Enhancing Occupational Safety and Health

Geoff Taylor, Kellie Easter and Roy Hegney

ELSEVIER
BUTTERWORTH
HEINEMANN

AMSTERDAM BOSTON HEIDELBERG LONDON NEW YORK OXFORD PARIS
SAN DIEGO SAN FRANCISCO SINGAPORE SYDNEY TOKYO

Elsevier Butterworth-Heinemann
Linacre House, Jordan Hill, Oxford OX2 8DP
30 Corporate Drive, Burlington, MA 01803

First published 2004

Copyright © 2004, Work Safety and Health Associates, Bdel Pty Ltd and
Roy Hegney. All rights reserved.

Roy Hegney is the original author of much of the material in Chapters 1, 3 and
5–7, and Best Practice in Chapter 2. Kellie Easter is the original author of much
of the material in Chapters 2 and 4. Geoff Taylor is the author of Chapters 8–11
and 13–14, and additional material in Chapters 1–7 and 12. There is some edited
material by Laraine Telfer in Chapter 8 (on Workplace Wellness) and by
Martin Peters in Chapter 12.

No part of this publication may be reproduced in any material form (including
photocopying or storing in any medium by electronic means and whether or not
transiently or incidentally to some other use of this publication) without the
written permission of the copyright holder except in accordance with the
provisions of the Copyright, Designs and Patents Act 1988 or under the terms of
a licence issued by the Copyright Licensing Agency Ltd, 90 Tottenham Court
Road, London, England W1T 4LP. Applications for the copyright holder's
written permission to reproduce any part of this publication should be addressed
to the publisher.

Permissions may be sought directly from Elsevier's Science and Technology
Rights Department in Oxford, UK; phone: +44-0-1865-843830;
fax: +44-0-1865-853333; e-mail: permissions@elsevier.co.uk.
You may also complete your request on-line via the Elsevier homepage
(http://www.elsevier.com), by selecting 'Customer Support' and then
'Obtaining Permissions'.

British Library Cataloguing in Publication Data
A catalogue record for this book is available from the British Library

ISBN 0 7506 6197 6

For information on all Elsevier Butterworth-Heinemann publications visit
our website at www.bh.com

Typeset by Charon Tec Pvt. Ltd, Chennai, India
Printed and bound in Great Britain

Working together to grow
libraries in developing countries

www.elsevier.com | www.bookaid.org | www.sabre.org

ELSEVIER   BOOK AID International   Sabre Foundation

Enhancing Occupational
Safety and Health

We have soothed ourselves into imagining sudden change as something that happens outside the normal order of things. An accident like a car crash, or beyond our control like a fatal illness. We do not conceive of sudden radical, irrational change as built into the very fabric of existence. Yet it is. And chaos theory teaches us that straight linearity, which we have come to take for granted in everything from physics to fiction, simply does not exist. Linearity is an artificial way of viewing the world. Real life isn't a series of interconnected events occurring one after another like beads strung on a necklace. Life is actually a series of encounters in which one event may change those that follow in a wholly unpredictable, even devastating way. That's a deep truth about the structure of our universe. But, for some reason, we insist on behaving as if it were not true.

(Michael Crichton, (1991). *Jurassic Park*, London, Century – reproduced by permission of Random House)

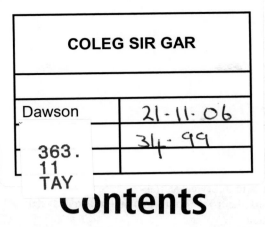

Contents

About the authors	*x*
Foreword	*xiii*
Preface	*xv*
Acknowledgements	*xvii*

1 Current concepts in work safety and health — 1
Modern trends — 1
Safety, health and risk — 4
Key factors in the development of workplace accident prevention — 8
Accident costs from workplace injuries and damage — 21
Issues in the work environment affecting employee well-being — 30
Further reading — 36
Activities — 37

2 Safety management — 38
Safety and health in risk management — 38
Formal and informal safety meetings — 47
Nature, occurrence and industrial relations implications of OHS issues — 52
Prioritizing and developing strategies to resolve OHS issues — 56
Budget planning, control of safety expenditure and supervision — 62
Best practice in work safety — 71
Contractor safety — 74
Behaviour-based safety — 76
Communication and meeting skills — 78
Further reading — 93
Activities — 94

3 Common and statute law — 95
Origins and types of law which influence occupational health and safety — 95
Duty of care — 100
Occupational health and safety legislation — 102
Legislation in accident prevention in the USA and European Union — 114

Further reading	118
Activities	119
Appendix 3.1 – A brief guide to some OHS legislation and organizations (to November 2003)	120

4 Hazard and risk management — 130

Risk concepts	130
Role of hazards in injury causation	132
Planned hazard identification, formal and informal systems	133
Risk assessment	139
Principles behind risk assessment, and importance and limitations of scientific assessment	144
Understanding risk	151
Risk control	155
Selecting and using personal protective equipment	157
Communicating risk	159
Further reading	161
Activities	162
Appendix 4.1 – Assessing risk with figures: Summary of US Navy risk assessment paper by Kinney and Wiruth	163

5 Workplace inspections — 168

Standard setting and formal and informal inspections	168
Structured and documented inspections	174
Format for inspections and reporting	178
Conducting a safety audit and preparing a report	185
Further reading	190
Activities	190

6 Accident prevention — 191

Accident causation factors	192
Accident investigation	201
Principal elements in developing a workplace health and safety plan	217
Accident, injury, compensation and safety data	227
Collecting, sorting, accessing and validating data	227
Data handling and analysis	229
Data reporting to management	234
Violence in the workplace	235
Further reading	238
Activities	239
Appendix 6.1 – Accident models	239

7 Risk engineering — 246

Risk and reliability	246
Systems engineering	250

Electrical safety	254
Fixed machinery hazards	259
Safe use of pressure vessels and lifting equipment	264
Fire hazard identification and extinguisher use	272
Fire safety managers	280
Building regulations and fire safety compliance	280
Determining building classifications, fire load and fire resistance	285
Preparing a technical brief of fire safety requirements for a building	286
The human element in fire causation and behaviour during fire emergencies	289
Locking systems – security versus safe egress	291
Fire prevention and emergency training programmes	292
Further reading	297
Activities	298

8 Health at work — 299

Development history of occupational health	299
Current developments in occupational health	301
Workplace stressors, processes, diseases and disabilities	303
Links between workplace stressors, processes and diseases	307
Worker health monitoring	324
Stressors inside and outside the workplace	326
Alcohol and drugs in the workplace	336
Workplace wellness	338
Further reading	342
Activities	344

9 Hazardous substance management — 346

Chemical elements, compounds, classes and physical state	346
Common chemical hazard classes and descriptors	351
Chemical reactions and structures	355
Classifying chemicals	359
Main factors in transport of hazardous chemicals	363
Dangerous goods vehicle and driver checks	367
Storage of hazardous substances	368
Information sources on chemical hazards	370
Procedures for receipt and dispatch of hazardous substances	376
Specialist facilities for the management of hazardous substances	382
Signage required for storage, handling and transport of hazardous substances, including dangerous goods	385
Controls to minimize employee exposure	386
Role of emergency personnel	390
Useful information on toxicity and confined spaces	392
Further reading	393
Activities	394

Appendix 9.1 – European Union and US material safety data
sheet requirements   395

10 Work environment   **399**
Major characteristics of noise   399
Control of excessive noise   402
Ergonomic principles and control of noise, vibration and
lighting hazards   404
Conducting a noise survey   405
Measuring light levels   410
Air contaminant measurement   414
Evaluation, exposure controls and reporting on chemical contaminant
monitoring   429
Thermal comfort and heat stress   433
Vibration   438
Non-ionizing radiation   439
Ionizing radiation   442
Further reading   446
Activities   448

11 Ergonomics   **449**
Origins and history   449
The person–machine model   452
Anthropometry   454
Relative merits of methods of collecting and applying anthropometric data   455
Common forms of occupational overuse syndrome (OOS) and
preventative ergonomic strategies   458
Methods for identification, assessment and control of manual
handling hazards   462
Assessment of manual handling tasks   466
Prevention of manual handling injuries   468
Assessing the energy cost of work   472
Ergonomic principles associated with integration of controls and displays   475
Job design structure   477
Ergonomic principles and design and redesign of work stations   480
Human error   484
Further reading   485
Activities   487

12 Workers' compensation and rehabilitation   **489**
Development of employer's liability   489
Employer's liability for compensation under workers'
compensation legislation   492
Process for resolving disputes in workers' compensation   493
Principles of rehabilitation applying to injured workers   495

	Role of rehabilitation in the workers' compensation system	498
	Factors in an effective injury management system	499
	Assistance in rehabilitation of employees	502
	Negotiation of premium level with the insurer	503
	Effective claims management	504
	Further reading	507
	Activities	508
13	**Health and safety training**	**509**
	Health and safety training programmes	509
	Training needs analysis, and design, conduct and evaluation of an OHS training programme	513
	Assessment methods	520
	Options for training delivery	523
	Evaluating a training session or programme	523
	The importance of procedures	524
	Further reading	531
	Activities	533
	Appendix 13.1 – OHS competencies and performance criteria for key workplace parties	534
14	**Health and safety management systems**	**542**
	Options for management of OHS in an organization	542
	Strategies to integrate OHS into organizational quality management systems	551
	Proposing and defending a strategy for management of change	558
	Reviewing an occupational health and safety management system	568
	Cost–benefit analysis for new acquisitions, refurbishments or maintenance	571
	Planning integration of the new risk management strategy with the organization's quality management (QM) programme	575
	Further reading	578
	Activities	580
	Information sources	**581**
	Some worldwide web addresses on OHS	581
	Index	585

About the authors

Geoff Taylor, M.Sc. (Occ. Hyg.), Grad. Dip. Bus. Admin, Chartered Fellow, Safety Institute of Australia, Dip. Occ. Hyg. (UK), C.I.H. (US), MAIOH, Director, Work Safety and Health Associates.

Mr Taylor has carried out field surveys of workplaces over 36 years. He has his own safety and environment consultancy, and is a sessional lecturer at the Australian Centre for Work Safety at Carlisle TAFE College, and at Curtin University. Geoff is a past president of the Safety Institute of Australia Western Australian Division, and was a board member of the International Commission on Occupational Health. He has been senior lecturer in OHS at Curtin University, a sessional lecturer at Edith Cowan University, OHS study area leader in TAFE-Western Australia and chief scientific officer with the Western Australian OHS authority. Geoff has developed and delivered OHS training at various management levels for the mining, construction and water industries and designed, coordinated and taught university and TAFE courses in OHS. He has an advisory role with the Mining and Resource Contractors Safety Training Association (MARCSTA).

About the authors

Kellie Easter, Grad. Dip. OHS (Curtin), RGN, MSIA, Lecturer, Australian Centre for Work Safety, Swan College of TAFE, Perth. Ms Easter has worked nationally in occupational health and safety and has held safety management positions in the health and manufacturing sectors. Kellie has worked as a consultant in the textile and mining sectors and is currently a coordinator at the Australian Centre for Work Safety where she manages the learning program, lectures and consults to industry. Her principal professional interests are in the application of quality management to safety and environmental practices in industry.

Roy Hegney, Grad. Dip. OHM (Ballarat), Grad. Dip. Ed. Training and Development (ECU), Lecturer, Australian Centre for Work Safety, Swan College of TAFE, Perth.

Mr Hegney has extensive operational and management experience in work safety. A graduate of Ballarat University and the Monash-Mt Eliza Staff College (1985), Roy lectured in OHS at Curtin University prior to working as a training consultant in the mining sector. He is currently the manager of the Australian Centre for Work Safety and has developed accredited courses in work safety including a nationally delivered course through Open Learning Australia (OLA).

Foreword

In June 1996 I was privileged to write the Foreword for the book *Enhancing Safety – An Australian Workplace Primer*. In the Foreword, I wrote among other things:

> *This book represents a most welcome addition to the library of the widest range of occupational safety and health practitioners.*
>
> *During the last 15 years, I have frequently been asked questions about suitable textbooks in the field of occupational safety and health. In the past, a number of excellent textbooks have been available, but each of them dealing only with one relatively narrow area, such as system safety or occupational hygiene. Even the books which attempted to cover a broader area, have not achieved such a coverage. They usually concentrated on the organizational aspects of the field, neglecting technical and scientific aspects, or vice versa.*
>
> *This book fills such a serious gap in the Australian safety and health literature, and perhaps is the first of its kind in safety and health literature published in the English language.*

It won a prize in the 1996 European Commission/International Social Security Association Health and Safety Training Competition.

Enhancing Safety has been accepted most enthusiastically by both technical and further education and university students of OSH, and by a wide range of occupational safety and health practitioners such as safety officers, occupational health nurses, risk managers and others. The book sold very quickly, and four more editions have followed.

A companion book to *Enhancing Safety* has now also been in use for four years. This book, *Advancing Safety*, builds on the information given in *Enhancing Safety*, and in a very successful way helps to further develop skills in a wide range of areas within the multidisciplinary occupational safety and health field. The treatment of both human and technical factors involved in the accident phenomenon is outstanding, and reflects well the practical experience of the three authors in the prevention of accidents and ill health at work. A number of principles and strategies discussed in the book have been tested by the authors, who have, among them, many years of practical experience in safety and health. The authors have used the elements of the contents of this book in teaching technical and further education and university students who aim to become OSH practitioners.

Geoff Taylor has now taken most of the two books, and with the assistance of Kellie Easter and Roy Hegney, has added a range of new material. The entire text has been reviewed to suit an international readership, and examples included from a range of countries with an English-speaking connection. It also recognizes that four of those countries operate within the EU OSH framework. The book attempts to bridge the gap between the Robens-style duty of care and the prescriptive approach used in for example the USA, via the bridge of self-regulation. It contains a guide to OSH legislation, administration and societies in over forty countries.

Enhancing Occupational Safety and Health then represents a most useful textbook for technical and further education and university students studying occupational safety and health. A very wide range of occupational safety and health professionals, such as safety managers, safety coordinators, occupational health nurses, occupational physicians, ergonomists, occupational hygienists, personnel officers and managers and many other professionals will find this book to be one of the most valuable resources for their work. This book will find its way into professional libraries of all organizations and individuals who strive towards the common goal of contributing to safety and health at work.

Dr Milos Nedved, FSIA
Associate Professor, Edith Cowan University
Perth, Australia, December 2003

Preface

This book has been written as a practical guide to occupational health and safety both for students of OHS and for people working in this field and for the many people in workplaces who need to be able to understand and apply OHS, in order to minimize injury, ill health, disease, disability and death among employees. Every attempt has been made to adopt a plain language, straightforward approach, and to relate the material to practical examples in the workplace at appropriate points.

This book arose from two books on OHS, *Enhancing Safety* and *Advancing Safety*, originally published for an Australian audience, which have gone to revised editions several times. They have found a ready acceptance in Australia by those studying OHS at technical and further education (community college or vocational) level and university level, as well as by those in workplaces undertaking short courses on generic OHS skills. Students include those studying to be OHS practitioners and those for whom OHS units are part of another course such as management.

This new book combines elements of those as well as some new material, and has been specifically written to appeal to anyone dealing with health and safety in any country which has a substantial English-speaking connection. The references to specific examples are taken from a number of those countries.

For that reason a table of those countries, their primary OHS legislation, administering agencies and national professional societies is provided in an Appendix at the back. There are some gaps in the last category which we invite readers to fill.

This book has a focus on self-regulation. While US legislation and that of some other countries in the Appendix still adopts a prescriptive approach, there is an increasing *de facto* dependence on self-regulation recognized in programmes such as the US Voluntary Protection Program and the adoption of Robens-based legislation in various countries worldwide. Also, the risk management approach, which the European Union has adopted for OHS, is woven throughout this book.

We make no apology for including a number of older references in the text. We consider they are either still valid (e.g. Kinney and Wiruth, Woodson) or are important in the development of work safety and health (e.g. Heinrich).

Preface

The demands on the OHS practitioner have not lessened. There are a number of growth industries such as tourism. Commercial outdoor adventure activities form a part of this, but outdoor adventure also poses duty of care requirements for schools, private associations with paid employees and leisure operators. The statutory duty of care to non-employees assumes much greater importance in this context, and the legal cases here and overseas suggest this area must receive much greater professional attention. Growth in tort litigation and public liability insurance rates is a rising concern.

However, this area raises new issues. Clearly there are a variety of long-term survival advantages for a society in which people are allowed to experience and manage risk. This also fulfils individual psychological needs, such as pitting the body and mind against a mountain. Paid football employees are expected to work in a climate of high intrinsic risk of injury. Despite good forecasting and planning, natural environments – air, sea, land (and for a few, space) – still carry elements of unforeseeable risk. It is suggested that most existing OHS law fails to adequately address the situation where non-employees pay to be exposed to the exhilaration of high risk or where employees deliberately expose themselves to high risk in organized professional sport.

In the area of violence at work, the usual techniques of risk management don't adequately address situations where one non-employee in a workplace very deliberately places the lives of employees at risk. The September 2001 attack on the World Trade Center using hijacked civilian aircraft is a gross example. There, however, good emergency response saved tens of thousands of lives. There is a slow spread of the risk management approach, formally embraced as I noted in the European Union's forward OHS strategy, into the whole area of ensuring a safe society. This presents new challenges for us all.

Two thoughts – it would be good to have one word in English for OHS; 'worksafety' suggests itself. It would also be good to have a word for unwanted events, free of the 'it just happened' nuance of 'accident'. We suggest 'accevent'.

'Section' in a reference to another part of a chapter refers to a section listed in the contents. The book uses the terms OHS and OSH interchangeably.

We hope you will find this new book useful and we welcome any suggestions for improvement.

Geoff Taylor, Editor, July 2004

Disclaimer

This book deals with occupational health and safety in an international context. While it contains references to certain acts, regulations, codes, guidance notes and standards, it is the responsibility of the reader to ascertain the particular acts, regulations, codes, guidance notes and standards as they apply to particular situations in particular jurisdictions. The UN Recommendations in Chapter 9, for example, may or may not have been adopted in your jurisdiction.

Acknowledgements

We are grateful for the support of Catherine Shaw, Commissioning Editor, Doris Funke, Editorial Assistant, and Deborah Puleston, Production Editor, Elsevier UK.

We acknowledge helpful comments from Ken Leeden, Claire Thomas, and Chris Harrison of Swan TAFE, assistance from Mariola Stanczak and staff of the Western Australian Department of Industry and Resources, Owen Wilson of ARPANSA, Anitha Arasu, Samantha Peace of the UK HSE, Bob White of the NZ DOL, library staff at WorkSafe Western Australia, Mandip Pabial-Parmar of IOSH, The Finnish Institute of Occupational Health, Rebecca Bunn of Lexisnexis UK, Peter Bateman of *Safeguard* (New Zealand), Scott Williams of the Industrial Accident Prevention Association of Ontario, the Canadian Centre for Occupational Health and Safety, and the US Occupational Safety and Health Administration.

Geoff wishes to thank his family and partners – Trish, Dan, Chris, Alana, Simon, Andy and Luke, for their support and assistance.

We are also grateful for the support of Associate Professors Milos Nedved, Edith Cowan University and Al Mims, University of Wisconsin.

We also acknowledge the support of Bob Stratton, Acting Managing Director of Swan TAFE College.

Throughout this book, every effort has been made to acknowledge the use of other people's work, and we apologize for any errors or omissions in acknowledgements. Please advise us of any.

We are particularly grateful to the Canadian Centre for Occupational Health and Safety for permission to reproduce material from 'A Basic OHS Program' in Chapter 6 (with minor changes). Acknowledgments also include: Australian Training Products, ACGIH, Alan Fox, Australian Building Codes Board, Australian Institute of Occupational Hygienists and David Grantham, Arbetsmiljofonden, Bodil Alteren, Steve Boster of the US Navy for work from G.F. Kinney and A.D. Wiruth, Professor Gunnar Breivik, Henry Cole of the University of Kentucky, Bonnie Claydon, Elaine Cullen of US NIOSH, Chamber of Commerce and Industry of Western Australia, Daryl Cooper, Professor Dennis Else, Elsevier UK for work from H.A. Waldron, Federal Office of Road Transport, Mike Gavin, Pat Gilroy of the Mining and Resource Contractors Safety Training

Acknowledgements

Association (MARCSTA), HMSO for work from the UK Health and Safety Executive, Human Kinetics for work from Lovato, Industrial Foundation for Accident Prevention, John Wiley and Co for work from Peter Bernstein and N.A. Leidel and K.A. Busch, Lippincott Williams and Wilkins for work in *Spine* from M.A. Adams and W.C. Hutton, Geoff McDonald, McGraw-Hill for work from R.S. Bridger, H.W. Heinrich, M. Sanders and E. McCormick, M.O. Amdur and J. Doull and A. McMichael in Egger, G. et al., Joe Maglizza, Moss Associates Ltd for work from Geoffrey Moss, Grant Newton of the NZ Wellington Tramping and Mountaineering Club, NOHSC Australia, the NZ Mountain Safety Council, Pearson Education for work from J. Stranks, Martin Peters for some edited material in Chapter 12, Professor Dan Petersen, Stephen Pheasant, The Random House Group for work from Michael Crichton, Professor Jens Rasmussen, University of California for work by Wesley Woodson, US Department of Health Education and Welfare-NIOSH, Lorraine Telfer for some edited material in Chapter 8, Pasi Toivonen for a photograph, Jim Whiting, WorkCover Victoria, and WorkSafe Western Australia.

Other external sources are acknowledged in the text. We have been unable to trace the source of the actual human figure in the chapter *Current Concepts*.

Dedication

This book is dedicated to Trish, Dan, Chris, Alana, Simon, Edith, Alan, Jean, Peter, Caitlin, Genevieve, Arthur, Rhys, Paige and Luke.

1

Current concepts in work safety and health

WORKPLACE EXAMPLE

An employee's leg was amputated after an accident in a sawmill in New Hampshire, USA. A 125 cm circular head saw was being adjusted by an operator. There was a mechanism to carry logs into the saw and this moved. It pushed the worker into the saw blade, which was rotating, and amputated his leg above the knee.

The subsequent Occupational Safety and Health Administration (OSHA) citations covered these wilful violations: failure to develop document and utilize procedures to control potentially hazardous energy, lack of guarding and positive means to prevent movement of the log carriage.

The citations also covered six serious violations. These included guarding, ladders, saw safety guides, and training in the bloodborne pathogen standard. Wilful violations are those involving intentional disregard or plain indifference to the requirements laid down in the Occupational Safety and Health Act, while serious violations are those where there is a substantial probability that death or serious physical harm could result and the employer knew or should have known of the hazard.

OSHA has a lock-out tag-out standard which means that before an employee starts work of this type, powered machinery must be shut down, positively prevented from movement, and the power source locked and tagged so that another worker cannot start the machinery accidentally.

(Source: OSHA, USA)

Modern trends

From past to present

The development of steam power, the introduction of devices such as mechanical textile looms, which steam made possible, and the increase in coal mining and iron ore smelting which steam demanded, all caused a rapid transformation in the conditions of many people's employment in the nineteenth century in Europe.

The same transformation shifted to America and is still in progress today in countries such as China. It was accompanied by moves by workers away from the land and towards industrial cities. Basic city sanitation was initially very poor in Europe and so environmental health problems overlay occupational health and safety issues.

Concern about workers' poor health, especially that of women and children, and about workplace accidents and community health, drove a process of legislative change. Health and safety legislation became commonplace throughout the world. It was largely prescriptive, that is, it specified a series of 'what to do's', and was generally limited to specific areas of work such as mining, construction, factories, shops and warehouses. It also relied on outside inspection of workplaces by a government agency.

There have been later developments in transport safety (rail, road, sea and air) and, in the last thirty-five years of the twentieth century, in environmental legislation. Both have a large component of health and safety of the public, not just that of workers. There was often a degree of fragmentation in the administration of health and safety legislation, with 'departments of health' involved in workplace health issues and, with a greater or lesser degree of cooperation, with 'departments of labour', although in some cases the inspectorate was directly managed by a medical practitioner.

Another key feature was the protection of workers' wages in the event of injury or ill health sustained at work, through the development of compulsory workers' compensation insurance. The worker was to be paid regardless of fault. This development had been preceded in some instances by cases in tort law which provided monetary damages to injured workers, although the legal obstacles were quite difficult to overcome and fault had to be established.

Health and safety law did provide for penalties if the law was breached, but too often the matter only went to court after an accident and even then judges were sympathetic to the company not the workers. Workers' compensation insurance premiums, which employers had to pay, were an important incentive to try to reduce injuries and ill health by improving safety and health standards in the workplace.

After World War II, the Scandinavian countries began writing health and safety legislation linked to industrial democracy. This was coupled with a new emphasis on self-regulation, and coverage of workers at all workplaces. This later spread to Britain and some of the countries of the former British Commonwealth. It is generally known as Robens-style legislation, after the British lord whose commission of inquiry recommended its adoption in Britain in the early 1970s.

In some jurisdictions, the government compensation and rehabilitation authorities, and the occupational health and safety authorities, have been merged. In British Columbia, for example, the workers' compensation premiums fund the occupational health and safety administrative function.

Current approach in the EU

There are some lessons in a Communication from the Commission of the European Communities on *Adapting to Change in Work and Society – a New Community Strategy on Health and Safety at Work 2002–2006*.

It notes that an ambitious social policy is a positive factor in competitiveness. The increasing percentage of women in the workplace, an older working population, and more social exclusion are considered by the Communication.

According to the Commission, promotion of well-being at work should include physical, moral and social well-being.

Response to the profound changes in society resulting from a knowledge-based economy should not ignore those existing industrial sectors which still have high injury rates, with even higher figures for the small and medium-sized businesses in those sectors. While women generally figure at the low end of occupational injuries, they suffer a disproportionate amount of certain types of occupational illness. The Communication also addresses the lack of worker motivation in insecure employment relationships and the issue of responsibility for telework (e.g. work done at home) as the distinction between employment and self-employment blurs.

The Commission notes that emerging issues such as stress, depression, anxiety, violence at work, harassment and intimidation are responsible for 18% of all health at work problems. Here it is not a specific risk but a raft of factors such as work organization, working time arrangements, hierarchical relations, transport-related fatigue, and the degree of acceptance of ethnic and cultural diversity within the firm, which counts. The Communication says that mainstreaming the gender dimension into risk evaluation, preventive measures and compensation arrangements, to take account of the specific characteristics of women, is vital.

Another area for greater emphasis is social risk – stress, harassment, depression and anxiety, and dependence on drugs, alcohol and medicines.

Inclusion of OHS in vocational training, dispensed regularly, and geared to the realities of day-to-day work, is also seen as a crucial element, as is making OHS part of the school curriculum. Awareness training for small and medium enterprises receives a special mention.

The need for enhanced protection of young people is recognized, while also responding to an ageing workforce.

Benchmarking; best practice and corporate social responsibility; integration of OHS compliance with OHS conditions in public contracts; and making OHS part of economic cooperation are also included.

The strategy to address this, in summary:

- Adopts a global approach to well-being at work, taking account of changes in the world of work and the emergence of new risks, especially of a psycho-social nature. As such, it is geared to enhancing the quality of work, and regards a safe and healthy working environment as one of the essential components.
- Is based on consolidating a culture of risk prevention, through combining a variety of political instruments – legislation, the social dialogue, progressive measures and best practices, corporate social responsibility and economic incentives – and on building partnerships between all the players on the safety and health scene.
- Points to the fact that an ambitious social policy is a factor in the competitiveness equation and that, on the other side of the coin, having a 'non-policy' engenders costs which weigh heavily on economies and societies.

Key elements of self-regulation style legislation

The key elements of self-regulation style legislation include:

- Usually, a policy-making body on which government, employers and industrial unions are represented.
- Unified administration of occupational health and safety law across the country, or in the case of federations such as Canada and Australia, across a province, state, or territory. This has not been fully realized in, for example, parts of Australia.
- Coverage of all workplaces and workers.
- Self-regulation, requiring commitment, occupational health and safety policies and procedures by employers, with occupational health and safety issues being resolved as far as possible in the workplace. The legislation is described as 'enabling' because it sets up a framework within which employers and employees can consult on occupational health and safety matters.
- A general duty of care by workplace parties such as employers, employees, contractors, the self-employed, occupiers, designers, importers, manufacturers, and suppliers.
- Consultation processes with the use of health and safety committees and, in some cases, health and safety representatives.
- 'Piercing the corporate veil'; that is, making the organization's managers and directors legally liable for accidents where appropriate, e.g. S. 52 of the Malaysian OSH Act.
- The use of administrative notices (e.g. prohibition notices, improvement notices) by inspectors to correct problems, starting immediately, in preference to prosecution for breaches or violations.
- Codes of practice explaining how to meet the duty of care in relation to particular matters, e.g. manual handling of goods and materials.
- A new approach to regulations – less 'what to do', more 'what to achieve' or even 'how to achieve it'.

In the USA in the 1970s there was also a renewed emphasis on occupational health and safety with the introduction of federal legislation governing both mining and non-mining workplaces. This was binding on all US states. Although it retained the prescriptive approach, more recently some companies have entered voluntary self-compliance arrangements.

You can read more about the legislative approach in Chapter 3.

Safety, health and risk

Common terms

In order for communication to be effective between those involved in occupational health and safety, the safety profession and other professions, it is important that we use common words such as 'accident, injury, hazard, safety, health and risk' with some consistency. Unless these words are given specific definitions in legislation, some effort must be made to give these words acceptable meanings.

Accident: An unplanned event that may or may not result in damage, loss or injury.

Injury: Damage to the body resulting from a delivery of energy to the body above the capacity of the body to cope with that energy *or* an interference with the normal function and systems within the body.

Hazard: A source of unwanted or excess energy with the capacity to cause damage, loss or injury.

Safety: An individual's perception of risk. Two alternative definitions are 'safety is a state of mind whereby workers are made aware of the possibility of injury at all times' (from Ted Davies, a mining safety expert, derived from Osborne, Canada), and 'safety is a state in which the risk of harm (to persons) or damage is limited to an acceptable level' (Australian Standard 4801). Some would argue for 'tolerable' not 'acceptable', saying no risk is acceptable.

Health: The degree of physiological and psychological well-being of the individual.

Risk: The combination of the likelihood that a hazard will actually result in an accident and the consequences of that accident, often expressed as the product of the two.

What is safety?

'Safety' is a word defined in the *Concise Macquarie Dictionary* as:

1. The state of being safe: freedom from injury or danger.
2. The quality of insuring against hurt, injury, danger or risk.

However, the definition is too narrow to derive any meaningful discussion or understanding about the nature of safety beyond a notional perspective. The words 'safety' and 'safe' give rise to an expectation that a state can exist which is free from risk. Nothing can be absolutely free from risk, consequently nothing can be absolutely safe. Rather, there are degrees of risk and consequently there are degrees of safety. In practice we say an activity, system, substance, etc., is *safe* if the associated risks are considered acceptable.

We may better define safety as an individual's judgement of the acceptability of risk. Safety is predicated upon two discrete activities:

1. The measurement of risk.
2. The value placed in the risk (judgement).

Measurement of risk

Risk is the potential for realization of unwanted, negative consequences from an event. Provided sufficient information is available it is possible to quantify the risk. Technically competent people such as chemists, doctors and engineers are trained to measure risks objectively. However, no accurate prediction can be made for the future. More relevant is the principle that the technically competent, while having the knowledge to quantify risks, are no more entitled than anyone else to decide who should be exposed to these risks, or the level of risk which is acceptable.

There are some measures of risk such as deaths per million people per year or loss of life expectancy from various events or activities which can be built up from careful studies. For example, one figure given for smoking (one pack per day) is 5000 deaths per million people per year.

Compare this with car accidents (300), snake bite (5) and lightning (0.5).

In the workplace we have to work not only on measures of risk like these, but also with people's perceptions of risk. Blue asbestos, in certain circumstances, has given a relatively high number of people in Western Australia a lung disease which is normally very rare. Unfortunately, however, this has led to extreme demands in relation to other types of asbestos in other situations. The problem is that when we spend money on health and safety, as a general rule we need to put the money into those areas where the greatest reduction in accidents or the more severe injuries can be achieved. Irrational perceptions can lead to scarce funds going to areas where little or nothing will be achieved, while other more important areas are neglected.

Nevertheless, the sensitivity of the issue cannot be ignored. If people are worried, irrationally or not, about a hazard in their workplace, particularly one which can't be seen, heard, felt, touched or smelt (like radon or a biological hazard), then it will affect their performance and stress levels and may be a source of costly high labour turnover. As a general rule, people may be prepared to accept a higher level of risk in the things they do of their own free will, e.g. recreation, than they are prepared to tolerate in the activities they carry out for an employer.

Shown below are three factors which need to be considered when assessing risk. See Fig. 1.1.

A fourth factor to consider is the legal situation. If perceptions of risk – even if you think they are wrong – lead to certain legal requirements, then those legal requirements cannot be ignored.

Various systems of evaluating risk have been devised, see the Appendix to Chapter 4 – for a system devised by Fine, Kinney and Wiruth.

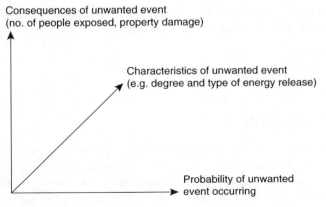

Figure 1.1 Size, magnitude or degree of risk

A second system of evaluating risk is given in John Ridley's *Safety at Work* (see Further Reading at the end of this chapter).

Value placed on the risk (judgement)

The decision to accept a particular level of risk is a matter of individual judgement. There are forces that create high levels of risk taking; an example is economic circumstances. It is worth noting that common law has come to recognize that workers are often forced to accept risks unwillingly in order to earn a living. This is reflected by the limitations placed on the doctrine of Voluntary Assumption of Risk (by employees) as a defence available to employers in the Tort of Negligence in many jurisdictions.

On the other hand, in some areas society, through parliament and the courts, decides what level of risk is acceptable. This can be spelt out in laws, regulations, standards or court judgements which set a precedent.

The approach in Robens-style occupational health and safety acts (see Chapter 3) is to require the amount spent on reducing the risk to balance the size of the risk. Failure to recognize and appreciate the process that determines safety results in much misdirected effort and, on occasions, leads to unnecessary and pointless conflict.

Work safety criteria

Despite what is often written, and said, about work safety not being based primarily on financial considerations, practical experience would suggest otherwise. Either from the perspective of the employee evaluating whether or not exposure to a particular risk is acceptable, to the capacity or intention of an employer to mitigate a particular risk, the criteria are primarily economic.

However, in addition to the economic concerns there are other factors that have an influence in establishing the workplace safety climate:

- the common law duties and liabilities of employers, as established and developed over the years by many court decisions
- the broad impact of the statutory duties imposed by self-regulation style legislation
- salient features of the workers' compensation and rehabilitation acts
- national or international standards and codes of practice
- codes of practice, guidance notes and national standards or rules issued by government occupational health and safety authorities
- safety policies determined by individual organizations stating the intentions of the corporate body towards the management of health and safety
- standards set down by industry groups and safe working methods that have evolved over time and have become acceptable practice in particular trades and occupations
- use of standards of exposure accepted internationally, e.g. the Threshold Limit Values (TLVs) for Chemical Substances in the Work Environment adopted by the American Conference of Government Industrial Hygienists (ACGIH)
- the conventions concerning occupational safety and health and the working environment developed by the International Labour Organization (ILO). Convention

No. 155 and Recommendation No. 164 set out the broad principles while other Conventions and Recommendations deal with specific issues
- collective bargaining or other industrial relations agreements.

Key factors in the development of workplace accident prevention

Principal contributors

It is difficult to clearly identify all the forces responsible for the wave of change in work safety and occupational health. The visible indicators of change are perhaps the changed role of the government inspectorate, the safety representative and media promotion. These signs of change represent only the curl of the wave. The real forces are far less obvious than inspectors, representatives and media promotion. Some examples are listed below.

Institutions

Many educational institutions offer studies in occupational health and safety or a particular aspect of this field, such as ergonomics or occupational hygiene. These courses have led to a large increase in highly skilled and career minded professional people working in the field.

Government administration

Government departments exist to administer the occupational health, safety and welfare legislation. These departments have an inspectorate role and may also provide information, education and training services. They are also given the responsibility to prosecute for a breach of particular legislation and the task of producing standards and codes of practice. Some also provide research grants for occupational health and safety studies.

Professional associations

The interest created by safety and health reform around the world has led to a growth in the number of people working in the field of accident prevention and health management.

Internationally, associations such as the International Social Security Association and the International Commission on Occupational Health have emerged as groups with potential for influence. A range of national professional associations also exists.

Trade unions

In many countries trade unions play an important role in securing and improving working conditions, often being represented on occupational health and safety policy-making bodies. Groups such as the International Federation of Free Trade Unions also bring pressure to bear on those countries where unions are weak or non-existent. A number of employee associations have appointed full-time health and safety officials.

This is likely to increase as the employee demand for safer and healthier work environments places increasing pressure on the traditional union services and skills.

Employer groups
Employer groups and associations are usually also represented in occupational health and safety policy-making bodies. Groups of employers in a particular industry sector may also develop particular approaches to health and safety relevant to that industry and provide health and safety assistance to employees in that industry sector.

International Labour Organization (ILO)
The governing body of the ILO is the International Labour Council on which employer and employee bodies and governments are represented. The ILO publishes a series of conventions and recommendations on occupational health and safety to be used as a basis for minimum standards. These documents are well regarded, as is the ILO *Encyclopaedia of Occupational Health and Safety* which has become a widely used reference source.

Public pressure
Many of the industrial processes and work activities that are carried out have an effect on people both inside and outside the workplace. Escaping chemicals, noise, dust and dumping of industrial wastes are just a few examples. There is undoubtedly a strong public and media influence in the drive for safer and healthier work.

The American scene and the birth of accident prevention

Developments
Although some steps were taking place in the United Kingdom to reduce accidents – for instance, the 1855 Factory Act provided for surgeons to investigate workplace accidents – the principal interest in the nineteenth century was in occupational disease, perhaps because the inspection system was headed by doctors. A scientific approach to the causes of accidents, and the acute physical trauma some produced, really started in the USA.

Progress in American industrial safety before 1911 was practically non-existent. With no workers' compensation laws, all states handled industrial injuries under the common law. In common law the legal defences available to the management of industry almost ensured that they would not have to pay for any accidents occurring on the job. Under this common law system employees did not automatically receive payments when injured on the job, as they do today. Before workers' compensation legislation, the injured employee had to sue the employer for recompense.

When the employee did sue, the employer had four legal defences. If any of the following could be shown, the injured employee would not be paid for the injuries suffered.

1. The employee contributed to the cause of the accident.
2. Another employee contributed to the accident cause.

3. The employee knew of the hazards involved in the accident before the injury occurred.
4. There was no employer negligence.

In 1908 the state of New York passed the first workers' compensation law which deemed, in effect, that regardless of fault, management would pay for injuries occurring on the job. Before this, most accidents fitted under one of the common law defences and hence management did not often have to pay for injuries resulting from accidents. The New York law was held to be unconstitutional. A similar law was passed in Wisconsin in 1911 and held to be constitutional. This Wisconsin law set the stage for all the other states, including New York, to resubmit or introduce similar laws. All states did, the last being enacted in 1947.

When management found itself in the position, by legislation, of having to pay for injuries on the job, they decided that it would be financially better to stop the injuries from happening. This decision by industry gave birth to the organized industrial safety movement.

Workers' compensation legislation provided the financial atmosphere for industrial safety. Without the legislation the safety movement would be far behind its position today. Workers' compensation laws in effect state that, regardless of fault, the injured employee will be compensated for injuries that occur on the job. Liberal interpretation of these laws in many jurisdictions today goes even beyond this original intent.

Improving working conditions

In the early years of the safety movement, management concentrated heavily, if not entirely, on correcting the hazardous physical conditions that existed. Doing this showed remarkable results during the first 20 years. In deaths alone the reduction in the USA was from an estimated 18 000 to 21 000 lives lost in 1912, to about 14 500 in 1933.

The death rate (deaths per million person-hours worked) for that period would show an even better reduction. This reduction came largely from cleaning up the working conditions. Cleaning up physical conditions came first – possibly because they were so obviously bad, and possibly because people believed that these conditions were actually the cause of injuries.

Contributions to accident prevention

Heinrich's safety philosophy

In 1931 the first edition of H.W. Heinrich's book *Industrial Accident Prevention* was published. This text in industrial safety was revolutionary, for in it Heinrich suggested that unsafe acts of people are the cause of a high percentage of accidents – that people cause far more accidents than unsafe conditions do. Heinrich's ideas were a departure from the safety thinking of the time. What he said, however, made sense to people in the field of safety, and his ideas were accepted. They were accepted so completely that even today we work largely within his framework. His work set the stage, in effect, for

Current concepts in work safety and health

all safety work since 1931. Perhaps it was because Heinrich proposed a philosophy for safety that his work was so important. Before the publication of his book, safety had no organized framework of thinking. It had been a mixture of ideas. Heinrich brought them all together and distilled some excellent principles out of previous uncertain practices. His 'Domino Theory' was based on the occurrence of an injury invariably resulting from a complicated sequence of factors, the last one of these being the injury itself. The accident which caused the injury was in turn invariably caused or permitted directly by the unsafe act of a person and/or by a mechanical or physical hazard.

He likened the sequence to a series of five dominoes standing on edge. These dominoes are labelled:

1. Ancestry or social environment
2. Fault of a person
3. Unsafe act or condition
4. Accident
5. Injury.

The following diagram, Fig. 1.2, illustrates Heinrich's theory.

The injury is caused by the action of preceding factors.

A great many safety people have preached his theory many times. Many have actually used dominoes to demonstrate it: as the first tips, it knocks down the other four dominoes, unless at some point a domino has been removed to stop the sequence.

Obviously the easiest and most effective domino to remove is the centre one – labelled 'unsafe act or condition'. This theory is quite clear; it is also quite practical

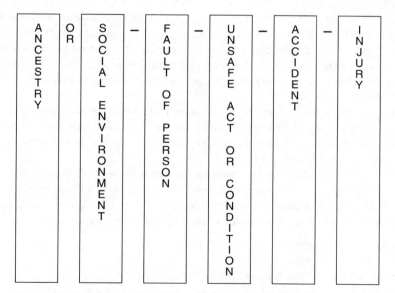

Figure 1.2 Heinrich's chain of multiple events model (with acknowledgement to Heinrich, H.W. (1950). *Industrial Accident Prevention*, 4th edn. New York: McGraw-Hill)

and pragmatic as an approach to loss control. Simply stated, it says, 'if you are to prevent loss, remove the unsafe act or the unsafe condition'. Heinrich's philosophy of accident causation also included the necessity for having properly educated and trained personnel within the workforce.

Disadvantages of the model

It should be realized that such unsafe acts or hazards as those referred to by Heinrich can be described in a multitude of ways and this is one of the difficulties in using this model. It does not provide a common description or classification which can serve as a basis for scientific study. It also involves both human behavioural factors and mechanical or physical factors within the same classification.

An adaptation of the Heinrich model is used in the concept of loss control as developed by Bird and Loftus.

Multiple causation model

The theory of multiple causation has its basis in epidemiology. In 1949, Gordon pointed out that accidental injuries could be considered with epidemiological techniques. His belief was that if the characteristics of the accident victim, together with the characteristics of the agent (or injury deliverer) and those of the supporting 'environment', could be described in detail, more understanding of accident causes could be achieved than by following the domino technique of looking for a single cause only. Essentially, Gordon's theory is that the accident is the result of a complex and random interaction between the victim, the agent and the environment, and cannot be explained by consideration of only one of the three.

Whilst there is little basically wrong with the domino theory introduced by Heinrich, there is however a need to apply a wider interpretation of its application, which has justifiably been criticized as being too narrow. For instance, when we identify an act and/or a condition that 'caused' the accident in the investigation procedures of today, how many other causes are we leaving unmentioned? When we remove the unsafe condition that we identify in our inspection, have we really dealt with the cause of the potential accident?

Today, we know that behind every accident there lie many contributing factors, causes, and subcauses. The theory of multiple causation is that these factors combine together in random fashion, causing accidents. If this is true, our investigation of accidents ought to identify some of these.

Let us briefly look at the contrast between the multiple causation theory and our too narrow interpretation of the domino theory. We shall look at a common accident: a person falls off a stepladder. If we investigate this accident using some current accident investigation forms, we are asked to identify one act or one condition:

- the unsafe act – climbing a defective ladder
- the unsafe condition – a defective ladder
- the correction – getting rid of the defective ladder.

This would be typical of a supervisor's investigation of this accident under the domino theory.

Let us look at the same accident in terms of multiple causation. Multiple causation asks: what are some of the contributing factors surrounding this incident? We might ask:

- Why was the defective ladder not found in normal inspections?
- Why did the supervisor allow its use?
- Didn't the injured employee know he/she shouldn't use it?
- Was he/she properly trained?
- Was he/she reminded?
- Did the supervisor examine the job first?

The answers to these and other questions would lead to these kinds of corrections:

- an improved inspection procedure
- improved training
- a better definition of responsibilities
- pre-job planning by supervisors.

Energy exchange model

Another approach in describing an accident is to consider the energy which is delivered and which produces the injury. The object delivering the energy is of secondary importance because of the difficulty in providing a suitable description of the specific object which caused the injury.

In this model the hazard is described not in terms of the object itself but rather in the type of energy exchange which caused the injury. This is the concept of the energy exchange model developed by Haddon in 1967 (see Further Reading at the end of this chapter). Table 1.1, based on one of Haddon's tables, shows the components of the energy exchange model.

Advantages and disadvantages of the model

All physical or chemical injuries are described as belonging to one of two classes. This model does have the advantage that an injury which has multi-causes can be described with two or more classifications. Perhaps a disadvantage of the model is that a large proportion of industrial accidents are due to mechanical energy exchange and hence this classification becomes too general.

Table 1.1 Components of the energy exchange model

		Cause of injury
Class 1	Injuries caused by delivery of energy in excess of local or whole-body injury thresholds	Type of energy delivered: • Mechanical • Electrical • Thermal • Chemical
Class 2	Injuries caused by interference with normal or whole-body energy exchange thresholds	Type of energy exchange interfered with: • Oxygen utilization • Thermal balance • Ionizing radiation

Haddon, an epidemiological researcher, was chosen by US President Lyndon Johnson as his head of highway safety. Crush zones, airbags and collapsible roadside 'furniture' are a result of Haddon's influence.

Injury causation model

Safety precautions have traditionally been oriented towards the removal or engineering control of potentially damaging energy, but the use of techniques for reducing non-culpable error is also potentially fruitful. These parallel approaches are combined into a simple model from which is derived a conceptual outline to guide in the selection of counter measures. From this model a course of formal study can be constructed in a manner suitable for incorporation into the curriculum of those who will be professionally concerned with the prevention of accidental injury.

Wigglesworth described this model in *A Teaching Model of Injury Causation and a Guide for Selecting Counter-Measures* in 1972 (see Further Reading at the end of this chapter). According to him, 'attempts to evolve a comprehensive theory of injury causation had been hampered by the lack of adequate relevant data and by the almost total lack of precision in the terminology used in the field'.

It should be appreciated from the above comment that before there can be an accident investigation there must firstly be a method of analysis which will describe the accident and all associated factors in such a way that it will provide data to assist in reducing accidents and injuries.

The model shown in Fig. 1.3 describes the accident sequence in terms of four factors:

1. Error
2. Accident
3. Hazard
4. Injury.

Such a model can form the basis of a scientific approach to examining an accident problem, as it provides a clear description of the component parts of the problem and hence enables the use of a systematic procedure for investigation.

The human engineering approach, designed by Wigglesworth to remove or reduce the likelihood of error, incorporates the traditional engineering approach of design to remove or reduce potentially damaging energy. The resultant combination, applied in the framework of Haddon, permits the construction of a model of injury mechanism, shown schematically in Fig. 1.3.

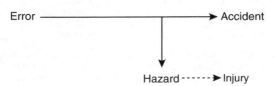

Figure 1.3 Injury causation model

Current concepts in work safety and health

In this model, some terms that are in common usage are given the more precise definitions listed below.

Definitions of components of the model

A 'hazard' is a potential source of bodily damage. Hazards may be classified in terms of the type of energy that they deliver – i.e. mechanical, thermal, electrical, chemical or ionizing radiation – or in terms of the energy exchange which they interrupt; i.e. thermal balance or oxygen utilization.

An 'accident' is an unplanned event that interrupts normal activity. It may or may not result in personal injury or property damage.

An 'injury' is a unit of bodily damage caused either by the delivery to the body of amounts of energy in excess of the corresponding local or whole-body injury threshold (as by impact or electric shock) or by interference with the normal whole-body or local energy exchange (as in suffocation).

An 'error' is a missing or inappropriate response to some stimulus.

This definition of 'accident' is one frequently used by safety engineers. The definitions of 'hazard' and 'injury' are based on the work of Haddon.

The teaching model demonstrates schematically the role of:

(a) missing or inappropriate response(s); and
(b) the source of potentially damaging energy.

The injury causal sequence always contains a hazard and an error. In most cases an error will lead to an unplanned event; that is, an accident. If no hazard is present the sequence will end at that point, but if the error–accident transition takes place in the presence of a hazard, or if the transition itself introduces a hazard, then, depending on a variety of factors, an injury may result.

The model shows that the hazard may be introduced in two ways. It may occur through an error on the part of the subsequent victim, for example, the inadvertent ingestion of a toxin. Alternatively, it may be introduced by an error on the part of another person, remote in time or space – for example, when a design fault causes a weakness in a structure that remains as an unsuspected hazard for a substantial period before eventual collapse. In both cases the causal sequence remains unchanged.

Wigglesworth listed typical human errors and their causes, involving:

- reception of information – 'failure of detection' including input overload, underload, or poor contrast, or 'incorrect identification' involving any of the first three causes or conflicting cues
- decision – 'failure to reach a decision' from lack of information, or an 'incorrect decision' from lack of information, incorrect priorities or inhibition of correct action
- action – 'failure to act' due to an underlearned sequence or inability to reach, 'incorrect control selection' due to lack of uniform layout, inadequate separation or coding or unintentional activation, or 'incorrect control adjustment' due to the presence of a non-controlling position or nonconformity with a stereotype.

The Wigglesworth model also demonstrates the role of probability in the sequence. The presence of a hazard does not automatically lead to an injury but depends upon some non-related factor. For example, when a brick is dropped from a height, the likelihood of injury depends on the number of people per square metre below and the distance the brick falls.

You will note that this model includes reference to the human factor involved, through the description of error. In addition, the hazard is identified and described in terms of energy exchange. This model specifically describes the factors involved in the accident sequence and, in particular, it highlights the contribution of error, as a human characteristic, to the sequence which may lead to injury.

Advantages of the model

In the past it has been usual for the safety practitioner to concentrate much effort on the hazard and the methods by which it can be removed or controlled. This approach has been very successfully applied in industry, but the injury causation model now provides the safety practitioner with another new field of investigation in determining the reasons for human error. The human error is not considered to have any element of blame. What is important is to determine the cause of the human error and to identify the stimulus that produced the behaviour pattern which resulted in the error.

An updated 'domino' sequence

Bird and Loftus extended the 'domino theory' in 1976 to reflect the influence of management in the cause and effect of all accidents which result in a waste of company assets. The modified sequence of events becomes:

(a) lack of control by management, permitting:
(b) basic causes (personal and job factors), that lead to:
(c) immediate causes (substandard practices – conditions – errors), which are the proximate (near) causes of:
(d) the accident, which results in:
(e) the loss (minor, serious or catastrophic).

This modified sequence can be applied to all accidents and is fundamental to loss control management.

The humanistic approach

Petersen, in 1978, developed the humanistic approach to management as it applied to safety. He considered that the traditional approach to accident causation was not keeping up with the changes and influences of the post-1960s management styles in industry. Petersen wrote extensively on the need to treat safety management in the same way business treated all other forms of management. He stressed the need to set objectives in safety performance and to make people accountable for meeting the objectives. This approach moved away from the Heinrich influences of the early safety philosophy based around unsafe acts and conditions and the pivotal role of the supervisor. While the supervisor remained a key player in the management role, the emphasis for performance outcomes should be driven by management accountability rather than supervisor responsibility, said Petersen. He also considered the concept of safety being part of a 'management system' and clearly understood the need to use accidents as a window into the work system.

Management oversight and risk tree

The Management Oversight and Risk Tree (MORT) is a system safety programme originally developed for the Nuclear Regulatory Commission in the USA by Johnson in 1980.

This programme is designed to accomplish the following objectives:

- prevent safety-related oversights, errors and omissions
- express risks in quantitative form (to the maximum possible degree), and refer these risks to proper management levels for appropriate actions
- make effective allocation of resources to individual hazard control efforts through the safety programme.

Rasmussen's SRK Model

The main thrust of this model is in the category of operator behaviour. It is based on Rasmussen's theoretical work on the analysis of operator tasks. According to his model, three levels of operator behaviour may be identified.

Skill-based behaviour

This refers to routine tasks requiring little or no conscious attention during task execution. In this way enough 'mental capacity' is left to perform other tasks in parallel. For example: an experienced car driver travelling a familiar route will control the vehicle on a skill-based level, enabling them to have an intelligent discussion with a passenger, parallel to the driving task.

Rule-based behaviour

This refers to familiar procedures applied to frequent decision-making situations. A car driver integrating the known rules for right-of-way at crossings with stop signs or traffic lights, deciding whether to stop the vehicle or pass the crossing, is functioning at this level also. The separate actions themselves (looking for other traffic, bringing the vehicle to a full stop, changing gears, etc.) will again be performed on a skill-based level. Making these familiar decisions and monitoring the execution of the skill-based actions requires some part of the total mental capacity available to the driver, but not all.

Knowledge-based behaviour

This refers to problem-solving activities, for instance when a person is confronted with new situations for which no readily available standard solutions exist. The same car driver approaching a crossing where the traffic lights have broken down during rush hour will first have to set their primary goal: do they want to proceed as fast as possible or do they want to minimize the chance of collision? Depending on this goal they will control the vehicle with varying degrees of risk taking (e.g. by ignoring some of the usual traffic rules whenever they see an opportunity to move ahead somewhat).

It is interesting to note that this model infers that an accident occurs as a consequence of goal setting followed by a human decision. Any error may be attributed to human behaviour, perception, cognitive skills and experience, factors which have been widely used in other accident models.

Rasmussen has also developed the boundary theory which has particular relevance to large-scale acccidents. In this theory organizations operate in a space of possibilities within the three boundaries of economic failure, unacceptable workload and functionally acceptable performance. Within this last boundary is the resulting perceived boundary of acceptable performance. The distance between these two is the margin for error. Rasmussen considers experiments to improve performance to create Brownian or random movements within this space. There are pressure gradients operating, such as the gradient towards least effort, and the management pressure towards efficiency, both driving the organization in the direction of the perceived boundary of acceptable performance. A counter gradient may exist in the form of safety culture campaigns. If corporate behaviour in the presence of strong gradients migrates past the perceived boundary and reaches the boundary of functionally acceptable performance, then an accident is likely.

Summary

These people have all made a significant contribution to the development of accident and injury prevention. Many other models have been put forward for consideration and, additionally, the work carried out by these early thinkers has been built on.

Today we have access to computer technologies and opportunities for research into the accident phenomenon. Through ever-increasing interest in the area, many more people are working in the field of occupational safety and health. The current thinking is that there is no single model or theory that is correct but rather that parts of all these models are relevant to particular workforces and industries. A list of references to models is given in the Appendix to Chapter 6.

Total loss control

Total loss control may be defined as those activities which are designed to reduce, prevent or otherwise control those events which produce economic or social loss. A total loss control programme may be designed to reduce the frequency and/or the severity of losses.

For example, the road freight industry, which involves both people and equipment, could institute a driver education programme in an attempt to reduce the frequency of accidents and incidents involving equipment. Larger organizations might provide their own fast emergency response at depots as a way of reducing the severity of accidents.

As the name implies, total loss control is concerned with controlling both the costs associated with losses which occur as a result of preventable events in the workplace, and the number of those events. Organizing and setting up a total loss control programme requires an organization to be very clear about the way in which its procedures operate.

Current concepts in work safety and health

A procedure is simply an organized scheme or plan of action for dealing with an identifiable body of knowledge. Your workplace already has procedures for dealing with various aspects of its operations. For example, there is a procedure for recording and paying wages.

What is needed to set up and operate a total loss control programme?

The first requirement is the recognition by management that a total loss control programme will pay dividends both in cost reduction and in a safer and more productive workplace. However, this recognition will not be obtained simply by asking for it. The need for such a programme has to be established by supplying management with convincing evidence. Such evidence might consist of:

- accident/incident investigation reports
- survey, inspection and audit reports
- cost analysis associated with accidents and incidents.

Having gained management approval, how and what do you require to put a total loss control programme into operation? You will need to consider all the following items:

- motivation of workers and management
- occupational health and safety requirements
- consultation with people in the workplace
- analysis of current operations
- establishing the costs of current operations
- setting up systems
- establishing procedures
- the results of system analysis
- injury and illness
- rehabilitation of workers
- training
- recruitment.

Motivation of workers and management

There are a range of strategies such as safety promotion activities, posters and films which help in motivating people to work safely and to report and investigate unsafe events. Ultimately, however, people will only cooperate if they are convinced that it is in their best interests. This will require constant feedback to people of the results of investigations as well as a demonstrable management commitment to fix problem areas.

Meeting occupational health and safety requirements

There are a range of specific legal requirements which must be met under the various pieces of legislation. You need to be aware of these generally and also of other requirements specific to your organization, and then act to meet them.

Consultation with people
A principal feature of the occupational health and safety acts is the setting up of a consultative mechanism in the workplace and the availability of people to consult. You should be aware of the requirements and how they apply to your workplace, and involve the people working in these areas.

Analysis of current operations
You need to establish an information base using the collection of facts about the workplace as the basis.

Establishing costs of current operations
In establishing costs, you may be required to make an 'intelligent guesstimate' of the costs required. After all, there may be little point in spending tens of thousands of dollars fixing a problem that is really a minor issue.

Setting up systems including accident reporting systems
In addition to the requirements set by local legislation (including an accident register) you will need to ensure that you have set up a detailed event-investigation system designed to educate staff as well as to collect data. There should also be an accident reporting system.

Appropriate procedures
The organization will need to establish procedures to deal with all the issues such as:

- the results of investigations
- safety analysis
- injury and illness management
- rehabilitation claims management
- local workers' compensation law.

This job will be made easier if procedures are written down. These written procedures can be used to train staff in appropriate ways to handle these matters.

Training
A basic principle of common law is that an organization's management must provide competent staff. In safety and associated training, as well as at induction sessions, the principles of accident prevention and safe behaviour should be emphasized at all levels of work.

Recruitment of staff
Those people recruiting staff should be familiar with the job and task specifications likely to affect health and safety. The organization should also ensure that it is complying with equal opportunity and anti-discrimination legislation.

Accident costs from workplace injuries and damage

Accident costs

Injury experience from workplace accidents can be found in many official publications. For a fuller picture, the average lost time injury frequency rate and the duration rate for the injuries need to be considered. You can usually obtain more information from your occupational health and safety authority.

It is useful to consider the statistics from different points of view:

- How do they look on a gender basis, considering the gender balance in different areas of industry?
- What do they tell you about age, especially in regard to duration rate?
- What types of injury have longer duration rates?
- Which age groups appear to be most at risk?

Note that the statistics are only as good as the injury reporting. The only nearly 100% reliable statistics, unfortunately, are fatalities.

The most common unit of workplace injury performance is the Frequency Rate. Other rates in common use are the Average Time Lost Rate (Duration Rate) and the Incidence Rate. Some organizations use other rates and indicators to measure their injury experience; examples being the Severity Rate and the Frequency–Severity Index. Severity Rate is based on those injuries which lead to more than a defined minimum of lost time.

The important consideration when selecting the appropriate measurement statistics is to ensure the time and effort required to collect the data are well spent. Producing statistics is a most important function of the safety and health professional, because decisions made by management are often based on the information supplied or inferred by the injury statistics; therefore, the message provided to management must be clear, accurate and relevant.

Various rates are used to identify several important factors, such as:

- the identification of accident/injury trends and hazards
- improvement or deterioration in safety performance in sectors of the workplace
- comparisons between like workplaces and industries
- where to allocate resources and support to achieve the best result for effort and expenditure
- the adequacy of existing policies, procedures and the work environment
- monitoring the overall organization performance, in particular the requirement to provide employees with safe systems of work
- compliance with statutory requirements, including working arrangements required by self-regulation style legislation.

The Frequency Rate

The Frequency Rate is used to identify the number of injuries experienced or expected in a period where one million person-hours of exposure occur. The formula to calculate the Frequency Rate involves two important pieces of information: firstly, the number of Lost-Time Injuries (LTIs) and, secondly, the total hours of exposure experienced by the group during the period of interest.

An LTI results through the loss of at least one full shift following the accident. Provided there is relevant information, the following formula will produce a Frequency Rate:

$$\text{Frequency Rate} = \frac{\text{Number of LTIs} \times 1\,000\,000}{\text{Total person-hours worked}}$$

Example A
An organization with 500 workers works a total of 1.15 million hours during one year. In the same period the workers experienced 46 LTIs.
Therefore

$$\text{Frequency Rate} = \frac{46 \times 1\,000\,000}{1\,150\,000}$$

$$= 40$$

This calculation demonstrates that, in the year, 40 LTIs occurred per million person-hours worked.

It is important to note that the Frequency Rate of 40 does not indicate how serious these LTIs were. It indicates that the injured worker was away from work for at least one shift following the accident. Where a worker is killed at work, one (1) is added to the LTIs for the purpose of calculating a Frequency Rate.

Example B
A small workshop of 10 workers works a total of 23 000 hours during one year. During this period the workers experienced two (2) LTIs. However, as we will see later in the section, no single rate is capable of identifying all the important factors.
Therefore

$$\text{Frequency Rate} = \frac{2 \times 1\,000\,000}{23\,000}$$

$$= 86.96\ (87)$$

Decimal figures are rounded to the nearest whole number; that is, 86.96 becomes 87. The calculation says that the group are experiencing a rate of 87 LTIs per million hours of exposure.

There must be some question over the value of knowing this Frequency Rate, as it will take the group of 10 workers, with the current exposure of 23 000 hours per year, over 43 years to amass the million person-hours. At which time, if nothing changes, they will have suffered 87 LTIs. What can we really do with this information?

Further comparisons are difficult between these two examples because the small workshop would have a Frequency Rate of 43.48 (43) if only one (1) LTI had occurred during the year. In summary, it could be considered that the Frequency Rate is insensitive

Current concepts in work safety and health

to small groups and because of the wide variations in the level of injury, ranging from one shift lost through to death, the Frequency Rate alone is at best suspect as a true performance guide. The Frequency Rate should not be disregarded, however, but used in combination with other indicators.

Average Time Lost Rate (ATLR)

This rate is often referred to as the Duration Rate, and is used to indicate the severity of injury. As we saw when discussing the Frequency Rate, we were unable to determine how serious the LTIs were because this data was not built into the formula. By using the ATLR in combination with the Frequency Rate a clearer picture of injury performance is possible. The ATLR is calculated by dividing the number of days lost through injury by the number of LTIs.

$$\text{Average Time Lost Rate} = \frac{\text{Number of days lost}}{\text{Number of LTIs}}$$

A death is counted as 220 standard working days lost.

Example C

In our organization with 500 workers, who experienced 46 LTIs (see Example A), assume the following injury outcomes:

```
10 LTIs resulted in 3 days each   =  30
 8 LTIs resulted in 6 days each   =  48
12 LTIs resulted in 14 days each  = 168
 4 LTIs resulted in 20 days each  =  80
10 LTIs resulted in 28 days each  = 280
 2 LTIs resulted in 42 days each  =  84
                           Total    690
```

$$\text{Average Time Lost Rate} = \frac{690 \text{ days lost}}{46 \text{ LTIs}}$$

$$= 15.$$

As a result of this calculation we are now in a better position to assess the injury performance. We now know the organization has had 40 LTIs per million hours exposure and each injury experienced during the year resulted in an average of 15 days off work. This additional information provides the safety and health professional with the necessary knowledge required to make decisions and recommendations on the areas of risk within the workplace where resources and effort should be directed.

Example D

Now let us consider the small workshop of 10 workers who experienced 2 LTIs (see Example B).

Enhancing occupational safety and health

1 LTI resulted in 2 days	2
1 LTI resulted in 4 days	4
Total	6

$$\text{Average Time Lost Rate} = \frac{6 \text{ days lost}}{2 \text{ LTIs}}$$

$$= 3.$$

The Incidence Rate

The Incidence Rate calculates in percentage terms the number of occurrences, such as LTIs, experienced by the work group.

$$\text{Incidence Rate} = \frac{\text{No. of occurrences} \times 100}{\text{No. of workers exposed}}$$

$$\text{Incidence Rate (large organization, i.e. Example C)} = \frac{46 \times 100}{500}$$

$$= 9.2\%$$

$$\text{Incidence Rate (small organization, i.e. Example D)} = \frac{2 \times 100}{10}$$

$$= 20\%$$

Summary

Both the large organization (Examples A and C) and the small organization (Examples B and D) carry out the same type of work – steel fabrication. The employees in both organizations work the same hours during the year, 2300 hours each. The results of our calculations, however, show vastly different performance outcomes. See Table 1.2.

The use of rates and other indicators are important factors in safety and health work. However, caution must always be shown when using these rates as the only means of measuring risk in the workplace. Remember that these particular rates do not identify the accidents in the workplace; they are only measuring Lost-Time Injury experience.

Types of workplace injuries

Some idea of the types of workplace injury can be had by studying the tables in various databases, some online. These can give a view on an occupation and industry basis

Table 1.2 Injury rates

	Frequency Rate	Average Time Lost Rate	Incidence Rate
Larger organization	40	15	9.2%
Smaller organization	87	3	20%

or a view of the duration based on the nature of injury. Other duration sheets relating to bodily location, mechanism of injury and breakdown agency (e.g. tool, substance) are also available.

Another way of looking at injuries, suggested by G.L. McDonald, a Queensland safety researcher, involves dividing injury into three classes:

Class 1   Accident permanently alters the future of the individual.
Class 2   Lost-time accident where individual fully recovers.
Class 3   Accidents which cause inconvenience to the individual but do not stop him/her from carrying out normal duties.

McDonald (see Further Reading at the end of this chapter) has produced line diagrams which graph class of injury and total cost of that class based on actual data for an OHS jurisdiction. The case for focusing prevention on Class 1 is very strong.

Distinguishing accidents from injury

Accident costs

In identifying accident costs, some attention needs to be focused on the 'non injury' part of the problem. H.W. Heinrich asserted in 1931, after original research in this area, that less than 10% of all accidents result in personal injury. From his studies the following information emerged: for every disabling injury there were 29 less serious injuries and 300 accidents involving no injury.

Further research undertaken by Simonds in the USA in 1954 showed that accidents gave rise to considerable indirect or uninsured costs due to overheads and delays. Refer to Simonds and Grimaldi (1989) in Further Reading at the end of this chapter.

By 1959 F.E. Bird Jr. had taken the accident loss potential one step further. He had identified the *damage* aspect inherent in accidents. His theory stated that for every one disabling injury at Lukens Steel Company in the USA there were 100 minor injuries and 500 property damage accidents.

Implications and application of the Bird–Heinrich triangle

Bird went on further to look at a wide range of accident reports across a large number of companies in America. In 1969 he analysed the information and generated the 'Bird Accident Ratio Triangle'. See Fig. 1.4 and Further Reading at the end of this chapter.

Conclusions that the study revealed

The 1–10–30–600 relationships in the ratio would seem to indicate quite clearly how foolish it is to direct our total effort at the relatively few events terminating in serious or disabling injury when there are 630 property damage or no-loss incidents occurring that provide a much larger basis for more effective control of total accident loss.

Heinrich's original data of 1–29–300 was based on 'accidents of the same kind and involving the same person'. The figures are averages of masses of people and all kinds

Enhancing occupational safety and health

Figure 1.4 The Bird accident ratio triangle

of different accident causes and types. It does not mean that these ratios apply to all situations. It does not mean, for instance, that there would be the same ratio for an office worker as for a steel worker. It might mean they could be averaged to this (or a similar) ratio, but certainly neither of these extremes would fit the ratio.

It also does not mean, as we have too often interpreted it, that the causes of frequent injuries are the same as the causes of severe injuries. We have typically believed a 1–29–300 ratio might apply to all kinds of accident types and causes. Obviously there are different ratios for different accident types, for different jobs, for different people, etc; for example, *electricity* would produce a different looking triangle to *handling materials*.

Common sense dictates totally different workplace relationships in different types of work. For instance, a steel worker would no doubt have a different set of relationships compared to those of an office worker. This very difference might lead us to a new conclusion. Perhaps circumstances which produce the severe accident are different from those that produce the minor accident.

For years, safety workers have been targeting frequency in the belief that severity would be reduced as a by-product. As a result, in Australia for example, Frequency Rates nationwide have been reduced much more than have Severity Rates. One Australian state reported a 33% reduction in all accidents over a period of 10 years while, during the same period, the number of permanent partial disability injuries actually increased. (This could have resulted from a statistical recording change rather than any real improvement in accident rates.)

With an infectious disease like sudden acute respiratory syndrome (SARS), a triangle standing on its apex may be more appropriate. The apex would represent one victim who flies to another country before serious symptoms appear, and the rest of the triangle the subsequent range of effects as the disease spreads in a new environment.

Direct and indirect cost of workplace accidents

All these studies serve to show the need for a detailed look into accident costs. They were all carried out in different industries and under different technologies and different social conditions than exist today.

Most of the formulas generated out of these studies will not apply to other organizations. For this reason, an accurate costing of loss experience must be conducted within individual organizations. Currently, most organizations do not use a proper cost analysis technique to identify major loss factors resulting from workplace accidents and injuries.

Insured costs are paid out by the employer's insurance company or the government authority concerned with providing the workers' compensation cover. Workers' compensation for employees is generally compulsory. However, uninsured costs are usually concealed as part of normal operating expenses. Most organizations pay attention to their workers' compensation premium. The annual premium is often used as a barometer of organizational performance, as it reflects quantifiable claims experience. However, this pay-out is not an accurate guide to total accident costs.

Often, little effort is put into undisclosed costs. These could be described as the uninsured costs. Information on these hidden costs is critical to planning strategies, decision making and control techniques. The logistics of such an undertaking provide a degree of doubt in the minds of some managers as to whether it would be a practicable exercise.

While it is agreed that the task would be monumental in an organization that is large and diverse, until accident costs are identified and loss potential considered there appears to be great difficulty in planning and implementing effective control strategies.

A closer look at accident costs

Should a serious effort be made to assess the total costs of accidents, it is soon realized that these costs come under two major headings:

1. Those covered by insurance.
2. Those not covered by insurance.

The costs covered by insurance are those which come within the scope of workers' compensation, motor vehicle, property, machinery damage, fire, public risk, etc.

All premium rates are based on the basic principles of insurance:

- the risk and extent of potential loss
- past experience.

The costs of accidents not covered by insurance are high, and can be several times the cost of those covered by insurance. Following on-the-job accidents, there will be disruption of the orderly processes, and these interruptions must be paid for. Let's consider some of the costs involved.

The costs of wages for time lost by people not injured

Ordinarily, employees near the scene of an accident which results in injury or damage will stop to watch, assist or discuss the accident. They may not be able to continue without the aid or output of the injured worker or damaged equipment.

It is reasonable to assume each employee will be producing a given amount in work value to the organization every hour. This will be at least as much as the wages the employee is paid; otherwise the employee would not be there. In other words, the person has to be producing (in US currency, for example) $20.00 before being paid $20.00. Obviously this is the absolute minimum. In addition, allowance has to be made for overheads, power, space, administration and the like. Close examination will show that this amount will often be two to two-and-a-half times the amount of wages paid before a start can be made on profits; that is, $40.00–$50.00 per hour.

The costs to repair or replace property damage

These relate to the costs of repairing or replacing machinery, raw materials, work in process and property not insured, inadequately insured or not insured for replacement value.

There are difficulties in estimating the real worth of an old machine destroyed by accident. There are many machines in use today in which all, or practically all, of the original price has been written-off in depreciation. This does not mean that the destruction of such machines is of no loss to the organization.

If a machine was bought ten years ago for $10 000 and at the date of destruction was still in good operating condition, even though the book value was nil, the purchase of a replacement machine to bring the plant to the same operating efficiency could now cost $16 000. The loss would be $16 000 and any additional taxation advantages such as depreciation, less the recoup of any salvage value and insurance payouts.

Loss of production by the injured worker

As in the case of the persons not injured, the work value of the injured person during time lost must be regarded as worth at least as much to that person's employer as the amount of wages paid for the employee's time. The loss also occurs when the employee is away from the job being treated for minor injuries, with the added cost of actual wages paid for the period.

Cost of supervisor's time

Supervisor's time is taken up following accidents, making adjustments, arranging assistance for the injured workers, preparing accident reports, training replacement workers and rearranging staff and workloads. The organization loses the value of the work the supervisor would have performed during the period of 'cleaning up' after an accident.

Supervisors do not just sit around waiting for an accident to occur to make themselves useful. In addition to getting work out on time, they are responsible for planning, instructing workers, improving methods, eliminating bottlenecks and the like. If supervisors had more than adequate time then the ratio of supervisors to workers could be reduced.

Decreased output of the returned injured worker

Often workers are not fully recovered when they return to work after being injured, and do not work at their normal rate. The percentage of pay which corresponds with the percentage reduction in output is a loss.

Uninsured medical costs borne by the organization
This cost is for the services supplied and equipment used in maintaining a first aid unit. Bandages, medication, vehicles and buildings have to be paid for, as well as the cost of maintaining the first aid centre and attendants.

The cost of time spent by senior staff
Accidents have to be investigated, particularly those involving serious injury or damage. As a result activities such as meetings, reviews, follow-ups and other discussions occur. These often involve senior staff. This represents a loss of productive work value.

Other costs
The costs mentioned above represent only a few of the valid elements of accident costs. Other costs include such items as hiring replacement staff and/or equipment, excess spoilage of product, damaged minor equipment, the cost of preparing reports for government departments and insurance companies, and many more. In the case of, for example, construction companies tendering for projects, serious accidents can result in loss of future contracts.

Implications of an injury on employers' liability, workers' compensation and rehabilitation

Premiums are not fixed; they are adjustable. If you reduce the risk of loss, and your claims experience is the main guide here, premiums will be reduced.

Workers' compensation insurance generally has maximum and minimum premiums for particular types of employment set by law. Good rehabilitation practices, with an early return to work for employees where possible, can reduce the overall costs of an injury.

In addition to the prescribed pay-outs set by workers' compensation law, employers may in some jurisdictions be liable to a common law claim (civil suit) for pain and suffering. Some pay-outs can be very high; for example, where an injured worker needs constant nursing care for the rest of a lifetime, or where back injury interferes with sexual function and hence may affect a marriage.

As unwanted events, accidents also include the undesirable effects of stress on workers. Some jurisdictions have moved to restrict claims based on stress, e.g. from retrenchment, even if the retrenchment is handled badly.

The other serious implication of a workplace injury, where it can be shown that the duty of care to employees was breached, is the possibility of a heavy fine under occupational health and safety legislation. The organization or an individual director, manager or supervisor could be charged. Some jurisdictions use manslaughter charges under criminal law for very serious breaches, and imprisonment is being considered as a possible penalty or can flow from the nature of the charge.

Issues in the work environment affecting employee well-being

Relationship between person, machine, environment, procedures and materials

There are five key overlapping elements in the work system: people, machines (or equipment), work environment, procedures, and materials. This section deals with these factors that make up a work system and how these factors have an effect on accident causation.

Before considering the delicate relationship between these factors, it is worth reflecting on the point that little thought was given to the relationship between persons and machines until relatively recent times. In this context the word 'machine' holds for any piece of equipment used to perform work. Consequently, the golf club and shovel are in this sense as much machines as the car we drive or the lathe and drill we work with.

When operating a machine a person receives information on the machine mainly through the senses of eyes, ears and touch. When information is received by a person it is processed to arrive at a decision. The information received will combine with the knowledge gained by past experiences that have been stored away in the memory. The decisions that result out of this combination of information will vary from simple responses, made automatically, to those that require a high degree of reasoning and care. Having arrived at a decision a person may then take action. The machine receives these commands and produces some work; the machine will then inform the operator of what is happening by displaying further information. The cycle begins again.

It is a reasonable fact that people are not perfect. However, some person–machine systems are designed on the premise that the human component will be error-free. It is most important that these systems are identified and modified to obtain a realistic reliability and monitored or modified to recognize and take account of the performance limitations of individuals. It is most important that we consider these three factors, person–machine–environment, as an integrated work process bound together by the procedures we adopt to actually perform tasks, and which, in some cases, requires attention to the materials used, e.g. flammable solvents. Figure 1.5 sums up the work system using slightly broader terms such as equipment and work methods.

Workplace relationships

Interface	Possible Hazard
Person–Procedure	Are the people familiar with the procedure to be followed for the job? Is the procedure able to be understood by the people doing the work?
Procedure–Machine	If the procedure calls for a mobile crane or some other piece of equipment to be used, is that equipment readily available? Is the equipment designed for the use spelled out in the procedure?

Current concepts in work safety and health

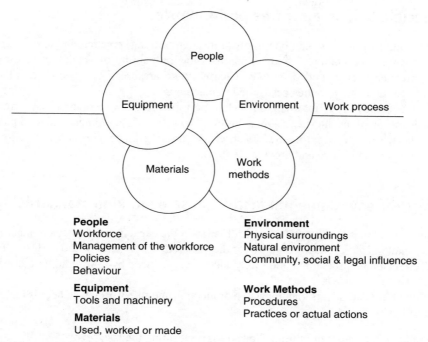

People
Workforce
Management of the workforce
Policies
Behaviour

Equipment
Tools and machinery

Materials
Used, worked or made

Environment
Physical surroundings
Natural environment
Community, social & legal influences

Work Methods
Procedures
Practices or actual actions

Figure 1.5 The work system

Interface	Possible Hazard
Machine–Environment	Does the environment prevent the use of certain machinery or equipment – for example, ground too soft for mobile crane, or induced voltage from nearby high voltage lines necessitating the use of conductive rather than electrically resistive footwear?
Person–Machine	Are the people trained and experienced with the equipment they are working with?
Person–Environment	Does the environment create hazards for the people working there – such as: wind, precipitation, temperature, persons working above or below other persons working at height and subject to falling hazard?
Procedures–Environment	Does the environment prevent procedures (safety rules) from being followed – for example, is adjacent live apparatus too close for workers to carry out their tasks without coming inside safety limits?
Materials–Equipment	Does the equipment used to mix flammable solvents have a spark-proof electric motor?
Person–Materials	Is there adequate distance and protection between personnel and explosives before a charge is detonated?

Principle behind error-free performance

While human reliability can be improved through training, practice, increased information and good communications, there is a low probability of obtaining error-free performance over extended periods. Clearly, there are areas that people excel in; similarly, there are activities best suited to machines.

A low error rate, if not totally error-free performance, can be obtained by careful consideration of the interfaces above, followed by good job, equipment and workplace design (see below), appropriate selection and training of people, and a system to assess error occurrence and provide feedback.

Effect of environmental factors on safe working standards

The environment has a significant part to play in the capacity of the person–machine combination to produce output at the desired level and without loss. The person–machine production and safety factors are reduced by downgrading the work environment.

Environmental factors that impact on production and safe working standards include the following:

- reducing the amount of time allowed for performing the task
- abnormal temperature conditions
- failing to provide adequate lighting or proper illumination
- restrictions to movement through special clothing and equipment
- failure to provide compatible workplace design
- excessive noise and vibration
- imposing stress: lack of rest, confinement, isolation
- emotional stresses: fear, anxiety, boredom or personal issues.

Generally, a machine is not designed to operate continuously at its maximum limit. Logically, we should not design work systems which demand constant maximum output and vigilance from humans. Both people and machines are less prone to error where environmental variations and extremes can be avoided. Also to be avoided is the belief that, because person and machine are seen to perform well in poor environmental conditions, improvement is not needed. The effectiveness of procedures in ensuring suitable practices depends on training, supervision and the culture of the organization.

Role of ergonomics in job design

Ergonomics is often referred to as 'human factors engineering' and practically applies knowledge of the strengths and limitations of people to the design of machines and jobs.

While the person displays a wide variation in factors such as height, weight, strength and endurance, there are some general principles applicable for most people.

Current concepts in work safety and health

These enable us to plan the work environment in such a way that we can optimize performance. Underlying the desire to see people perform to optimum level is an appreciation that the mental and physical well-being of the individual is of the utmost concern.

It must be clearly understood by those responsible for the design of work methods, and those who demonstrate and supervise work practices, that normal variations in people place a limitation on the capacity for all individuals to perform at the same level. The failure, by those who plan and programme work systems, to grasp an understanding that individuals have different performance levels, sits at the core of many workforce accidents. See Fig. 1.6.

Technical specifications

The person, the human machine, is not a very efficient unit when compared against most mechanical machines; however, the person has greater flexibility in decision making and is often required to work in areas where mechanical machines cannot function. The following examples of ergonomic parameters are general indicators only; the exact requirements will vary with the individual. Before using this type of information to assess a work system, always refer to the latest technical publications and any relevant standards.

Human factor parameters

Power Output: 0.1 to 0.2 kw per day.
Efficiency: 20–30%, i.e. 70–80% heat loss.
Lighting requirements: General work areas – 200 lux.
                 Medium machinery and assembly work – 400 lux.
                 Fine machinery and assembly – 800 lux.

Optimum temperature:

Type of Work	Temperature (°C)
Clerical	21
Sitting – light manual	19
Standing – light manual	18
Standing – heavy manual	17
Severe (hard) work	5–16

Influence of perception, memory and risk on decision making

Perception

Some factors affecting perception by workers are:

- hearing range 20–20 000 Hertz
- vision takes time to adapt to the dark after exposure to bright light
- expectancy causes people to see what they want to see
- people cannot concentrate continuously on the same task
- reaction time is about 0.3 seconds, even for a simple reaction.

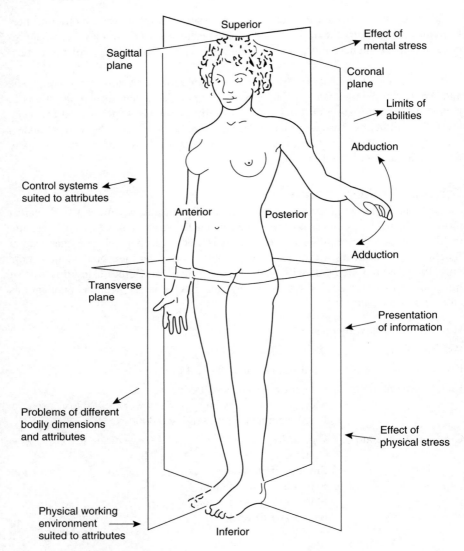

Figure 1.6 The person and the work environment

Memory
Long-term memory capacity is vast and readily accessible, but can be unreliable. Short-term memory is limited and expecting people to have full retention of many items of information (greater than seven) is unwise.

Risk
The person has the ability to make reliable accurate judgements about the probability of events. The exceptions are the high and low levels where people tend to fix on the

outcomes of 0 (no chance) and 1 (certainty). Since risk is quite frequently at very low levels of probability, people's estimation of risk is quite poor.

Environment

The person works best over a narrow range of environmental conditions. High temperatures produce heat stress, low temperatures cold stress. High noise levels cause pain and hearing loss. High and low frequency vibration can lead to discomfort and injury. A person cannot be exposed to ionizing radiation, toxic fumes or hazardous chemicals above permissible threshold limits without threat of injury.

Information capacity

There is a limit to the amount and rate of information which people can receive and process. When the system is overloaded, the coping mechanism is to accept the important information and ignore the rest. Generally, a person cannot do too many tasks simultaneously without error.

The safety precedence sequence

A thorough review of the work process, combined with an appreciation of human limitations, provides the opportunity to better understand the high risk areas within the work system. Having identified areas within the system that need improvement, we are in a position to consider an appropriate strategy. The strategy consists of the following sequence:

MOST EFFECTIVE

1. *Design to eliminate hazards*
   - substitute a safer process or product.
2. *Design to minimize energies*
   - reduce energies to the minimum, e.g. valves installed at convenient heights for repair.
3. *Install physical barriers or safety devices*
   - interlocks, pressure relief valves, fall-arresting devices.
4. *Install warning devices*
   - warn personnel of a potential and/or actual hazard, e.g. fire alarm, roped-off area, warning signs.
5. *Design to minimize human error potential*
   - human factors review – reduce: number of error-provoking situations; for example, UP button on top; DOWN button on bottom.
6. *Write-protective procedures*
   - most frequently used, though of limited effectiveness, e.g. operating manuals, safe operating procedures.
7. *Select, train, motivate and supervise personnel*
8. *Identify residual risk to line management*
   - when all the above controls have been considered, either accept the **residual risk** or cease that particular task.

LEAST EFFECTIVE

The safety precedence sequence is a method of reviewing safety options. Clearly, the most effective option to reduce hazards is to eliminate them altogether. This is rarely possible. Most options are controlled by cost, technical capability and information. As we select options lower down the sequence, the controls become less reliable and less effective. In most cases, some compromise has to be reached in order for work to continue.

Further reading

Bird, F.E. (Jr.) and Germain, G.L. (1969). *Damage Control*. New York: American Management Association.

CCH, Australia. (2000). *Planning Occupational Safety and Health*. 5th edn. Sydney: CCH, Australia.

Guarnieri, M. (1992). Landmarks in the History of Safety. *Journal of Safety Research*, **23(3)**, 151–8.

Haddon, W. (Jr.) (1963). A Note Concerning Accident Theory and Research with Special Reference to Motor Vehicle Accidents. *Annals of the NY Academy of Sciences*, **107**, 635–46.

Haddon, W. (Jr.) (1967). The Prevention of Accidents, Chapter 33 in *Textbook of Preventive Medicine* (Clark and MacMahon (eds)), Boston, Mass.: Little Brown.

Haddon, W. (Jr.) (1973). Energy Damage and the Ten Countermeasure Strategies, *Journal of Trauma*, **13(4)**, 321–31.

Haddon, W. (Jr.), Suchman, E.A. and Klein, D. (1964). *Accident Research: Methods and Approaches*. New York: Harper and Row.

Heinrich, H. (1959). *Industrial Accident Prevention,* 4th edn. New York: McGraw-Hill Company, Inc.

Heinrich, H.W., Petersen, D. and Roos, N.R. (1980). *Industrial Accident Prevention*. New York: McGraw-Hill.

Holt, A. St.J. (1998). *Principles of Health and Safety at Work*. 2nd edn. Wigston Leicestershire: IOSH.

ILO Convention 155, Recommendation 164. *Occupational Health and Safety and the Working Environment*. Geneva: ILO.

Kinney, G.F. and Wiruth, A.D. (1976). *Practical Risk Analysis and Safety Management*. AD/A–027189 Washington, US Dept. of Commerce National Technical Information Service.

Manuele, F. (2002). *Heinrich Revisited: Truisms or Myths?* Itasca Ill: NSC Press.

Marsh, S.M. and Lane L.A. (2001). *Fatal Injuries to Workers in the United States 1980–1995*. Cincinnati: US DHHS NIOSH.

Mathews, J. (1993). *Health and Safety at Work*, 2nd edn. Sydney: Pluto Press.

McDonald, G.L. (1995). *Safety, Consignorance or Information*. Technical Papers of the Asia Pacific Conference on Occupational Health, Brisbane, September. Toowoomba, University of Southern Queensland.

McDonald, G.L. (1998). *The Cost of Damaging Occurrences. Work and Road*. Proceedings of Safety in Action Conference, Melbourne, 25–28 February. Melbourne, Safety Institute of Australia.

Ridley, J.R. (1998). *Safety at Work*, 5th edn. London: Butterworth-Heinemann.
Sauter, S.L. et al. (2002). *The Changing Organization of Work and the Safety and Health of Working People*. Cincinnati, US DHHS NIOSH.
Simonds, R.H. and Grimaldi, J.V. (1989). *Safety Management*. 5th edn. Homewood, Ill. (Illinois): Richard D. Irwin.
Standards Australia. AS1885.1–1990. *Workplace Injury and Disease Recording Standard*. Sydney: Standards Australia.
Stellman, J.M. (ed.) (1998). *Encyclopaedia of Occupational Health and Safety*. 4th edn. Geneva: ILO.
Taylor, G.A. (1998). *Swirling Mists – Where is the Narrow Way?* Safety in Australia, **21(3)**, 22–26. (A review of some OHS journals under 32 topic areas reflecting demands on OHS practitioners.)
Wigglesworth, E.C. (1972). A Teaching Model of Injury Causation and a Guide for Selecting Counter Measures. *Occupational Psychology*, **46**, 69–78.
Wigglesworth, E.C. (1978). The Fault Doctrine and Injury Control. *Journal of Occupational Trauma*, 18, **(12)**, 37–42.

Activities

1. Ask three fellow workers to explain to you their understanding of 'self-regulation'.
2. Determine how accidents are identified in a workplace.
3. Identify the most common form of injury in an industry or workplace.
4. Identify an accident with no injury outcome.
5. Review an accident report and consider the information value as a guide to future prevention.
6. Identify three environmental factors that influence safety performance in a positive way.
7. Inspect a work environment and identify a good ergonomic job design.

2

Safety management

WORKPLACE EXAMPLE

An Ontario mining company was fined a total of CAN $225 000 for the death of an employee. The worker was using a stick to scrape off finely crushed rock on a pulley while a conveyor in an ore-crushing house was running. The stick got caught between the pulley and belt and was pulled in. A screw on the end of the stick caught the worker's glove and he too was pulled in. This occurred even though a guarding device was there, but this still allowed access to the pinch point. The worker, who was on the night shift, was not being directly supervised.

The lack of supervision on the night shift and the lack of sufficient guarding were offences under the Ontario OSH legislation.

(Source: Ontario Ministry of Labour, Canada)

Safety and health in risk management

Introduction

There are many processes and tools used in safety management. Most of the processes used to manage safety aim to eliminate or control costs and the exposure of people to harm.

The management strategy

When we look at individual processes used in safety management, it becomes very clear that each process is linked and is usually not used in isolation. Two common forms of safety management include:

- risk management
- loss control.

However, before we look more closely at the two safety management processes, it is important to consider that both risk management and loss control are linked by controlling or managing hazards within the workplace.

Definitions

Definitions which will assist you include:

Hazard
When a hazard is present, the possibility of adverse effects exists. A hazard may be:

- any thing or any condition which has the potential to cause injury or harm to health
- a source of potentially damaging energy
- a situation with the potential for harm to life, health and property.

Risk
This can be viewed as one of the following:

- the probability that an event may occur resulting in personal injury or loss to the organization
- the probability that a hazard is realized, i.e. leads to an untoward event
- the measure of how likely it is that injury will occur in a given situation.

Measures of risk usually also include the consequences of the untoward event actually occurring.

Risk management, loss control and acceptable risk

Risk management
Fundamentally, this is a process of planning, organizing, directing and controlling activities which lead to the identification, assessment and control of risks in an organization.

Loss control
This involves identifying and controlling weaknesses in a workplace system where there is potential for loss to people, plant and equipment.

Acceptable risk
A risk is considered to be acceptable when the money value of further control measures far outweighs the expected reduction in risk those measures would provide. It is a value judgement in which a number of parties have a stake. The higher the risk, the more should be spent on reducing it.

Elements of risk management

Risk management involves the identification, assessment and control of all areas of risk in an organization and aims to minimize loss or wastage of business assets. When managing safety, remember the Pareto 80/20 rule which states that 80% of accidents are the result of 20% of activities. Targeting the most dangerous 20% of activities and preventing these accidents makes the most effective use of time.

The control of risks will enable an organization to operate efficiently and safely. On this understanding, risks can be managed through five basic principles:

1. Risk evaluation
   Once a list of risks within an organization has been compiled, the impact of each risk on the organization – assuming no control action has been taken – requires evaluation. Risks may be put in order of priority to decide when control action is required.
2. Risk avoidance
   In this stage, risks are eliminated. An example of this is a manual handling hazard in which lifting heavy boxes may result in back strain. The risk may be avoided through the use of a mechanical aid (fork-lift truck), therefore eliminating the risk of back injury with the task. Another example is where a fire hazard exists in a particular area of a plant. Switching to non-combustible materials can eliminate this hazard, and so prevent it being realized.
3. Risk reduction
   One method of reducing a risk is through engineering. For example, a conveyor belt nip point may present a hazard of trapping hands or fingers at that point. Engineering controls, in which guarding prevents contact with the moving parts in equipment, will control the hazard – for example, the use of electrical motors instead of shafts and belts in powering machinery.
4. Risk retention
   In cases where risk cannot be avoided, or reduced, risk may be retained and the cost absorbed by the organization, taking an insurer's point of view of risk retention. From an OHS practitioner's viewpoint, insurance such as workers' compensation or fire can be taken out for some retained risk.
5. Risk transfer
   Risk transfer involves shifting the point of risk to other parties who are better equipped to control the risk. This might be done because of inadequate equipment, experience or skill with a particular task. In an organization an example of this is the use of emergency management teams which are trained to combat disasters. Insurers tend to use the term 'risk transfer' to mean that the cost of the retained risk has been transferred to the insurer.

Techniques used in risk management

Risk management should involve identifying and controlling risks before loss occurs. A workplace or organization may be looked at through a systems approach when considering risk management. A systems approach looks closely at areas within the workplace through the following steps.

Step one

One of the first practical steps that you can take with risk management is to consider the work systems within an organization and then at each department level. Work systems include all the components required to perform a particular function, for example:

- human resources – people
- equipment – tools, plant
- environment – physical surroundings, lighting
- procedures – the way or method in which work is carried out
- supervision – control, steering.

All of these components are required to function together; however, if a risk or subsequently a failure occurs in any one area, efficiency is reduced and loss occurs.

Risks associated with the above example may include:

- equipment not regularly serviced or maintained
- ineffective safe operating procedures
- manual handling risks
- floor surface, e.g. slippery
- people not properly trained.

Step two

Identify the work areas or departments within your organization. Your organization may, depending upon its structure, include these six areas:

Office	Shop floor
Stores	Dispatch
Maintenance	On the road

Once you have established the work areas and work systems, risks may be identified using this grouped approach.

Step three

Risks may be identified through employee identification of them, safety inspections, analysis or accident statistics, and near miss incident audits, checklists and reports. This is not an exhaustive list, however. Once risks are identified, review the process involved from Step 1, within the selected department (Step 2), then evaluate and apply risk control principles in consultation with the department supervisor, employees and health and safety representatives (where they exist).

Other methods of risk identification include the use of structured programmes such as:

- fault tree analysis, and
- hazard and operability study.

Major areas of risk

Areas of risk may be divided into:

- employees and management
- equipment, machinery and plant
- materials
- the environment
- work practices.

Employees and management

Management failure in the work system can take various forms. Therefore, consideration should be given in the preparation and implementation of such a system to the following:

- formalized procedures and correct use of PPE
- enforcement of the health and safety policy and rules
- regular reviews of all written systems of work
- establishment of health and safety committees and election, where law provides, of health and safety representatives ('delegates' under US mine safety legislation)
- industrial relations
- adequate training and instruction
- safe plant and equipment
- clear and effective communication pathways
- checks that procedures are followed and review of these on a regular basis
- ongoing education and training
- competent supervision.

Equipment, machinery and plant

As employees are often in direct or indirect contact with items of equipment, machinery or plant, questions that need to be asked are:

- Is there adequate guarding on all plant and equipment?
- Is equipment used for its designed purpose?
- Is equipment regularly maintained?
- Do selection criteria for equipment meet standards and legislative requirements?
- Is there a safe working procedure documented and practised with plant, machinery and equipment?
- Has consideration been given to the ergonomic principle of fitting the task to the person?

Materials

Material handling and processing may involve risks such as manual handling, thermal conditions and dust exposure. In consideration of this, risks may be associated with the:

- poor selection of materials
- work practices unsuited to the materials
- poor housekeeping
- poor cleaning, handling and disposal of waste.

The environment

The environment in which we work presents hazards which can be grouped into six main categories:

- chemical — risks from dust, fumes and vapours or flammable, corrosive, irritant, carcinogenic or teratogenic substances
- physical — lighting (inadequate lighting to attend to tasks), noise (hearing damage from loud noise), radiation, temperature (temperature extremes, e.g. too hot or cold), gravitational, hydraulic, kinetic (moving parts) energy or stored energy (e.g. springs, flywheel), poor layout, lack of storage, poor housekeeping
- biological — waste disposal, syringes, animal-borne diseases, cooling towers
- ergonomic — floor surfaces (floor surfaces that are slippery or hard), poorly designed work stations (e.g. for VDUs)
- electrical — wiring, tools, control panels, circuits, fuses, etc.
- psychological — quality of work, fatigue, pace and length of work, quality of management.

Work practices

Work practices are the way in which tasks or procedures are carried out. It is not only the physical aspects of work practices which may present a risk but also the layout of the work areas. Work practice problems include:

- bad timing with process
- safe work procedures may not be adhered to or developed
- work schedules may result in employees rushing
- poor ergonomic design of work stations
- inadequate storage space and housekeeping problems
- lack of job rotation
- repetitive actions in individual work tasks.

Risk control

In order for risk control to be successful in the workplace, we must manage the risk associated with the underlying hazards. This should be done on a priority basis; in other words, priority should be given to the hazards or activities which have the potential for greatest adverse effects (put 80% of the effort into this 20% of hazards).

An appropriate response to a hazard should consist of selecting the highest measure of control possible. Figure 2.1 outlines how hazard control measures from high to low order will assist in eliminating or reducing risk within your organization.

Function of a loss control programme

Loss control involves identifying and controlling weakness in a workplace system where there is potential for loss to affect people, plant and equipment. In practice,

44 Enhancing occupational safety and health

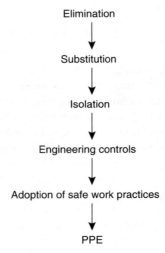

Figure 2.1 Hazard control hierarchy

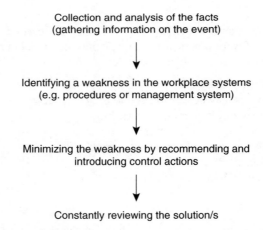

Figure 2.2 Principles of controlling loss

when an accident occurs, or accident potential exists, the principles of controlling loss are shown in Fig. 2.2.

If we look at general components within a workplace, weakness may exist at many points in procedures and practices. For example, if a procedure for purchasing chemicals does not take into account safety requirements or its toxicity, an accident or loss may result. Many chemicals can be substituted by less toxic ones and personal protective equipment may need to be ordered along with the chemicals. Other safety considerations which should be included when purchasing chemicals are:

- the need to obtain and make available a material safety data sheet
- the maintenance of records for quantity and type of chemicals used

Safety management

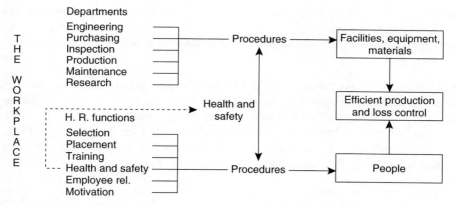

Figure 2.3 Loss control areas (Adapted from Petersen, D. (1978). *Techniques of Management*, 2nd edn., New York, McGraw-Hill reproduced by permission)

- a chemical register detailing chemical identification, quantity and location on site
- placards to alert the public of hazardous material and to assist emergency teams
- control strategies to prevent adverse effects of the chemical on health, prosperity or life
- health surveillance
- emergency procedures for the particular chemical
- educating and training of personnel in correct use and application of a chemical
- correct labelling of containers.

This is only one example of how safety needs to be incorporated in just one area of a workplace. Your workplace, depending on its structure, may include other departments as outlined in Fig. 2.3.

Controlling areas of loss

Loss control is mainly directed at areas such as personnel safety, property conservation, environmental protection, security and product safety. To control loss at any level of the workplace, procedures which integrate safety into each function need to be developed, practised and reviewed on an ongoing and regular basis. Efficient procedures are, however, only one part of loss control.

When we talk about loss control, some important points must be considered:

- An unsafe act, an unsafe condition, and an accident (leaving aside certain natural disasters) are all symptoms of something wrong in a management system.
- We can predict that certain sets of circumstances will produce severe injuries. These circumstances can be identified and controlled through consultation mechanisms, regular inspections and accident or incident investigations, to name a few examples. Safety should be managed like any other company function.
- Management should direct the safety effort by setting achievable goals and by planning, organizing and controlling to achieve them.

- The key to effective line safety performance lies in management procedures that fix accountability.
- The function of safety is to locate and define the operational errors that allow accidents to occur.

In loss control, areas of weakness or poor control may account either directly or indirectly for an accident. Therefore, through safety practices, weakness in the organization's system may be controlled; that is, loss control is achieved.

To go one step further, loss control should also aim to identify weaknesses in the main management system. Components of the management system which reflect, or should reflect, in-built safety include:

- management organization
- accountability for safety
- systems to identify hazards

Table 2.1 Management of labour sources of loss

Problem	Remedy	Responsibility
Poor selection	Correct job description and thorough interview and selection procedures	Line managers Personnel (HR) Dept.

Labour strength:
  Too high
  Too low
	Human resources planning in conjunction with workplace activity levels	Line managers
		Personnel Dept.

Inadequate training	Training needs analysis and training plans	Line managers Training managers
Insufficient instruction	Job instruction	Line managers Supervisors
Incorrect method	Improve job method	Line managers Engineering Dept. Development Dept.
Absenteeism	Determine cause	Line managers Personnel (HR) Dept. Nurse/medical practitioner Industrial chaplain
Injury	Investigate Determine cause Correct	Line managers Safety coordinators Engineering Dept.
Discipline and accountability	Enforce regulations	Line managers Supervisors
Ineffective supervision	Education/training Accountability	Line managers Personnel/training
Industrial disputes	Good industrial relations	Line managers Supervisors Personnel (HR) Dept.

(Adapted from IFAP, Western Australia, with permission)

Safety management

- selection and placement of employees
- training and supervision
- motivation
- accident records and analysis
- medical assistance
- alcohol and drug abuse programmes
- approaches to absenteeism and rehabilitation
- fire and explosion control
- transport activities.

Ideally, loss control measures should be based on the principles of, firstly, setting standards for every activity and area; secondly, providing resources and organization to achieve set goals; and, lastly, assigning responsibility, authority and accountability to ensure desired levels of performance. Table 2.1 shows labour sources of potential loss and the area of responsibility which can influence each source of loss.

Formal and informal safety meetings

Preparing a technical report on an aspect of risk management

A suitably prepared technical report can be of considerable assistance to a safety and health committee in a number of ways:

- if it is circulated before the meeting so that the committee have adequate time to read it, it can save a lot of time in discussion and avoid the committee having to meet again because the members find the report lacks sufficient information
- it can allow committee members to seek their own information and consult with other interested people before the meeting
- the consultation from circulation of the report can bring out 'hidden' issues which might not have been addressed by the committee because they were not aware of them
- the technical report, if properly prepared, will adopt a systematic approach to the assessment of risk.

What should the technical report include and how should it be structured? To some extent this is going to depend on the subject area it deals with, and any existing formats that your organization works to, but here are some suggestions:

- date the report
- identify the writer/s of the report
- explain why the report was written
- show how information was collected
- detail the method used to make an assessment of risk (for more information refer to Chapter 4)

- put forward conclusions and recommendations
- put an *executive summary*, which covers the essential details of the report, at the front of the report.

Getting information for the report

Information you will need to write the report includes details from the workplace; for example, becoming familiar with the area of the workplace concerned, the nature of the problem and the number and type of workers affected by the problem and their possible degree of exposure.

You then need reference information; for example, an MSDS for a chemical problem, a supplier manual for a machine tool, a standard on forklift (industrial truck) use, or a guidance note on Legionnaire's disease.

Using pictorial and graphical information

The old saying that one picture is worth a thousand words remains true. However, it is also possible to present information in a quite misleading way – through graphs, for example. Look at the two graphs in Figs 2.4 and 2.5.

This is the same information but notice what happens when the position of the x-axis is moved and the y-axis scale changed. The drop in LTIFR looks quite dramatic in the left-hand graph, so ensure that your graphical data is presented correctly. Computer programs offer you a relatively easy way to draw up graphs, bar charts and pie charts and even to present data from three variables at once using 3-D representation. For example you can graph LTIFR, month and work area.

Many of us are not dab hands as artists, but we can take advantage of photocopiers with cut and paste, and clip-art on computers to improve our reports and presentations. Ensure that you are not in breach of copyright and acknowledge diagrams drawn by others. With computer tools like PowerPoint, interesting overheads can be made or a presentation can be fully supported with laptop and video projection, providing the equipment is available. Tables have become much easier to construct using the

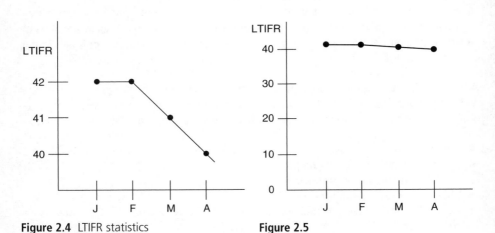

Figure 2.4 LTIFR statistics

Figure 2.5

Safety management

appropriate menu items on a computer. Computers also offer software containing project planning programs and charts.

Remember that just as you explain a table or graph in a presentation, in a written report you must do the same, not just leave it to the reader. In a presentation, avoid the mistake of putting up a complex table on overhead and giving the viewer 10 seconds to digest it. Processes and procedures with a number of steps and decisions are often most easily absorbed by others if you draw up a flow chart. A number of standard *icons* are used in flowcharts – look them up.

Finally, you will find Gantt charts, or similar, of use for sequencing OHS plans (= OHS programmes in the USA). This indicates time along the top and activities down the left-hand side. An example is given in Fig. 2.6.

Barriers to effective risk communication

Risk communication is an important aspect of the OHS programme in any organization, and in fact is also important in the public arena generally, because many people in the workforce will be influenced by risk communication taking place in the wider community.

But there are barriers to risk communication. These include:

- *Cultural barriers.* The cultural effect can come from, for example, ethnic or religious sources. For example, the idea that a higher source of wisdom is protecting a person may well be very important for giving them the confidence to do things which possess inherently greater hazards, as long as it does not lead them to believe that they don't need to do everything humanly possible to minimize the risk. Certain

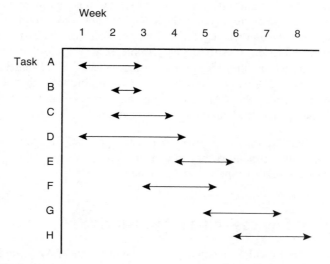

Figure 2.6 Gantt chart

groups may have a 'fatalistic' attitude – 'que sera sera' – what will be, will be. Another possibility is that the culture may relate more to the social group a worker belongs to; for instance, those where drug taking or other substance abuse is commonplace. There may be a 'to hell with it' attitude to heavy drinking or unsuitable diet.
- *Belief models.* So-called 'health belief models' play a part in the way people respond or don't respond to promotion of better health.
- *Unbalanced presentation of risk.* Ionizing radiation has had a number of benefits for society, for example in medicine. But it has been subject to a 'what you can't see and understand will get you' attitude by many. So the disposal of even low-level radioactive waste has been a source of debate for many years. When we look at figures for risk, many people accept the risk in motor vehicle travel or smoking, but object to the far lower risk of some uses of ionizing radiation. It is true that risk which is *voluntarily assumed* or taken on, or is commonplace, is often more easily accepted especially if the benefits are plain. Motor vehicles offer people mobility so they trade that against risk. However, it is only 100 years since a red warning flag had to be carried ahead of a motor car.
- *Media influence.* This can be positive or negative. Quite often the media will warn of a risk before parliamentarians and judges, as social arbiters, act to increase people's protection. But the media can also create unbalanced attitudes, panic or hysteria.
- *Responses to low probability/high consequence risks versus high probability/low consequence risks.* What you can easily foresee injuring you in the next five minutes is more likely to result in steps to avoid injury, than what may be very serious if it happens but you perceive as very unlikely. On a geological survey in the tropics you keep an eye out for crocodiles while swimming, but ignore the chances of a widow-maker bough from a self-pruning tree falling on your tent, because it offers the benefits of shade and something to fasten the ridge-rope to.
- *The 'what they don't know won't hurt them' view.* It is to be hoped that consultation in occupational health and safety is doing away with this, but there may be a fear in some management circles that too much honesty about risk will make people too frightened to behave and respond rationally to it. There will be extreme situations where such an approach is justified. For example, some surgeons limit information given to patients because they know that on balance the benefit of surgery outweighs the potential risks. The worker who survived by jumping through smoke into the water from the burning Piper Alpha oil rig in the North Sea couldn't afford to consider too many questions about the near future at that point.

Studies have shown that people's perception of risk is generally not well developed, so an important role in occupational health and safety is to assist people to overcome some of the barriers, and so approach risk rationally.

Preparing a safety case on a particular risk

Safety cases have been widely adopted in the wake of the Piper Alpha and Alexander Kielland oil rig disasters. The effort in writing them is justified in workplaces with high consequence risks, and a large *technical safety* requirement. They are a requirement under petroleum safety legislation in Australia and the UK, for example, and a guide to them is available from the administering authority. The UK also requires them under the COMAH regulations.

On-shore safety cases would be justified in a chemical process plant or one employing large amounts of hydrogen; for example, in oil refining or direct reduction of metallic ores, or in a nickel and cobalt refining process using ammonia and hydrogen sulfide. It was a hydrogen leak past an eroded O-ring due to flexing in the joints of the rocket, which was the immediate cause of the *Challenger* space-launch disaster. However, managerial factors led to acceptance of a certain level of erosion in the O-rings. Not following up on auditor's warnings about technical safety, such as fire sprinklers affected by salt crystallization from seawater, led to their failure to operate on Piper Alpha. It is the interaction of these managerial factors with technical safety which forms an important part of a safety case.

The principles used in a safety case can be applied to any significant risk in the workplace. They include:

- Thorough specification, design and design review, construction and commissioning using techniques such as HAZOP (hazard and operability study) which asks 'what if' questions and FMEA (failure mode and effects analysis) which looks at likely sources of engineering failure and the consequences of it. The 'what if' questions can include:
  - effects of under pressure or over pressure
  - effects of underfilling or overfilling
  - effects of surge or backflow
  - too hot or too cold.
- Use of the analysis above to develop appropriate methods of risk control including maintenance and repair, and control during maintenance and repair.
- Running likely scenarios for various types of failure and developing recovery or emergency response strategies. Both the Flixborough (UK nylon plant) and Piper Alpha disasters started during maintenance procedures involving modifications to process pipework. The Chernobyl nuclear plant disaster in the Ukraine in 1986 began during system test procedures.
- Developing an appropriate management structure and organizational design and linking policies, procedures, consultation, communication and training to the 'engineering' aspects of the area for which the risk is being considered. A review of the *Columbia* space shuttle break-up identified possible communications gaps between operational engineers and management.
- Considering ongoing audit, review and follow-up.

A full safety case approach involves three parts:

- a facility description
- a safety management system with fourteen key areas
- a formal safety assessment.

For workplaces placed close to the community or involving non-employees (such as shoppers) in a workplace, this approach includes appropriate measures to protect them. A safety case has been run, for example, for a public swimming pool using cylinders of chlorine for water treatment. A safety case run on the Bhopal pesticide plant in India would have identified the problem of a lack of buffer zone due to poor, if not non-existent, enforcement of local government residential area planning. As it was, in 1986 too many residents were living too close to the plant.

Presenting accurate and sensitive information on risk to exposed groups

Information on risk needs to be presented accurately to working groups, but that information may need to be handled in quite a sensitive way.

Some examples of this are:

- Manufacture of a formulated chemical where the market is highly competitive. Information on ingredients must be disclosed to workers making the formulation, but the trade secret must be kept. This is sometimes done by giving each ingredient a company code name and providing workers with adequate OHS information on the ingredient. Secondly, only workers who need to know the formulation process, and emergency and certain other staff, are provided with that information.
- A number of workgroups can potentially encounter violence in the workplace. These include police, people in banks, shops or service stations handling cash, railway patrol officers, taxi drivers, prison staff, hospital emergency staff, those caring for the mentally ill, bar and nightclub staff. There may be a temptation to shy away from this in some workplaces, but it must be put to the people exposed honestly and in such a way that they are reassured about the steps the employer has taken for their safety, and the steps to take in an emergency. Where an incident has occurred, professional counselling of the staff concerned may be needed.
- Situations where women, especially those who may be pregnant, can be exposed to chemicals (drycleaning), anaesthetics (surgery), metals (lead in mineral assay), or radiation (radiography). The question of balancing equal opportunity against avoiding exposure is a sensitive one. It is not a matter where the employee can 'voluntarily assume the risk', as present day common law and OHS statute law does not recognize the employee's right to make an informed choice to accept the risk.

Nature, occurrence and industrial relations implications of OHS issues

Labour relations

The relationship enjoyed between the employer and the employee will have a significant bearing on the effectiveness of safety programmes.

The essential factor is the credibility both parties perceive in the objectives of the particular programme. It could be argued that in matters affecting the well-being of the workforce, safety is quarantined from the often adversarial nature of labour relations. This is not the case, and industrial politics often bring an influence to the management of safety and safety programmes.

However, industrial issues over choice, empowerment, perception and commitment are likely to influence most workplaces and effective safety programmes must be designed to function in all environments.

An employer in the power industry introduced a safety programme aimed at eliminating death through electrocution in its workforce. The safety programme had a number of objectives including staff training, new protective equipment, revamped isolation procedures and a review of the earthing methods. All of these initiatives were supported by an electrical safety handbook, safety posters, safety promotion and remedial medical checks for people experiencing electrical contact.

The organization spent a large amount of money on the programme and management were truly committed to the safety programme. Death through electrocution was significantly reduced over the following five years.

During this period a survey was conducted to identify the attitude of the workforce towards the way safety was dealt with in the organization. The overwhelming view was that the organization displayed little interest in the safety of the workforce. This view was not consistent with the employer view and further analysis was required.

It was discovered that the management initiative on reducing electrical fatalities had been largely undervalued and even ignored. The workforce believed that management were not focused on the risks of most concern to the workers.

Death by electrocution was ranked behind some other worker safety concerns. Workers were more concerned at a lack of resources and support to better manage the following risks:

- structural collapse resulting in brain damage or quadriplegia
- oil and electrical burns resulting in disfigurement, particularly facial burns
- cancer resulting from the handling of, or exposure to chemical substances
- loss of limbs or vision
- falls from heights onto sharp or pointed objects
- working alone, being injured, with no immediate response
- lack of traffic control at worksites.

The employer was not seen as taking an active enough interest in these safety issues.

A review of safety minute meetings, going back nine years, showed only the occasional reference to electrical safety, including fatalities. However, matters relating to the integrity of working structures, in particular, and the exposure to chemicals were matters of continual discussion and concern.

The employer is charged with the provision of a 'safe system of work'. The efforts made by the employer in this case, in seeking to eliminate electrical fatalities were proper. However, the delicate nature of personal perceptions and the subsequent implementation of safety programmes are always going to be difficult areas to match.

Labour relations will benefit from improving risk communications and the development of a better understanding of self-regulation. These are the goals both employer and employees should work towards.

The funding for safety programmes will always have to compete with other organizational requirements. It is imperative that money is spent on safety initiatives so as to gain the maximum benefit, and to minimize the risk of occupational health and safety becoming an issue fought on the industrial relations front.

Overlaps often occur between occupational health and safety issues which arise in a workplace and the industrial relations of that workplace. Quite often, of course, unions may play a background or even a statutory part in who is nominated to be a health and safety representative or elected committee member, and may influence the election process.

With some issues, if resolution through representatives or committees following the agreed procedure is difficult, employees or a union, or unions, might make the issue an industrial relations dispute. A safety issue like a crane dropping something three times at the same site within weeks is likely to become an industrial relations issue.

Examples of OHS issues and their occurrence

Clearly, the range of possible OHS issues and the reasons why they arise is huge, if we consider the variety of workplaces, the types of processes and the variations in people's training and personalities.

Someone may have seen an ergonomic keyboard or chair in another firm and decide that they would like to have (or should have) one. The media may play a role by publicizing a particular suspected hazard, such as electromagnetic radiation from mobile phones.

As a result of a bad accident, a union may be warning its members to be on the lookout for similar problems elsewhere, for instance the use of nylon slings with a crane instead of the preferred steel rope slings. The equipment may simply have been badly designed, or it may have been badly maintained. Employers who did not have the accident may as a result have to respond with improved equipment, even though their maintenance programme is tip-top.

Changes to legal requirements in OHS legislation can catch employers unawares, until someone in the workplace raises the issue. This may occur because a health and safety representative has just been to a training course. He or she comes back more aware of hazards, and usually acts out of genuine concern, although the odd individual may act primarily to demonstrate that fellow workers were correct in electing them.

However, properly trained representatives and committee members are likely to solve many more problems than they create.

Hidden agendas which may trigger OHS issues

On some occasions what appears to be an OHS issue arises from other separate concerns. For example, workers on a construction project which is due to end may try to prolong the project by slowing or stopping work over OHS issues, such as dust suppression. It is sometimes difficult to distinguish these 'intentional' disputes from those which arise because supervisory control slackens as people are pulled off site towards the end of the construction phase.

Sometimes the apparent reason for the issue is an OHS matter, but there is really a deeper problem. For instance, a site working under great pressure to complete a project by a set date can build up high shared stress levels. A maggot found in canteen food, for example, can then trigger a mass tools-down or walk-off.

Downsizing affects the number of personnel assigned to particular areas (what was called 'manning levels'). Disputes then arise as to whether the reduced staff levels raise safety issues. So staffing levels then bring OHS face to face with industrial relations. Quite often there is a genuine OHS aspect to staffing levels. One fatality in New South Wales arose out of regular inspections of a dam. Originally done by two men, it was cut to one. When he fell in there was no one about to rescue him. Fumigation of grain is another situation where such concerns have arisen.

Another area where industrial relations meet OHS is in hours on the job. An estimate in 2000 by the National Sleep Foundation says that 47 million US workers lack sufficient sleep and are irritable at work.

Role of management and workers in the occurrence of OHS issues

OHS issues will certainly arise from time to time in any workplace, and can be handled through the consultative process which Robens-style OHS law sets up – employers talking with health and safety representatives, or in health and safety committees. In small workplaces, or those without representatives, the employer can discuss matters one-on-one with employees. Issues only become disputes if there is a difference between an employee's request, demand or point of view and that of the employer. The dispute may be for a number of reasons – one, of course, is inability to pay for risk control measures. Another may be that the employer is dependent on others to implement solutions. For example, a contractor at a mine site may operate machinery belonging to the principal which needs new capital investment for safe operation rather than just maintenance. An accident may make employees aware of a risk they weren't aware of before.

Staging the introduction of necessary control measures can make them affordable and may well be accepted by a health and safety committee as a way forward.

Implications of OHS issues for industrial relations in the workplace

Certainly in some parts of the world much of the impetus for change towards modern OHS legislation came from the union movement. So it is not surprising that OHS issues do play a part in, or form part of, industrial relations.

This sometimes offends some OHS practitioners who believe that OHS issues should be dealt with 'scientifically'. This is often a good approach, but power relationships and community concerns and perceptions can cut across this. Asbestos in asbestos-cement sheet in school roofs became an issue for teachers' unions and school parent groups, for example.

Sometimes unions act on OHS issues which are outside their own immediate sphere. A wharf union may declare a foreign-flagged vessel black because of conditions they discover on-board affecting crew safety, such as inoperative or rusted lifeboats.

It is safe to say that OHS issues will continue to overlap with industrial relations, and so on occasions become mainly industrial relations issues (for those who see them separately). Equally, industrial relations issues which start as concerns about wages or shifts can engender bad feeling. In that organizational climate, relatively minor safety or health issues may be picked on consciously or unconsciously by employees to indicate their dissatisfaction and become the tinder for a major industrial relations dispute.

Prioritizing and developing strategies to resolve OHS issues

Factors determining priority for resolution of OHS issues

At the outset it is important to point out that proper attention to management of, and consultation in OHS will avoid many OHS problems becoming *issues*. Many people believe it is important to distinguish OHS issues from industrial relations, but in reality they do interact because working conditions are a legitimate part of industrial relations. A number of factors come into play in relation to resolving occupational health and safety issues. These include:

- Whether the issue presents an imminent and serious risk. Most OHS legislation provides for a worker to refuse work in these circumstances. It is also a situation where in some OHS jurisdictions a health and safety representative may issue a default or provisional improvement notice (PIN).
- A refusal to continue with work in an area after a serious injury or near miss.
- A situation where an inspector (or 'investigator' under some OHS legislation) issues an improvement or prohibition notice.
- The discovery of a serious but previously unnoticed hazard.
- Unreasonable or irresponsible directions by management.
- Unreasonable or irresponsible demands by health and safety representatives, employees or unions.
- Critical deadlines for production and supply of goods or completion of construction work.

It is important to balance issues involving the immediate risk of physical injury against those which may affect health, even if not in the immediate future. For example, scaffolding without a kickboard versus use of a solvent which damages the liver over time.

Identifying strategies for resolving a particular OHS issue

There are really two parts to this process. The first involves ensuring that a suitable procedure for issues resolution is decided by consultation in the workplace before the need for it arises. The second is to utilize suitable problem-solving, conflict resolution and negotiating techniques when an issue does arise. One of the keys to most successful negotiation is to agree on the process for negotiation first, which really goes back to the first point. More information on this can be found later in this chapter.

Some Robens-based OHS legislation provides a process for issue resolution where a workplace has not agreed on its own process. This is often presented in the form of a flowchart. The usual advice to employees if they have an OHS problem is to raise it with their supervisor in the first instance. Where an atmosphere of trust prevails this is good advice. However, if the employee feels that raising an issue may result in discrimination, the employee may rightly prefer to go through the health and safety representative, or a union. The health and safety representative or a union may resolve the issue with management or it may be dealt with by the health and safety committee. If it cannot be resolved, an OHS inspector may be asked for a determination and issue a notice. Further review by heads of OHS authorities, or courts varies with the particular jurisdiction. In New Zealand, for example, the District Court resolves disputed notices.

Major issues affecting many stakeholders may be resolved by the tripartite body set up in Robens-based legislation. In those jurisdictions with PINs or default notices, resolution proceeds further automatically because an inspector may be required to confirm or disallow the notice.

Methodology for resolving an OHS issue

The basic strategies for resolution of issues have already been given. The actual methodology includes:

- Accurately identifying what the real issue is, as distinct from what appears to be the issue. For example, the apparent issue may be housekeeping or use of mineral fibre-based material on a construction site, while the real issue is a desire to prolong the contract because construction is nearly complete.

    As noted, people may criticize health and safety issues which refer to staffing levels, saying that the levels are an industrial relations issue. However, it is not that simple – staffing is often directly related to health or safety. For example, there is a need for a second person in a job involving boating, a second person who can clearly see the rear-door exit of a bus, and a second person in the cockpit of a commercial aircraft.
- Working with the other party to agree on the method to be used for resolution (refer to the information on strategies above).
- Seeking out any further information required. This could include an MSDS; incident and accident statistics; information from expert bodies; standards; methods of control of the risk used in other companies and industries; or interviewing other employees.

- Deciding on an approach to control with, if necessary, an action plan.
- Altering procedures and providing any necessary additional training.
- Monitoring the results of the resolution to see if the solutions are working. If this is done actively it prevents a poorly resolved issue continuing to simmer.

Problem solving

Problem-solving skills are essential in reaching the desired outcomes in all instances where there is a need to come to an agreement with a number of people. The following are a series of steps that may be useful for efficient problem solution.

1. Define goals and objectives
   - Exactly what do I want from this negotiation?
   - What do I have to get to meet my needs?
   - What am I willing to give up to get what I want?
   - What are my time and economic requirements for this negotiation?
2. Clarify the issues
   - What are the issues as I see them?
   - What is the supporting framework for my position?
   - How will I present it to the other party?
   - What are the issues as seen by the other party?
   - How will they support their position?
   - What are the significant differences in the way the parties view the issues?
3. Gather information
   - Who will I be negotiating with and what do I know about them? How do they approach a negotiation? What are their ego needs?
   - When and where will the negotiation take place? What advantages or disadvantages do the alternatives have for me or for the other party?
   - What are the economic, political and human implications of the issues?
   - What personal power do I have that can be used constructively in this negotiation?
4. Humanize and set the climate
   - How can I best establish rapport with the other party?
   - How can I establish a win/win climate?
5. Prepare for conflict
   - What will be the major points of conflict?
   - How will I determine the other party's needs compared to their wants?
6. Compromise/resolution of the issues
   - How will I attempt to resolve conflict? How will I respond to the other party's attempts to resolve conflict?
   - What concessions am I prepared to make? Under what conditions?
   - What do I expect in return for my concessions?
7. Agreement and confirmation
   - How formal must it be?
   - What approval process will be required? How long will it take?
   - What implementation steps will be needed?

Identifying key stakeholders in the resolution of an OHS issue

The stakeholders are going to vary depending on the issue but they could include:

- company shareholders
- company customers, who are often the general community
- contractors supplying a company with products, expertise or labour.

For example, prolonged industrial action at a mine over staffing levels and length and structure of shifts can affect all these stakeholders. It can also mean loss of markets to competitors both locally and internationally.

However, the key stakeholder is the individual employee. It is he or she who stands to suffer the most direct consequences from the risk and to be injured or develop ill health. Other stakeholders are the health and safety representatives and committee members, the various levels of management and the company directors. In some major issues government is also a stakeholder. The unions are stakeholders, particularly where the issue becomes identified with a win or loss by employees, unions, representatives or management.

Some issues in one company can involve many other companies as stakeholders. For instance, a health and safety representative in a casino was dismissed, but ostensibly not because of his campaign on passive smoking. This issue stood to have widespread ramifications for an organization attracting many less-smoking-conscious overseas patrons and for other organizations in the food and entertainment sector. A second example which went before the Australian Human Rights and Equal Opportunity Commission was the right of women of child-bearing age to work with lead, so sometimes EEO goes head to head with OHS.

The community can be a stakeholder, too. Safety incidents such as an ammonia leak from refrigeration in a brewery or from ammonia being transported by rail can affect people within several kilometres.

Government responsibilities in OHS issues

You will need to refer to the legislation under which you work for the approach taken to resolve issues which are not or cannot be resolved at workplace level. These generally involve seeking a determination from an inspector who may uphold the employee view, and issue an improvement or prohibition notice, or uphold the employer view. Further appeal by parties can then be to the head of the OHS administering authority or a court (depending on what type of notice the appeal involves). Further appeal against the head of the OHS authority may be to a court or industrial commission.

As mentioned earlier, issues of widespread concern may be dealt with by the tripartite advisory body set up at a national, state, province or territory level, referring them to a specialist industry or other committee where necessary.

Readers are referred to their own legislation for the details of OHS issue resolution mechanisms.

The variable approaches present challenges for anyone involved in managing OHS in organizations operating in more than one jurisdiction.

OHS issue resolution

An OHS policy will usually contain a general reference to consulting with employees on OHS matters. However, to be useful, it needs to be fleshed out in guidelines or a specific procedure. Workplace or enterprise agreements can also contain reference to means of resolving industrial relations issues and these, as we have seen, can include OHS issues. Usually, any procedure will require initial referral of a concern to a supervisor, who may confer with a health and safety representative, and then provide for further referral, if necessary, to the health and safety committee, if one exists. If not, it may provide for the matter to go further up the management chain.

Roles, responsibilities, knowledge and skills of workplace players relating to issue resolution

Health and safety officers/advisors/coordinators, health and safety representatives, health and safety committee members, union representatives, supervisors and managers all have a role in the resolution of OHS issues.

Health and safety officers/advisors/coordinators

Well-trained health and safety officers/advisors/coordinators can play an important role in issue resolution by providing an impartial and skilled source of advice on the hazards and risks involved in the issue, on applicable standards and possible control measures. They may do this through informal contact with management, health and safety representatives or employees. They may also operate as one of the management nominees on the health and safety committee and be asked by the committee to research the issue and so assist the committee to resolve it.

Health and safety representatives

During issue resolution, health and safety representatives, where they exist, may be involved in liaison with employees and the employer, and referral to the health and safety committee of any matters that should be considered by the committee. An inspector may be notified where issues are unresolved and the issue is considered a serious or imminent risk.

Health and safety committee members

Where health and safety committees exist, the committee members probably have a crucial role to play in preventing OHS issues arising in the first place. Effective review of the workplace, its processes, people, equipment and hazards, using a systematic approach, will largely achieve this. Well-trained and well-informed committee members, be they those elected by employees, or those nominated by management, can do a lot to promote an effective health and safety climate in the organization. Keeping abreast of improvements in comparable organizations or industries (benchmarking) will certainly assist the safety effort and lead to best practice.

Union representatives

Union representatives may be given no formal role in the resolution of OHS issues under OHS legislation. In reality in many workplaces, if the procedures for resolving an OHS issue fail, pressure may develop from the union in that workplace, if there is one. Negotiating skills combined with a good knowledge of the background to the OHS matter itself will undoubtedly assist a union representative in effectively representing the needs of union members.

Supervisors

First line supervisors may well be the key to the way OHS issues are handled in an organization. Employees are advised to raise their concerns with their supervisor in the first instance or raise them with the health and safety representative, who then raises them with the supervisor. If there is an imminent risk of harm or adverse health effects, the first line supervisor needs to make a prompt decision to pull his or her team out and stop the process until the matter can be examined. If it is not an issue of that kind, the supervisor needs to have the authority to address less serious issues there and then. For issues which he or she can't resolve, the supervisor will need to show that they can confidently raise workers' concerns with higher management, and be prepared to implement solutions or, on the other hand, explain why no change has been agreed to.

First line supervisors again play an important role in issue prevention by setting an example of safe work and by being proactive about safety in toolbox meetings, in their day-to-day contact with team members and in ongoing hazard identification. Clearly a supervisor may have difficulty if they are less well trained in OHS than a health and safety representative.

Managers

Managers need to be thoroughly familiar with the issue resolution procedures agreed on by the organization. Ideally, they will have had some training in this and have run through some role plays or case studies.

Managers, depending on their level, are in a position to determine or change health and safety policy. They also play a part in setting budgets, including budgets for improving safety. They hold, or should hold, the purse-strings when money is needed to implement a solution to an OHS issue, such as renewed training after an accident has given rise to an OHS issue.

Role and responsibilities of union and employer associations in resolution of OHS issues

While occasionally an OHS issue is of such immediate importance that peak employer associations and peak union bodies become involved, this is rare. More commonly, it is likely that an OHS issue arises which is common to many workplaces. It may be an old issue, such as fatigue at work, but with a new face, due to changes in workplace organizational structures, staffing and economics. The issue can then be dealt with at the tripartite OHS policy body level (where such a body exists). An expert working party may be set up to examine the issue and make recommendations. As a result, the body may issue a new code of practice designed to address the risks involved in the issue. Comparatively recent codes of practice on violence in the workplace have arisen out of this issue of common concern to many workplaces dealing with cash and the public.

Union and employer associations, also, usually have some OHS research and information gathering capacity and this can be used by employers or unions to assist in resolving issues in their own workplaces.

The peak bodies also have the capacity to communicate directly on some OHS issues through newsletters and interaction with the press. They also run training courses and seminars. Some union organizations improve the effectiveness of health and safety representatives in resolving issues by being one of the organizations accredited to train those representatives, where accredited training is mandated.

Role and responsibilities of inspectors or investigators in resolution of OHS issues

In the self-regulating OHS framework we have now, inspectors may form an important second force in issue resolution. The inspector's role can arise in two ways.

Firstly, in jurisdictions with a provisional improvement notice (PIN) or default notice system, the inspector may be called in to affirm, vary or not affirm the notice. The PIN, in some cases, will indicate that the issue resolution procedure has not worked. In other cases it will indicate that there is an imminent and serious risk of harm to health or injury, such that issue resolution procedures are too cumbersome to offer immediate protection. What is a reasonable period of time before a PIN is issued will depend on the nature of the issue.

In jurisdictions without a PIN system, the inspector really should only be called in at the point where the internal procedures, such as referral of the issue to the health and safety committee, have not produced a result. Management may be unwilling to spend the money needed on risk controls, for example. It will be up to the inspector to make a decision about the employees' concerns. If he or she finds in favour or partly in favour of the employees, an improvement, prohibition notice or similar may then be issued.

Sometimes the boot is on the other foot, and employees or unions resist something which contributes to safety, such as a drug and alcohol policy.

Budget planning, control of safety expenditure and supervision

Preparing a written case for expenditure on a specific safety initiative

An area which you may be involved in is justifying expenditure on a specific safety initiative. Recommendations for a safety initiative may come from various sources such as the safety committee, safety representatives, unions and the management.

Safety management

The key thing to appreciate when recommendations are being made is to establish clearly that:

- baseline statutory requirements are being met
- the organization's policy statement is being followed
- the cost associated with implementing the recommendation is known
- the benefits associated with implementing the recommendation are identified.

Your justification should clearly identify that the benefits outweigh the costs of the expenditure. This is a key factor in gaining approval for safety initiatives from senior management. Cost/benefit is difficult to establish for diseases which are a long time showing up. A different justification for prevention must be used.

Operating and capital cost records

Organizations usually keep a variety of records which document operating and capital costs. These records assist the organization in determining the efficiencies and effectiveness of their business sections. They also assist with determining budget planning.

A portion of these records may involve costs associated with occupational health and safety. The costs of various elements of loss control – for example, security, workers compensation or fire – can be broken down for each centre or department to assess performance.

General records which may be kept include the following:

- an asset register listing capital items purchased, which are usually displayed on a balance sheet
- profit and loss statements:
  - premium (workers compensation)
  - staff training
  - employee equipment item costs.

Other reports may include absenteeism, department compensation costs, etc. These reports will permit performance ratios to be developed to monitor efficiency – for example, the ratios of workers compensation costs to the total wages bill. Uninsured cost calculations for accidents allow the effect of accidents on profitability to be demonstrated. This can assist in seeking allocations for health and safety initiatives.

Preparing a budget for operating and capital expenditure

Budget planning in occupational health and safety requires that all departments within an organization consider income and expenditure needs in advance for any given financial year. Items that require projected cost planning within each department may include:

- training in occupational health and safety
  - course numbers and type
  - cost of courses

- number of employees to be trained
- cost of temporary, or replacement personnel
- pre-placement health assessment costs
  - based on average recruitment numbers
- loss control costs
  - based on previous years' experience
- workers' compensation insurance costs.

An organization will have an insurance premium calculated annually for the coverage of its employees; however, with the assistance of the organization's accountant, the premium may be divided between departments. Departmental workers' compensation costs may then be offset against the allocated premium each month. This method will allow departments to recognize their costs and plan safety management in a pro-active fashion, to reduce the loss from injury experience. For example, the total workers' compensation premiums for company XYZ may be 100 000 currency units (e.g. euro). This may be allocated to the following departments:

- Administration   20 000
- Production   50 000
- Distribution   30 000

This type of budget planning will assist each department to manage their costs more efficiently. The insurer can usually provide data on claims experience details for departments or work areas and this assists with safety budget allocations. Targets for a reduced budget may then be introduced to each department to improve financial management.

If you are in the position of safety advisor with an organization, department managers may request your assistance in the budget planning phase. This will ensure that, through a team approach, safety costs are managed effectively, thus benefiting safety performance.

Writing a draft health and safety policy

The ownership of health and safety strategies usually originates with management's commitment to safety, which may be communicated in a policy statement. Strategies may then be formulated at each level of the corporate ladder in order to achieve common goals. Achievement of these goals can then be evaluated and new strategies developed to maximize safety performance, in line with the idea of continuous improvement.

An example of a health and safety policy is given in Fig. 2.7.

Health and safety policies may outline objectives in key health and safety areas, such as mobile equipment hazards, plant or noise. They may outline the key occupational health and safety accountabilities of the different levels of staff.

Key roles of supervision in health and safety management

The success of a health and safety management programme will depend on how functions are perceived by management and how responsibility for health and safety is delegated throughout an organization. Responsibility for managing health and safety requires

This example of an Occupational Safety and Health Policy should be used only as a guide in the development of a specific policy appropriate for the organization. Guidance is in italics.

Occupational Safety and Health Policy

General Commitment Statement

This policy recognizes that the safety and health of all employees with [Company name] is the responsibility of company management. In fulfilling this responsibility, management has a duty to provide and maintain, so far as is practicable, a working environment that is safe and without risk to health. This responsibility includes:

- providing and maintaining safe plant and systems of work
- making and monitoring arrangements for the safe use, handling, storage and transport of plant and substances
- maintaining the workplace in a safe and healthy condition; and
- providing information, training and supervision for all employees enabling them to work in a safe and healthy manner.

[The General Manager] is responsible for the implementation and monitoring of this policy. The safety and health duties of management at all levels will be detailed and company procedures for training and back-up support will be followed. In fulfilling the objectives of this policy, management is committed to regular consultation with employees to ensure that the policy operates effectively, and that safety and health measures are regularly reviewed.

Duties

Recognizing the hazards occurring in the [ ] industry, this company will take every practicable step to provide and maintain a safe and healthy work environment for all employees. To this end:

Management

- is responsible for the effective implementation of the company safety and health policy
- must observe, implement and fulfil its responsibilities under the Acts and Regulations which apply to the [ ] industry
- must ensure that the agreed procedures for regular consultation between management and those with designated and elected safety and health responsibilities are followed

clear statement of responsibility

requirements of relevant OHS legislation

specific provisions and objectives

nominated individual with responsibility for implementation of policy

acknowledgement of benefits of consultation with employees

need to demonstrate an understanding of the hazards in this industry

commitment to provide resources

- must make regular assessments of safety and health performance and resources in co-operation with those with designated and elected safety and health functions
- must ensure that all specific policies operating within this company are periodically revised and are consistent with company health and safety objectives
- must provide information, training and supervision for all employees in the correct use of plant, equipment and substances used throughout the company; and
- must be informed of incidents and accidents occurring on the company premises or to company employees so that safety and health performance can accurately be gauged.

Employees

- have a duty to take care for their own health and safety and that of others affected by their actions at work
- must comply so far as they are reasonably able, with the safety procedures and directions given by [Company name]
- must not wilfully interfere with or misuse items or facilities provided in the interests of safety and health of company employees; and
- must, in accordance with agreed company procedures for accident and incident reporting, report potential and actual hazards.

This policy will be regularly reviewed in the light of legislation and company changes.

Management seeks co-operation from all employees in realizing our safety and health objectives and creating a safe work environment.

All employees will be advised, in writing, of agreed changes and arrangements for their implementation.

Signed   **(Managing Director)**   Date

Signed   **(Manager)**   Date

need for separate policies and procedures for specific areas which will support overall policy

company procedures and training arrangements to be followed

requirements of the relevant OHS legislation

acknowledgement of agreed company procedures for dealing with those with nominated or elected health and safety functions

Figure 2.7 Example of health and safety policy (With acknowledgement to the Chamber of Commerce and Industry of Western Australia, *Handbook of Occupational Safety and Health*, Update 26 September 1999, pp. 6001–6002).

the support of the same people who manage or are in charge of any other aspect of the organization's operation.

Poor management of health and safety may, in many cases, be attributed to perceiving safety as a separate function to other functions within an organization. As a result, many health and safety programmes have tried to promote safety through slogans such as 'Safety First'. Ask yourself the question 'why should safety be first, or last for that matter?'.

In fact, effective and efficient health and safety management requires a systems approach whereby health and safety is an integrated part of every operation and work process. If health and safety management is perceived as an integral part of every job, safety performance may be built in to the work system.

Safety management functions

Safety management is a process of planning, organizing, leading and controlling (and, where relevant, staffing) to enable the integration of safety into every part of the organization's processes.

Functions include:

- planning
  - determining what is to be achieved
  - deciding what is to be done
- organizing
  - allocating human and material resources
  - deciding how it is to be done and who is to do it
- leading
  - guiding the work efforts of other people
  - deciding how to make sure it gets done
- controlling
  - monitoring performance, comparing results
  - deciding if it is, or is not, getting done and what to do if it is not
- staffing
  - what type of staff are needed?
  - will they be employed or contracted-in?

(*With acknowledgement to Claydon, B.* Organisation and Management Study Guide. *Perth, Australia, TAFE Publications.*)

Effective management and supervision requires that authority, responsibility and accountability relationships are clearly defined and executed throughout the organization, and that they are matched by appropriate training and availability of resources and funds.

Concepts of responsibility, authority and accountability

Definitions

'Authority' – is the legal power or right which is based on recognition of the legitimate right of individuals or groups to exert influence, with recognized boundaries, through their formal position in an organization. (Claydon, see above.)

'Responsibility' – being answerable or accountable for something within one's power, control or management.

'Accountability' – liable to be called to account.

Who is allocated 'authority' in the workplace?

Accountability for managing occupational health and safety is assigned to the employer in legislation. Inspectors also have specific authority to assess workplace safety systems.

If a health and safety advisor or coordinator is appointed to an organization, authority will be given by management to perform certain functions and activities. Health and safety representatives (or delegates in US mines) are also given authority through statute law to participate in safety management within a workplace.

Who is allocated 'responsibility' in the workplace?

Every person within the workplace has some responsibility for health and safety. Responsibility is generally allocated to positions within an organization once specific activities in safety have been identified for each department or work section. Responsibilities may be divided as follows.

Senior management responsibilities

These include:

- providing:
    - a working environment that is safe and without risks to health
    - safe equipment and systems of work
    - adequate resources, information, training and supervision
    - ongoing and effective health and safety promotion
- establishing a process to identify, assess and eliminate hazards or control risks
- ensuring that:
    - relevant laws are complied with
    - workplace rules, procedures and methods are developed and maintained
- auditing and reviewing safety.

Manager and supervisor responsibilities

These include assisting with:

- implementation of policies and procedures
- identification of workplace hazards
- ensuring that effective consultation occurs
- investigating accidents and incidents
- induction and ongoing training for employees
- responding to issues raised by health and safety representatives
- submitting statistics and reports.

(*With acknowledgement to OHS Authority (Victoria, Australia.) 1993. SHARE,* Responsibilities for Health and Safety in the Workplace.)

A supervisor, for example, may be responsible for a variety of management functions including:

- meeting production requirements
- meeting quality targets
- budget control
- safety management.

As safety management will be a high priority, merging with all management functions, the responsibilities incurred can be further examined. A supervisor may promote safety management through initiating and participating in a whole range of activities such as conducting toolbox meetings, induction training programmes, hazard inspections and review of work practices. All of these actions can be viewed as proactive safety initiatives which will enhance productivity and efficiency.

Once management has decided what aspects of safety management the supervisor will be responsible for, documentation can be designed which will allow the supervisor to document practices performed which meet the designated responsibilities. An example of this can be seen in Fig. 2.8 – the Supervisor Health and Safety Report. This report can be completed on a weekly basis by a designated supervisor, allowing review of departmental and shift performance. Requirements for supervisor training and specialist advice on safety issues may be identified from such a report.

Employee responsibilities

These require an employee to:

- follow work practices, procedures, instructions and rules
- perform all duties in a manner which ensures their health and safety and that of others in the workplace
- wear protective equipment supplied
- report hazards.

Who is held accountable in the workplace?

Once responsibilities have been allocated throughout the workplace, individuals or work teams become accountable for meeting the expectations these responsibilities entail.

If we look at the allocated responsibility of a supervisor to submit statistics and reports, it will be relatively easy to see if the supervisor has met their responsibility. If they have submitted the required statistics and reports, the accountability that responsibility involves has been met for a given time-frame.

Good management requires that before being held accountable, a manager, supervisor or employee must be assigned the level of authority needed to match the responsibilities assigned.

How will management change affect safety management?

The traditional allocation of safety management through levels of management is changing. Evolution of a team approach is gaining momentum in association with Quality Management (QM), best practice and continuous improvement strategies. As we evolve from a hierarchical management system to a more integrated approach involving 'work teams' in decision making, safety practice can benefit from the dynamics in the organization. Work teams where health and safety are foremost add value by:

- planning the change
- creating the right climate
- anticipating resistance
- seeing the benefits
- listening
- providing follow-up or evaluation.

SUPERVISOR HEALTH AND SAFETY REPORT

Department: _____ Supervisor: _____

Week Ending: _____ Av. Employee No: _____

INSPECTIONS

DATE	AREA	HAZARD	CONTROL ACTION/OUTCOME

Total No. of Inspections: _____ Total No. of Controls: _____

HEALTH AND SAFETY PROMOTION

TYPE	DATE	TOPIC
Training Staff		
Team Briefing		
Individual Talks		
Induction		
Other:		

Total Safety Education: _____ Number of Employees: _____

ACCIDENT INCIDENT DETAILS

DATE	NAME	INCIDENT DESCRIPTION (WORK TASK INCLUDED)	CONTROL MEASURE	TIME LOST HRS

Total No. of Accidents: _____ Total Lost Time: _____

DEPARTMENTAL MANAGER COMMENTS:

SIGNATURE: _____ DATE: _____

HEALTH & SAFETY COORDINATOR COMMENTS:

DATE RECEIVED: _____ SIGNATURE: _____ DATE: _____

Figure 2.8 Supervisor health and safety report

This supports the QM approach whereby the traditional management hierarchy is breaking down, hence recognizing the role of individuals who work at the coal face as key change agents. Such people may have a great deal of knowledge of, and expertise in, work processes. Employees have a vital role to play in goal setting and developing strategies for implementation, particularly in their own part of the workplace.

In the future we may see a greater number of workplaces supporting this approach of empowerment. Providing there are adequate support structures in place, such as training, shared values and vision, and shared benefits, safety management has the potential to benefit greatly from the skills of those who often know the workplace best – first level supervisors and shop-floor employees.

Critical elements in change include:

- commitment from senior executives to workplace culture change
- senior management developing a vision of where they want the organization to be
- selling the reason for the vision to all
- involving all levels in planning the best way to achieve the vision
- preparing middle management for new roles and ensuring people skills are developed
- ensuring a broad agenda is developed with a single bargaining unit.

The next aspect is the concept known as benchmarking. The health and safety records of organizations with similar activities can be looked at by you and, if they are better than those of your organization, the ways they achieve this can be examined and implemented by your organization.

Benchmarking, however, can be very industry-specific, meaning that a benchmark which is used by a mining company may not be appropriate for a manufacturing company. Having said this, care also needs to be taken with similar industries using the same benchmarks. For example, if a mining company in Canada sets a benchmark associated with a particular operation, adaptation and the use of the benchmark must be carefully considered if applied to a minesite in Indonesia – because of cultural and environmental variations.

Safety benchmarks, if used, should be developed by employees for operations associated with their specific departments. In this instance, they may represent very site-specific requirements, which can be measured and monitored in association with performance indicators.

As mentioned earlier, the process of continuous improvement can incorporate a variety of activities designed to achieve even better safety. This process involves developing key elements from which performance can be measured. Work teams are encouraged to develop strategies which are time limited, to meet the given key elements. This process should be evaluated at frequent intervals and progress in continuous improvement fed back to the work team for strategy review.

Many organizations have used a combination quality-safety management system to enhance productivity and competitiveness. There are standards for safety management (see Further Reading), and these can be linked to the ISO 9000 to 9004 quality standards to establish quality safety systems for this purpose. According to Rothery, the use of the ISO standards 'provides the essential information needed to take management policy or quality assurance and convert it into action' (see Further Reading at the end of this chapter). Further linking with the ISO 14000 environmental management standards may also apply in your workplace.

The health and safety plan

The first and most important aspect of health and safety management is a health and safety plan. This is described in detail in Chapter 14.

To examine the adequacy of a preventive health and safety plan, ask whether it addresses the following:

- commitment
- communication
- consultation
- hazard identification and risk management
- auditing
- procedures
- accountability
- training
- investigation of accidents
- selection of personnel.

Best practice in work safety

Introduction

No respectable health and safety policy would be complete today without the CEO assuring stakeholders that the management team is seeking to emulate best practice. Usually the challenge for management is to convert policy statements into workable procedures and practices. However, in the case of work safety, what constitutes best practice is not that clear. Unlike, for example, engineering, finance and stores management, where reasonably established performance indicators exist, work safety remains a difficult area to manage. The two most common performance indicators are injury measurement and workers' compensation costs. Both are of limited value as a basis on which to determine progress towards achieving a best practice safety model.

Defining best practice

Best practice management systems in work safety are formal, sustainable systems of work, which are integrated into the operational culture of an organization. The system must be based on the principles of self-regulation and self-assessment and support a high degree of employee involvement and participation.

Best practice systems embrace the hazard management model of:

- identification
- evaluation
- assessment
- control.

These hazard management elements are supported by an efficient programme of monitoring, training, communication and reporting.

It is inherent in the model that all hazards are translated into reliable, quantified bands of risk and rated accordingly. This function is critical to the safety system as the rating forms the basis for both determining action priorities and measuring the impact of any risk reduction. This process provides the benchmark for continuous improvement.

Best practice must also demonstrate a methodology to ensure compliance, notwithstanding that compliance is merely a building block for best practice.

In an organization aspiring to best practice, accidents are viewed as the result of an error in the system of work and as a positive opportunity to modify a system that has failed. Accidents are investigated in terms of non-culpable error with a clear focus on the people/environment/machinery/material/work practices relationship.

A best practice organization is subject to external auditing and reporting, and the results of audits are used as risk management data to monitor performance.

Where do you find best practice?

In attempting to discover and define best practice one approach as noted earlier would be to look at successful organizations in various industry sectors, analyse their work safety strategy and then duplicate their techniques and systems.

If successful, this approach would have the basic attraction of being able to implement a system that has been tested, shown to work elsewhere, and which offers cost savings in programme design and development time. There is also the added benefit in that the programme elements can be readily imported and a safety management system installed relatively quickly. This strategy will always have appeal with contractors, particularly those involved in high risk and fast track projects typically found in the mining and construction sectors, two areas where effective safety management is critical.

Using this methodology could lead to the discovery of a best practice model, but regrettably the imported system will almost certainly resist replication when imposed on a new environment.

This was a lesson learned the hard way. The US industrial giant Union Carbide reportedly believed that its operation in Bhopal, India was a replica of its plant in Virginia, USA. Investigation and litigation following an isocyanate escape at their Bhopal plant showed that was not the case.

The decision to take an engineering design, a set of management safety systems and corporate policies from one work environment and impose them on another environment, without careful thought as to their adaptability, invites a repeat of the Union Carbide experience.

Another temptation is to simply 'copycat' a procedure on the premise that, if we are doing the same thing as a leading player, we are achieving best practice.

An example is drug testing in the workplace. Without any real evidence as to the effectiveness of such testing on safety performance and having generally established no realistic standards, some people are being tested and their lives affected in programmes that lack validity, reliability and credibility.

While this practice continues with mixed support, a performance-limiting factor such as fatigue is rarely addressed.

Adapting these types of strategies to the identification and implementation of best practice demonstrates a lack of understanding of the basic principles of taking a systems management approach to work safety.

Performance indicators

The common view is that an accident must involve some form of injury. This is not correct and is counter-productive. The focus on injury or outcome probably resulted from our over-dependence on the Lost-Time Injury (LTI) as a performance indicator. Regrettably, the mindset is well entrenched and the paradigm shift needed to redress the present thinking will not happen until we replace our dependence on the LTI with alternative performance indicators which are understood, accepted and relevant to industry needs.

On the positive side, in most workplaces there already exists formal documentation available to report accidents, incidents, near misses, hazards and injuries. However, few workpeople appear able to clearly understand exactly what is to be reported. Even fewer appear to know the role each report plays in their safety. This must cause concern that the effort put into information collection, and its analysis and interpretation, is of questionable value.

Clearly, the information available has the potential to be of great assistance to those exposed to the risks, those that create the risks and others working in the prevention area. Regrettably, the Lost-Time Injury (LTI) experience is of minimal benefit in developing a base for a prevention strategy.

The LTI has traditionally been generally viewed as a statistical function reflecting the overall failure rate and safety climate at the department, organizational and industry levels. While there is some conventional wisdom and historical foundation in the relationship between the LTI and overall performance, there are far more accurate performance measures available. As the LTI became a standardized performance indicator for safety performance, some organizations became adept in avoiding the recording of LTIs. The LTI is now being used as a performance indicator in contract specifications. This may encourage under-reporting if contracts can be breached by a high recording of LTIs.

The LTI is only one factor in determining the safety management performance of an organization, and even then only a minor factor. It is debatable whether the LTI has any direct bearing on the safety climate in the organization. Safety is an individual's perception of risk. This perception is affected by many factors, mostly outside the direct control of the workplace. Performance indicators must reflect a wide range of variables in order to gain a clearer and more accurate picture of safety management in the work system.

In order to gain any real appreciation of organizational capacity to manage a safe system of work, accurate and timely information is necessary on critical performance variables.

These variables include control over error management, accident prevention techniques, hazard management, risk engineering, and corporate arrangements for employee consultation and reporting.

Clearly, we must not marginalize the use of the LTI as a means of identifying a need to improve performance. However, we must take advantage of all the available performance indicators if we are to maximize the prevention opportunities.

Some of the more progressive key performance indicators include the following:

- the match between procedures and practices
- the variation in the organizational safety profile
- risk communication systems
- information and data traps
- action response rates
- self-regulation arrangements
- integrity of policy
- performance monitoring standards
- decision-making arrangements
- problem-solving opportunities.

All these indicators are quantifiable and able to reflect the preparedness of an organization to manage risk in the workplace.

The indicators are proactive and chosen to support the control of errors in the workplace and the prevention of accidents.

Contractor safety

Introduction

There has been a marked shift in many areas of work from using the organization's own employees for nearly all of its work to outsourcing a range of areas of work through the use of contractors. Recent evidence from actual studies suggests that there is a high correlation between this and a decline in occupational health and safety. Contractors may:

- come from very small contract firms or be self-employed
- larger firms with an office on the principal's construction or mine site
- work from home or a location remote from the principal's operations using electronic links to the principal
- be a labour hire firm providing the principal with workers with a wide range of skills as needed.

The contractors in turn may be bringing in subcontractors. The work contracted out can include operations such as call centres to answer inquiries, generate business or solicit donations, or a myriad of other tasks.

The principal organization, if it is to ensure high standards of OHS within its area of legal responsibility, must know that any contractors or subcontractors have an adequate system to ensure OHS. The contractor must meet certain levels of achievement and aim at consistency with the overall OHS strategy of the principal.

Participation

Contract workers on a site or in a workplace belonging to the principal need to feel that they are part of the site reporting and participatory mechanisms, so that they can make an effective contribution to safety and health on site. Longer-term contractors need to be part of site health and safety committees. Even short-term contractors need to participate in site safety meetings.

Induction and training

The employees of contractors either need to attend induction programmes run by the principal or to attend an induction programme run by the contractor which has the principal's approval. In some industries such as mining, where contract employees may go to many sites and work for many principals, a common basic induction, for which a ticket is issued, has been introduced in some places. This avoids the need to constantly receive the same training.

It is important that the contractor's workers are familiar with the procedures applying to a particular site for items such as lock-out and tag-out, other permit systems and emergency response. So even if there is common industry-wide induction, site induction training on specific issues should supplement this, ensuring it is a good fit to the basic industry-wide induction.

Special efforts need to be made for temporary or part-time workers in particular to overcome the fewer opportunities they would otherwise have to receive training.

Labour hire

Labour hire firms face quite large logistical problems because while they retain legal responsibility for the health and safety of their workers on site in many jurisdictions, they are almost entirely dependent on the principal to ensure that health and safety.

In the nature of things, those delivering a product or service offsite are seen in a different light to those working on the principal's site alongside the principal's workers or working for the principal's supervisors.

Some OHS legislation says, on the one hand, that the principal is responsible for the health and safety of the contractor and any person employed or engaged by the contractor in relation to matters over which the principal has control, but then says that this in no way reduces the duty of the contractor to workers they employ or engage. It further mentions that the principal cannot vary this duty in the contract.

This makes for a degree of confusion about responsibility on a site, and also fails to recognize the possibility that specialist contractors, rather than the principal, may know best about the safety controls their type of work requires.

Tenders

An important step in ensuring health and safety in contract situations lies at the tender stage. The principal needs to ensure that there is evidence that the contractor has an adequate OHS system, and a good accident record. (The latter can be difficult for contractors just starting out.) Principals using a prequalifications system which screens contractors out before they can even submit a bid, make it even more difficult for a new company.

It is important that the key performance indicators (KPIs) for OHS (and for quality and environmental management) are set out in the tender, and that the contractor demonstrates how it will meet these. The achievement of KPIs by the contractor needs to be monitored by the principal for the life of the contract. The principal needs to explicitly recognize the costs associated with having and operating a health and safety system to avoid competitive underbidding, which compromises safety.

Subcontractors

One approach used to address the health and safety of subcontractors is for an industry such as the construction industry to set up minimum requirements, e.g. for welders. These subcontractors can then be required to undergo suitable training and assessment to acquire an industry passport. Contractors can then be required by the principal to hire only subcontractors who hold the passport.

Behaviour-based safety

The concept of controlling behaviour in the workplace grew with enthusiasm and behaviour-based safety programmes have become a key focus in some organizations following the publication in 1931 of H.W. Heinrich's book *Industrial Accident Prevention*. Heinrich's research suggested that unsafe acts by people had a direct causal relationship to accidents. He further indicated that accident causation was the result of a series of preceding factors including: ancestry and social environment, fault of the person, unsafe acts or conditions which lead to an accident and injury.

Heinrich, at the time, was an insurance investigator, who based his research on examining accident reports completed by company supervisors prior to 1931. He concluded from his research that approximately 90% of all accidents were caused by the unsafe acts of workers. It would be interesting to know if those company supervisors had adequate skills and knowledge of the work system to identify the true accident causation factors, or if they simply adopted a blame mentality when accidents occurred.

Through the evolution of time, coupled with Heinrich's accident causation theory, behaviour-based safety programmes emerged as the key to harness the unwanted outcomes or accidents. Behaviour-based safety is essentially about identifying and eliminating unsafe worker behaviours, often referred to as risk behaviours, and promoting conformity and the practice of safe work behaviours or critical behaviours.

When implementing a behaviour-based programme, the methodology varies between workplaces; however, there are some common elements:

- Establishing a critical behaviour list
  Job tasks are reviewed by the work team and a list of critical behaviours or safe work behaviours are documented.
- Monitoring critical behaviours
  Work behaviours are observed to determine if they have met the standards set in the critical behaviour standards or if risk behaviours are occurring. The observation is carried out by supervisors and employee colleagues.
- Correcting risk behaviours
  Education and training can assist with correcting risk behaviours; however, there is also a tendency to use formal warning procedures and dismissal if these behaviours are not corrected.
- Incentive schemes to promote safe work behaviours
  These schemes are varied; however they aim to encourage worker participation in the programme. They may include monetary or material rewards for demonstration of safe behaviours and outcomes such as reduced accident numbers and increased productivity.

As we enter a new millennium, behaviour and accident causation models from the 1930s remain with us. Age does not necessarily destroy their value. Some organizations, including Dupont, embrace these concepts and apply 'behaviour-based safety'. Krause and Hidley in California (see Further Reading) have also strongly espoused them. Others shun the idea based on evidence suggesting failure, and a growth in alienation between workers, employers and unions.

Work and the methods used for completing tasks, education and experience levels of employees, and environmental factors associated with work have evolved over time. Based on this, we could then suggest that the application of 'behaviour-based safety' as a stand alone practice in safety management can only fail, as we have evolved from the 1930s and the needs and functions within the workplace have significantly altered.

The Hegney–Lawson System Risk Model (see Chapter 6) is applied to the workplace system as opposed to one element within the system. The model supports the theory that for work to function at any level, and function safely, input or resources involving people, equipment, the environment, the materials used, and the methods used, are fundamental. Work or the interaction of these five ingredients can result in performance levels that range between stable and unstable, fluctuating over time.

Performance fluctuations are influenced by many factors including work schedules, the condition of equipment and raw resources, staffing numbers and experience, lighting and thermal control. Therefore, to influence work output we must have some control over the many variables which form the ingredients of the work process. This is where tools such as hazard management and quality control become essential. If errors in the work system can be identified, and hazard identification and risk management applied – i.e. the hierarchy of control, the work system is likely to remain in a stable state more often than an unstable one.

Alternatively, if we were to rely solely on behaviour-based safety to keep our work system in balance, many other variable factors that influence work and efficiency could be overlooked. For example, while carrying out his job John is identified as not wearing the required protective glasses. Action will be taken to ensure that this behaviour does not occur again. However, if the employee is asked why they were not wearing protective glasses, we may begin to focus on a more comprehensive systems approach which identifies system failures instead of blaming employees for bad behaviour.

In a systems approach, the system failure is identified through risk management and the hierarchy of controls is applied to minimize exposure to the risk source.

We may in fact find that by applying a systems approach there is no need to wear safety glasses because the risk can be eliminated or controlled through other measures.

Behaviour-based safety does have some value where the risks in other aspects of the work system have been dealt with through the comprehensive approach suggested above.

Communication and meeting skills

Communication

Effective health and safety communication skills will become very important to you when liaising with management and/or employees. Communication is an interpersonal process of sending and receiving symbols with meanings attached. The giver of the message presents the message based on their experiences, values and skills. The values of the giver in part will be shaped by the culture of their family and the other groups in which they live and work. The same is true for the receiver of their message. Effective communication occurs when the intended meaning of the sender and the perceived meaning of the receiver are one and the same.

Efficient communication results in minimum cost in terms of resources expended, e.g. use of memos, group meetings.

Communication, to be effective, requires that:

- the message is accepted
- the giver and receiver of the message both understand it clearly
- appropriate action results from the communication of the message.

For an employee, it is important for them to develop skills to represent the views of their work group, not just their own views. On occasion, the views of a certain member of the work group may not be shared by others. It will take good judgement to decide when to accept group views and when to accept that an individual may, in fact, have a view worth supporting even if it is not supported by other work group members.

If you are health and safety facilitator, you will need to develop skills in:

- interviewing (counselling fellow workers in your work group and asking them for information)

- reporting to meetings and discussions (e.g. health and safety committee, management)
- note taking
- problem solving
- report writing.

Interviews

Interviewing is an important part of all aspects of management including health and safety management. It can involve people in work teams, supervisors and managers. It may include:

- hiring the person most suitable for the job
- appraisal of a person's performance including their safe and healthy work skills
- discussing options for return to work after an injury
- raising the issue of the need for an employee assistance programme
- terminating employment, including termination where an employee has repeatedly breached conditions of employment
- investigating an accident or breach of a rule, standard or policy.

Planning

Planning an interview is important, especially if it involves hiring or termination, because of the equal opportunity and unfair dismissal laws, which apply in many jurisdictions. Even if there is to be an on-the-spot 'step into my office' interview for some reason, it can pay to spend a few minutes thinking beforehand about how to conduct the interview.

Some interviews are less formal, such as those which take place when an audit team questions different workers about aspects of their work.

An interview should be:

- conducted in a quiet place
- conducted in a situation in which the exchange is confidential to the parties concerned
- conducted without interruption such as fixed or mobile phone calls or while the interviewer is dealing with email or flipping through an in-tray. (For this reason a separate room away from the office desk, such as one which can be booked for the purpose, makes good sense.)

Seating

Attention needs to be given to the seating positions, both in relation to relative height and layout (if the interview is conducted while seated). An interviewer can interpose a desk or choose not to. What difference does this make to the atmosphere of an interview?

If a number of people have to interview a person, a circle or a round table is often preferable to having the interviewee sit facing a line of people behind a long table. This is less threatening for the interviewee.

Preparation of questions

Preparation is important. Each type of interview will have its own requirements. If it is to select someone to hire, ensure you have summaries of the reasons for shortlisting and all the shortlisted applications properly assembled. You will need to draw up a list of questions you or the interviewing panel will ask. In some types of employment the same questions must be asked of each applicant. Questions should relate to the selection criteria and be phrased to encourage the interviewee to give an example of their experience or how they would handle a situation. Take a moment to imagine yourself in the other's shoes – switch roles and ask 'How would I react to this question? Would I be defensive, secretive, cooperative, or appealing to be believed?'

If there is a 'live' segment where a candidate has been advised in advance to prepare to give a short training session, the training aids should be ready and in working order. Or you may wish to take a candidate to a piece of equipment and ask questions, or have he or she demonstrate its correct and safe use. This requires pre-arrangement with the work area concerned – embarrassing the interviewee in front of other workers is not the aim of the exercise.

Rehabilitation

Interviews with an injured person in regard to rehabilitation can involve a group and may include the supervisor, occupational health nurse, health and safety officer and rehabilitation specialist. A friendly, informal, positively charged atmosphere is important, as light duties may involve perceived loss of status and concerns about ability to work with impaired function, or the possibility of pain, and a different working area and work group. Care must be taken to preserve medical confidentiality and to obtain the rehabilitee's consent to release of information.

Employee assistance

When it comes to employee assistance programmes, such an interview often arises out of realization by a supervisor or fellow team members that an employee's performance has lessened. There are many influences in our lives which can affect the way in which we perform our work. These may include domestic concerns, financial difficulty, substance abuse and peer group pressure. Such an interview may be one-on-one and kindly but firm. The interviewer needs to be well prepared and have a clear basis for what they can offer by way of further assistance. For instance, external counselling may be offered. If the employee is able to return to feeling happy, healthy and secure, work performance should be enhanced. If performance doesn't improve after assistance, the employer may need to follow a formal warning process which could lead to termination.

Termination

In some employment there is a clearly stated policy about which acts or omissions are unacceptable, and that two or three breaches are grounds for dismissal. If the employee has reached the limit and has received adequate formal warning, the interview should be firm and formal. The interviewer should be prepared for the interviewee to be evasive, despite the findings, and should have made preparations, if necessary, to call others in to back up what they are saying, e.g. 'This is the third time you have entered an area marked off with bunting and with a sign indicating entry is only allowed with a permit.' Particularly in jurisdictions with unfair dismissal laws, such interviews require careful handling. But even if there is no legal requirement, the employee should be left feeling that the process was fair and accorded with the natural justice principle 'hear the other side'.

In this type of interview as in others it is important for the interviewer to listen carefully to the responses and allow for adequate time to give the interviewee time to think and add to what they have said.

Try and conclude all interviews on a positive note, as ongoing dialogue may still be required after the interview.

Recording

Records of all interviews should be kept. Depending on the nature of the interview, different types of records apply. These can range from a formal write-up of a job selection interview to a note in a supervisor's diary about a caution or formal warning issued to an employee, initialled by the employee if desired. Note taking or a checklist during an interview assists with short-term notes; however writing a more thorough record should be attended to as soon as possible whilst the memory is still clear. Where the record summarizes the thoughts of a group, it needs to be circulated so that the members of the group can read it and sign it if they agree, or request amendments.

Questioning techniques

The techniques used in questioning are important. Be clear about the purpose of your questions before you ask them. Use link questions to move smoothly from one type of question to another, e.g. 'You were saying earlier on that ...' or 'How does this affect your work?'

A question should be clear and well thought out especially if it has been written in advance for use by a panel in an interview. Supplementary questions should be used to clarify or probe an answer the interviewer has received. Open-ended questions, rather then yes/no or true/false style, often give rise to better responses. However, yes/no style questions are still useful simply to verify something.

In another approach, the interviewer can make a statement and wait for a reaction, and can use facial expression alone to obtain a further response. Comparison is also useful, e.g. 'How does the way that you did this today compare with the way in which you did it yesterday?' Focusing is another technique – here the interviewee is asked to be more specific about something they have said. On the other hand an interviewer may wish to expand on a matter, seeking more detail, e.g. 'What was Jane doing just before she restarted the lathe?'.

Do not use counter-productive questions. The aim is to get the other person talking – not to suggest 'right' answers, discourage, confuse or mislead. As noted in Chapter 6, avoid blame when interviewing people during an accident investigation.

The emphasis in interviews on eye contact is sometimes pushed too far – looking out of a window to reflect for a moment is normal and acceptable for an interviewee and prevents what is sometimes an almost hypnotic effect in a long interview.

Listening and taking notes

Listening is an important skill, whether it is during training by the trainer and trainee, as an interviewee or interviewer, or while attending a presentation at a seminar or conference. Many committees such as health and safety committees need accurate listening to provide minutes. It begins with the ability to hear clearly, so the surroundings are important. (No training or meetings alongside a minesite crusher!)

Listening is also aided by the other senses, e.g. a talk on chemical odours as a warning sign (if the odour threshold is lower than the permissible exposure standard) would be aided by allowing a sniff of the chemical if given safely. Sight is the dominant sense (in those who aren't sight-impaired) and so example objects or pictures assist the listening process.

Attention span can be relatively short, so the listener must be conscious of the need to concentrate because it is often not possible to do anything about the standard of presentation (e.g. monotonous, no light or shade or emphasis, poor structure) on a topic the listener simply must find out about. The listener needs to listen carefully for key points often given at the beginning of, e.g. a training session, and focus on the points during the presentation which expand on these. There is a need to weed out what is irrelevant. It is important for the listener to try and develop a concept map or mental map or image which is similar to that of the presenter, sometimes clarifying it with a question.

A listener coming from a different experience base to that of the presenter will relate what is heard to his or her own experience base, and may as a result fail to understand what it is the presenter is trying to get across. If the listener then uses what is called cognition to convert the received message into action, problems can arise. For example, if an instructor (who had worked on a lathe with a left-handed cutting head advance mechanism) told a trainee who was trying to use a lathe with a right-handed advance mechanism to turn the head advance mechanism two turns right, thinking the head would be withdrawn from the workpiece, an accident could result.

In many situations the listener needs to take notes. Listening for key points and using abbreviations (as long as they can be understood later) is vital. At the end, writing up the notes promptly, highlighting key issues, filling in the spaces and clarifying issues by going to the references or discussing those issues with the other attendees or trainees is important. Always ask the speaker if notes are available or make sure you get a copy if they are offered. Voice recording can be valuable if permitted but is often tedious to turn into written notes later. For minute takers especially though, a voice recording can be useful as an important check on the accuracy of details. With minutes especially, listen beyond the discussion for the actions decided on, who is to implement them and by when. An action list can then be developed and circulated to those who have been assigned tasks.

As noted, listening is important to interviewers and interviewees. Some interviewers will need to overcome a tendency to stereotype people, e.g. in some occupations, a man with a pony tail and a denim jacket is the norm; and some women must wear a head covering for religious reasons. The interviewer must sometimes overcome dress prejudices to find the best person for the job – so the interviewer needs to give extra attention to listening. Sometimes an answer can be clarified by the interviewer repeating it back as they understood it, saying something like 'You said you were near the factory door when you heard Chandra yell for help, is that correct?'.

From the interviewee's perspective, the need to listen carefully to the question is paramount. Although the response must be reasonably quick, it is permissible to ask for a moment or two to think about some tougher questions, without giving an appearance of uncertainty. The interviewee, just like a person in a written examination, must answer the question asked (not another one) and if they are not clear they should ask for it to be repeated or clarified.

In general, a shared view of an exchange between people (i.e. people who are on the same wavelength) is the aim of many interviewing or listening situations.

Effective presentations

Communication

Promotion of workplace health and safety requires good communications between people. A person technically proficient in health and safety must communicate well to be effective.

In the Robens-style consultative approach to health and safety, a lot of importance is placed on negotiation by health and safety representatives, as well as discussions in health and safety committees. Training is a key feature. Effective presentations of information and points of view to other people are essential.

The source or sender of the message may have poor public speaking skills and be unable to address even a small group of people confidently. So when asked to do so, their attitude towards themselves, the subject matter and even the person who asked them to speak may interfere with the message to be conveyed. Often, they know too little, and sometimes too much about a topic to do justice to it when presenting information. Knowledge on a particular topic as well as knowledge of the audience are important determinants of how successful a communication will be, and how readily a message will be accepted. So, if you are presenting, make sure that you get the necessary information, you understand it and you put it into a form that others can digest. You must know and understand what you are talking about.

As all communication takes place in some situation, it is important that we consider the social system or setting in which the communication takes place. (As an example of this, you would not tell a group of five employees at the local bar on a Friday afternoon that their services are no longer required.) We are continually reminded in the press just how important culture is as a barrier to effective communication. Race, religion and politics in themselves influence our attitude towards others and, therefore, will influence our communication.

Planning

Here are some issues to consider when planning a presentation:

- intended outcome
- needs of the audience
- size of the audience
- nature of the audience

- time available
- physical facilities available
- aids to presentation desirable
- timing and duration of breaks
- cross-cultural communication
- wearing appropriate attire.

The needs of the audience will allow some objectives for the presentation to be set, so the presenter needs to consider a range of issues, including:

- who it is made up of, how diverse they are – in age, education, gender, nationality
- what they are expecting
- what they want to know or what they are seeking (you may think they want a safer workplace, but in fact they may want more money – although 'dirt money' is now an outdated concept, some people are still paid for taking extra risk)
- whether they are there by choice or there because they have to be (e.g. have they just come off night shift?)
- any language difficulties (in some cases people's written comprehension is better than their spoken comprehension and so despite the 'death by overhead' idea, written overheads or distributed copies of the speech can help – 'overhead' here refers to any projected image)
- perceptions, e.g. 'You're talking to us disabled people but you're lucky, because you're not disabled'.

Speaking

Apart from straight presentations, there are many other ways of getting the message across. But let's look at these straight presentations, where you are not able to actually deliver them on site, so the audience cannot directly experience the topic during its presentation. Here are some pointers:

- Generally speaking, a large audience requires a more formal, less interactive style.
- Select your key points.
- Write the body of your presentation on these points.
- Provide an introduction (try to include something attention grabbing) and a conclusion. You should now have a beginning, middle and end.
- Check that the order of presentation is logical.
- Decide whether you are going to memorize, use speakers cards, notes or speak from overheads or slides. Select and make your supporting visual material. Do you need photos, films, videos or other demonstration items? Obtain, make or order them. A computer-based program such as PowerPoint allows you to create notes and overheads, either for overhead projector or video projector. Ensure that it can be used at the venue. Ensure that the laptop low power and screensaver settings are off.
- Deliver the talk to your mirror or someone long suffering (your dog?).
- Whittle it down to time allowing spots for the audience to absorb any visual material and to ask questions.

- Plan your humour, or any special demonstrations or effects. Rehearse the joke or demonstration.
- In relation to slides and overheads:
  - key them on your cards or notes.
  - have them in the right order (check run slides to see they are right way up and not mirror image).
  - ensure that diagrams, print and details in pictures which you want to emphasize are clearly visible to the audience without effort. Keep detail down and allow time for it to be absorbed.
  - check the operation of the projector. (Do you do it from the rostrum, or will an operator do it on request?). In a PowerPoint presentation, once again, have you disabled the computer screensaver?
  - ensure that the lighting allows you, the overheads, and demonstrations, to be clearly seen. Can lights near the screen be dimmed?
  - ensure that you and the visual aids are clearly visible to all (no obstructions such as the lens holder on an overhead projector).
- Check the sound level of amplified systems so that all the audience will be able to hear.
- If you intend to use a whiteboard, ensure that the pens work, have thick points, dark ink and that writing can clearly be seen in the back row.
- Give yourself enough time to have everything ready before you begin the presentation.
- Put down your watch when you start and stick to time. You may want to ask for questions to be left to a question session at the end.
- Greet your audience at the start. Thank your audience at the end.
- Use body language to convey messages.

The senses as mentioned earlier have an important role to play in the value the listener or participant gets from your presentation. Taste and touch generally cannot aid us a lot (although they would be valuable to a nutrition lecturer or a fabrics presenter respectively). So we must make the best use of sight and hearing. Smell can also be used to add reality as noted earlier in 'Listening and taking notes'.

A speaker needs to pay attention to a number of items – these include body language, dress, movement, and voice. At the same time they need to be conscious of the nature of their audience and how the audience relates to those items.

It is important to decide how to tackle such issues, so asking questions when commencing allows the speaker to be more aware of them (e.g. even the simple question 'Can you hear me at the back?').

Dress

Dress, touched on earlier in 'Listening and taking notes', can be a factor which stimulates the prejudices or intrudes on the comfort zone of a person or audience. An unforgettable scene involved a supreme court judge in wig and robes being honoured in a dance by an Australian aboriginal leader clad in a loincloth, skin daubed with ochre, with a tassled coloured ornament held in his mouth. The appearance of each indicated high status within their own group, but not automatically in the other.

A tie and white shirt on a minesite may set up an immediate barrier. It may be unavoidable but the visitor in the white shirt and tie should be aware of it.

Movement

Movement and gesture are also important. Effective use of hands and body can do much to aid a presentation. Moving out from behind a table or lectern towards the audience can help to reduce barriers and engage that audience. But be careful that any sound system supports this and learn to switch off the lapel microphone when you move back to the fixed one. Confidence and attitude can be judged from how a speaker sits, stands and moves, and also from eye contact. Rehearsing in front of a mirror is a well-tried technique.

Voice

Apart from the visual aspect, the voice also needs attention. A few deep breaths will reduce any sound of nervousness at the outset. It is important not to speak too fast and for a speaker to deliberately slow the delivery and watch pronunciation if an audience has trouble understanding because of language, accent or social background. The speaker should choose words to suit the audience (e.g. dust disease of the lungs – not pneumoconiosis – for some audiences), and relate what is said to the needs of the audience. Use of the vernacular to suit the audience needs care – it can offend some listeners and may even reduce the authority of a presenter, although the listeners use the vernacular themselves. Care is needed in trying to be 'one of the boys or girls'.

Pausing for effect, using light, shade and where appropriate, humour, are all important. Some presenters excel at *ad hoc* humour based on the situation at hand, others find it better to prepare humour. As noted earlier, it is useful to try out a presentation before delivery to a friendly listener. The listener can suggest clarification and the length can be adjusted. Deciding in advance about what can be omitted is good practice, in case the speech time is cut by a chairperson or by other circumstances.

Humour is often used to break the ice or find common ground with an audience, and so develop a measure of acceptance for a speaker. Establishing credibility can be vital, so as important first steps, the speaker should describe who they are (some introductions do this), what their background is and how they will help the audience. To do this they may need to check that they have gauged the needs of the audience correctly. At the outset it is also useful for a speaker to explain how and when questions will be taken. Once underway, outlining the structure of a presentation is vital. This, and recapping the main points at the end, are 'givens' and yet are often forgotten.

In some circumstances the speaker should be prepared to revise the presentation on the spot if unusual circumstances arise. (This is much more easily done if they are well prepared.) And a speaker should always be prepared to work without overheads or PowerPoint if circumstances make it unavoidable.

These few tips are no substitute for experience in dealing with audiences, but they will allow a speaker to avoid the reputation of never learning from their mistakes, which some speakers sadly acquire.

Reporting and reporting back

Before going to interviews or meetings it is important that you make use of some of the skills above. Listen carefully, take effective notes, and plan how you intend to report on an issue (present your case). You will be more effective if you use an assertive (but not aggressive) approach. Make sure you cover all the key points.

In reporting back to a work group on a meeting or interview, use the notes you have taken, and be prepared to explain points of view put at the meeting. Even if they are likely to be unpopular with the work group, it is important to do this, so that the work group is better able to think about what action they want you to take next. You may think public speaking skills are confined to more formal occasions, but a number of the points set out in the section on effective presentations can apply to any situation in which you are speaking to, or with, a group of people.

Barriers to communication

Non-assertive behaviour
This occurs when we don't express our feelings and needs, and we ignore our own rights, allowing others to infringe on them. We avoid conflict at all costs. We sacrifice our wishes for those of another person.

This is emotionally dishonest, indirect, inhibited, and self-denying. We often feel anxious and disappointed with ourselves and bottle up all the anger and frustration, which often leads to health problems. We suffer in silence to avoid confrontation, tension and conflict.

Aggressive behaviour
This occurs when you force your ideas and feelings on other people without any regard for them, and usually at other people's expense. You will stand up for the things you want but overlook the rights and needs of others, even to the stage of humiliating others and dominating them. You certainly don't keep things bottled up, and you make sure you tell others exactly what you think of them! This behaviour is hostile, defensive and self-defeating, simply because we lose the support and cooperation of others who will try to avoid us and our bad-tempered outbreaks. Aggressive people end up feeling more angry and frustrated, self-righteous and possibly very guilty and embarrassed later. Aggression leads to loneliness and defeat.

If an aggressive person is requested to wipe up an oil spill from the floor to reduce a slipping hazard, the response may be 'Why should I wipe up the spill when I didn't put it there?'.

Assertive behaviour

You can express your feelings in ways that don't violate the rights of others. You can say how you feel and what you think, and stand up for your rights. This behaviour is usually emotionally honest, direct and expressive. Assertive people make their own choices, with thought and often research, so they feel confident and good about themselves. They are interested in what others experience as it can be a guide as to what the consequences of a choice are.

Usually, assertive people reach their goals, and if they don't they still feel good about the effort they put into what they did, and what was achieved in the attempt. This improves self-confidence, as the more willing a person is to step out and try new things and learn from others, the more they grow and extend their skills.

Assertive people don't always get their own way, but they have acted with dignity and earned others' respect. They do not make choices for others, and so treat other people's right to think and plan for themselves with respect. This means they are welcome anywhere.

If an assertive person is requested to clean up an oil spill, the response may be 'I have already been asked by "X" to readjust the clearances on the machine urgently. While I would normally be only too happy to clean up an oil spill, can I ask that you ask someone else this time, or check with "X" that it's OK to attend to the machine later?'.

Meeting arrangements

The following points are provided to assist health and safety facilitators establish procedures for the conduct of meetings, such as meetings of the health and safety committee.

Who will chair the meeting?

The committee members select a chairperson. The position could be rotated between employer representatives and employee representatives at each meeting or for a specified period. The chair should allocate times and order of consideration for each discussion item in consultation with other members.

Will there be a quorum?

The fixed number of members that must be present for any meeting to properly proceed should be decided. All parties need to be represented and, when setting the quorum, the parties should allow for absences.

Who will take the notes or minutes?

Health and safety committees must take minutes of meetings and they should be retained for not less than three years. This could be done by a committee member or a minute taker provided for this purpose. Typing and photocopying facilities need to be available.

Timekeeper

A timekeeper can ensure that:

- the fixed time for a meeting is adhered to.
- the allocated time for each item is not exceeded without the further consent of the meeting.

Meeting arrangements

Issues that may be relevant when determining the frequency of meetings include the:

- likely volume of work to be handled by the committee
- size of the workplace or area covered by the committee
- number of employees and work groups covered
- kind of work carried out
- nature and degree of risk across the workplace or area covered by the committee.

Meeting dates should, as far as practicable, be arranged well in advance, preferably on a regular day that is suitable to all concerned. By adopting this arrangement, meetings can easily be planned months in advance. Notices of health and safety meeting dates should be posted where all employees can see them.

For meetings to operate effectively, every effort should be made to ensure scheduled meetings take place. Where postponement is absolutely necessary, an agreed alternative meeting date should be made and announced as soon as possible.

A copy of the agenda and accompanying papers, with a notice of the time and place of the meeting, should be sent to all participants in sufficient time for them to be considered before the meeting. Doing this by email, not hard copy, has its drawbacks.

Reasonable time should be allowed during each meeting to ensure discussion of all business.

Effective meetings

The communications aspect of meetings should not be underestimated as this sort of interaction provides an excellent forum for planning and preparing people, resources and time; solving problems; presenting financial and business reports; and generating motivation and enthusiasm among staff members.

The value gained from meetings will depend on the combinations of personnel and planning, place, documentation process, and the implementation of decisions reached.

Success of a meeting

The success of a meeting is measured by the way in which the purpose of a meeting was accomplished.

Factors contributing to a good meeting include:

- ample notice of time and location for meeting
- prior circulation of an agenda
- good chairpersonship
- punctuality in beginning and finishing on time
- observation of rules governing the meeting

- ability of members to accomplish assigned tasks
- well-directed discussion
- provision of solutions
- wise use of time
- competent recording of proceedings, including action items and responsibilities and timelines for them.

Documentation for and of meetings

There are three main items included in meeting documentation.

1. Notice of meeting
   This could take many forms: telephone call, facsimile or email, message, typed or printed form. However, when verbal advice of an important meeting is given, it is always best to provided written backup.
   A notice of meeting will include:
   - date, time, place and type of meeting
   - items to be discussed.
   Usually the document is sent out well ahead of time so that there is an opportunity for preparation.
2. Agenda
   This is a more detailed list of the items to be considered at a meeting. The chairperson and secretary may have expanded versions to serve as a guide to the conduct of the meeting. Spacing between items allows for the insertion of notes as the meeting proceeds.
3. Minutes
   These are the official record of proceedings at the meeting. They should be written in the past tense and in sentence form.

A court of law will accept signed minutes as prima facie evidence, so care in recording and safe-keeping of minutes is most important. It is usual to show only decisions and actions, and maybe some key points. Key documents should also be attached for future reference.

Who will issue minutes?
A person should be made responsible for this task.

Who will draw up and issue the agenda?
A member of the committee should be made responsible for drawing up the agenda. Adequate notice of items to be discussed should be given to this person to ensure timely distribution of the agenda.

How will decisions be made?
When discussing health and safety issues, decisions are best made by consensus rather than majority vote.

Who sees the minutes?
A copy of the minutes should be provided to each member of the committee as soon as possible after the meeting. Copies of the minutes should be displayed, or made

XYZ COMPANY LIMITED

Agenda

Date: Location:

Time: Meeting Place:

1. Opening

2. Attendance and apologies

3. Minutes – adopt minutes of the previous meeting as circulated correcting any error omissions

4. Matters arising from the minutes

5. New business (items for consideration)

6. Accident review

7. Training and education

8. Inspections and investigations

9. Other business

10. Next meeting

Figure 2.9 Example of committee agenda

available by other means, for the information of employees. A member of the committee should be responsible for providing the employer with recommendations of the meeting.

Figures 2.9 and 2.10 are examples of a committee agenda and committee minutes (in action format).

Enhancing occupational safety and health

Health and Safety Committee Meeting Minutes

Date:

Employer: Present:

Address:

Site Location:

Address:

No.	Item for Consideration	Recommendations	Action by	Target Date

Figure 2.10 Health and safety committee meeting minutes

Further reading

Alteren, B. and Hovden, J. (1997). The Safety Element Method – A User Developed Tool for Improvement of Safety Management, *Safety Science Monitor*, **(1)**, 3 at www.ipso.asn.au. See also *Safety Science* (1999), **31**, 231–64.

Bird, F.E. (Jr) (c.1980) *Mine Safety and Loss Control*. Loganville, Ga: Institute Press.

Bone, D. (1988). *Four Key Elements of Good Listening*. The Business of Listening, pp:12–21, 60–2. California: Crisp Publications Inc.

British Standards Institute (1999). *British Standard 18001 OHS Management Systems (Requirements), 18002 (Guidelines)*. London: British Standards Institute (developed with 13 other standards bodies).

Brooman, G. in Williams, S. (ed.) (1994). *Successful Training Strategies*. Ch. 1, Sect. 6, pp. 1–9. Ontario: Southam Information and Technology Group.

Brown-Haysom, J. (2001). Men Behaving Safely (behavioural safety in mines). *Safeguard (New Zealand)*, July/August, pp. 40–3.

Department of Industry Science and Resources. (2000). *Guidelines for Preparation and Submission of Safety Cases*. Canberra: DISR.

Editorial (2001). A Guide to Successful Meetings (health and safety committees). *Accident Prevention (Canada)*, March/April, p. 10.

Emmett, E. and Hickling, C. (1995). Integrating Management Systems and Risk Management Approaches. *Journal of Occupational Health and Safety – Aust. NZ*, **11(6)**, 617–24.

Everest-Hill, D. (2002). Purchasing Power (integrating H&S into purchasing). *Accident Prevention (Canada)*, March/April, p. 14.

Fettig, A. (1990). *Ten Ways to Make Your Meetings Sizzle!* The World's Greatest Safety Meeting Idea Book, pp. 10–16, 20–7. USA: Growth Unlimited Inc.

Health and Safety Executive. (1998). *Successful Health and Safety Management*. 2nd edn. Sudbury, Suffolk: HSE.

Keith, N.A. (2002). What Happened to Prestart Reviews? *Accident Prevention (Canada)*, July/August, p. 10.

Kowalski, K.M. and Rethi, L.M. (2003). Out of Box Approach to Mine Safety. *Professional Safety*, **48(1)**, 21–7.

Krause, T.R. (2000). *The Role of Behaviour-based Safety in the Workplace*. Proceedings of Minesafe International Conference, pp. 475–82. Perth, Chamber of Minerals and Energy of Western Australia (Inc).

Mayhew, C. et al. (1997). The Effects of Subcontracting/Outsourcing on Occupational Health and Safety. Survey Evidence from Four Australian Industries. *Safety Science*, **25(1–3)**, 163–78.

Metzger, C.R. (2002). Evidence-based Loss Control. If Not Now, When? *Professional Safety*, **47(8)**, 39–43.

Mol, T. (2003). *Productive Safety Management*. Oxford, Butterworth-Heinemann.

NIOSH. (2000). *Safety and Health Resource Guide for Small Business*. Cincinnati: US DHHS NIOSH.

NSCA. (1995). OH&S Performance Measures. *Australian Safety News*, **66(1)**, 50–1.

Petersen, D. (2003). *Techniques of Safety Management – A Systems Approach*. Des Plaines Ill.: ASSE. (Also separate book. ... *A Human Approach*.)

Rothery, B. (1996). *ISO 14000 and ISO 9000*. Brookfield, Vt.: Gower Publishing Co.

Shaw, A. and Blewett, V. (1995). Measuring Performance in OHS: Positive Performance Indicators. *Journal of Occupational Health and Safety – Aust. NZ*, **11(4)**, 353–58.

Standards Australia. (1999). Australian/NZ Standard 4360. *Risk Management*. Sydney: Standards Australia.

Stephenson, J. (1991). *System Safety 2000*. New York: Van Nostrand Reinhold.

Standards Australia. (1998). Australian Standard 4804. *Occupational Health And Safety Management Systems – General Guidelines On Principles, Systems And Supporting Techniques*: Sydney: Standards Australia.

Standards Australia. (2000). Australian Standard 4801. *Occupational Health And Safety Management Systems. Specifications With Guidance for Use*. Sydney: Standards Australia.

Townsend, A. (2001). A Cure for all Ills? *The Safety and Health Practitioner*, **19(1)**, 24–6. (re. behavioural safety).

Vassie, L. (2000). Effectiveness of Safety Improvement Processes: Behavioural-Based Approaches to Safety. *The Safety and Health Practitioner*, **18(5)**, 28–31.

Whiting, J. (1995). Proof Positive. *Australian Safety News*, **66(9)**, 34–42 (PPIs).

Williams, J.H. (2003). People-based Safety. *Professional Safety*, **48(2)**, 32–6.

Worksafe Australia. (1994). *Positive Performance Indicators. Issues and Applications*. Canberra: AGPS.

Worksafe Australia. (1995). *Occupational Health and Safety Considerations for Workplace Agreements*. Sydney: Worksafe Australia.

Worksafe Australia. (1998). *National Guidelines for Integrating Occupational Health and Safety Competencies into National Industry Competency Standards*. (NOHSC:7025). Canberra: AGPS.

plus OHS authority guidance notes or advice on election of health and safety representatives, operation of health and safety representatives and committees, and resolution of health and safety issues.

Activities

1. In a workplace of your choice, identify three elements of the management system which will impact on loss control.
2. Review the minutes from a safety meeting and identify if any barriers to effective communication occurred.
3. Develop or review a procedure for issue (dispute) resolution to promote the management of occupational safety and health in a workplace.
4. Select an occupational health and safety issue published in a newspaper and determine the difference in employer and employee viewpoints on the issue.
5. Write a presentation on a health and safety topic.
6. Write a draft health and safety policy.
7. Prepare a written case to justify expenditure on a specific safety initiative.

3

Common and statute law

WORKPLACE EXAMPLE

The US Mine Safety and Health Administration (MSHA) shows on its website a page covering the twenty most frequently cited standards, that is the standards on which the most frequent violations are based. It shows the period covered and the number and percentage of each type of violation; in most cases the reader can click on a *Get Tip* button to find suggestions to avoid a citation. The top five violations in 2002 involved guarding of moving machine parts, mobile equipment safety defects, electrical conductors, housekeeping, and inspection and cover plates.

Origins and types of law which influence occupational health and safety

Common and statute law

The term 'common law' refers to a system of law originally derived in England and imported into many other countries at the time of colonization. Common law is based on sets of rules developed by judges usually hearing cases in superior courts. Common law is also referred to as 'case law' and can be viewed as legal rules not contained in legislation.

'Statute law' is legislation. The word 'legislation' refers to both statutes and delegated legislation such as regulations and by-laws. Statutes or acts are made by legislatures.

There are important differences between common and statute law. Judges, through the common law court system, could be seen as making decisions in order to resolve specific disputes between various parties. If, for example, a worker is injured in the workplace a court will decide on the question of blame and what compensation is owed. This may set a precedent for future cases. In some places, action by the state against a person may also occur under common law.

The courts, however, cannot establish rules that will prevent future accidents occurring. Common law is re-active, whereas legislation has the potential to be pro-active and may serve to prevent problems rather than simply provide a means of settling disputes.

Legislation is much more flexible than common law and always takes precedence over it. Parliament (or Congress or a state or provincial legislature) within the limits of its power can change any principles deriving from case law that may be considered undesirable.

The courts have an important role in ensuring that legislative bodies follow proper procedures in making legislation and are charged with the responsibility of resolving disputes which may arise over the meaning of words in the legislation.

Common law is derived from custom as decided by the courts. Following the conquest of England by the Normans in 1066, William the Conqueror allowed the law as it existed to continue. At that time each shire had a small tribunal, usually made up of local community leaders, who met to determine disputes between the local people and to administer justice. The King did, however, establish a court, the Privy Council, to consider the more important cases of the day. Further, unlike the shire tribunals which were a local court system, the King's court had jurisdiction throughout the land. Although superior to the local tribunals, the King's court was required to administer the same customary laws which were based mainly on tradition and varied from shire to shire. Over time the King's (or Queen's) court became a centralized body of law and reduced local practices to a common level throughout the whole of England, the common law.

The system served the people well for some time; however, the common law or judge-made law tended to be inflexible. The people, frustrated by injustice, petitioned the King for personal justice or 'equity' in the law. The King responded through the establishment of a court system that provided equity and compensated for the inflexibility of the common law. Today, the courts of common law and equity are combined in some countries; however, equity still enjoys reasonable flexibility.

Origins and development of law

In many of the countries of the former British Commonwealth, the system of common law and a considerable body of statute law were inherited from the United Kingdom. The decisions of the UK's higher courts were given a lot of weight in many of these countries. These courts included the Judicial Committee of the House of Lords, the Court of Appeal and the Privy Council. The United States system of law also derives from the United Kingdom, although, like Australia's for instance, it is heavily conditioned since 1776 by a written constitution, and a federal system of government, unlike the United Kingdom. Increasingly, law in countries like the United Kingdom, Eire, Cyprus, Malta and the other countries in (or about to be in) the European Union is influenced by EU law. This brings influences derived from Roman law into the judicial process, and from France, for example, the perceived benefits of codification of the law.

Divisions and hierarchy of courts

The specific details of these vary widely between jurisdictions. However, there is a hierarchy of courts, so that appeal from an initial decision, and in some cases further appeal from a first appeal decision, is possible. In countries with a federal structure there are distinctions between matters falling within state or provincial jurisdiction and those falling under federal jurisdiction, although generally final appeal is to a court set up in the federal jurisdiction.

Civil and criminal jurisdictions

The legal system is generally divided into two major areas – civil law and criminal law.

(a) Civil law deals with disputes between individuals. The purpose of this law is to provide compensation.
(b) Criminal law is the law that operates between the state and the individual. The purpose of criminal law involves punishment, usually either a fine or imprisonment.

When a natural person, corporation or government body is prosecuted for a breach of legislation, including OHS legislation, the case is heard under the rules governing criminal matters. When the employee sues the employer for damages arising out of a work injury, the case is heard in the civil jurisdiction.

The US Occupational Safety and Health Act (OSH Act) applies the word '*civil*' in a different way. There are civil penalties for some offences and criminal penalties for others.

How work safety and health civil actions and breaches of statutes are dealt with in the court system

The victim of an injury sustained at work can, in many jurisdictions, sue in the courts for pain and suffering, and if the court finds the employer was negligent, receive an award of damages. This does not address loss of earnings, which the workers' compensation system is designed to cover. The possibility of launching a civil action for damages, in addition to receiving workers' compensation benefits has been ruled out by statute law in some jurisdictions. However, this can have the effect of transferring the cost of people with serious long-term work-caused disability to the social security system.

Prosecutions under occupational health and safety legislation can take a number of forms. They can be based on:

(a) failure to obey an administrative notice, e.g. prohibition or improvement notice.
(b) breach of the general duty of care (such prosecutions are usually brought after an injury or serious near miss).
(c) breach of a specific section of the act or regulations (or also in the USA, rules or standards), and called in the USA, a citation for a violation.

Penalty is by way of a fine, which for a serious breach by an organization can be very large. Some jurisdictions have introduced a charge of industrial manslaughter with the potential for a gaol sentence. Some jurisdictions, e.g. Alberta, have introduced compulsory training orders as an alternative to fines. The South African Department of Minerals and Energy has been investigating the use of 'no penalty' sections in the Mine Safety and Health Act to facilitate accident investigation.

Interpretation

Any person called upon to interpret an *act of parliament* (or of Congress or of a state or provincial legislature) should be able to do so based on the information contained within the *act*. The act must speak for itself. A *statute* usually contains an interpretation section which explains the meaning of words in that statute.

For example, the US Occupational Safety and Health Act lists in Section 3 the meanings given to the terms 'occupational safety and health standard', 'national consensus standard' and 'established Federal standard'.

So it is clear what is intended by these definitions. However, uncertainty in legislation is not uncommon and, to provide the courts with guidance in how to interpret legislation, parliaments or congresses have generally passed 'interpretation acts' to establish certain meanings. An important meaning concerns the matter of gender. Under the law, unless otherwise specified in the particular statute, words using the masculine gender include women. Further, the expression 'person' includes a body corporate; the term 'natural person' is used to consider a person in the living form.

Where a statute is not clear on the meaning of words and expressions as set out in specific legislation, and where statutory guides are not available, the common law established through the cases which have come before the courts provides the judge with some rules to follow to enable proper interpretation.

Some of the rules used in some jurisdictions are the *literal rule*, the *golden rule* and the *mischief rule*.

Other grammatical rules of construction may be used to interpret words within a phrase of a statute or phrases within a statute.

For example in the UK, the Sunday Observance Act 1677 (still in force when *Gregor v. Fearn* was heard in England in 1953) provided that:

> '... *No tradesman, artificer, workman, labourer or other person whatsoever shall do any worldly labour, business or work of their ordinary callings upon the Lord's Day* ...'

It was held in that case that the words 'or other person whatsoever' must be construed *ejusdem generis* (of the same kind) with those before them. Thus an estate agent (realtor) is not within the section, which appears to be limited to work of a manual nature.

Also, in some jurisdictions, courts are required to refer to the speeches given in the legislature, by those introducing the bill on which the act was based, so as to give due weight to the intent of the legislator.

Making an act of parliament, congress or legislature

An act of parliament, congress or legislature (we will use 'legislature' for simplicity) begins life as a policy, an electoral mandate or pressure brought through public opinion or community needs and pressures. The 'bill' is the formal documentation presented to the legislative body for its consideration. The minister, parliamentary secretary or congressman responsible for the legislative process may release a 'discussion document' (sometimes called a white paper or green paper) for public comment and the acceptance of 'submissions'. The US congressional process has the feature that two members of Congress may also take responsibility for developing and presenting a bill. Hence the OSH Act is also known as the Williams-Steiger Act. The Congressmen originating the bill may even belong to different parties.

Once the results of submissions are known and considered, following further steps a bill may then be drafted by 'parliamentary' or 'congressional counsel'. After debate, if passed, notification of the act is published in an official gazette (in the USA the Federal Register) which indicates when it will become law.

Delegated legislation

When passing an act, the legislature may make provision for what is known as delegated legislation. Delegated legislation is used to grant some person or body the power to make orders, regulations and rules which have the force of law. The reasons for using delegated legislation arise out of the limitations placed on the legislature during and after the formal process of passing legislation. They are:

(a) Lack of time – the legislature has insufficient time to deal with and debate all issues necessary for efficient government.
(b) Urgency – the legislature is not always in session. Furthermore, the process is often slow.
(c) Flexibility – a ministerial or secretarial order can be made speedily if legislation needs clarification. The legislature would need to pass another statute to achieve the same effect.
(d) Technical matter – much legislation today is of a technical nature and requires detailed expert input. This is best left to ministers or secretaries of state advised by their 'departments' or by statutory boards or commissions.
(e) Future needs – the legislature cannot foresee all the difficulties associated with health, safety and welfare issues. Future difficulties are best dealt with by *delegated legislation*.

Forms of delegated legislation

These vary with the country but usually include regulations. Regulations are rules made under each act to make sure that the requirements of each act are met. Regulations are used to establish minimum standards in the workplace. In the USA these occupational safety and health standards appear in the CFR, the Consolidated Federal Register (the US government gazette), primarily in 29 CFR 1910. Mining safety is found in 30 CFR, chemical transportation safety in 49 CFR.

Status of codes of practice, and standards

Codes of practice

Codes of practice are used in some countries. They are not legislation, but may be given some legal status in relation to defences in court. They are developed to provide practical assistance to people to comply with the requirements of OHS and other acts and regulations on a workable basis.

National government standards

These are documents issued by an OHS authority which set standards in particular areas, e.g. hazardous chemicals. Their legal effect varies between jurisdictions. See earlier re their effect in the USA. They should not be confused with standards issued by standards organizations.

Standards issued by standards organizations

There are a large number of national standards relevant to OHS issued by organizations such as the ANSI, BSI, Canadian Standards Association, Standards Australia, Standards New Zealand and so on, sometimes referred to as consensus standards. They may be given some legal status by the OHS laws or regulations – if so, they are said to be 'picked up' by the legislation. International standards may also be referred to, e.g. ISO, CIE, IEC and ICRP.

Guidance notes

Some OHS authorities may issue a *guidance note* or *guideline* which is an explanatory document providing detailed information on the requirements of legislation, regulations, standards, codes of practice or matters relating to occupational health and safety.

Industrial agreements

Industrial or collective bargaining agreements, or in Australia awards made by an industrial commission, detail the conditions of work and often include occupational health and safety provisions. There may also be agreements on issues such as the resolution of disputes over dangerous work and the right to stop work with pay, although this issue is also covered in some OHS legislation.

Duty of care

Evolution of common law principles and precedence

The practice of judge-made law is to decide on a particular set of facts. The decision becomes the general rule or precedent to be followed in similar future cases. A judge is bound by the decision of a superior court. The exception is that many final courts have the power to overrule their own earlier decisions.

Nature and elements of the tort of negligence

The word 'tort' is used to indicate certain civil wrongs as distinct from criminal wrongs. Here, we are primarily interested in the 'tort of negligence' where the term 'negligence' refers to a breach of a 'duty of care' owed by one person to another. For OHS purposes we can refer to the duty owed by employers towards their employee(s).

For those countries which until recently accepted the authority of English decisions, the duty of care test was defined in the case of *Donoghue v. Stevenson* (1932). In that case it was held that a duty of care was owed to one's 'neighbours'. The meaning of the word 'neighbour' was defined. The rule that you are to love your neighbour becomes in law – you must not injure your neighbour. You must take reasonable care to avoid acts or omissions which you can reasonably foresee would be likely to injure your neighbour.

Who is my neighbour? 'Persons who are so closely and directly affected by my actions that I ought reasonably to have them in my contemplation as being so affected when I am directing my mind to the acts and omissions which are called into question.'

However, earlier decisions in England and elsewhere had also developed a doctrine of care owed to others including that owed by employers to employees.

Employer's duty of care

The employer's duty, as recognized by law in many jurisdictions deriving from English law, is to take 'reasonable care' for the safety of his/her employees in all circumstances of the employment. This duty of care is usually summarized as follows:

(a) provision of a safe place of work.
(b) provision of a safe system of work.
(c) provision and maintenance of safe plant and equipment.
(d) provision of competent staff to manage and supervise the employer's business.

How is the standard of 'reasonable care' defined?

The standard of care required might be said to be that amount of care that would be exercised in the same circumstances by a reasonable person. In fact, the degree will vary greatly according to the circumstances.

Standard of care

The standard adopted by the courts is an objective one – what would a reasonable employer have done in the same situation? In determining this standard of reasonable care several questions arise. When considering the employer's duty of care, should consideration include that the defendant, whilst an employer, may also be an entrepreneur and businessperson (and is therefore concerned to make a profit), or should safety override these considerations? Where an employer's business involves risks to workers by the nature or method of operation, should the employer be subject to claims for damages on each occasion an employee is injured? As an ordinary, reasonable person, can the employer be allowed to indulge in occasional oversight, slips or errors common to most members of society, or should the employer be considered to be 'infallible'?

In examining the nature of the employer's duty of care and the standard of the ordinary, prudent, reasonable employer, it is necessary to consider principles enunciated by courts, and to consider the extent to which they are likely to provide a reliable guide.

Occupational health and safety legislation

Work safety and health reforms – Robens' principles and current approach

Readers need to familiarize themselves with the OHS legislation which affects their particular environment and need. However, readers are advised to read all the material provided to gain an appreciation of the variations between the legislation in different jurisdictions.

History

It is not intended to dissect the various pieces of legislation here but rather to develop a common framework that is able to be used in the wider context for the general application of most, if not all, of the self-regulation style legislation. The original occupational health and safety legislation was based on the English Factory Acts (1833–1894) and the Workers' Compensation Act of 1897. Compensation legislation had been introduced by Bismarck in Germany in 1884. These factory acts were developed to help improve the poor working conditions in factories and mines that had developed during the Industrial Revolution. Many members of the old British Commonwealth inherited much of this style of occupational health and safety legislation, as did the USA. The English Factory Act style of legislation is prescriptive and requires the attainment of minimum standards and the periodic inspection of the workplace.

In 1972 a new style of legislation was introduced in South Australia, and in 1974 in the United Kingdom, as the result of the findings of the Robens Committee – the Health and Safety at Work Act. The Robens Committee criticized a system of too much law administered by too many authorities based on the enforcement of minimum standards by an inspectorate. The Health and Safety at Work Act provides for self-regulation by industry and does so by involving those most at risk in the workplace – the workers. An emphasis was placed on prevention, rather than inflicting heavy penalties as punishment, although serious penalties were retained for gross breaches of the law. The Health and Safety at Work Act also consolidated many scattered pieces of legislation, enabling occupational health and safety to be focused into one act. Occupational health and safety legislation in many parts of the world has adopted the Robens model in whole or in part. Standards and regulatory requirements must still be met, but with the opportunity for worker participation and involvement. Legislation in Scandinavia which predated Robens contained a number of its features, particularly consultation with workers. The UK is now considering a further change of approach with a proposed Safety Act.

In 1970 the USA took a different tack with the Occupational Safety and Health Act. Some comparisons and contrasts with that appear in the discussion below, and the US legislation is discussed in a section at the end of the chapter.

Principal features of the Robens or self-regulation approach

Tripartite policy-making body

As part of the Robens approach, policy making in OHS has passed from a traditional public service department to a body with representation from government, the unions, employers, and, in some cases, independent experts. In the Canadian Province of Alberta, for example, S.6 of the OHS Act sets up an OHS Council. The USA has not generally adopted the Robens principles, but does have under S.7 of the OSH Act a multipartite National Advisory Committee on OSH.

Unified administration

The Robens recommendation for a unified administration has not been accepted in all those jurisdictions which have generally accepted other Robens features. Specialist areas such as the nuclear industry may operate under the prime OHS legislation but be regulated by a specialist inspectorate under an agency agreement with the prime OHS authority. Another such area would be air safety. Note that the USA, not Robens-based, has two principal OHS administrations – the Occupational Safety and Health Administration, OSHA and the Mine Safety and Health Administration, MSHA. See 'Legislation in accident prevention in the USA and EU' later in this chapter (p. 114).

There is a trend in some parts of the world for OHS authorities to adopt positive performance indicators and contract the operation of these systems to accredited consultants.

Coverage of all workplaces

If all OHS legislation is considered, nearly all workers are now covered by one piece of OHS legislation or another. Special provisos may apply to coverage of military personnel. For example, under the Australian OHS (Commonwealth Employment) Act, the Chief of the Defence Forces can declare exemptions. Most jurisdictions, however, now have no problem including police within the coverage of the OHS legislation.

One growth area which presents special problems is adventure tourism or school-based adventure activities. For example, twenty-eight people died when heavy rain flooded a Swiss canyon in which they were canyoning and in the UK, Lyme Bay was the site of a canoeing tragedy. OHS legislation varies in the protection it affords non-employees in a workplace. This issue is compounded when those non-employees are in a workplace which is the natural environment, often highly variable in the risks it poses. The employees of the tourism company are leading, guiding or mentoring a party of tourists (often novices and often young) in a situation similar to the training situation many teachers and vocational lecturers face (and not just outdoors). Those employees are also working in that environment, which may lie outside the jurisdiction in which the company resides. The adventure leader may be a contractor to the tour company, not an employee, further diluting the line of responsibility. Some jurisdictions are taking a twofold approach to this, under the pressure of prohibitive insurance premiums, by re-applying the concept of voluntary assumption of risk in tort law to people who pay for adventure in high-risk situations, and by tightening up the training certification for

Figure 3.1 Adventure tourism in New Zealand (Photo courtesy of Grant Newton, Wellington, NZ)

leaders. It will be interesting to see if the change to tort law flows through into the OHS legislation. The NZ Mountain Safety Council (see Haddock in Further Reading, Chapter 4) has a very useful book on risk control in such activities.

Gunnar Breivik (see Further Reading in Chapter 4) offers an interesting comment on this: 'Another distinction in relation to risk that is of vital importance, is the distinction between action and arena. Risk theorists often focus upon individual or collective acts or actions, but I think arena is more important. I was first surprised when we tested climbers, parachutists and other risk sport groups and found that they were moderate risk takers. But it makes sense. When you are climbing you try to remain as safety conscious as possible and not take unnecessary risks. Since you are in a dangerous arena, you may have accidents and problems. The most important decision point is whether you are willing to enter the arena, to take up climbing, or not. On the other hand, people in safe arenas may be more willing to take risks. They may show a more risk-taking attitude. But it matters less, since the risks are objectively very small, whatever the line of action they take.'(*Reprinted by permission.*)

Another company presenting special problems in OHS, and where an employer may be far from the arena in which its employees operate, is the labour hire company.

Self-regulation

Self-regulation has been chosen as the main strategy in improving occupational health and safety standards in the workplace. There are strong logical arguments that point us towards self-regulation; however, for governments to have committed their industry and citizens to this approach, based on our limited knowledge and understanding of the processes, has required a leap of faith. Self-regulation moved us away from the inspectorate-driven, prescriptive legislation of the past century and three quarters, to a new model based primarily around a general duty of care by various stakeholders in the workplace, and to consultation, cooperation, information and training. Fines and penalties for breach of duty under Robens legislation remain and, in fact, the level of penalty is being continually increased. A jurisdiction may also have on-the-spot fines, e.g. Queensland has this and Alberta introduced it in 2002. Robens-style legislation is described as 'enabling' because, by providing for health and safety representatives (in many jurisdictions) and for health and safety committees, it sets up structures to facilitate consultation in OSH. In Canada this concept is referred to as internal responsibility. The USA has adopted Voluntary Protection Programs (see 'Legislation in accident prevention in the USA and European Union' at the end of this chapter).

General duty of care

A key concept in Robens-style legislation is to take the concept of the general duty of care developed through common law tort cases (and already in some factories legislation) and apply it to a variety of stakeholders in the workplace. Canada now views this as part of an organization's due diligence.

The key section in Robens-style legislation is the duty of employers. Employers are required to provide a work environment, plant and system of work which, so far as is reasonably practicable (the wording of this varies between jurisdictions), do not present a threat to the health and safety of employees. They must also consult with workers and provide adequate information, instruction and training, and report accidents of certain types (in some cases also critical incidents) to the relevant OSH authority. Employees are also subject to a number of duties which are enforceable in law.

Depending on the particular jurisdiction, duties of care may be imposed on principals (in respect of contractors), on occupiers, on the self-employed, and on manufacturers, designers, importers and suppliers of plant and chemicals. The legislation may include a duty of care on employers (and in some cases self-employed and employees) in relation to the health and safety of non-employees – for example, passers-by near a construction site.

Consultation and health and safety representatives

The process for initiating and electing health and safety representatives, where it is in the enabling legislation, such as for example in S.17 of the South African OHS Act, varies somewhat in each jurisdiction. Some provide accredited training for such representatives.

For the consultation process to begin to have effect, there must be more than good will on the part of both the employer and the employee. The worker must be able to fully participate in matters that affect his/her safety and health, and the employer, generally represented through management, must be able to provide the guidance and resources to support the necessary growth of the individual. There has been an opportunity over the past thirty years to demonstrate a commitment to this process. Refer to the relevant legislation to identify the role and powers of the representative.

It is generally believed that the role of the health and safety representative has made a significant contribution to the reduction of injury in the workplace. As to the consultative process between the worker and the employer, it is unclear if the role of the health and safety representative enhances consultation in occupational health and safety or detracts from the process through the creation of another communication barrier.

There is clearly a part for health and safety representatives to perform in the facilitation of occupational health and safety. Legislation establishes functions for health and safety representatives in those jurisdictions where they exist. Properly structured training and development programmes, supported by progressive safety policies designed to complement self-regulation style legislation, are central to the effectiveness of consultation by management, workers and health and safety representatives, and in health and safety committees.

A major problem has emerged in some jurisdictions because, while accredited training may be required for representatives, no similar mandated requirement exists for management training in OHS.

Health and safety committees
Most Robens-based jurisdictions have health and safety committees made up of management representatives and a number of people selected by the employees, usually the health and safety representatives. For a clearer outline of the committee's role, refer to the relevant legislation. For instance, in the Canadian Province of Manitoba, it is set up under S.2 of the Workplace Safety and Health Act.

Individual workers
Most jurisdictions allow an employee to refuse to undertake unsafe work, and to continue to be paid, if certain conditions are met.

Resolution of issues
All new legislation provides a mechanism to promote the resolution of occupational health and safety issues at the workplace level. The purpose of these arrangements is to reinforce the Robens principle that the solutions for accident prevention are essentially a matter to be resolved between those who create the risks and those who work with them. There is also a clear intention on the part of governments to reduce the industrial relations conflicts often associated with occupational health and safety issues.

Generally, organizations are expected to set up their own issue-resolution procedure, with provision, where this is not effective, to involve, progressively, the OHS authority and/or the courts.

National uniformity
A key element in the transition to self-regulation is the achievement of, or at least the objective of, rationalizing the legal arrangements, and hence the administrative arrangements too. In federations such as the USA, Canada and Australia, national companies have to operate under several pieces of state or territorial OHS legislation. So it is important that national OHS authorities take the lead in setting common standards which can be adopted at the state, provincial or territorial level. In Canada, for example,

hazardous substances control is based on the national *Workplace Hazardous Materials Information System* (WHMIS). In the EU similar aims exist, with national OHS authorities making the EU-wide OHS directives law in their respective countries. In the USA, while it is not, as noted, Robens-based, national OHS standards apply throughout the country, under S.6 of the OSH Act. States have the option of accepting federal administration of workplaces, or not doing so, but the national standards still apply if they don't. Just as the US OSHA covers federal employees, in both Canada and Australia federal employees are governed by specific national OHS law and regulations, the Canada Labour Code Part II and Canada OSH Regulations, and the OHS (Commonwealth Employees) Act and Regulations respectively. The Canada Code also covers employees of companies or sectors that operate across provincial or international borders. These businesses include airports, banks, canals, exploration and development of petroleum on lands subject to federal jurisdiction, ferries, tunnels and bridges, grain elevators licensed by the Canadian Grain Commission and certain feed mills and feed warehouses, flour mills and grain seed cleaning plants, highway transport, pipelines, radio and television broadcasting and cable systems, railways, shipping and shipping services, and telephone and telegraph systems.
(*With acknowledgement to the Canadian Centre for OHS.*)

Corrective procedures

The Robens Committee avoided a preoccupation with the negative outcomes of failing to adhere to safe working practice. Rather, the Committee recognized the need to enforce the spirit of the law through a mechanism based on administrative rather than judicial procedures.

In much Robens-based legislation this progressive principle is now embedded in the form of various *notices* available to correct the work system prior to the occurrence of an event that is likely to injure or cause damage. This is adopting a proactive position on the management of risk rather than taking action only after loss has occurred – the reactive approach. The US *citation* approach, if used before an accident, is similar in some respects.

In principle, there are two types of notice – one that seeks to improve the work system by compliance with some section of the legislation, while the other type is used to limit or prohibit an activity that is likely to give rise to imminent or serious harm. For example, the New Zealand Health and Safety in Employment Act provides for four types of notices, including improvement notices and prohibition notices. A prohibition notice can be issued by an inspector and work ceases or use of equipment stops until required steps are taken. In some jurisdictions, there is provision for elected health and safety representatives to issue notices on a provisional basis. This applies until an inspector can attend and either support or not support the notice.

In New Zealand a trained health and safety representative can issue a 'hazard notice'. An inspector there can issue an 'infringement notice' with an on-the-spot fine if there has previously been a hazard notice or an improvement notice issued which was not acted on.

An improvement notice issued by an inspector allows a specified period of time for correction of the problem.

A relatively new approach being used in some jurisdictions is for the inspectorate to conduct audits of health and safety management systems. This function is, as noted earlier, being contracted out in some of those jurisdictions.

Examining the process used to improve health and safety, and bringing in a system which recognizes this, emphasizes a positive approach rather than just fault finding.

Safety and health policies

Not all Robens-based legislation specifies that the employer prepare and promulgate a health and safety policy. (The UK Health and Safety at Work Act and the South Australian Occupational Health Safety and Welfare Act do.) There are advantages in issuing a properly stated policy because such a document assists in identifying the direction an organization is taking, through its management plan, to satisfy the duty of care towards employees, clients and the public.

Where legislation prescribes arrangements for the provision of a health and safety policy, the requirement is usually conditioned by the size of the workforce. There is no logical reason why this should be the case.

When establishing a health and safety policy, consideration must be given to two specific stages in policy development.

Stage one   The general policy statement. A statement that sets out the philosophy of the organization; management commitment; levels of accountability; and employee responsibilities; all written in clear and precise terms.

The policy should establish organizational targets and goals, indicate how the policy will be resourced, and stipulate an agreed process for review.

Stage two   The policy must also include an implementation strategy. This includes the clear identification of the management representative responsible for introducing the policy. This will generally be the most senior accountable person in the organization.

Piercing the corporate veil

Robens legislation carries with it the intention to extend the onus of individual liability beyond the workplace to the boardroom. It can be credited with some success in achieving this. However, ever-increasing penalties will not make the ability to gain a successful prosecution against a recalcitrant employer or manager an easy task. Historically, the burden of proof on the prosecution seems to grow relative to the size of the penalty. Perhaps the strategy should be to reduce the penalties, and focus on education and training as the means to bring about compliance. This, after all, was a basic concept in the Robens recommendations. One jurisdiction has introduced compulsory orders to provide training as an alternative to penalties. Notwithstanding the dubious benefit of increased penalties in the management of occupational health and safety, it is likely that the penalty mentality will continue to drive reform.

A further move has been to introduce some form of manslaughter charge into the occupational health and safety legislation, to compensate for a reluctance in some jurisdictions to prosecute under the general criminal law for OHS offences. There is also a need to address the question of who will be charged and to redefine the penalty if the charge is to be laid against a company (a company cannot be gaoled) and not merely some transient member of the management team. On the other hand, companies can often afford to pay monetary fines for their directors but the threat of gaol, which is an option being discussed, is a different matter. However, it also introduces issues such as to what extent the individual person, collectively working within the managerial system and within a given economic environment, is to blame. Canada has recently introduced measures to modernize corporate liability after the Westray mine disaster in Nova Scotia. Careless disregard of the duty to ensure worker and public safety, where this disregard leads to bodily harm or death, could result in a charge of criminal negligence. This builds on recent reforms to Part II of the Canada Labour Code.

Requiring reporting of OHS and environmental performance to the stock exchange by listed companies (the triple or quadruple bottom line) is another step.

Codes of practice

The use of codes of practice as a preferred model for the application of regulations has been incorporated into the self-regulation style legislation. The use of codes allows for greater flexibility in the management of occupational health and safety at the operator level. The method of establishing an acceptable code of practice can be somewhat cumbersome, because of the three social partners involved. The legal status of codes of practice varies from jurisdiction to jurisdiction.

New regulations

While change has affected the way occupational health and safety legislation deals with the duty of care arrangements, so that it now includes consultation, training and information responsibilities, regulations have also undergone a significant change in style. Regulations have previously been referred to as prescriptive, in that they set out in some detail the required condition, or state, of the system or equipment. Following the introduction of the Robens-style legislation, regulations became more oriented towards performance; that is, end results. There is now debate about whether regulations should focus on performance or process.

To provide an example, let us consider the erection of scaffolding in terms of regulations:

Prescriptive: Scaffolding is to be erected where people are required to work at a distance greater than 2 metres above the ground.
Performance: People required to work above ground level will be provided with scaffolding to enable them to work safely.
Process: Scaffolding will be erected in accordance with the appropriate standard and, after discussion with users, to determine the most effective way of carrying out the task.

The US OSH Act specifies that OHS standards should be expressed in terms of objective criteria and desired standards of performance where practicable.

Some Robens OHS jurisdictions, e.g. Western Australia, have replaced all pre-Robens regulations with a new comprehensive set (Alberta is currently doing this) and define the relationship of compliance with the regulations to the general duty of care.

Links to compensation and rehabilitation

There has been a trend in recent years to tie the occupational health and safety legislation to workers' compensation and rehabilitation reforms. In some jurisdictions, for example British Columbia, the administration of OSH and of compensation and rehabilitation have long been given to a common authority.

Although there is a certain administrative appeal for the bureaucrats to deal with these matters collectively, the solutions required to redress the incidence of industrial injury and disease bear no relationship to the problems associated with the management of either compensation or rehabilitation. The obvious danger for people concerned with the occupational health and safety agenda will be the attention given to the management of controls over ever-increasing compensation costs at the expense of urgently needed resources in prevention work.

One example is the use of legislation to prevent affected workers making certain types of claims – for example, for management/workplace-induced stress. This can have the effect of reducing the driving force (moral hazard) which increased compensation premiums have on the preventive effort.

However, cooperation between the OHS authority and the compensation and rehabilitation authority can enable the collection of better statistics, and this in turn can be used to optimize the use of an OHS authority's resources, using, for example, inspections targeted at high risk organizations or areas of activity.

Duties of employees

These vary between jurisdictions but generally include some or all of the following (however, be sure to check your own OHS legislation):

- be careful and look after your own safety
- do not put the safety of workmates and others at risk
- wear protective equipment supplied if asked to do so, or if the job normally requires it
- follow the safety rules and procedures of your workplace
- report hazards, accidents and injuries
- cooperate with your employer on health and safety matters
- do not work under the influence of alcohol or drugs
- do not misuse things provided for health and safety, such as machine guarding.

Duties of employers, occupiers and manufacturers

Employers

Again these vary between jurisdictions, but generally include some or all of the following:

- Employers must give the employee a safe and healthy place to do their work. It is impossible to make all workplaces clean and totally safe and to keep them that way.

However, the employer must make it as safe as reasonably practicable. An employee should be able to get to their workplace easily and safely.
- The way an employee does their work (work practices) must be organized in a safe way. They cannot be asked to carry out a task in a way that is dangerous or risky. They must be properly supervised at work.
- Machinery must be safe to use, installed properly and used with safe work practices.
- Materials and substances must be used, handled, stored and transported safely.
- Information on health and safety (including substances and materials) must be provided.
- Employees have a right to lunch areas, changing rooms, toilets, wash basins and other facilities.
- Employees should be trained and instructed in how to use machinery and materials safely. Information on the machines they use, and any material that they use, should be available. Materials and training must be in the employee's own language if they are not an English speaker.
- An employer must keep records on an employee's health and any accidents that the employee has at work. An employee's health should be checked regularly.
- Health and safety policy must be developed in consultation with employees' representatives.
- First aid and medical services must be provided.
- Where hazards cannot be avoided, protective equipment must be provided free to the employee, and training in its use given.
- Injuries and accidents above a certain severity must be reported to the OHS authority.
- Regular checks must be made on noisy areas and those where chemicals are used, plus appropriate health checks.
- An OHS policy must be developed consultatively.
- Records of employee health and injuries must be kept.
- Non-employees in the workplace, or affected by the work, must be protected.

Some employers, as principal contractors, may be responsible for contractors and their employees in some jurisdictions, so check your relevant act.

Occupiers

This duty must also be checked against your relevant legislation but, generally, an occupier or a person who has control of non-domestic premises in which people work must ensure:

- premises which are safe and without risks to health
- adequate access and egress
- plant or substances provided for use by the workers are safe or without risks to health.

Manufacturers (and designers, importers, suppliers)

The duties vary from jurisdiction to jurisdiction but generally the people above must ensure that plant or substances for use in a workplace are:

- safe and without risks to health if properly used
- properly tested in order to ensure such safety or health

- accompanied by adequate information on safe use and, in the case of substances, results of tests on them (at a practical level a material safety data sheet – MSDS).

Mechanisms for resolution of issues

It is important to look at the requirements of particular OHS acts and regulations affecting you. A flow chart for resolving issues may be given. The general requirement, in line with the concept of self-regulation, is for each workplace to develop an issue-resolution procedure. This may involve the manager together with the health and safety representative, the health and safety committee or the employees directly.

If this fails to resolve the matter, in some jurisdictions an inspector, as noted, can be called in and may issue a prohibition or improvement notice. Further appeal against the inspector's decision is possible. This may be to the relevant OHS authority, such as, for example, in Alberta to the OHS Council, or to a court of law.

Some OHS acts provide for work to stop through a joint employer/health and safety representative decision pending resolution of an issue, or by a decision of the employer or the health and safety representative.

Role of government inspectors

All OHS authorities appoint inspectors to assist with the implementation of the legislation, the enforcement of it, the investigation of serious injuries and accidents, and the resolution (if necessary) of OHS issues.

Direct regular 'shop-floor' inspections of workplaces in a number of jurisdictions are being reduced. This type of primary inspection is still used in targeted workplaces based on, for example, known high risk or poor OHS performance, based on indicators such as workers' compensation claims. Increasingly, as noted earlier, the inspection force in some jurisdictions is involved in auditing OHS management in the workplace. Some of the auditing may be contracted out. Inspectors are clothed with adequate legal powers to investigate accidents and serious incidents, and in some jurisdictions can call on police assistance. Check your relevant legislation but generally they can:

- enter a workplace
- inspect the workplace and issue notices if necessary
- seal off an accident scene
- seize evidence
- require the production of records
- interview witnesses
- examine plant, substances and other relevant things
- take photographs and make measurements
- exercise other powers given in the regulations.

Inspectors also present evidence as part of the prosecution of an OHS case in court.

Again, check your relevant legislation, but generally, if an OHS issue cannot be resolved at the workplace, an inspector can be called in to make a decision. He/she may issue a prohibition notice or an improvement notice (see 'Mechanisms for resolution of issues' above) as part of the decision.

In some jurisdictions inspectors have the power noted earlier to levy an on-the-spot fine as part of an infringement notice for a breach of OHS law.

Role of health and safety representatives

Many Robens-based jurisdictions as we have noted provide for the election of health and safety representatives. The method of election of the representative is generally given in the legislation.

Once again check the relevant legislation but, generally, the health and safety representative's role allows them to do the following:

- inspect workplaces if the employer has been given reasonable notice or if there has been an accident, incident or if there is an immediate risk
- request an inspector to accompany an inspector on an inspection tour
- represent employees on health and safety issues and to be present at interviews, with the permission of the employee
- investigate complaints
- access all information except confidential records
- be assisted by consultants if the employer agrees
- request that the employer set up a health and safety committee and examine subsequent records
- issue provisional improvement notices (PINs) (in some jurisdictions) or in New Zealand, hazard notices
- initiate emergency stop-work procedures in the case of an immediate threat to health and safety
- report hazards or unsafe work practices to the employer and health and safety committee (if there is one)
- undertake approved OHS training
- consult with the employer on planned changes to the workplace
- act as an employee member on the health and safety committee.

Role of a health and safety committee

Most Robens-based jurisdictions provide for the setting up of a health and safety committee (or safety committee), depending on the number of employees in a workplace. Generally, the committee has an equal balance of employer and employee committee members with at least one employer representative who has the power to give effect to committee decisions. In some jurisdictions, the committee may include both elected employee committee members and health and safety representatives.

You need to check the arrangements in your relevant act and also refer to that act, and in some cases the regulations, for the role of the committee, but it will do some or all of the following:

- meet regularly and keep minutes
- make recommendations on health and safety issues to the employer
- establish and review work practices and procedures, and planned workplace changes
- provide information on health and safety to employees
- keep itself informed about OHS standards
- recommend OHS training
- review accidents and incidents
- consider matters referred to it by a health and safety representative
- consider issues relating to rehabilitation of injured employees.

Legislation in accident prevention in the USA and European Union

United States

The approach involving cooperation in OHS between employers and employees which was introduced in Britain, Canada and Australia in the 1970s had its origin in Sweden twenty years earlier. The USA took a different approach, which still envisages cooperation between employers and employees but does not formalize this with joint safety and health committees or elected safety and health representatives. The US approach, like that in Robens-based OHS jurisdictions, involves employer, labour and expert representation at the policy level. One important difference is that criminal violations and penalties cannot apply to charges brought under the general duty clause in the Act; they are classed as civil charges. Criminal charges and penalties can apply for breach of gazetted standards.

Until 1970 occupational health and safety requirements in the United States were enforced at a state level. However, the Walsh-Healy Act of 1936 had given the US Department of Labor the authority to set OHS standards to be met by employers seeking public contracts over $10 000. Further acts later included smaller contracts, and federal or federally financed construction projects.

In 1970 the US Congress, as noted earlier, passed the Williams-Steiger Occupational Safety and Health Act (OSH act). A new administration was set up, the Occupational Safety and Health Administration (OSHA), part of the Department of Labor. This works in conjunction with the US Department of Health and Human Services which is responsible for the National Institute of Occupational Safety and Health (NIOSH), part of the Centres for Disease Control. The act allowed national collection of OHS statistics for the first time. Occupational safety and health standards are put out by the Department of Labor and take legal precedence over state laws and regulations. However, states can continue to administer OHS legislation as long as it meets federal standards. Some standards are keenly fought in the US courts, for example the benzene standard was fought by industry for ten years. Unlike the British legislation, the US legislation has adopted, until now, a prescriptive approach, with a strong emphasis on compliance and *citations* for *violations*.

Where there is no penalty applied, a citation may operate like an improvement notice under Robens legislation, although even some Robens-based jurisdictions provide for on-the-spot fines.

The concept of health and safety representatives is not in the legislation. In 1982 a form of self-regulation was introduced. The Voluntary Protection Program allowed best practice employers reduced penalties and exemption from inspection.

As noted earlier, MSHA, also part of the Department of Labor, enforces OHS in US mines under the Mine Safety and Health Act, but again with research and other support from NIOSH, such as miners' medical surveillance.

A fuller description follows.

US OSH law and administration

The OSH Act asserts federal jurisdiction on OSH matters. States can submit plans to assert jurisdiction on OSH issues which are covered federally. These can be approved subject to certain conditions. States can also assert jurisdiction on OSH issues where no federal standard exists.

The Act extends to the District of Columbia, Guam, Puerto Rico, Outer Continental Shelf Lands, Wake Islands, Johnston Islands, the (US) Virgin Islands and North Mariana Islands.

As noted earlier, OSHA is responsible for the administration of the Act and its enforcement. Small business is specifically considered in parts of the Act.

The general duty

The Act places a general duty on an employer to furnish employment and a place of employment which are free from recognized hazards that are causing or likely to cause death or serious physical harm to his or her employees. Employers must also comply with OSH standards promulgated under the Act.

Federal government agencies must set up safety and health programmes consistent with the standards.

Employees' duties and rights

Every employee must also comply with all applicable OSH standards and all rules, regulations and orders issued under the Act. An employer and an employee representative may accompany any inspector during an inspection of a workplace.

Employees or representatives of employees can request an inspection by OSHA if they believe a standard is being violated which threatens physical harm, or where imminent danger exists. The employees or representatives of employees can notify the Secretary of Labor or the inspector, in writing, of any violation. Procedures exist following this for informal review of any refusal by an inspector to issue a citation, and the Secretary must give reasons in writing for his or her final decision.

In the case of imminent danger, if the Secretary fails to seek a court order after an inspector has concluded that there is imminent danger, and other procedures under the act cannot immediately deal with it, an employee can seek a writ of mandamus compelling the Secretary to seek a court order.

A refusal to work is not normally protected under the OSH act, but under certain conditions it would be. The employee would need to advise the employer of the hazard and not leave the workplace, although he or she could refuse to be further exposed to the imminent hazard. He or she could not leave the workplace to contact OSHA if the employer didn't respond.

Employees also have the right to know about hazards in general and hazardous substances, through provisions of the act and its accompanying rules.

Accurate records of exposure to toxic materials or harmful physical agents, and monitoring of them, must be kept by an employer and the records must be available to employees or their representatives.

The act provides protection from discrimination to employees who lodge a complaint, cause proceedings to be instituted, or are required to testify.

The district courts can order their rehiring or reinstatement with back pay.

Standards

The OSH act describes the rulemaking process by which occupational safety and health standards become law. A standard may be a national consensus standard produced, for example, by the American National Standards Institute (ANSI), or it may be an established standard produced by a federal agency. Promulgating the standards is known as rulemaking. A rule published in the federal register (government gazette) promulgates, modifies or revokes an OSH standard.

The rulemaking follows recommendations from an advisory committee set up after the standard is referred to the Secretary of Labor by an interested person, an organization of employers or employees, a nationally recognized standards-producing organization, the Secretary of Health and Human Services (HHS), the National Institute of Occupational Safety and Health (NIOSH), or a state or political subdivision.

The Secretary of Labor promulgates a proposed rule and interested parties can comment for up to thirty days. They can also file written objections which must be followed by a public hearing. They can also apply for a temporary variance from the standard or a provision of it.

An employer can seek temporary variance from a standard. A revised standard must be accompanied by a statement of explanation. A rule promulgated which is substantially different to a national consensus standard must be accompanied by a statement of explanation by the Secretary of Labor in the Federal Register. Regulatory impact of assessment of new standards may be undertaken. A proposed rule can be challenged by any person, and a public hearing follows. After that the rule can be promulgated, modified, or revoked. Its effective date may be postponed to give employers time to adjust. A standard can be varied to allow approved experiments to demonstrate or validate new or improved safety and health techniques.

There is a National Advisory Committee on OSH of twelve people, four appointed by the DHHS, comprised of representatives of management, labour, occupational safety and occupational health professions and of the public. The others are appointed by the Secretary of Labor.

The Advisory Committee on Standards can be up to fifteen people, and includes someone designated by the Secretary of HHS and also equal numbers of persons with experience and affiliation to represent employers and workers involved, together with representatives of the state OSH agencies. The Secretary may also include representatives of professional OSH organizations, nationally recognized standards organizations, but no more than the number of representatives of federal and state agencies. Meetings and committee records are public.

A standard must provide for labels or other forms of warning about hazards, symptoms, emergency treatment and proper conditions and precautions for safe use and exposure. Personal protective equipment or control or technological procedures should be provided in the standard, as well as type and frequency of medical examinations or other tests of employees at the employer's expense. If the tests are for research, the Secretary of HHS may pay for them.

Even if a standard has been promulgated it may be challenged in court by any affected person.

Decisions by the Secretary of Labor on standards, rules, orders, variances, extensions and reduction of penalties must have the reasons published in the Federal Register.

The Secretary of Labor may apply to a court to put a stop work injunction on an employer where there is imminent danger. The court can then prohibit the employment of any individual at that workplace, but with due allowance to allow key personnel to remain to deal with the risk or to manage issues involved with continuous process operations.

There is also an OSH Review Commission of three members appointed by the President with the advice and consent of the Senate. The Commission determines matters if a citation is contested. An administrative law judge appointed by the Commission hears any proceeding instituted before the Commission. Judicial review of Commission decisions is available. The Secretaries of Labor and HHS can prescribe rules and regulations relevant to their responsibilities, including inspection of an employer's workplace.

The OSH act among other things also sets up NIOSH, mentioned earlier, which develops and establishes OSH standards and carries out research. The act also provides for training grants and collection of statistics.

Citations and enforcement

Where an inspector believes an employer has violated either the general duty, standard, rule or order under the OSH act, he or she can issue a citation. The citation fixes a reasonable time to abate the violation, and must be posted in the workplace. There is provision for a notice for minimal violations with no direct or immediate effect on safety or health. The citation is followed by a notice of the penalty, if any. If neither the employer nor employees contest the citation within fifteen days, the citation and penalty is not reviewable by any court or agency.

If the employer fails to correct the violation within the specified time for abatement (allowing for any genuine appeal in progress), he or she will be advised of the penalty. If they do not contest the penalty within fifteen days, again it is not reviewable. Any appeal for review goes to the OSH Review Commission and it can decide to alter or

Enhancing occupational safety and health

not alter the citation and penalty. Judicial Review of the Commission decision is a further option under the act, both for the employer and the Secretary of Labor.

It is an offence with a criminal penalty to give advance notice of an OSHA inspection.

Penalties

There are violations, serious violations and willful violations. For the meaning of 'serious' and 'willful' see Workplace Example of Chapter 1. A civil penalty of up to US $70 000 (2003) may apply for each violation, and a minimum of $5000 for each wilful (US – willful) violation.

Certain wilful violations, under the Sentencing Reform Act, become Class B criminal misdemeanours, and if they cause death, the sentence is $250 000 for an individual and $500 000 for an organization. Both can be put on probation too, and a prison term is a possibility.

An employer who wilfully violates any OSHA standard, rule or order, where that violation causes the death of an employee, is liable, if convicted, for a $10 000 fine or imprisonment for up to six months, or both. If it is a second offence, the penalties double.

Civil penalties are assessed by the OSH Review Commission taking into account the size of the business, good faith, previous record, and the gravity of the violation.

A criminal or wilful violation cannot be based on the general duty clause in the act.

Note: in the USA the word 'docket' means an official public record and can include, for example, proposed standards, regulations, or adjudication information.

Europe

In the European Union, 25 member countries continue to be responsible for passing and administering their own health and safety legislation. However, they are obliged to apply the requirements of the European health and safety directives adopted by the European Commission. For example, the so-called European 'six-pack' included issues such as manual handling, and optometric tests for VDU operators.

The headquarters of the European Agency for Safety and Health is in Bilbao, Spain.

Further reading

Abrams, A.L. (2002). OSHA and Ergonomics Enforcement Under the General Duty Clause. *Professional Safety*, **47(9)**, 50–2.

ABSG Consulting. (2003). *OSHA Compliance Self Study Course*. Rockville, MD: ABSG Consulting.

Cooper, S. (2002). The Inquest Experience. *Accident Prevention (Canada)*, March/April, p. 27.

Delpo, A. and Guerin, L. (2002). *Federal Employment Laws – A Desk Reference*. Berkeley, CA: Nolo Press.

Editorial. (2002). OSHA's Voluntary Protection Program Turns 20. *Professional Safety*, **47(9)**, 12.

Feitshans, I.L. (1998). Self Regulation – An American Route to Safety and Health on the Cheap? *The Safety and Health Practitioner*, June, 34–6.

Harrington, J.M. and Gill, F.S. (1999). *Occupational Health*, 4th edn. Oxford: Blackwell Science.

Keith, N.A. (2002). Benchmarking Due Diligence. *Accident Prevention (Canada)*, November/December, p. 10.

Keith, N.A. (2002). Westray's Enduring Legal Legacy. (Nova Scotia mine disaster). *Accident Prevention (Canada)*, March/April, p. 10.

Seo, D.C. and Blair, E. (2003). Dissecting OSHA's Cost-benefit Analysis Offers Insight for Future Rulemaking. *Professional Safety*, **48(4)**, 37–43.

Taylor, G.A. (1988). Dicey v. Robens – An Analysis of Some of the Consequences of the Modern Approach to Occupational Health and Safety Legislation. *Australian Industrial Safety, Health and Welfare*, 17.10.88, **44**, 549–54, 578. Sydney: CCH Australia.

Wren, J. (2000). The Right To Say No (workers' rights). *Safeguard (New Zealand)*, May/June, p. 18.

Young, S.L. et al. (2002). Safety Signs and Labels. Does Compliance with ANSI Z535 Increase Compliance with Warnings. *Professional Safety*, **47(9)**, 18–23.

plus legislation on occupational health and safety; mines safety and health; petroleum safety; maritime safety; transport safety; hazardous materials/dangerous goods, nuclear and radiation safety; Workers' compensation and guides from government authorities and others explaining these, e.g. NZ DOL (2003): *Guide to the Health and Safety in Employment Act*. 2nd edn, and IAPA: *Guide to the Ontario OHS Act*.

US OSH law and standards can be easily accessed on the US OSHA and MSHA websites, www.osha.gov and www.msha.gov, Australian, New Zealand and some UK OHS law on www.austlii.edu.au, and Canadian OHS law on laws.justice.gc.ca/en/L-2 (no www). Some information on Pacific states is in www.paclii.org. See also www.worldlii.org.

Activities

1. Find three examples of areas of human activity covered by common law and three covered by statute law.
2. Find an example of a rule, order, ministerial direction, or US standard relating to OHS and explain it.
3. Draw a flowchart for the process of creating and passing an Act in your provincial, state or national legislature.
4. Outline issues a court trying a common law case would look at in deciding whether a system of work was unsafe.
5. Access legal case reports and briefly descibe a leading tort case in the area of OHS.
6. Explain 'onus of proof' and 'standard of proof'.
7. Describe the current penalties applying under your OHS legislation.
8. List twelve key features of the OHS legislation in your area or covering your place of work.
9. In your OHS jurisdiction describe the rights and powers of health and safety representatives (or in the USA, employee representatives).

Appendix 3.1 – A brief guide to some OHS legislation and organizations (to November 2003)

International	Conventions	Administration	Professional organizations	Professional qualifying organizations	Other organizations
United Nations – International Labour Conference	ILO Conventions esp 155 and 161	International Labour Organization			
United Nations		World Health Organization			
United Nations		International Social Security Association			
European Union	European Commission Directives	European Agency for Safety and Health			
			International Commission on Occupational Health, International Occupational Hygiene Association, International Ergonomics Association	International Society of Mine Safety Professionals	

Country	Legislation (not including workers' compensation)	Administration	Professional organizations	Professional qualifying organizations	Other organizations
Australia	National Occupational Health and Safety Commission Act, OHS (Commonwealth Employment) Act, Seacare Act, and State and Territory OHS and Mine Safety Acts, plus Regulations	National standards, etc.: National Occupational Health and Safety Commission Enforcement: State and territory OHS departments, and in three states – mines departments	Safety Institute of Australia, Australian Institute of Occupational Hygienists, Ergonomics Society, Australian Acoustical Society, College of Occupational Nursing, Australia and New Zealand Society of Occupational Medicine	Safety Institute of Australia, College of Occupational Nursing, Australasian Faculty of Occupational Medicine	
Bahrain	Labour Law, Amiri Legislative Decree	Ministry of Labour and Social Affairs			
Bangladesh	Factories Act, Shops and Establishments Act	Department of Labour – Inspectorate of Factories			Bangladesh Occupational Safety Association
Botswana	Factories Act				

(*Continued*)

Country	Legislation (not including workers' compensation)	Administration	Professional organizations	Professional qualifying organizations	Other organizations
Canada	Federal provincial and territorial Acts and Regulations, Canada Labour Code	Administration – provincial and territory OHS departments National information and Workplace Hazardous Materials Information System – Canadian Centre for Occupational Health and Safety	Canadian Society of Safety Engineering, Association of Canadian Ergonomists, Canadian Occupational Health Nurses Association, Occupational and Environmental Medical Association of Canada		
Caribbean	21 countries – see www.ilocarib.org.tt – see also Jamaica				
Eire	Safety Health and Welfare at Work Act, SH&W (Offshore Installations) Act, Mines and Quarries Act	Health and Safety Authority	Irish Society of Occupational Medicine		National Irish Safety Organization

Fiji	Health and Safety at Work Act	OHS Service Organization	
Gambia	Public Health Act, Hazardous Chemicals and Pesticides Control and Management Act	Ministry of Health, and Factories and General Labour Laws Inspection Department	
Ghana	Factories Offices and Shops Act, Mining Act	Labour Department	
Hong Kong SAR	OHS Council Ordinance and OHS Ordinance	Department of Labour	Hong Kong Occupational Safety and Health Association
India	Factories Act and 15 other acts, including Mines Act and Oil Mines Regulations	State government inspectorates, and for mines, Directorate General of Mines Safety	National Safety Council of India
Jamaica	Factories Act and Regulations, Mining (Safety and Health) Regulations	Ministry of Labour Social Security and Sports, Department of Mines and Geology	Jamaican Association of Safety Professionals, Occupational Health Nurses Association of Jamaica
Kenya	Factories Act	Occupational Health and Safety Department	
Kuwait	Decision No 1 on Scaffold Work	Ministry of Social Affairs and Labour	

(Continued)

Country	Legislation (not including workers' compensation)	Administration	Professional organizations	Professional qualifying organizations	Other organizations
Lesotho	Employment Amendment Act	Ministry of Labour			
Malawi	Occupational Health Safety and Welfare Act, Mines and Minerals Act, Mines Quarries Works and Machinery Act	Ministry of Labour and Vocational Training, Ministry of Mines and Energy			
Malaysia	Occupational Safety and Health Act and Regulations	Department of Occupational Safety and Health			Malaysia Society for Occupational Safety and Health
Mauritius	Occupational Health Safety and Welfare Act 1988	Ministry of Labour and Industrial Relations			
Namibia	Factories Machinery and Building Work Ordinance, Labour Act, Minerals Act	Ministry of Labour, Ministry of Mines and Energy			
New Zealand	Health and Safety in Employment	Occupational Safety and Health Service of	New Zealand Occupational Health Nurses		New Zealand Safety Council

	Act, Hazardous Substances and New Organisms Act	Department of Labour	Association, Australia and New Zealand Society of Occupational Medicine
Nigeria	Factories Act, Minerals Act, Coal Mine Act	Ministry of Employment Labour and Productivity, Ministry of Energy and Mines	
Oman	Sultani Decree No 40 (and a law on workers' compensation)	Ministry of Social Affairs and Labour	
Pakistan	Labour Code of Pakistan – Factories Act, Mine Act and other Acts	Ministry of Labour Immigration and Overseas Pakistanis	
Papua New Guinea	Industrial Safety Health and Welfare Act	Ministry for Labour and Employment	PNG Occupational Health and Safety Association Inc
Philippines	Several Acts collated in Philippine Labor Code Book IV	Department of Labor	Safety Organization of the Philippines Inc

(*Continued*)

Country	Legislation (not including workers' compensation)	Administration	Professional organizations	Professional qualifying organizations	Other organizations
Seychelles	Occupational Health and Safety Decree (Amendment) Act	Ministry of Health			
Singapore	Factories Act and Regulations and orders	Ministry of Manpower	Academy of Medicine, Public Health and Occupational Physicians Chapter		National Safety Council of Singapore
Solomon Islands	Safety at Work Act				
South Africa	OHS Act and draft Regulations	Department of Labour and Department of Minerals and Energy	South African Society of Occupational Health Nursing Practitioners, Southern African Acoustics Institute, Ergonomics Society of South Africa		
Sri Lanka	Factories Acts, Laws and Ordinances, Shop and Office Employees Act, Mines and Minerals Law	Department of Labour			Ceylon Society for Prevention of Accidents
Sudan	Not available	Ministry of Labour and Social Security			

Swaziland	Factories Machinery and Construction Works Act	Ministry of Enterprise and Employment	
Tonga	Not available	Ministry of Labour Commerce and Industries – Factories Inspectorate	
Tanzania	Occupational Health and Safety Act, Industrial and Consumer Chemicals Management and Control Act	Ministry of Labour – Factories Inspectorate	
Uganda	Factories Act	Department of Labour	
United Arab Emirates	Not available	Ministry of Labour and Social Affairs	
United Kingdom	Health and Safety at Work Act, Nuclear Installations Act, and others, and Regulations	Health and Safety Executive assisted by specialist agencies (e.g. nuclear) and local government Policy: Health and Safety Commission	Institute of Occupational Safety and Health, Institute of Occupational Hygiene, Society of Occupational Medicine, British Occupational Hygiene Society, The Ergonomics Society, Royal College of	National Board of Occupational Safety and Health, British Examining and Registration Board in Occupational Hygiene, Faculty of Occupational Medicine of the Royal College of Physicians	Royal Society for the Prevention of Accidents

(*Continued*)

Country	Legislation (not including workers' compensation)	Administration	Professional organizations	Professional qualifying organizations	Other organizations
United States	Occupational Safety and Health Act, Mine Safety and Health Act, Toxic Substances Control Act, and standards gazetted in the Consolidated Federal Register	Department of Labor – Occupational Safety and Health Administration and Mine Safety and Health Administration Environmental Protection Agency, and in some states, state OHS departments National research and standards recommendations: National Institute of Occupational Safety and Health, and Bureau of Mines	American Society of Safety Engineers, American College of Occupational and Environmental Medicine, American Conference of Governmental Industrial Hygienists, American Industrial Hygiene Association, Human Factors and Ergonomics Society, American Association of Occupational Health Nurses, Acoustical Society of America	Board of Certified Safety Professionals, American Board of Industrial Hygiene, American Board for Occupational Health Nursing	Nursing Society of Occupational Health Nursing

Zambia	Factories Act, Mines and Minerals Act	Ministry of Labour and Social Security – Factories Inspectorate, Ministry of Mines and Minerals Development
Zimbabwe	Factories and Works Act, Mining and Minerals Act	National Social Security Administration OHS Division

Note: Some US- and UK-based professional organizations have a large international membership, and more of the professional organizations may also be qualifying organizations. The list does not include workers' compensation legislation.

4

Hazard and risk management

WORKPLACE EXAMPLE

A 41-year-old electrician was working in an aerial lift truck (cherrypicker) and died after falling from the basket. Three men including the victim were installing electrical wiring for a message sign on the side of a highway bridge. The electrician had entered the basket and swung it up between two of the bridge beams. He called down about how much pipe he needed for the electrical wiring. The basket was above two closed traffic lanes, but the electrician then moved it to a point above the nearest open traffic lane while the others cut the pipe.

A semitrailer was approaching and followed the warning cones (witches hats) and signs directing traffic to merge right. The driver proceeded down the lane closest to the closed lanes and when about 16 metres from the workers' vehicles, he noticed the cherrypicker basket was over that open lane. Before he had time to slow or move safely to another lane, the truck hit the basket and the electrician was thrown to the road.

A key recommendation from the subsequent investigation was to keep equipment within the boundaries of established work zones.

(With acknowledgement to DHHS-CDC-NIOSH, US (2001): *Building Safer Highway Work Zones*.)

Risk concepts

The word 'hazard' comes from the Arabic 'az zahr' (a die, plural dice). Risk comes from the Italian 'risicare' to dare. Blaise Pascal (of the pressure units) developed the idea of risk in the seventeenth century with the statement: 'Fear of harm ought to be proportional not merely to the gravity of the harm but also to the probability of the event'.

Daniel Bernoulli in the eighteenth century followed with the idea that when future events are uncertain, 'the *utility* of an outcome is dependent on the particular circumstances of the person making the estimate. There is no reason to assume that the risks anticipated by different individuals must be deemed equal in value'. Different people place different values on risk. If they are only concerned about consequences, and fail to allow for low probability, they are too risk-averse. If they are only concerned about probability, even though consequences may be extreme, they are foolhardy.

Risk, economics, psychology, adventure work and insurance

It is important to view risk from both a positive and negative perspective. Some people put a high utility on a small probability of big gains. It follows that they give low utility to a possible loss, even though the probability of that is greater. So we are also influenced by the 'disutility' of a loss event. Utility is linked to the psychology of motivation. Utility is a concept which has been very much used in economics to describe the different values people assign to different *goods*. These include goods and services in the public arena, such as availability of medical treatment. This explains the demand for a good or service and so how much people are prepared to pay for it. It is a useful concept to consider when discussing the different utilities different people assign to safe work, and the relative utilities they assign to safety and scale of earnings. A person desperate for food or money will accept a lower degree of safety (higher degree of risk). That is, for them, safety as a good has a lower utility than food.

Predicting the future can only ever involve deciding, out of all *known* possibilities, those which are more likely to occur. So unless a person consigns themselves entirely to fate (i.e. makes no positive decisions designed to influence their future), moving into the future involves the conscious taking of risk. The degree of risk depends on the individual. John Keynes, possibly the most influential economist in the first half of the twentieth century, realized that there was a difference between risk and uncertainty. Risk allows the calculation of probabilities. Not so with uncertainty – for example, no one can actually predict the price of nickel or gold in seven years based on past knowledge. Uncertainty, if you like, is 'the gods', risk is 'the odds'.

This gives us freedom to try new things, to influence the future – we are not locked in. This contrasts with earlier notions of a protective saint for occupations who could be asked to alter fate. Keynes was concerned that too great an emphasis on always trying to anticipate the future could handicap the exercise of our animal spirits – our initiative. Decisions often have to be made on limited data, with constant correction as new data comes to hand – the *plan, do, check, act* cycle. Another economist, Kenneth Arrow, worked as a long range weather forecaster in World War II. He realized that predictions could not be accurate and offered to resign. He was told to continue, because the predictions were needed for planning purposes!

In Chapter 3 we quoted from Gunnar Breivik, who is a professor of sport and physical education. He is interested in the tug-of-war which can exist between people's desire for excitement and challenge, and society's drive towards security and safety. There is a need for challenge because it hones our survival skills as human beings and influences our health. There is a fine line between the state ensuring safety and the 'nanny state'. For instance, a former Western Australian Minister for Labour responsible for OHS was at the same time opposed to compulsory bike helmets. The same debate arises in relation to the 'safety net' in an economic rationalist economy.

Breivik realizes that in the area of physical and mental challenges in natural environments, all reasonable steps should be taken to ensure avoidance of injury or death. However, there is still an area of uncertainty, often associated with weather or water conditions. It is therefore inevitable that even with careful planning, proper gear and good training, injury and death are still possible. This should not be a reason for prohibiting or making such activities too difficult to carry out because of the threat of being sued – an important consideration for work activities such as adventure tourism and school wilderness and kayaking expeditions. Branches sometimes do fall on campers, in spite of an intelligent choice of where to spend the night. Moves to require boats to carry EPIRBs if going more than a little offshore in unprotected waters are justified, because they reduce the overall risk which rescuers need to take. However, this does not mean that organizers of such activities should expect rescue to be borne completely by the public purse.

Finally, a brief mention of the links between risk and insurance. John Graunt, a button and needle merchant, and the astronomer Edmund Halley, produced tables of mortality in the seventeenth century, although these were not used by insurers to make life assurance more predictable for over a century. Occupational guilds had common funds to cover the death of a member. 'Bottomry' originated in 1800 BC, in the Code of Hammurabi – also noted for its severe penalties for poor builders – as an early form of insurance for shipping. This was followed many years later in 1696 by a shipping list provided for marine underwriters who met in a coffee house owned by Edward Lloyd, hence the 'Lloyd's List'. The policy issued by the underwriter derived from the Italian 'polizza', meaning evidence or a setting forth. Futures trading which later developed allowed farmers to insure against natural events by accepting a set future price for their product. Companies hedge on currency values to protect themselves against the unpredictability of currency movements.

(Partly adapted from Bernstein, P.L. in *Against the Gods – the Remarkable Story of Risk*, copyright © 1997 John Wiley and Sons, Inc., by permission of John Wiley and Sons, Inc.)

Role of hazards in injury causation

Interaction between an individual and a hazard can have various types of outcomes. It could be an incident which results in no loss or damage to life and equipment; or only to property; or only personal; or result in both personal and property damage.

Definitions

Definitions which will assist you include:
Hazard:
- any thing or any condition which has the potential to cause injury or harm to health, or
- a source of potentially damaging energy.

Hazard and risk management

Risk
- the probability that a hazard will actually result in an accident, considered together with the consequences of the accident

Hazard identification
- the process of recognizing the presence of hazards

Risk assessment
- the process of deciding how likely it is that a particular hazard will affect a person in varying circumstances

Risk control
- the process of selecting and applying measures which will reduce the risk to acceptable levels

Risk monitoring
- the process of checking that the selected control measures are effective.

Hazard identification and risk management

This is the interactive process used by an organization to identify hazards, evaluate the risk associated with them, and control the risk to reduce the incidence of injury.

The process of hazard and risk management in the workplace is a tool available to all people which, when applied correctly, leads to a safer work environment. Through applying the process set out below, the risks associated with hazards can be managed.

The four steps in the management of hazards and their associated risks are:

1. Hazard identification
2. Risk assessment
3. Risk control
4. Monitoring the effectiveness of the controls.

Some replace the word 'risk' with 'hazard' and include an evaluation step. There are six Es (good) and five Is (bad) of risk control or management: engineering, education, encouragement, example, enthusiasm and excellence; and impunity, impatience, improvise, impulsive, and impossible (Ted Davies) – the switch from noun to verb to adjective with the Is is deliberate.

Planned hazard identification, formal and informal systems

Identification

Many organizations have policies and procedures in place to identify potential risks and hazards within their work environment.

Hazard identification may occur by the use of various tools, strategies, and information sources. Some information sources include:

- material safety data sheets (MSDS)
- national, state or provincial injury statistics
- chemicals notification and assessment scheme documents under the OECD Protocol
- standards and codes.

MSDS

These particular sheets referring to chemical-based products will provide information on materials before they are purchased and used on site. They also detail the potential health effects, etc. This particular tool can help in reducing hazards at the assessment stage by assessing the hazard and risk and allowing a less hazardous replacement.

National injury statistics

Information on injuries released by national OHS authorities can predict injury trends and develop precise, preventative strategies when comparisons are drawn with organizations similar to the one where you are working. State, provincial or local authority documents may also be useful.

OECD-based assessments for new and existing products and chemicals

Assessment guidelines for new chemicals under an OECD protocol includes a toxicology assessment based on environmental and workplace concerns. This assessment is available from national authorities. In the EU there is the European List of Notified Chemical Substances (ELINCS) and for existing substances EINECS. Some US states, e.g. New Jersey and Pennsylvania, have right-to-know lists, and Canada has domestic and non-domestic substances lists. Australia has assessments under the ICNAS legislation implementing the OECD protocol.

OHS standards and codes

These consist of practical advice on issues and include preventative strategies to assist with hazard control. They are a baseline for comparison and a check for organizations to see if they conform with relevant acts and regulations and meet the duty of care.

Other forms of hazard identification

These include:

- workplace inspections
- supervisor/worker discussions
- independent audits
- job safety analysis
- hazard and operability studies
- health and safety representatives/employer discussions
- your organization's workers' compensation data.

Hazard and risk management

Task				
Step no.	Hazard	Risk from the hazard	Control measure	Step in procedure

Procedure:

Step 1

Step 2....

Training and assessment in procedure:

Performance Criterion 1

Performance Criterion 2....

Figure 4.1 JSA form

Job safety analysis has four steps:

1. Break the job down into steps.
2. Identify the hazards for each step.
3. Identify risk controls for each step.
4. Write safe job procedures.

A form to assist you is given in Fig. 4.1.

Formal and informal identification

Hazards, as noted, may be identified in the workplace from inspections. Two main types are used.

Formal identification

This involves the use of checklists which will assist with the identification process. Checklists are used as a tool to assist with distinguishing between acceptable and unacceptable standards of workplace practices, materials, equipment and environment. Checklists are usually used on routine inspections by safety personnel and employees.

Informal identification

Any person in the workplace can participate in informal inspections. It simply requires a person to identify a hazard and report the finding to the appropriate person for risk management. Hazards in the workplace may have direct relevance to an accident sequence. This may include unsafe acts or conditions. If we review some of the models which have been devised to understand accidents, the concept of hazards being present is a common theme.

Role of hazards in injury causation

All of the items listed below are examples of readily identifiable hazards and the associated disease or injury which can occur.

Hazards	Disease or injury
Biological causes	
● animal-borne	anthrax, Q fever
● human-borne	AIDS
● vegetable-borne	aspergillosis (farmer's lung)
Chemical causes	
● acids, alkalis	dermatitis, eye injury
● metals	cadmium, lead, or mercury poisoning
● non-metals	arsenic or cyanide poisoning
● gases	carbon monoxide poisoning
● some organic compounds	cancer
● dusts	lead poisoning, silicosis
● vapours	mercury poisoning
Electrical causes	
● electricity	burns, heart failure
Mechanical causes	
● unguarded machines	acute physical injury
● hydraulic pressure	acute injury
Overexertion/postural causes	
● overexertion, lifting or pulling	musculo-skeletal injury
● repetitive use	tenosynovitis
Physical causes	
● heat	heatstroke, cramp
● lighting	headaches, eyestrain
● noise	noise-induced hearing loss
● vibration	white finger disease
● ionizing radiation	lung cancer
● pressure	bends (decompression sickness)
● gravitational energy, falls of people or objects	acute injury
Psychological causes	
● shiftwork, stress	fatigue

Hazard and risk management

The New Zealand Health and Safety in Employment Act, for example, now specifically addresses stress. Psychological effects may also result from violence, harassment, or abuse.

Wanted and unwanted energies

In any inspection it is useful to look at the various forms of energy which exist in the workplace. There are many forms of energy that exist in our community environment. Of particular interest are those energies that are associated with the work environment. 'Wanted or necessary energy' is the critical element needed in the production process. 'Unwanted or excess energy' is the element that is least desired and is the cause of most losses in production. It is therefore important to distinguish between wanted and unwanted energies.

It is important to firstly list the energies that exist in the workplace:

- potential energies
- kinetic energies
- mechanical power
- acoustic and mechanical vibrations
- electrical
- nuclear particle radiation
- thermal energy
- chemical
- microbiological energy
- muscle energy.

It is apparent that all of these energies when used appropriately (wanted) have a necessary function in our work environment and community. However, once these energies are freed from their wanted path, by deliberate or accidental release, they become unwanted energies or hazards. If they travel the wanted path in excess they also present a problem.

So, emphasis must be placed on managing the control measures which prevent energies of a wanted nature becoming unwanted, or energies at a desirable level becoming energies at an undesirable level. Injury may occur if risk control has not been properly performed.

Hazard identification and reporting provisions in OHS legislation

Within the occupational health and safety legislation there are generally provisions for identification and reporting of hazards in the workplace. Each of these sections in your relevant legislation should be looked up and read thoroughly.

Inspecting the workplace

Inspection requires a complete understanding of the processes at work, procedures, policies and the overall work system. Records should be taken which are suited specifically to the workplace. This prevents collecting useless and irrelevant information.

There are three procedures best suited to an inspection:

1. Develop a plan of the workplace.
   This includes a floor plan and can include locations of machinery, doors, fire escapes, storage areas, work stations, windows, etc. A plan of work practices/flows highlighting raw material to product completion will assist you in developing a systematic approach to workplace inspections.
2. Prepare a written description of the workplace.
   This includes documentation of employee numbers, current state of accident/injury rates, shiftwork, downtime, etc.
3. Develop and use checklists.
   This aids in ensuring hazards have been identified on the basic/immediate level. They can be done by safety representatives, safety coordinators, employees or union representatives. Once a checklist has been filled out, assessment decides the priorities and the best control measures can be selected.

The above steps can help:

- provide a systematic approach to identifying safety problems
- health and safety representatives from a number of departments to work together to identify/solve common problems
- provide permanent records for the organization.

Risk assessment will be covered in detail later in the chapter, but it is briefly mentioned here because it forms part of the report mentioned below.

Risk assessment

A risk assessment should be undertaken by employer representatives in consultation with health and safety representatives and employees, where relevant.

Once a hazard has been identified, assessment of the risk that the hazard presents will depend on the hazard source and standards, codes and other information which may exist. An example of this can be found in chemical exposure hazards in the workplace. Environmental monitoring may be used to assess the exposure risk within the workplace. Through the measurement of air samples, the result or measurement value may be compared against standards such as the legislated or recommended exposure standard for the substance in question, for example, a US OSHA PEL or an American Conference of Governmental Industrial Hygienists (ACGIH) TLV, or a UK HSE REL. This will indicate the risk associated with the hazard by comparing the risk assessment result with regulations, standards and codes of practice. You may also wish to refer to Appendix 4.1 at the end of this chapter.

Hazard and risk management

Risk control

Hazards, and the risks they present, can be prioritized and control measures adopted based on the safety precedence sequence (see Chapter 1) or the hierarchy of controls (see Chapter 2).

Hazard and risk report forms

If hazards or accidents are not reported at a workplace, the risk of a serious injury resulting from these increases. For a safer workplace, it is imperative that a reporting system of hazards be implemented. This allows for adequate identification, assessment, evaluation and control. Hazards can be entered on a hazard register.

Hazard report forms vary from one organization to the next. It is important to have a hazard report form suited to the work performed at a particular workplace, so accurate, pertinent information is collated.

Fig. 4.2 is an example of a hazard reporting and risk control form.

Risk assessment

Requirements for proper risk assessment with employee participation

A risk assessment involves careful examination of the hazards in a workplace and the factors involved in them, such as the type of hazard, the type of equipment, training and number of operations and existing systems in place in relation to control of the risk. In the assessment, the degree of risk associated with each hazard is decided based on the use of standards where available. The 'control' step involves developing measures to reduce that risk.

Assessment criteria used to influence employers' and employees' decision making

These criteria will potentially include:
For employees:
- lack of information on which to base decisions and adequate training to make use of the information
- knowledge from the media or previous injuries to self, workmates, other people
- concern that control measures could cost jobs
- information or misinformation from friends, media, community
- ideological attitudes (i.e. the boss is against us)
- distrust of expert report or opinion
- potential controls seen as too much trouble.

Many of these can be addressed in a constructive way.

Enhancing occupational safety and health

HAZARD IDENTIFICATION	*DESCRIPTION OF HAZARD:*
RAISED BY: _____ DATE: _____	_____ _____ _____
RISK ASSESSMENT AND EVALUATION	1. INDIVIDUAL: _____ 2. OTHER PERSONS: _____
POSSIBLE LEVEL OF RISK. INDICATE THIS LEVEL AGAINST EACH SECTION (1–5) ASSESSMENT COMPLETED BY: _____ DATE: _____	3. COMPANY: _____ 4. ENVIRONMENT: _____ 5. OTHER: _____
RISK CONTROLS	1. ELIMINATION: _____
DETAILS OF CONTROLS IDENTIFIED FOLLOWING ASSESSMENT CONTROLS IMPLEMENTED BY: _____ DATE: _____ BRANCH/PROJECT/SITE MANAGER _____ DATE: _____	2. SUBSTITUTION: _____ 3. ISOLATION: _____ 4. ENGINEERING: _____ 5. ADMINISTRATION: _____ 6. P.P.E.* _____ *(Personal Protective Equipment)

Figure 4.2 Hazard reporting and risk control form

Hazard and risk management

For employers, decisions may involve these criteria:

- reliance on expert opinion, but a concern about its cost
- better access to trained resources to consider the evaluation results
- concerns about the cost of injuries or disease if the hazard evaluation doesn't lead to needed changes
- concerns about the cost of controls if the evaluation indicates problems which aren't easily tackled
- concerns that standards used in evaluation don't reflect 'real world' demands
- industry standards.

A constructive approach to agreement on evaluation criteria by both employees and employers relies on building up an effective climate of consultation and trust in the workplace before, where possible, addressing problems.

Effect of media information on risk assessment

Advertising campaigns, media articles and programmes can raise many emotive issues which can alter the perception of safety.

First and foremost is the threat to the family unit by the creation of a *death-like* image with the worker's family; for example, that asbestos threatens to take a person away from the family if personal protective equipment is not worn. This may be necessary and accurate. For example, campaigns of this nature include AIDS, driving safety, and smoking.

It is the current affairs programmes which may have the most emotive effect on an individual's perception of a hazard, thus influencing risk assessment.

While media reports can be valuable in raising issues, difficulty arises when the media supplies biased information to a greater population than perhaps an organization can reach. As a result, organizations may find it difficult to provide information that is believed, particularly if it contradicts what the media has outlined.

See also 'Barriers to effective risk communication' and 'Presenting accurate and sensitive information on risk to exposed groups' in Chapter 2.

Critical factors in the relationship between safety and production

Safety and productivity

The critical factors in the relationship between safety and productivity can be determined by understanding what productivity is. Productivity is that component in an organization which attempts to produce as much of something as possible in the least expensive way. Safety can become involved in this process.

To reduce cost and time, additional risk to an organization's resources is incurred (usually by management). Managed correctly, resources can be maximized through safety, and safety can be involved with production in the following ways:

- maximizing use of human resources (by reduction in injuries)
- maximizing use of capital resources (reduction of machine breakdowns)
- optimizing the person–machine interface (combination of 1 & 2)
- reducing potential risk areas by identification of hazards, assessment, and control of risks
- maximizing financial resources (insurance premiums)
- reducing potential for litigation
- increasing morale.

Roles of individuals, safety representatives, safety committees, supervisors and management in the assessment process

Responsibility may be defined as an individual's obligation to carry out assigned duties. Authority implies the right to make decisions and the power to direct others. Responsibility and authority can be delegated to others, giving them the right to act for the employer. It is important to note that, while some responsibilities can be delegated, the management remains accountable for seeing that they are carried out. The responsibilities of an employee apply to everybody in the workplace. First line supervisors have additional responsibilities, and senior management, including the CEO have extra responsibilities again.

Employee responsibility

Employees bear some responsibility for risk assessment at work. Employees must follow the established work practices and procedures as instructed. They must be also be involved in the identification of hazards and in assessment of the risk associated with those hazards through active participation in meetings and other communication opportunities.

Health and safety representatives

In those jurisdictions where the relevant OHS legislation sets out arrangements for the appointment of representatives, their role in relation to risk assessment is clearly set out in the appropriate act or regulations.

Health and safety committees

Likewise health and safety committees can be established in most Robens-based jurisdictions and are a critical element in occupational health and safety at a workplace. These committees are made up of employees and management and become the ideal forum for the exchange of information essential in risk assessment.

The hazard and risk issues in question can be dealt with in terms of identification, evaluation, assessment, control and review by employers and employees looking at the many perspectives.

A joint health and safety committee brings together labour's in-depth, practical knowledge of specific jobs and management's larger overview of job interrelationship, general company policies and procedures and financial position. The appropriate sections in your relevant local OHS legislation may outline the purpose, duties and standard procedures for meetings of the committee.

Management responsibility

Organizations often select a manager who is responsible for safety and health, rather than delegating the responsibility for OHS throughout all positions. This function may be incorporated within another portfolio such as a human resource manager or personnel manager and is hence overloaded. As a result, OHS issues usually end up at a committee level and the issue of risk assessment is often tabled at the meetings, when they could have easily been dealt with at levels closer to the departments. Responsibility for health and safety for all management levels is imperative. Everyone, including employees, has rights and responsibilities in this area.

Senior managers

Senior managers are responsible for developing and maintaining a safe workplace, and this includes adequate and correct risk assessment. Senior management must:

- ensure all legislative requirements are met as a base level
- ensure adequate budgeting is established for implementing control strategies
- work in coordination with other members listed above
- lead commitment by example
- liaise with unions and other industry groups
- maintain emphasis on health and safety.

Line manager/supervisors

It is important that line manager/supervisors fully understand what risk management is and understand the task of risk assessment.

Health and safety coordinator/officer/manager

These officers have specific responsibility for the risk assessment process in the organization and for reviewing its progress. Adequate qualifications should be evident with experience in the field of work. The assessment should be performed fully and be part of a complete risk management programme, i.e. not just a light-hearted token gesture. The coordinator should not be solely responsible for decisions and actions taken.

The specific responsibilities of these managers include:

- organizing meetings
- selecting information sources for assessment
- inspection of the workplace

- monitoring compliance with legislation
- assisting people at all levels of the organization.

The management of OHS risk assessments

For occupational health and safety assessments to have any positive effect in the workplace, several steps should be considered.

These include:

- training
- establishment of safe work practices
- elimination of hazards.

and many others.

The key factor is commitment, particularly management's commitment. Without such commitment and management involvement the assessment process will fail and the hazard will be inadequately controlled.

Risk management and assessment are not an extra part of an employee's job but an integral, full-time component of each individual's responsibilities. Everyone from senior management down is an employee.

Principles behind risk assessment, and importance and limitations of scientific assessment

Introduction

When the risk from a hazard is evaluated by people within a work environment, a variety of methods may be used involving one person or many persons. It is important to remember that people working in the area of a hazard may have an expert working knowledge of usual work practices and possible implications of a hazard. You may also find that employees involved in maintenance, area supervisors/team leaders and safety representatives can be a valuable resource during the assessment phase.

Application of standards, codes and industry practice to assist in problem solving

Problem solving

The technique of problem solving can be applied at the evaluation stage of risk management. Problem solving involves evaluating possibilities for management of a situation, leading to control of a hazard where possible. To assist with this, you may refer to tools such as national standards, standards from OHS authorities, industry standards, and codes of practice, to provide a guideline on acceptable standards and practices.

A code of practice, for example, may provide practical advice on preventative strategies and/or a practical means of achieving provisions and specifications. Although, on their own, national standards and codes of practice are not a legislative requirement

Hazard and risk management

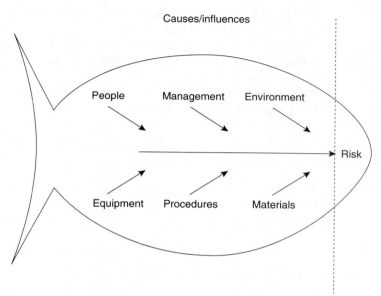

Figure 4.3 Ishikawa fishbone diagram

unless the legislation makes them so, they are a very good information source on acceptable industry requirements and practices.

The three types of standards, and codes of practice, should be listed as reference material from a consultant in the specialized area. The consultant should be up to date with the available information. These documents provide an immediate source of information for the particular hazard in question, and are available in hard copy or electronic format.

Another method which can be used when assessing a risk is to assess the relationship between the causes and effects of a risk. The process for doing this is represented in the 'Fishbone' or 'Ishikawa' diagram above (see Fig. 4.3).

This 'cause-and-effect' diagram was developed to represent the relationship between an 'effect' and all the possible 'causes' influencing it. The effect or problem is stated on the right side of the chart and the major influences or 'causes' listed to the left. It is an effective tool for studying processes and situations and for planning.

A cause-and-effect diagram is a means by which a list can be displayed pictorially. Arranging lists in this way leads to a greater understanding of a problem and possible contributing factors. These diagrams are most effective *after* the process has been described and the problem well defined because, by then, team members will have a good idea of which factors to include.

Steps in constructing a cause-and-effect diagram

Construct the actual cause-and-effect diagram by:

1. Placing the problem statement in the box on the right.
2. Organizing potential causes into categories: people, policies, procedures, equipment, etc.

(Note: human resources, machines, materials, and methods are alternative categories that are more relevant to manufacturing organizations.)
3. Identifying the problem and brainstorming all possible causes of the problem.
4. Entering minor causes and detail. Look for causes or influences. Keep asking:
   - What causes that to happen or occur?
   - Why is that?
   - Who, what, where, when, why, how?
   - Ask *WHY* 5 times!

A well-detailed cause-and-effect diagram will take on the shape of fishbones.
In order to find the most basic causes of the problem:

- look for causes that appear repeatedly.
- reach a team consensus.

The diagram shows us what we *think* are causes; we need to validate that with data.

Limitations on field testing arrangements

When we consider the range of environmental monitoring which can be attended to both in-house and through consultant employment, we need to consider the meaning and limitations of field testing of some risks.

A good example is noise testing. Due to legislative requirements in some jurisdictions, a noise survey is required every five years in particular workplaces. In addition to this, the noise survey must be carried out by a competent person who may also need to be accredited.

If we consider initially the possible limitations of a noise survey, the following variables need to be considered:

- is all equipment operating and representing usual noise levels in the work environment at the time of the survey?
- employee numbers and noise exposure patterns may change throughout periods of the day and with different shift cycles
- equipment used to record the noise levels must be calibrated to strict specifications
- are employees exposed to high noise levels on a regular basis outside the workplace? (This may contribute to hearing loss and permanent damage.)

When the results of a noise survey have been collated, noise areas in excess of the required standard, 85 dB(A) for example, should be the target of control strategies. There will, however, be a percentage of the population that will receive hearing damage from noise levels below 85 dB(A).

What is the possibility of noise control strategies which result in possible risks not being controlled?

Although we recognize that environmental and biological monitoring have some inherent limitations, it can be assumed that by monitoring and controlling hazard exposure within a limited range, safety will be maintained for most of the employees. There will always be employees who are more sensitive to hazard exposure and this should be considered when matching the right person to the right task.

There are many positives and negatives associated with conducting an assessment of a risk. Some of the associated limitations can be strictly related to in-house assessments and others to the introduction of a consultant. As with the type of assessment, so too there are limitations with information collected.

The following is a list of factors to consider:

- cost
- equipment failure
- human failure
- sample group
- inherent errors
- bias
- measurement reliability/reproducibility
- hidden agendas
- interpretation of results
- familiarity with the work system
- relevance of information, results, data, to the problem.

Professional assistance in the risk assessment process

The practice of occupational health and safety is multidisciplinary and, for this reason, situations in the evaluation of hazards may be outside the expertise of the organization and better suited to an expert. These are often consultants in a particular field such as noise, manual handling, auditing, to name a few.

When can a consultant assist?

Hazards in the workplace are best solved in-house where there are adequate resources to accommodate and control the risks. The cost and time of professionals (engineers, doctors, ergonomists) have to be considered when professional assistance is required. However, there are times when special problems or limitations in resources prohibit the effective management of hazards in the workplace and a consultant is needed.

Situations where a consultant may be required include:

- when a second opinion is needed.
- when several solutions to a problem need to be assessed.
- when you don't know if there is a problem in the workplace.
- relieving in-house resources for other tasks.

When selecting a consultant keep the following in mind:

1. Define the hazard.
    Once you have clearly identified the hazard, the selection of a consultant in a particular field will be easier. Professionals usually specialize in such areas as industrial

hygiene, ergonomics, noise and medicine; therefore, your consultant's speciality should match your hazard.
2. List who may be able to assist with the hazard.
   Considerable information is available from government providers which may reduce the cost involved.
3. Compile a short list of appropriate consultants.
   Consultants may be sourced through previous experience, industry or employee peak bodies, government agencies and the telephone book.
4. Call the consultants for a proposal.
   How a consultant may measure, assess and report findings can vary. It is a good idea to discuss the hazard with potential consultants to determine the possible consultant's time required, resources and their past experience with similar hazards.
5. Develop contract specifications.
   This will propose costs and resource requirements for the consultant's scope of work.
6. Obtain a preliminary and final report and recommendations.
   This will allow some planning in terms of hazard control measures in initial stages. A final report should provide you with clear recommendations which are likely to adequately control the hazard when implemented. It will also act as a record of assessment which should be kept by the organization.

Advantages and disadvantages of independent assessment

Some advantages of a consultant include:

- will hire or bring any necessary evaluation equipment
- in-depth approach to a particular hazard
- cross-industry experience, i.e. lateral thinking
- may save the organization valuable time
- can be seen as unbiased by management and workforce
- sees things in a new light.

Some disadvantages include:

- unfamiliarity with workplace and organizational culture
- may take time to gain cooperation of key stakeholders
- may not come up with solutions with a local fit
- in some cases – cost, availability, e.g. distance, demand for services.

The safety factor in risk assessment

The risk associated with a hazard is considered using a number of factors. These include:

- the probability of the hazard leading to an accident and, possibly, injury
- the consequences if the hazard results in an accident
- the degree of exposure of individuals or, in critical incidents without personal injury, the degree of exposure of – for example – equipment.

Table 4.1 Responses to risk

Probability	Consequence	Response
Low	Low	Retain risk
Low	High	Insure and have a major hazard control plan
High	Low	Loss control management. Quick treatment
High	High	Manage out the risk or insure

Accidents may have low to high probability of occurring (low to high risk) and low to high consequences. The risk factor is the combination of the two. So a potential accident may be 'low probability – high consequence'. For example, the Ukrainian nuclear reactor disaster at Chernobyl in 1986 might fit here, or the disasters at the Mayak nuclear facility near Chelyabinsk 1949–1967; or it could be 'high probability – low consequence' e.g. fingers cut on copier paper. Both of these may involve the exposure of many people. But if you were the operator of a nuclear reactor which would you give priority to? One approach to this by W.D. Rowe is shown in Table 4.1.

There is another important consideration and that is the individual's perception of risk. This can vary depending on a person's experiences, on media information or misinformation, on lack of information (ignorance) and on culture – for example, belonging to a group involved in higher inherent risk, such as divers, or a particular social or ethnic group.

Statistical information may be available to assist in evaluating risk. Within a society this can be compared, for example, risks of smoking with risks of diving. Within an organization, good accident statistics can assist in giving a risk ranking to different hazards.

Sometimes sufficient is known about risks to talk of 'safety factors'. For instance, we may know that particular frequency of maintenance of pipework (say once per month) means that a spray of hot liquid under pressure only happens once in five years; and this is decided by an organization to be acceptable. If the organization decides to increase the frequency to once per fortnight, thus reducing the chance of an escape to once in 20 years, we say they have adopted a safety factor of four (20 over 5).

Exposure standards for airborne chemical contaminants are usually set somewhat lower than the levels which cause observable effects – that is, a safety factor is built in.

Most Robens-based OHS legislation refers to 'reasonable practicability', usually defined as balancing risks against the cost and technical feasibility of controls.

Assessing a risk in the work environment

Once a hazard has been recognized or identified decisions need to be made about:

- the nature of the problem – is it relatively straightforward and simple, or does it require more detailed consideration?
- who is to conduct the evaluation?
- when and over what time period is evaluation going to take place?

- what method of evaluation is to be used?
- what information is needed to support the evaluation – i.e. methods, standards, codes, information from similar worksites, injury data?
- what equipment if any is needed for the evaluation, e.g. sound level meter for noise, electronic gas detector for chlorine levels?
- who is to be consulted before the evaluation occurs, e.g. shopfloor employees, health and safety representative?
- in which particular areas the evaluation should be carried out
- what form of documentation is needed – i.e. what style of report is needed to use the evaluation in the assessment phase?
- who will do the assessment, what they will want and how they will want the evaluation information presented?

From this a procedure, plan of attack or flow chart can be developed.

Steps that may be taken to assess a risk involve evaluation of the risk that the hazard poses to the workplace, personnel and environment. During an in-house evaluation the following may be considered.

Assess the risk associated with the hazard by determining:

- Is the risk associated with the hazard acceptable? Information may be required at this stage including relevant standards or codes of practice, and the relevant occupational health and safety legislation.
- What are the controls necessary to reduce the risk? This may involve introducing short-term controls. For example, if an oil spill on a floor surface presents a slip risk, placing paper towels over the area and then washing the floor may be required before long-term hazard management through equipment maintenance is arranged.

The next step is to prioritize hazards identified in order from high risk to low risk, viz:

'A' class hazard
Any condition, practice or procedure requiring immediate attention to prevent the potential for the loss of life, loss of a limb and/or extensive losses or damage to equipment and/or property.

'B' class hazard
Any condition, practice or procedure which has the potential for serious injury and/or equipment/property loss or damage, but less severe than an 'A' class hazard.

'C' class hazard
Any condition, practice or procedure which has the potential for minor injury or minor equipment/property loss or damage.

Hazards with the highest risk rating should receive immediate control priority.

The system developed by Fine, Kinney and Wiruth in Appendix 4.1 at the end of this chapter gives more guidance on assessing risk.

Understanding risk

An explanation

Risk can be considered as the potential for adverse effects to result from an activity or an event. The risk in placing a bet on a horse race or buying a Lotto ticket is not in how much you are likely to win, but rather how much you are likely to lose. Generally, the acceptable level of risk is determined by what is prepared to be lost, balanced against possible gains. People have different levels of acceptable risk across the full range of their interests and activities. This includes risks that are taken in the workplace.

We often hear comparisons being made between the death or injury experience associated with workplace hazards and those associated with road traffic accidents. In almost all circumstances, the road traffic death and injury rates will be higher. This comparison is used to indicate that the workplace is safer than the highway by demonstrating that the risks are lower.

This statement is probably correct; however, the point is often lost because comparisons are not made between like objects. Exposure to road traffic accidents is generally considered a matter of free choice. (This is a questionable assumption.) Exposure to hazards in the workplace is not considered a matter of free choice. (The argument of freedom to leave if you don't like it is invalid.) There is an apparent principle in risk-taking that the acceptance of risk varies greatly based on whether the risk taken is voluntary or involuntary. There is resistance by workers to readily accept even low levels of involuntary risk.

Some risks people are exposed to are not as obvious as those associated with grinding discs, running machines or falls. Risks such as exposure to asbestos in school rooms, radiation from electrical equipment or from noise hazards are less obvious and generally require professional expertise in order to quantify the risk levels.

Risk quantification of the above hazards requires the use of technical equipment and skill to determine the level of exposure which can then be compared against a relevant standard. The people carrying out the measurement work must use their judgement to ensure the final results of exposure to the hazards are communicated as accurately as possible.

Few measurements taken in the field can be considered 100% correct. They must, however, reflect a reasonable evaluation of the exposure level. Experience shows that risk information is better accepted when those exposed are involved in the analytical process. The building of trust and confidence in the minds of people makes the risk communication process much simpler.

Measuring risk

Before a judgement can be made about the safety of a process or activity, it is necessary for the individual to be aware of the risks involved. Knowledge about risks can be gathered from several sources; however, acceptance of risk information is most likely gained through personal experience. For example, information on the risks associated with grinding discs is readily available. Workers are aware that grinding discs can shatter if not properly treated. However, very few workers will ever see a grinding disc shatter. Consequently, a full appreciation of the risks associated with this particular activity may not be possible for every worker. As a result, there are several ways in which we try to communicate risk to those people who are not exposed to rare events, such as shattering discs. Some of these ways have been successful, others less so.

LEAST SUCCESSFUL	MOST SUCCESSFUL
– tradition or folklore	– formal and informal training
– use of common sense	– application of recognized standards
– fear (e.g. gory films)	– safety meetings
– punishment	– tool box meetings
– trial and error	– posters
	– encouraging worker participation

In some jurisdictions, as noted in an earlier chapter, health and safety representatives are able to stop employees operating in an environment when they have reasonable grounds to believe that to continue would expose employees, or any other people, to a risk of imminent and serious injury to their health. It is reasonable to expect that most employees, at all levels of organizations, are not familiar with all risks to which all other employees are exposed. Additionally, it would be impossible to remain constantly aware of the exact levels of all risks in the workplace.

For this reason, it is important to have a mechanism in place that enables employees to determine risk levels quickly and with confidence, so that proper assessments can be made about the nature of a particular risk. Failure to have such a mechanism can cause unnecessary delays or even stoppages in the workplace because employees are unable to assess, without assistance, if a particular exposure causes them to be at risk of imminent injury or harm to their health.

There is a saying that goes something like this:

It is too late to remember you should have drained the swamp when you are up to your armpits in crocodiles.

Similarly, it is too late to remember you should have developed a system to solve problems productively when faced with a group of workers concerned that some new, strange smell is making them feel ill and causing drowsiness to operations personnel.

When people express concerns about safety in the workplace, it is often caused by uncertainty about the nature of the risk. This ignorance is often expressed by task avoidance or, in extreme cases, complete refusal to perform certain work. The mechanism to remove uncertainty begins with the willingness of management to address the issues that concern workers with a proactive rather than reactive approach.

Once a procedure has been developed that encourages supervisors and workers to openly discuss questions of risk, support structures such as training, information sources and communications systems can be developed to complement the purely technical aspects of risk identification. Commitment in the workplace to a problem-solving mechanism is likely to be determined by the level of involvement of those actually exposed to the risks. This requires planning on the part of management and supervision to provide workers with adequate time to participate in problem-solving activities. Often those exposed to the greatest risks are those closely involved with the production process.

Risk evaluation

Risk is determined through the combination of two variables. Firstly, the probability of an adverse event occurring (an accident) and secondly, the severity of that event (the damage or injury or number of people affected). The probability variable considers the chances, odds or likelihood of an accident occurring, while severity is concerned with the extent of damage resulting from the accident. A result is found somewhere between no damage and death, using the damage or injury approach.

Case study

A mechanical fitter has been offered a job in the maintenance section of the Saline Gold Mine. The mine has been producing for four years and currently employs a total of 100 people. Part of the fitter's duties includes working on plant underground. The fitter has no underground experience, but is assured that she will be provided with the proper training and supervision to make her underground work as safe as possible. The fitter wants to know more about the risks associated with working at Saline and asks about the safety record. She is told that there have only been eight accidents in the past 12 months at the mine. This was a 50% improvement on the previous 12 months and a further 50% improvement is expected over the next 12 months. (The company uses the term 'accidents' to imply Lost Time Injuries.)

Has the fitter been given enough information to satisfy herself about the risk of working underground at Saline Gold Mine?

What does the fitter know?
- that 8% of the people at the mine had lost time from injuries during the past year (2002)
- that 16 people had lost time injuries in the mine in the year before last (2001)
- that the company expects only four accidents to result in lost time injury in the year ahead (2003).

What doesn't the fitter know?
- where the injuries occurred
- the extent of the injuries (severity)
- who had the injuries
- how many people worked at the mine in 2002.

All are important factors in risk determination.

When the fitter departed to consider the job offer she was given a copy of the company's annual safety report, a document provided to prospective employees to promote the image of the company as a good employer.

From information contained in the report it was possible to construct the following Tables 4.2, 4.3 and 4.4.

Table 4.2 Injury experience 1999–2002 all employees

Year	No. Emp	Sex (M)	Sex (F)	LTIs	Hrs/Emp	Risk Ratio %	Exp × 1000
1999	200	196	4	11	2460	5.5	44
2000	250	243	7	12	2500	4.5	52
2001	320	293	27	16	2420	5.0	48
2002	100	86	14	8	2384	8.5	29

Table 4.3 Injury experience 2002 × classification

Classification	No. Emp.	Sex (M)	Sex (F)	LTIs	Hrs/Emp	Risk Ratio%	Exp × 1000
Plant Operator	20	16	4	1	2300	5.0	44
Maintenance	18	15	3	3	2800	16.7	16
Miner	40	40	–	3	2400	7.5	32
Supervision	10	9	1	1	2200	10.0	22
Administration	12	6	6	–	2000	0	24
Total	100	86	14	8	2384	12.5	28

Table 4.4 Injury experience 2001 × classification

Classification	No. Emp.	Sex (M)	Sex (F)	LTIs	Hrs/Emp	Risk Ratio %	Exp × 1000
Plant Operator	64	56	8	1	2400	1.6	153
Maintenance	64	62	2	2	2500	3.1	80
Miner	128	128	–	8	2400	6.2	38
Supervision	32	29	3	2	2600	6.2	83
Administration	32	18	14	3	2200	9.2	23
Total	320	293	27	16	2420	5.0	75

Code:

Classification	–	Job or function.
No. Emp	–	The number of employees in each classification.
Sex	–	(M)-Male; (F)-Female employees.
LTIs	–	Lost-Time Injuries (as defined by a national standard, e.g. AS 1885.1).
Hrs/Emp	–	The average number of hours worked by employees in that classification over the period, i.e. 12 months.
Risk Ratio %	–	The probability of an employee in that classification being injured, expressed as a percentage, e.g. Maintenance Employees 2002 $\frac{\text{LTIs} \times 100}{\text{No. of employees}} = \frac{3 \times 100}{18} = 16.7\%$
EXP × 1000	–	The average number of hours worked between or without an LTI in that classification ×1000.

Note: the numbers used in this case study have been designed to highlight particular points and are likely to be unrepresentative of the real workplace experience. Further, some numbers have been rounded to provide easy calculation and are not intended to be exact.

Risk assessment

It is correct to say that Saline had 16 LTIs in 2001 and eight LTIs in 2002, representing a reduction of 50%. However, Saline employed 320 people in 2001 and only 100 in 2002. Consequently, to gain a true appreciation of the injury experience, the fitter calculated the injuries per 100 employees. The results were interesting.

In 2001 the probability of injury at Saline was five injuries per 100 employees, or one chance in 20, or about a 5% chance in percentage terms. The 2002 total was only 8 LTIs, 50% less than 2000. The fact that there were 220 less employees makes no statistical difference because the fitter was making comparisons using a common base; that is, injuries per 100 employees. In 2002 there were eight injuries per 100 employees, or one chance in 12, about an 8% chance. Clearly, the number of LTIs had halved at Saline, but at the same time the risk of injury at Saline had increased.

This all proved very interesting, but the fitter was specifically interested in risk associated with maintenance work. The Injury Experience Table 4.4 showed that the maintenance group had two LTIs in 2001. There were 64 in the group; two LTIs represented one chance in 32. In 2002 the group was reduced to 18 and it experienced three LTIs; the chance of injury in the maintenance group increased fivefold.

The chance of lost-time injury in 2002 was one in six, making maintenance workers the highest risk group, according to the safety report. The fitter made a few discreet inquiries about the injuries to the maintenance workers and about other issues not covered in the Safety Report. The result of her inquiries led her to conclude that she should accept the offer to work at Saline Gold Mine as a fitter in the maintenance section. She considered the associated risk acceptable.

Risk control

Hierarchy of risk control options and principles, and selection of countermeasures to provide a reduction in risk

Links between risk engineering and risk control

Risk is the probability that an event could occur in the workplace resulting in personal injury or loss to the organization. Risk management involves the identification, assessment and control of all areas of risk in an organization and aims to minimize loss or wastage of business assets. From an insurance viewpoint, risk avoidance, risk control, risk transfer and risk retention are some options in risk management. Risk control is a strategy which allows the use of a hierarchy of control measures. Risk engineering generally involves the use of engineering measures to reduce or eliminate risk.

Enhancing occupational safety and health

The link between risk control and risk engineering involves taking into account the control strategies used by management and the fact that if hazards are controlled then the risk associated is also reduced.

Countermeasures to lower employee exposure to hazards

A range of countermeasures is available to lower employee risk. They involve reviewing:

- task
- engineering
- guarding
- methods
- training
- PPE
- policy
- substitution
- shielding
- practices
- information
- worker behaviour.

Risk control options for reducing work activity injury

The main purpose of control methods is to eliminate or reduce risk in the workplace. It is essential that a thorough examination of the workplace be carried out to reveal the types of hazards and their extent. With this information, a comparison can be made with the requirements of appropriate acts, regulations, codes and standards. This evaluation and assessment stage is essential because it will determine the types of control methods used.

Physical control strategies are often used. There are three basic principles to be recognized when using these control strategies.

If you cannot eliminate the hazard:

- the next most effective results are achieved with permanent physical barriers
- generally, the least effective are non-physical barriers
- multiple barriers are often needed to provide adequate control.

Risk control status is achieved through field observations, regular and independent inspections, statistical analysis and in discussion with all levels of the workforce. The task is to ensure, as accurately as possible, that the barriers in place to control hazards are appropriate and effective. The 'safety precedence sequence' mentioned in Chapter 1 is useful as a guide in establishing the adequacy of risk control strategies:

- eliminate hazards
- minimize energies
- apply physical barriers
- use warning devices
- minimize the potential for error

Hazard and risk management

- revise and reinforce procedures
- select, train, motivate and supervise personnel
- identify residual risks.

In making a decision on the adequacy of risk control, it is necessary to consider the risk priority. Those hazards with the greatest worker exposure and potential for harm should require closest control. Following determination of control status, it may be decided that the control is not adequate. At this point, a decision must be made to accept the risk associated with the hazard as presently controlled, or to improve control.

The hazard control principles can be readily integrated into management systems such as quality management.

Selecting appropriate methods to monitor hazard controls

Once the control methods have been chosen and either installed or introduced, a programme of maintenance and monitoring must be put in place. This makes sure that the control equipment and techniques meet the requirements of the workplace. A number of monitoring methods can be used to check the reduction of risk:

- health monitoring
- environmental monitoring
- supervisor's observations
- checks on maintenance of control measures to make sure that they operate to the correct standards and limits.

Selecting and using personal protective equipment

Introduction

Personal protective equipment (PPE) has a twofold role in the workplace. In some cases it is a passive safety measure, used to minimize injury if other protective measures fail. But it is also used as the main means of injury prevention in situations where it is impossible to guarantee the effectiveness of other safety measures, e.g. penetration-resistant boot soles on a construction site, fall arrest harnesses where falls are likely to cause serious injury or death.

In some cases the equipment is the primary means of preserving life, e.g. self-contained breathing apparatus in an inert gas atmosphere, or while working underwater, or in a vacuum.

PPE includes:

- respiratory protection designed to prevent the absorption of gas or vapour
- clothing for the body designed to prevent skin contact with toxic, corrosive, or irritant substances, biological hazards such as bacteria, or the effects of solar radiation
- gloves to prevent the same, or to prevent electric shock, laceration, skin abrasion, amputation of a finger by a knife, or the effects of vibration

- boots to prevent the same as in bullet point two above, to prevent pentration of the sole or crushing of the toes, or to reduce the chances of slipping
- face shields, glasses or goggles for eye protection, which may involve radiation (from the sun, welding, or lasers), dusts gases or vapours, penetrating or projecting objects, or flying objects or particles
- head protection (hard hats where there are low roofs or openings or the likelihood of falling objects) or broad brimmed hats or attachable broad brims on hard hats to provide solar protection
- fall restraint harnesses
- solar protection factor rated sunscreen cream
- leaded clothing for X-ray and radiation workers and for sensitive areas of patients
- chainmesh abdominal guards for meat workers
- high visibility clothing and vests for road or construction work
- reflective or electrically illuminated clothing for construction and road work at night
- clothing designed for work in cold environments, either natural or manmade (freezer compartments)
- clothing designed for work near heat radiation or to improve comfort and keep body temperature down in hot conditions
- ear muffs or plugs to reduce the possibility of hearing loss.

Selection of the PPE

Many items of PPE are the subject of testing to an accepted national or international standard. In addition, flow charts within some standards assist in selecting the specific type of a particular category of PPE, e.g. choice of self-contained breathing apparatus, positive pressure air-supplied respirator, powered air purifying respirator, or passive respirator (and which type of cartridge or canister), if respiratory protection is being sourced.

Within this envelope of equipment which meets the standards, the purchaser in the workplace should take note of price, availability, degree of comfort, fit and acceptability to the workforce, spare parts and technical backup, and ease of cleaning and maintenance. Acceptability ('the look') and comfort are important because they will affect the degree to which workers comply with instructions to use PPE.

PPE necessary for particular work must be made available at the employer's expense. A risk assessment of particular tasks, as part of a job safety analysis, is a key step in this process.

Employees must have a direct role in selection of equipment such as boots, where fit and comfort are critical.

Responsibility

Responsibility for purchasing, maintenance (e.g. regular respirator cartridge changeover), testing, storage and cleaning of PPE must be clearly established, as well as responsibility for any necessary record keeping, e.g. last test of self-contained breathing apparatus. Training in PPE use is another area of responsibilty.

Induction and training

Where employees are expected to wear or use PPE, they should be properly trained from the outset in how to do this correctly. Comfort and fit are crucial, and if equipment is personal issue, this must be dealt with from the outset of employment.

Training should cover the situations where conflict arises from the need to wear PPE versus comfort, for example a full protective suit for certain chemicals creates problems from metabolic heat generated from bodily activity, and this is compounded if the environmental conditions are hot.

Cold weather gear can cause conflict in areas like manual dexterity and tool use.

Heavy work such as shovelling can be difficult if a respirator creates serious breathing resistance.

PPE use areas

Signs and lines on floors may be used to define areas where PPE usage is compulsory. It is important that everyone entering that area wears the equipment if the requirements are to have credibilty with the shopfloor workers. This means senior management setting the example, even if on strict grounds of exposure = concentration × time, it could be argued that they don't need to.

It needs to be remembered that certain tasks such as welding or grinding (producing flying objects, dust, noise) can affect not just the person doing the task but others nearby.

Screens should thus be used, plus mechanical ventilation if needed, and/or the others issued with appropriate PPE themselves.

Communicating risk

Negative dominance

Perception is reality in the area of risk communication. Research has shown that it takes three positive messages about risk to neutralize one negative message – the theory of negative dominance. It needs four positive messages to actually leave a positive impression after that one negative message.

This is a first key feature of the 'rule of three'. The second application of the rule of three comes in constructing a response to questions about a risk. This can be done with a message map, in which questions are anticipated and answers thought out in advance. Three areas of the issue with three questions each are anticipated.

Credibility transfer

A low credibility source can be improved if it is supported by a source of higher credibility. Different groups rank differently in terms of public credibility. In the USA the highest groups are health professionals, safety professionals and university scientists. The environmental professions, media and activist groups rank next, and industry, federal government and paid external consultants last.

It has to be accepted that 5% of people will never accept a positive message about risk. Communication with the public needs to be set at four levels below the average school education grade of the people concerned.

Factors in risk acceptance

If there are no perceived benefits associated with a risk, the perceived risk level rises. Sufficient information to make an informed decision reduces anxiety. If a solution is imposed, the perception of risk is heightened, and if people aren't listening, they become very much more upset. Trust is the strongest factor in people accepting a risk, with the benefit next, voluntary acceptance next and fairness (sharing the risk) next.

Primacy and recency

Constructing a message about risk should have three parts. What is said first and last is remembered last. The principles of primacy and recency were known by the ancient Greeks. It is important to first of all learn what it is that people want to know. Each third of the message needs to have three supporting statements, the 'rule of three' again.

The media

The media tend to cast people as either villains, victims or heroes, with the media often the heroes. A 'villain' can respond in a way which partially improves their image. Personalizing an issue (e.g. Kim Phuc, the little girl running with napalm on her back in Vietnam 26 years-or-so ago) can markedly offset attempts to depersonalize an issue.

Homework

It is important to research the perceptions of one's opposition. For example, to some in Islam the concept of luck in relation to life and death may be anathema because it is blasphemy to see it as anything but God's will.

Gaining attention over mental noise

If you are speaking to people who are upset, you will only hold their attention initially for two minutes. The maximum briefing should be no more than 20 minutes. Trust must be established and in a media interview nine seconds and 27 words may be all you have. The vital ingredient is to show that you care. Women convey this impression more readily. When people are upset they have difficulty processing information, they tend to be distrustful and to think negatively. This mental noise affects hearing, understanding and remembering.

Poor processing of information can be overcome by listening and responding in threes, and distrust can be overcome with credibility and mirroring restating the person's concerns. People who are upset also focus on negative words and body language. (Leaning back carries an idea of 'casualness'). So try not to use 'no, not, can't, don't, never, nothing, none' as they establish a negative frame for communication.

Trust in high stress situations

The factors creating trust in high stress situations are:

- listening, caring and empathy which is worth 50% in getting a positive response from the listener (trust must be established in the first 30 seconds)
- competence and expertise which provide 15–20% of a positive response
- honesty and openness: 15–20%
- dedication and commitment: 15–20%.

(With acknowledgement for this section to Dr Vincent Covello, Columbia University.)

Further reading

Breivik, G. (1998). (Norwegian University of Sport and Physical Education). *The Quest for Excitement and the Safe Society*. Safety Institute of Australia – Safety in Action Conference and Expo, Melbourne 25–28 February.

Brown-Haysom, J. (1999). Floor Safety: Getting A Grip, (housekeeping). *Safeguard (New Zealand)*, November/December, pp. 38–42.

CCH Australia. (2000). *Planning Occupational Safety and Health*. 5th edn. Sydney: CCH Australia.

Editorial. (2002). Liberty Mutual Study Reveals Leading Injury Causes. *Professional Safety*, **47(6)**, 19.

Emmett, E. and Hickling, C. (1995). Integrating Management Systems and Risk Management Approaches. *Journal of Occupational Health and Safety – Aust. NZ*, **11(6)**, 617–24.

Everley, M. (1995). The Price of Cost Benefit. *Health and Safety at Work*, **17(9)**, 13–15.

Haddock, C. (2003). *Managing Risks in Outdoor Activities*. 2nd edn. Wellington: NZ Mountain Safety Council Inc.

Health and Safety Executive. (2003). *Risk Perception Leading To Risk Taking Among Farmers In England And Wales*. Bootle UK, Health and Safety Executive.

Kinney, G.F. and Wiruth, A.D. (1976). *Practical Risk Analysis and Safety Management*. AD/A–027189 Washington, US Dept. of Commerce National Technical Information Service.

Low, I. (1988). An Introduction to Accident Damage Control in the Workplace – Part 7. System Failure Analysis. *Journal of Occupational Health and Safety – Aust. NZ*, **4(6)**, 555–65.

Macfie, R. (2002). Pain In The Paddocks (farm culture). *Safeguard (New Zealand)*, May/June, pp. 36–41.

Main, B.W. (2002). Risk Assessment is Coming. Are You Ready? *Professional Safety*, **47(7)**, 32–7.

McDonald, G. (1995). Safety, Consignorance or Information. *Technical Papers of the Asia-Pacific Conference on OHS*. Brisbane.

McDonald, G.L. (1993) *The Nature of the Conflict and the Need to Resolve It. (Injury Categories)*. Proceedings of the Minesafe International Conference, Perth, March 1993, pp. 371–88. Perth: Chamber of Mines and Energy.

Rasmussen, J. (1997). Risk Management in a Dynamic Society – A Modelling Problem. *Safety Science*, **27(2–3)**, 183–213.

Rasmussen, J. and Svedung, I. (2000). *Proactive Risk Management in a Dynamic Society*. Karlstad, Swedish Rescue Services Agency.

Standards Australia. (1999). *AS/NZS4360. Risk Management*. Sydney: Standards Australia.

Swartz, G. (2002). Job Hazard Analysis. A Primer on Identifying Hazards. *Professional Safety*, **47(11)**, 27–33.

Taylor, G.A. (2001). *Odds, Gods and Accidents*. Perth: Cindynics Applications Press.

Tweeddale, M. (2003). *Risk and Reliability of Process Plants*. Houston, Gulf Professional Publishing.

Wigglesworth, E.C. (1978). The Fault Doctrine and Injury Control. *Journal of Occupational Trauma*, **18(12)**, 37–42.

plus standards and relevant OHS legislation, codes of practice, guidance notes and national, state or province workplace injury statistics (metrics).

Activities

1. List the hazards likely to be faced by the following: Maria Fallucci – a mineral assay laboratory technician, Fred Kennedy – a worker in a sheetmetal/light engineering workshop, Faluso Osagyefo – an outside construction worker, Vo Thuc Do – an underground coal miner, Miriam Steinepreis – a clerical and keyboard worker, Nana Vouzinis – a general nurse. Use the headings 'chemical, biological, physical, electrical, overexertion/postural, mechanical, psychological'.
2. In a workplace of your choice, conduct a formal inspection to identify any hazards that pose a risk to people, the environment and/or equipment.
3. Select one hazard identified in the workplace and evaluate the risk from the hazard using a risk assessment tool.
4. What class of hazard have you identified?
5. Identify risk control strategies for the hazard identified, noting the cost/benefit of actions recommended.

Hazard and risk management

6. List the types of personal protective equipment used at the workplace and evaluate if there are any other control options that could be used in preference.
7. Outline how hazard identification and risk control can impact on the work system.

Appendix 4.1 – Assessing risk with figures

Providing you can recognize the hazard, you can start to run through the rest of the process. In any situation, you will need to consider:

- how likely is it the hazard will lead to an accident?
- how likely is it anyone will be exposed to the hazard?
- how serious are the consequences?

G.F. Kinney and A.D. Wiruth from the US Naval Weapons Research Centre, looking at safety with explosives in 1976, gave us some safety maxims:

- We cannot completely avoid the many hazards in life and we cannot eliminate all the risks these hazards create.
- In ordinary life careful thought and effort can reduce risks to acceptable levels.
- Use our limited resources to get the best benefits from risk reduction; don't use them in hopeless attempts to completely eliminate some risks.

Kinney and Wiruth developed a risk assessment system based partly on work in 1971 by William Fine, also from the US Navy. The system gives us some ways of describing the likelihood of a hazard giving rise to an unwanted event. The descriptors of 'likelihood' are:

Might well be expected	10
Quite possible	6
Unusual but possible	3
Only remotely possible	1
Conceivable but very unlikely	0.5
Practically impossible	0.2
Virtually impossible	0.1

You will see numbers have been put on the classes, with the first one hundred times higher than the last.

Next, Fine added a consideration of the 'likelihood of people being exposed to the hazard':

Continuous	10
Daily (frequent)	6
Weekly (occasional)	3
Monthly (unusual)	2
Two or three times (rare)	1

Once per year (very rare)   0.5

Lastly, Fine considered the 'consequences' if the hazard gave rise to an unwanted event:

Catastrophe (many fatalities)   100
Disaster (several fatalities)   40
Very serious (fatality)   15
Serious (serious injury)   7
Important (disability)   3
Noticeable (minor first aid)   1

He also gave an alternate dollar value choice where property was damaged, but no one was hurt. As dollar values change, we will ignore that here.

You may have noticed that there is another factor which is going to affect any assessment. Twenty people may be exposed continuously to a hazard of a type which *might well be expected* to result in an untoward event. If that unwanted event is capable of producing at least one fatality with one person exposed, with twenty people exposed, it may result in many fatalities.

For the last three consequence descriptors, we could up the rating shown if there are injuries to several people, not just one, because the rating given for that level takes no account of the numbers of people injured.

Geoff McDonald, an Australian accident specialist, looks at injury in a slightly different way. You could use his descriptors, as given in Chapter 1, for the last three types of injury:

- person's future permanently altered (dead or permanently disabled) – Class 1
- person is not fit for work for a length of time but fully recovers and adopts normal work and lifestyle – Class 2
- person is inconvenienced by discomfort, pain or limited motion but can carry on – Class 3.

We can obtain a risk score by multiplying the 'likelihood of a hazard giving rise to an unwanted event' times 'likelihood of exposure' times 'consequences'.

A score of no more than 20 is desirable for risks at work. Interestingly, this is lower than the risk score for driving to work, or riding a bicycle for exercise, Kinney and Wiruth note. Up to 20 would be considered low risk, 20–70 moderate risk, 70–200 substantial risk (correction needed), and 200–400 high risk (urgent correction needed).

Over 400, whatever is being done should stop, or if the risk is being assessed in advance, shouldn't proceed until corrections can be made.

Values under 20 suggest acceptable risk. The fact that some everyday situations have higher scores reinforces the points which will be made about methods of reducing risk. Let's try applying Fine's ideas to a familiar risk scenario. A contractor decides to clean the gutters using a two-legged ladder. There is nowhere on the gutter to tie off (fasten) the top of the ladder.

Hazard and risk management

Let's assume the person on the ladder tries to reach just that extra bit sideways, rather than getting down and moving the ladder. The ladder tips sideways and the person half jumps to save themself. How would we rate this?

Firstly, how likely is it that this would happen? We decide it's 'quite possible', score 6.

How often does the person do the job? Let's say it's two times a year, that's 'rare', score 1.

What is likely to happen after jumping? If our person is lucky, he or she lands in a bush – a rosebush or a geranium. Consequence – 'noticeable', if a rosebush – score 1.

Or they twist an ankle – we'll give that 'important', score 3. What then are our overall risk scores?

$6 \times 1 \times 1 = 6$ for the scratches. This seems OK.

Or $6 \times 1 \times 3 = 18$ for the ankle. This is still within the low risk area.

What if we decide it's likely to lead to a broken leg? Depending on the person's age, and the type of fracture, this might still only get a rating of 'important', score 3, or it could be 'serious' (permanently alter the person's future), score 7.

If 'serious' is chosen, this gives us $6 \times 1 \times 7 = 42$, that is, it's in the moderate risk area. (Kinney and Wiruth call that 'possible' risk, but that's a confusing descriptor).

You decide that moderate risk is not acceptable to you, and use a four-legged ladder instead to reduce the chances of the ladder moving from under you, although this may present greater difficulty in placing it over garden beds underneath the gutter.

The assessment can also be carried out using a nomogram, Fig. A4.1.

Select the likelihood and exposure, connect them and extend the line to the tie line. From here draw a line through the selected level of consequences and extend to give a risk value.

A second nomogram, Fig. A4.2, can be used to determine the cost effectiveness of control measures.

Connect the risk value found in Fig. A4.1 to the degree of risk reduction desired and extend to the tie line. Connect this to correction cost and extend to the justification factor line. This will establish how worthwhile the control measure is. Obviously you may need to adjust the cash values of the third line to reflect local conditions.

(*The information in this Appendix is reproduced courtesy of the US Navy, China Lake, California – see Further Reading in this chapter.*)

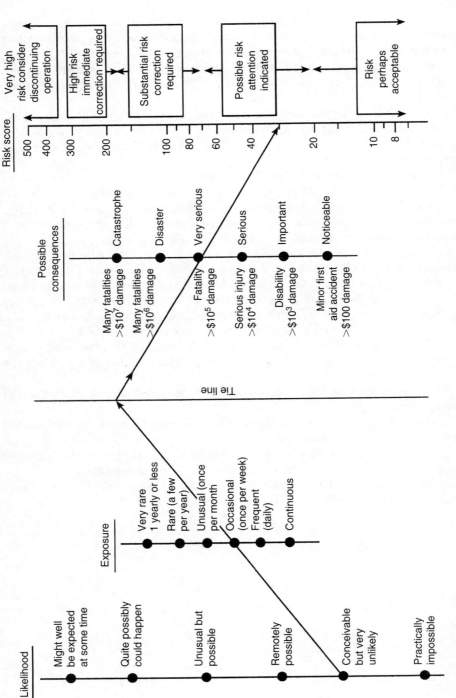

Figure A4.1 Risk analysis

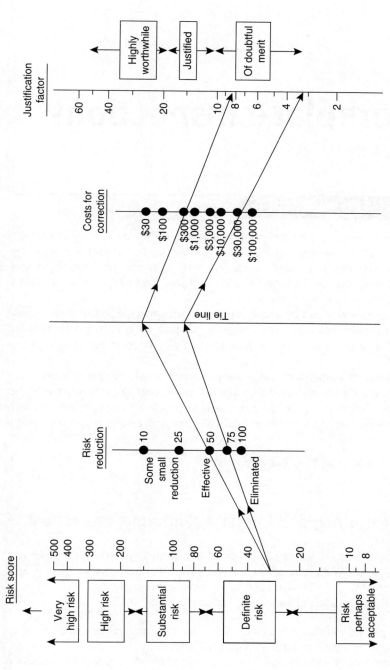

Figure A4.2 Cost effectiveness factors

5

Workplace inspections

WORKPLACE EXAMPLE

A New Zealand company paid a hefty fine after being prosecuted by the Health and Safety Service. During a concrete pour a floor collapsed and a worker fell about nine metres to the ground, sustaining serious injuries and requiring nine months to recover. Injuries included leg, rib and jaw fractures, abrasions, dental injuries and difficulty in breathing.

The worker was employed by a labour hire agency engaged by the construction company. The mould into which the concrete was poured did not have adequate support, and whether the concrete had set had not been checked before the next pour.

The OSH Service commented that a company with a safe system of work must ensure that its staff are committed to making it work, with appropriate checks on that. The construction company, said the OSH Service, had not followed its own quality control procedures before the pour, which required an inspection using a checklist.

(Source: the OSH Service, New Zealand.)

Standard setting and formal and informal inspections

Workplace inspections are an important part of any accident prevention or risk management programme. Workplace inspections are carried out by a wide range of people from management through to health and safety representatives.

The inspection is conducted for two basic reasons:

1. To check specific conditions within the workplace, to measure performance and to ensure that acceptable standards are being achieved.
2. To monitor the work environment for the purpose of identifying accident causation factors and hazards.

Inspections can also be divided into classifications such as:

- formal or informal
- planned or unplanned
- general or critical.

All inspections have the potential to be beneficial provided it is clear in the minds of those doing the inspection and those being inspected, as to what is to be achieved from the inspection process.

Many workplace inspections tend to concentrate on the physical or visual aspects of the work environment, and in so doing limit the effectiveness of the time and effort put into the inspection process.

Workplace inspections must not be limited to identifying the origin of oil leaks or checking the integrity of lifting devices but must include the less obvious aspects such as the:

- communication system
- flow of information
- identification of training needs
- effects of technological change.

All are equally important in determining a safe work environment.

Large and small sections of organizations are coming under increasing pressure to formalize and document their work safety and health activities. Examples include employee pre-employment checks, induction and training records, hazard reporting and the issuing of personal protective equipment. The use and application of inspection techniques can be viewed as an extension of the trend to identify and document work safety and health initiatives and activities. In workplaces of all sizes, there is a place for some type of formal and regular inspection system.

Definitions – types of inspections

This chapter considers three types of inspection techniques:

1. Workplace inspections
2. Safety audits
3. Safety surveys.

Workplace inspections

The workplace inspection addresses the work process and primarily considers aspects of the work activity and how they compare with a set of predetermined standards. Workplace inspections must look at the relationships between the people, the equipment and the procedures in determining if standards are being met and maintained. It is of limited benefit to merely consider what can be seen in the workplace, as this may be in a continual state of change.

The workplace inspection should be conducted in such a way that it is possible to identify variations from established work procedures. The workplace inspection should determine the appropriateness of existing safety policies and whether these need to be enforced, modified or changed in the light of changes in technology, employee skills, industry knowledge and legislation.

An inspection that is carried out on a job while it goes on in a particular place with particular equipment, tools and people is sometimes called Job Observation or Job Performance Sampling (JPS).

Having stressed the need to inspect all elements of the work system, the work environment itself must be inspected. Work is not carried out in a vacuum; consequently, it is important to identify areas within the work environment which may be having an adverse affect on the people, equipment, materials and procedure relationship. Inadequate work environment conditions result, at best, in employee discomfort and in some cases, contribute significantly to the creation of hazards, accidents, and injury.

Safety surveys

The safety survey is a detailed inspection, generally into one aspect of the work system. A survey could be used to examine a specific component in either the work environment or the management system. Usually, the safety survey requires consideration across the whole range of organizational responsibilities. The need to carry out a safety survey can arise out of a workplace inspection or a safety audit.

For example, it could be discovered during a routine workplace inspection that some areas of the plant are experiencing difficulties with discs during grinding operations. The difficulties may be concentrated in one section of the workplace or throughout all sections. Some employees feel that quality control is lacking. Alternatively, some supervisors believe the discs are excellent and last longer than the previous brand. The purchasing department is pleased with the discs, supplies are always available and they are cheaper than previous models. The safety representative is sick of people complaining to her. The union representative is having a meeting on Friday to consider a ban on the discs. The matter has been raised at the last three safety committee meetings.

The safety of the grinding discs, their suitability or otherwise, clearly must be addressed. The most productive method of conducting the survey is to carry out an inquiry or investigation in consultation with those most affected; that is, the users of the discs.

Safety audits

The health and safety audit subjects the whole work system to closer examination. In this context, the work system includes both the work environment and the management environment. The management environment includes responsibilities that are reflected at the workplace in specific quantifiable terms. The need to establish standards that measure variables, such as the provision of information, the adequacy of instruction and training, and the levels of cooperation and consultation, are now imperative as these matters are factors in the legislation of most jurisdictions.

Standard setting

In conducting an inspection, there is an understanding that whatever is being inspected is to be compared against some established standard. If there is no criterion for comparison, you cannot inspect – you merely observe or look. A set of standards needs to be developed, if they do not already exist, on which comparisons can be made. This is the most difficult part in preparing for the workplace inspection – identifying the standard.

The majority of things inspected do have parameters, usually determined through reference to an appropriate guideline such as a national standard, regulation, code of practice, guidance note, etc. The challenge in workplace inspections is to predetermine acceptable benchmarks for the measurement of those factors that influence accident prevention and hazard management for which there are no guidelines.

Conventional wisdom suggests that standard setting within the work environment is a matter for joint decision making between the employer and the employees. Consequently, legislation introduced to provide for self-regulation in the workplace clearly identifies a role for health and safety representatives and health and safety committees in both workplace inspections and standard setting.

It could be argued that there are more than the two basic reasons previously stated for conducting workplace inspections. However, in the context of this subject all other criteria will be considered as part of these two basic reasons.

The broad view

The monitoring of work safety and health standards within many workplaces is conducted on an informal, unplanned and unrecorded basis. This results from the fact that most workplaces are managed by owner/employers who are likely to employ less than 10 people. The need to undertake regular documented safety inspections is generally not fully appreciated. Workplaces with small numbers of employees are often lacking in many of the safety and health initiatives considered as normal practice in larger workplaces. This also extends to the provision of safety and health expenditure on things such as extraction systems for dust and fumes, lifting equipment, workplace layouts and adequate working space. The reasons why smaller workplaces adopt different standards from larger workplaces is not out of a lack of concern for the employees, but rather a need to spend available resources differently. However, the economic rationalism of business decisions taken to support the growth, or survival, of the small enterprise must remain balanced with the needs of the employees. Because some standards may be different between the small and large workplace, this does not excuse the small employer from establishing and maintaining appropriate standards and controls. In workplaces of all sizes, there is a place for some type of formal and regular inspection system.

How are standards identified?

In many cases, identifying a standard that can be adopted to measure workplace performance is straightforward. For example, Regulation 7 of the UK Dangerous Substances and Explosive Atmospheres Regulations 2002 requires certain places at workplaces to be classified in relation to the subject of the regulations as per Schedule 2, and to further divide the places classified as hazardous into different hazard zones. By referring to Schedule 2, places within a workplace where explosive atmospheres may occur can be identified; the regulations also provide a basis for their zoning. Similarly, the Planning (Control of Major Accident Hazards)(Scotland) Regulations 2000, implementing 96/82/EC, set out controlled quantities of substances in Schedule 1, e.g. for hydrogen it is 2 tonnes. So the Schedules provide standards as a basis for decision making. Regulations have the effect of spelling out the specific requirements of the legislation. Regulations may prescribe minimum standards and have a general application or they may define specific requirements related to a particular hazard or a particular type of activity.

Common law

Decisions handed down by the courts offer a guide to determining acceptable workplace standards. Decisions arising out of criminal, civil and coroner's courts all have an effect on standard setting. Smoking in the workplace, originally raised as a health issue, has taken on a much wider perspective following court awards of damages against an employer who neglected to protect an employee from the effects of passive smoking. While the issue of smoking remains a health issue, the court decisions will focus the attention of employers on their duty to provide a safe place of work for those affected by the smoking of other employees. The motivation for a future possible total ban on smoking in the workplace is likely to result from a fear of litigation more than any associated health issues.

Standards

A published national or international standard must be complied with where a regulation requires that particular standard to be met. The compliance with any standard must not be inconsistent with the regulations' general intentions. The vast majority of national or international standards are not mentioned in regulations. They do, however, provide an acceptable guide to standard setting in the workplace. In South Africa, the OHS act defines a standard to include any useful standard, not just those with the force of law.

Other standards

Codes of practice, guidance notes and industry standards are sometimes referred to in legislation. Most standards contained in statutory documents and associated support material call for the minimum acceptable conditions.

Identifying standards relevant to a particular work environment

It is often the case that particular organizations will establish internal policies, rules and work systems which call for a level of performance above statutory or industry standards. Self-regulation legislation provides the impetus for employers and employees to establish and maintain the highest standards for work safety and health in every workplace. However, it is an industrial reality that the workplace standards will vary greatly between, and even within, organizations. The principle to adopt is that the acceptable standards should always be as high as reasonably practicable in each particular circumstance and with an eye to what is generally considered best practice within an industry section. Standards must never be allowed to fall below the level prescribed by law. There may also be industry-wide standards which can be used.

Some standards for items such as ventilation, access and egress and fire safety can be found in building codes.

The dangerous goods codes based on the UN Recommendations set many of the standards relating to dangerous goods transport by road and rail.

The regulations for hazardous substances labelling in workplaces in many countries tend to be based on either European Union or US requirements.

Listing known risk areas in a particular work environment

If an inspection, survey or audit has not been done before, it is useful, in line with the Pareto Principle (80% of the problems come from 20% of places and 80% of the control effort should be directed at that 20%) to tackle known risk areas first.

Information to do this may come from:

- asking employees, health and safety representatives or the health and safety committee
- examination of accident and injury statistics for the workplace, if these detail particular areas
- background information and study about likely problems in particular workplaces (This could include a study of statistics collected by an OHS authority.)
- reports of accident investigations.

Your list may either be based on physical (geographic) areas in the workplace, specific types of jobs (e.g. manual handling) or specific types of machinery (e.g. drill press).

Conducting workplace inspections

Formal and informal inspections

Formal inspections are generally planned and scheduled checks, which occur at regular intervals throughout the workplace. Usually these are conducted on a monthly basis and some type of official documentation is used to communicate the inspection results to the employees.

Informal inspections can be valuable; however, they tend to be less controlled and as a result often cover too many or too few issues. The results of informal inspections are often communicated verbally and much worthwhile information can be overlooked or forgotten. Typically the supervisor or manager will walk around the workplace and point out or discuss issues with workers. The informal inspection has the benefit of creating awareness and is often seen as less threatening than formal inspections.

A willingness to develop a team atmosphere of trust is required. Open-ended questions can often provide better information, e.g. not 'do you do this while setting up the machine?' but 'tell me about setting up the machine'.

Structured and documented inspections

Reasons for workplace inspections

Prior to conducting any type of inspection there needs to be a plan developed that establishes the reason for the inspection process and the method to be used when carrying the inspection out. Most workplaces are complex environments that do not lend themselves readily to the inspection process. The workplace is generally in a constant state of change. It is dynamic. This is particularly true of mines and construction sites. Consequently, workplace inspections provide only a general overview of the working conditions and are used primarily to highlight areas where special attention may be necessary. Inspections are conducted to satisfy the following basic requirements:

- identification of hazards
- assessing the potential losses from these hazards
- selecting control measures designed to eliminate or reduce the hazards to an acceptable level
- monitoring the effectiveness of control measures
- reviewing compliance with established standards.

Reasons for formal inspections

The difference between formal and informal inspections was described in 'Conducting workplace inspections' above. While informal inspections are intended to occur in any health and safety management system, it is the formal inspections which will actually be specified.

They may be one of the tools used in the overall audit of health and safety. A standard form of reporting will be used which has been developed beforehand. Formal inspections are intended to ensure that structured inspections take place as a normal part of the management system. They also assist in complying with the law.

Some inspection reports must be kept and be available to OHS authority inspectors as a legal requirement.

The workplace inspection team

The workplace inspection may be carried out by a wide range of people. Using the same checklist, a different emphasis could be placed on the same item; for example, ventilation is an item that appears on checklists from almost all workplaces. However, a health and safety representative or mechanical fitter may look at the ventilation system in a vastly different way from an industrial hygienist or a ventilation engineer. The important issue in a workplace inspection is to identify if the ventilation system is working properly and receives regular maintenance. It is not a requirement to be a specialist in the area of ventilation to determine if the system is functioning properly in terms of providing worker comfort. This principle of identification will apply to the majority of inspection items. On occasions, it will be necessary to use professional specialists to carry out inspections and prepare reports. This is the exception rather than the rule. All efforts should be made to develop the skills of the organization's own people, enabling them to conduct inspections and make reports. In considering who should conduct inspections, thought needs to be given to the benefit of cross inspections. This is where people swap around inspection duties between different areas of the workplace to provide 'a new set of eyes' for the purpose of identifying possible hazards and accident causation factors that could be overlooked or underrated through familiarity or acceptance.

Most large organizations have some form of inspection procedure. Self-regulation legislation provides the right for health and safety representatives to inspect the workplace, providing certain conditions are complied with. Further, in some workplaces, inspections could be conducted by members of the health and safety committee. Certainly, the role of supervisors and managers includes inspection activities, and in some organizations there will be safety staff or first aid people who also conduct inspections of one form or another. Clearly, there is a need to coordinate the activities of all these employees for the purpose of maximizing the benefit of workplace inspections and to minimize duplication. An agreement on how, where and when inspections will be carried out seems to be the best way of establishing a credible procedure. The agreement is worked out among all those involved and requires the support of both employees and management. Results of workplace inspections need to be readily available to all employees. Inspections provide an excellent reference point for safety meeting discussions and toolbox meetings.

Types of inspections – planned and unplanned

Inspections may be considered under different classifications and each has a positive benefit provided a proper inspection plan has been developed and objectives defined prior to commencement.

Formal and informal inspections have already been explained in 'Conducting workplace inspections' above.

Planned and unplanned inspections

This classification refers specifically to scheduling of the inspection. For example, should the inspection take place on the last Friday in the month at 10.00 am (planned) or, alternatively, be carried out without prior notice (unplanned)? Again, both planned and unplanned inspections have the potential to be beneficial. Planned inspections are most common because they are less stressful – workers can prepare. Unplanned does not mean it is unplanned by management.

Two main points of view exist with planned inspections. The first is that knowing an inspection is to be conducted at a predetermined time and knowing what is going to be inspected provides the workers and supervisors with the opportunity to 'make things right' before any checks are carried out. The main reason is often to avoid criticism of the work group by other employees. The concern is that, in carrying out the inspection, an unrealistic appraisal of that particular aspect of the workplace could be gained if the conditions were different from those that existed during normal operating procedures.

Experience has shown that there have been occasions when hazards and potential accident causation factors have gone undetected because the workplace was not inspected under normal working conditions. There should not be an underestimation of the possible benefits that can be realized from having workplace inspections conducted by people who do not regularly work in the area. People carrying out inspections who are unfamiliar with a particular workplace or work system are forced to ask questions of the locals. The resulting exchange of information can lead to the identification of opportunities to improve safety levels overlooked by the locals because of either familiarity or complacency with their own work environment.

The second main point of view is that, even if the planned inspection gives a false impression of the normal workplace conditions, at least once a month some action is taken to correct things. The view has a good deal of merit. However, if an assessment report indicates that conditions are good when workers and supervisors know otherwise, then distrust arises. The whole effort put into the inspection process can be discredited as a result of what the workers or supervisors see as the inability of the workplace inspection methods to identify, with reasonable accuracy, true measures of performance and risk.

Although unplanned inspections are often seen as disruptive, this need not be correct. It is, however, necessary to determine whether the advantages gained by unplanned inspections outweigh the disadvantages caused by any possible disruptions. The benefits of unplanned inspections are gained from the opportunity for the inspection team to check the workplace under normal operating conditions, or as close to normal as possible. The inspection results are likely to be far more reliable and valid. It follows that decisions made as a result of observations made under normal work conditions will find greater support among the workforce.

General and critical inspections

General inspections relate to checks on matters such as housekeeping, chemical use, handling techniques, use of personal protective equipment, machine guarding and noise exposure. General inspections tend to cast a wide net and require careful attention in defining the scope of the inspection for the results of the inspection to be meaningful. For example, knowing that personal protective equipment is not being used to the standard required is too broad an observation. To develop any practical solutions for improvement it is necessary to identify personal protective equipment in more specific terms; that is, eye protection, hearing protection, gloves, hard hats, etc.

Critical inspections are more specific and often of a more technical nature than general inspections. A critical inspection would be used to check cranes, lifting devices, electrical systems, noise levels from a particular source(s), explosives handling, chemical storage, lighting levels around walkways and a wide range of other factors that have a direct influence on the employee's safety. Most workplace inspections will contain elements of both general and critical inspections.

When is the best time to conduct inspections?

The timing of workplace inspections is most important and, at best, the only picture you are likely to get from the effort is a snapshot of the conditions as they existed at the time of the check. One hour earlier or later a different rating could result. Consequently, carrying out inspections after the area has been cleaned up, limits the opportunities to identify risks and reduces the effectiveness of the exercise.

The use of spot checks (unplanned inspections) is one method of overcoming the bias that appears in the analysis of some reports on workplace inspections. However, the use of this technique creates its own administrative problems and the stress, created by the uncertainty of when checks will be carried out, appears to have little effect on motivating people to change their behaviour. The best approach is to gain acceptance of the workplace inspection process through the promotion of positive outcomes. People at all levels of the organization are concerned about criticism. If inspection procedures are going to be effective, they must be seen as beneficial and be accepted as part of the normal operational system. If this can be achieved, the timing of inspections becomes a simple matter of scheduling.

Documentation for a safety survey

A safety survey is also a form of workplace inspection. However, the safety survey is used to focus attention on a specific activity or condition. The application of the safety survey identifies the level of a specific hazard in a specific environment.

Examples of workplace issues which require safety survey techniques include:

- levels of heat stress experienced by workers employed around furnaces
- repeated occurrences of guards being removed and not replaced on running machinery
- the exposure levels of workers to asbestos fibre during the removal of asbestos roofing
- the adequacy of *Out of Service* procedures and why they are not always used on night shift
- eye protection not being used by electrical fitters outside their workshop
- identification of manual handling techniques in the stores section that could cause injury
- employees' attitudes on safety programmes
- the effectiveness of induction training on first year apprentices.

Safety surveys are usually carried out in response to some negative outcome such as:

- failure to reach a standard
- an increase in accidents or injuries
- employee unrest.

These should not be the only reasons for surveying. Consideration could be given to conducting a safety survey in an area of the workplace that has a good level of performance, the purpose being to determine how success is being achieved and working out how techniques and principles adopted in the good areas can be transmitted to areas where an improvement in performance is required.

Safety surveys are widely used to establish reasons for other areas of concern with far less technical requirements. The safety survey may also follow from some type of preliminary inspection and the nature of the survey may be such that technical skills and equipment are needed to determine the risk level, e.g. for airborne vapour.

Format for inspections and reporting

Development of a recording and reporting system

Inspection documentation

The checklists used in workplace inspections and the resultant reports and supporting materials are important records that provide valuable information to assist in decision making.

The documentation should be seen as an organizational asset. As there is always considerable effort and expense associated with workplace inspections, not to fully utilize the information is wasteful. The reports generated through workplace inspections could identify opportunities beyond the identification of hazards and accident causation factors. Inspections can also identify production improvements, better maintenance techniques, means of reducing wastage, and increased efficiency, all of which can contribute to an improved, more productive and safer work environment.

Use of checklists

The checklist is a widely used instrument and is ideal as a starting point in workplace inspections. The important criterion when using checklists is that they must be flexible enough to allow items that are not listed on the check sheet to be considered. Checklists are only a guide to assist in the process of inspections; they must not set the agenda, but rather be part of it. The checklist is beneficial in providing a means of measuring performance against a predetermined standard. In order to achieve some reliability, there must be consistency. Consistency is gained with practice by conducting workplace inspections and comparing results with others or against previous inspection checklists. The results attained through workplace inspections are often more beneficial than what is reflected simply by the information collected on checklists.

Reporting details

The manner in which inspection results are reported will depend primarily on the reason for the inspection and the person conducting the inspection. The organization's safety officer, the supervisor, the fitter and the health and safety representative could all use the same workplace inspection checklist and produce different reports. This is a normal and expected outcome as the reasons for conducting the inspection may be different in each case. The important point is that the performance criteria are correctly applied. There are two main methods of reporting workplace inspection results:

1. Evaluating the outcome of the findings and listing recommendations as to how improvements could best be achieved.
2. Tabulating inspection results and presenting the conclusions and any comments made by the employees involved – for example, the workgroups involved or safety committees – for the purpose of enabling others to determine appropriate actions based on findings.

Producing the results of workplace inspections

The production of a report or summary of the results of the workplace inspection is often a delicate undertaking. At this stage in the process, the need to have an agreed set of standards becomes obvious. The essence of an inspection outcome is to have those involved focus their attention on the action required to meet the targets established by the agreed standards, not to become defensive or begin to argue over what standard should apply.

In order to avoid any additional disagreement about the inspection results, it is imperative that the person conducting the inspection provides a report that adheres strictly to the agreed standards. It may well prove that the standards set are inadequate or too rigid for certain work groups. The report needs to reflect this so future inspections can be conducted using modified standards. This is, after all, the principle behind the concept of self-regulation. However, it is important not to create a standard of convenience. Particular attention must always be given to maintaining minimum standards consistent with any existing legislation, industry standard and the ever-present common law onus on employers to provide the appropriate duty of care towards their employees.

In selecting items for inclusion in a checklist, it is beneficial to group the inspection items under convenient headings. In addition to making the inspection process easier to organize, the groupings assist in making the results easier to summarize when reporting the inspection findings. The following are examples of how inspection items could be grouped. However, each workplace is different and groupings should be selected that are suitable for particular locations.

Group 1 – Guarding and protection devices
a. missing or inadequate guards against being 'struck by'
b. missing or inadequate guards against 'striking against'
c. missing or inadequate guards against being 'caught on, in, or between'
d. missing or inadequate guards against 'falling from or onto'
e. lack of, or faulty support, bracing or shoring
f. missing or faulty warning or signal device
g. missing or faulty automatic control device
h. missing or faulty safety device.

Group 2 – Structural defects and hazardous conditions
a. sharp edges, jagged, splintery, etc., conditions
b. worn, frayed, cracked, broken, etc., conditions

c. slippery conditions (for gripping or walking)
d. dull, irregular, mutilated, etc., conditions
e. uneven, rough, pocked, or with holes
f. decomposed or contaminated conditions
g. flammable or explosive characteristics
h. poisonous characteristics (by swallowing, breathing, or contacting)
i. corroded or eroded conditions.

Group 3 – Functional defects
a. susceptibility to breakage, collapse, etc.
b. susceptibility to tipping, falling, etc.
c. susceptibility to rolling, sliding, slipping, etc.
d. leakage of gases, fumes or fluids
e. excessive heat, noise, vibration, fumes, sparking, etc.
f. failure to operate
g. erratic, unpredictable performance
h. lack of adequate electrical grounding
i. operation that is too fast or too slow
j. low voltage leaks
k. signs of excessively high or low pressure
l. throwing off of parts, particles, materials, etc.
m. indication of need for special attention.

Group 4 – Work environment
a. noxious fumes or gases
b. flammable or explosive fumes or gases
c. insufficient illumination
d. excessive glare from light source
e. hazardous dusts or atmospheric particles
f. hazardous or uncomfortable temperature conditions
g. excessive noise.

Group 5 – Material storage and personnel exposure
a. improperly secured against sudden movement such as falling, slipping, rolling, tipping, sliding, etc.
b. unsafe storage that permits easy contact by persons or equipment
c. unsafe exposure to heat, moisture, vibration, flame, sparks, chemical action, electric current, etc.
d. congestion of traffic or working space
e. unsafe attachment of object to agent
f. unsafe placement of objects in agent
g. unsafe distribution of objects around agent
h. protruding objects
i. use of unsafe storage containers
j. faulty ventilation of stored materials

k. unsafe traffic layout
l. poor housekeeping.

Group 6 – Practice and procedures
a. correct use of plant, tools and equipment
b. safety rules and safe operating procedures being followed
c. correct use and maintenance of personal protective equipment
d. relevant information available on hazardous materials
e. employees aware of the need to report hazards
f. all workplace hazards clearly defined in written procedures
g. employees provided with proper training for the job.

Note that Group 6 is positively phrased. It is useful to use either all positive or all negative phrasing. The items contained in these groups are a 'guide' to setting out a safety inspection format. There are special needs in every workplace, as there are special risks, such as the handling of cyanide in some process activities, the need to work in confined spaces, working at heights or operating mobile plant. All of these have their own associated hazards and the checklist design should reflect the individual requirements associated with the different workplaces and activities. It is difficult to import into a workplace a checklist designed for another environment without modification.

Key elements in conducting an inspection

How should an inspection be conducted?
The inspection process requires the support of all sectors of the workforce in order to be effective. The best way to achieve support is to clearly communicate to the workforce what is to be gained out of an open and honest appraisal of the workplace and the work systems. Communication, by definition, implies a two-way exchange of ideas and information. Workers and supervisors, particularly, have shown a dislike for open criticism. The workforce must be provided with an opportunity to adjust to the likelihood of having their workplace, their job and even their behaviour monitored and recorded, albeit by fellow workers. Prior to the inspection taking place, people will want to know what is to be achieved by their participation and what the plans are to deal with the results of the inspection. There is a need to think out a strategy which includes the arrangements for any action plans to follow the inspection. There is nothing more threatening to the success of the inspection than to be asked by somebody 'what do they do with the information you are collecting?', and having to reply, 'I don't know'. There is a need to develop credibility in the inspection process. The more planning that occurs, the more confidence the workforce are going to have in the process and the greater should be their commitment.

Encouraging workforce involvement
When inspections are being carried out, every opportunity should be taken to involve as many people as possible in the process. The input of those involved directly in the

work activity is crucial to the development of any necessary corrective action which may be recommended. Gaining the support of workers and supervisors makes the change process much simpler and more acceptable than trying to force new work methods or ideas onto people.

For example, if a person is observed not to be wearing required eye protection when performing a task where some form of protective equipment is needed (for example, operating the grinding machine), how is the issue best dealt with? Rather than making some comment on an inspection sheet such as – 'eye protection not being worn in the machine shop', consider asking the employee why the required protection is not being used. The person may have simply forgotten to use the eye protection or may find it uncomfortable to wear. However, the reason for non compliance with this standard industrial practice could be more complex. Perhaps the worker has been provided with eye protection that 'fogs up' and interferes with visual ability. Whatever the reason, asking 'why is the eye protection not being used?' provides some important information and allows for the proper consideration necessary to find workable solutions to such issues.

When conducting the inspection, the results should always be discussed with the people who work in an area under review. The input from those involved directly at the workface is crucial to gaining acceptance of any comments made as a result of the observations.

Steps in conducting a workplace inspection

The key steps are:

- There must be agreement among all those involved as to what is being observed. Everybody participating in the workplace inspection process has to be aware of the inspection objectives and the performance rating criteria.
- Determine the time available to conduct the workplace inspection. The time available to carry out the actual physical inspection is not open-ended so by defining an acceptable time-frame, such as four hours each week, it becomes easier to set priorities and draw up a checklist.
- Design a format that is clear and easy to use. Select a suitable scale and define each reference point to avoid confusion and to clarify performance standards.
- Conduct the inspection openly and assess items honestly.
- Evaluate the information gathered from the inspection and report the results clearly and accurately.

Preparing a measuring scale for results

Information for the checklist is usually gathered by means of a rating system. There are many rating systems used in conjunction with checklists and inspections, for example:

- POOR – FAIR – GOOD – VERY GOOD – EXCELLENT
- 1 . 2 . 3 . 4 . 5 . 6 . 7 . 8 . 9 . 10

- STANDARD NOT MET – MEETS STANDARD – STANDARD EXCEEDED
- 0% 10% 20% 30% 40% 50% 60% 70% 80% 90% 100%

Some checklists use only two categories such as 'YES/NO'.
An example would be:

- Is the housekeeping satisfactory?   YES/NO
- Are people wearing correct eye protection?   YES/NO

This type of rating provides little information, as the options are often too limiting to determine a true account of the issue under review. The two-category system should only be used when there is a clear and specific matter to be considered.

For example, if the inspection included the need to identify particular safety controls on a piece of equipment, it may be appropriate to use a two-category approach. Such a case could be the inspection of a 'bench grinder' where the interest is in identifying specific safety features, for example:

- Is the grinder properly located?   YES/NO
- Is the observation shield in place?   YES/NO
- Is the observation shield clean/unpitted? YES/NO
- Is the workrest gap set correctly?   YES/NO
- Is the correct type of wheel fitted?   YES/NO

The knowledge that can be built up about this particular 'bench grinder' is sufficient to gain an appreciation of the necessary safety requirements through the use of a simple 'YES/NO' checklist. However, if the category checklist provided a 'NO' response to the question 'Is the bench grinder in a safe condition?', little would be known! Whatever the rating system chosen to measure performance, it must reflect as closely as possible the true state of the item under inspection. If a checklist is good, a second person using it will get much the same results as the first person.

Safety parameters

It is difficult to use a checklist if there is no clear definition as to what can be expected at either end of the rating scale. Ideally, each position or point on the scale should be defined.

When using the popular rating scale 'Poor to Excellent', it is necessary to define both ends of the scale for the outcome of any inspection to mean anything worthwhile to people who are affected by the results. Most people feel comfortable with a result that shows up as 'excellent' but are usually not pleased with a result expressed as 'poor'. The displeasure is usually expressed as 'what do you mean by poor?'. This is a valid question to ask by somebody who could be criticized as a result of the workplace inspection. To avoid creating unnecessary ill feeling through a lack of understanding, the points on the rating scale need to be defined. See Fig. 5.1.

The other popular scale uses a rating of 1–5, 1–8 or 1–10, for example. Again, each point needs to be defined in order to avoid confusion when results are known. The other benefit from defining the scale is that people are immediately aware of what needs to be done in order to improve performance. See Fig. 5.2.

Enhancing occupational safety and health

Workplace inspection sheets				
	Poor	**Fair**	**Good**	**Excellent**
1. Lighting	Inadequately illuminated work areas and walkways.	General lighting minimal, some areas require upgrading and maintenance.	Lighting in work areas and walkways satisfactory, some maintenance required.	Lighting standards above minimal requirements, regular maintenance carried out.
Comments:				
2. Ventilation	Ventilation poor, presence of dust, fumes, vapours, etc.	Ventilation system ineffective, irregular monitoring checks carried out.	Ventilation system effective, regular monitoring checks carried out.	System effective, properly maintained and monitored regularly.
Comments:				

Figure 5.1 Example of ratings for a workplace inspection

QUESTION	POOR		FAIR		GOOD		EXCELLENT		COMMENTS
	1	2	3	4	5	6	7	8	
1. LIGHTING									
2. VENTILATION									
3. NOISE									

Figure 5.2 Inspection results rating scale

Limitations in conducting inspections

The application of checklists and the carrying out of workplace inspections is only a form of identification. Further, it is only one form of identification aimed at uncovering hazards and detecting accident causation factors. The other opportunities for identification include daily contact between worker and supervisor, regular safety meetings, hazard reporting methods and accident investigation.

The workplace inspection is an important part of the identification process; however, it is the action taken by those responsible after the inspection has been completed and assessed that determines the real effectiveness.

Conducting a safety audit and preparing a report

Difference between audits and inspections

General

The safety audit expands the concept of inspections beyond the readily visible aspects of the work environment to include the qualitative elements that are not easily measured; that is, policy development, information flow, and consultation process. It is normal business practice to initiate and conduct checks on critical aspects of the work system. The financial controls of an organization come under close scrutiny through the audit process. Stores and materials are constantly under review through regular stocktaking checks. Equipment is monitored through planned maintenance schedules. These audit checks and schedules are all part of a control system necessary to ensure an efficient and profitable enterprise. Should exerting greater control over potential losses created by accidents and injuries be seen as any different to exerting control over losses created by poor financial or stores management? The safety audit functions in a similar way to a financial audit. Both look at the underlying strengths and weaknesses of the system, at the controls and at the performance indicators. The system is checked and evaluated to identify compliance with standards and to highlight any hidden risks or hazards that could threaten the organization or the employees' well-being.

Safety audit objectives

The objectives of safety auditing centre on the need to evaluate the effectiveness of the organization's health and safety management strategy.

For example:

(i) to assess the operational risks, including the identification of hazards, potential hazards and accident causation factors
(ii) to carry out a critical review of the organization's administrative arrangements, to ensure compliance with any legal requirements and to measure performance against a set of standards.

The safety audit is designed to take account of quantitative information available through data on injuries, incident reports, insurance claims, accident reports, first aid records and any other documents that can provide an indication of performance. In looking at the administrative structure, attention is drawn to the support mechanisms that provide the foundation for hazard management and accident prevention.

These include:

- policy development
- training strategies

- planning schedules
- target and objective setting
- consultation arrangements
- communications and information systems
- issue resolution techniques.

The application of safety audits

Employers and management require a system that identifies and measures the performance of all the organization's activities. It is often the case in organizations that all the responsibilities and controls are not clearly defined. Regrettably, occupational health and safety is such an area. Few examples exist, even among sophisticated companies, of well defined responsibilities. This can be clearly seen in enterprises where the organizational structure has the occupational health and safety function reporting through to the personnel (human resources) manager, while the bulk of the hazards and accidents are occurring in the operations and maintenance areas. Risk communication is difficult at the best of times and even more complex if the lines of communication are unsatisfactory.

There is another order of difficulty gained when you try to force into the line process some external influences such as occupational health and safety initiatives conceived by the human resources department. It is best if the occupational health and safety function as such reports directly to the head of operations. In addition the responsibilities and accountabilities of line management for health and safety must be clearly defined. The application of safety audits is well suited to identifying the strengths and weaknesses in an organization's approach to safety management. The safety audit is used to identify *how* hazard and accident causation factors are being recognized, reported and controlled; the effectiveness of policies and rules; information systems; reporting techniques; and training.

Effective use of safety audits

The safety audit includes the application of workplace inspections but goes much further to not only inquire into the visible components of a particular activity or piece of equipment, but to look at relationships in terms of the whole system, i.e.:

MANAGEMENT – PEOPLE – EQUIPMENT – PROCEDURES – MATERIALS – ENVIRONMENT – COMMUNITY

For example, portable fire extinguishers are to be seen in most workplaces and almost every checklist generated will make some reference to them. Typical checkpoints are likely to include:

	YES	NO
- Correct type of extinguisher for location	√	
- Correctly located on wall, etc.	√	
- Within maintenance period (Service Date ../../..)	√	
- Correct sign in proper position	√	
- Clear area around extinguisher	√	

This type of inspection regime is very important; however, it tells little about the capacity of the organization to cope with a fire. Fire extinguishers are first response appliances and are only of benefit if used quickly and properly by a person familiar with their correct use.

Satisfied that we have the correct equipment in the correct place and in working order (a fact that cannot be proven until the device is activated), there is now a need to know how effective an employee would be in using the equipment to extinguish a fire. Unless there has been an opportunity to assess the skill of the employee under real conditions, any evaluation will only be an estimate of the likely outcome.

Steps in conducting a safety audit

The essential steps in conducting a safety audit are:

- Decide the scope of the audit. Is it to cover all the workplace, a particular part or particular operations? Is it to tackle long-range or short-range issues?
- Compile a set of major headings for the audit. These can include policy, procedures, consultation, management, training, monitoring, for example.
- Construct checklists for each heading. There may be one checklist or there may be several to allow for greater detail.
- Decide reference standards for questions.
- Ensure that the questions in each checklist are clear and precise.
- Consult with stakeholders in the workplace on the audit. These include safety and health representatives and committees.
- Put together the documents or data input forms and computer software to handle the audit. Ensure that these include a rating system, provision to act on findings and bring-up on action taken or not taken.
- Line up people who will carry out the audit.
- Run a pilot audit to iron out problems.
- Run the audit; this may be periodic or continuous.
- Properly record all data collected.
- Initiate action where the audit determines it is required, ensuring that someone is nominated for each action required and that time lines are set as appropriate.
- Ensure that recommended action is followed up to ensure it has occurred.
- Review the audit through the consultation mechanism to ensure that it is adequate to pick up the problems which are leading to accidents and injury, i.e. is valid.

Defining performance indicators

The audit requires clear thinking about the performance indicators which will be set. Some examples in relation to the fire preparedness example are given here, using some of the headings given a little earlier for fire preparedness.

Policy: Do *all* employees who are likely to move in the vicinity of the extinguisher understand the organizational policy on fire control?

Consultation:	Have *all* employees who are expected to use portable fire extinguishers agreed to do so?
Training:	Has training in the use of the equipment been provided to *all* employees – are there regular opportunities to practise?
Information:	Are records kept on *all* fire outbreaks and on *all* occasions when fire extinguishers are used?
Communication:	How aware are *all* employees about evacuation procedures, i.e. what to do when they first see a fire? Who they tell and how?

System reliability

There is a need to stress the point that *any* employee using the area under audit could be in the position of having to take corrective action by being first at the scene of a fire. This could include employees who only use the area at irregular intervals such as a walk-through. For the system to be considered reliable, it is necessary to assess the knowledge of all employees who could be in the area under review at the outbreak of a fire.

Testing the system

The opportunity to test fire protection systems under real situations is, thankfully, a rare event in most organizations. Consequently, risk determination is difficult and the best assessment can only be an approximation of future outcomes. Nevertheless, assessments and approximations are critical factors in setting standards and measuring performance. To position a fire extinguisher on a wall and expect some untrained employee to use the device is unacceptable. To have the extinguisher, maintain it, and not expect it to be used to protect life and property is nonsense. The well conducted safety audit will discover the strengths and weaknesses in the system and provide a blueprint for any action plans necessary for improvement.

Producing a list of assessment criteria

Each of the performance indicators given in the last section, for example, for 'Policy', needs some criteria on which to base an assessment.

The assessment criteria should include an active verb, for example, for policy on fire control, one assessment criterion might be: 'Employees can explain the organizational policy on fire control', or for training, one criterion might be: 'Employees can correctly select and use the fire control equipment.' Whether these criteria are met can be ascertained in appropriate ways, for example, by discussion with the employee or demonstration of equipment selection and use.

Developing a model safety audit for a particular work environment

Developing a full audit for a particular work environment can be quite a large and complex task. This depends obviously on the size and degree of variation in the work environment.

However, a model can be constructed for a suitably small and not too complex work environment and this can serve as an example for any future full-scale exercises. (Refer again to the earlier parts of this section on auditing.) Develop the key headings for the selected workplace, e.g. fire, electrical, machinery, manual handling, vehicles. Develop performance indicators for each of these, e.g. 'training' – are employees properly trained for the class of crane they operate? Then for each of these create suitable assessment criteria, e.g. 'employees can produce a suitable certificate for the class of crane operated'.

This may be enough – there may be no need to go into detail in every case. However, if the 'cheeseborer' approach is used, that is, selecting some areas and going into detail in those, optional additional criteria can be included to allow checks 'on the ground', that is, actually in the workplace, rather than simply interviewing people, e.g. demonstrating correct performance of a particular task using that class of crane.

Conducting a safety audit and producing a report

You may conduct the audit alone or using a team. (See the section 'Structured and documented inspections' earlier.) Just as for inspections, it is important to consult with employees, employee representatives, and health and safety committees on the audit process, design, development, workplace approach, evaluation and reporting. That does not mean they physically conduct the audit.

While discussion with managers and supervisors is an important aspect of auditing, ensure that sufficient 'cheeseboring on the ground' is used to validate claims made by people. Clearly this requires skilled handling if people are not to be put offside. (See the section 'Format for inspections and reporting' earlier.)

With good starting documentation, after results are entered, you can evaluate the audit results. Your report should identify key strengths and weaknesses, and suggest remedial actions and deadlines for these. Depending on who does the audit, it may also identify who is responsible for these actions.

To sum up, audits:

- are a continuous measure
- measure the success of preventive strategies
- may use numerical tools for summing up, in which case different performance indicators and/or headings may be given different weightings
- emphasize key performance indicators such as management, policy, and procedures
- identify courses of action at the reporting stage
- can use action scores to set priorities for follow-up
- utilize checklists as a tool
- must be followed up to see that remedial action has taken place
- can lend themselves to graphical reporting
- assist with clear setting of further goals.

Further reading

Anonymous. (1999). Safety MAP Means New Era in OHS. *The Quality Magazine (Australia)*, **8(4)**, 28–9.
Editorial. (2001). Prestart Health and Safety Review Guideline. *Accident Prevention (Canada)*, July/August p. 5.
Everley, M. (1995). The Price of Cost Benefit. *Health and Safety at Work*, **17(9)**, 13–15.
Glendon, I. (1995). Safety Auditing. *Journal of Occupational Health and Safety – Aust. NZ*, **11(6)**, 569–75.
ILO. (1995). *Safety, Health and Welfare on Construction Sites – A Training Manual*. Geneva: ILO.
Kase, D.W. and Wiese, K.J. (1996). *Safety Auditing: A Management Tool*. New York: Van Nostrand Reinhold.
Kesling, R.L. (1992). Auditors Team up for Safety. *Safety and Health*, **145(5)**, 50–2.
Loud, J.J. (1989). Are Your Safety Inspections a Waste of Time? *Professional Safety*, **34(1)**, 30–2.
Occupational Health and Safety Authority (Victoria). *Safety Map. A Guide to Health and Safety Management*. Melbourne: OHS Authority.
Read, J.A. and Law, G.A. (1992). How to Design an Effective Health and Safety Audit. *The Safety and Health Practitioner*, **10(1)**, 19–22.
Rotheray, M. (1996). Aviation Ground Safety. *Safeguard (New Zealand)*, March/April, pp. 40–1.
Victorian Workcover Authority (VWA) and JAS-ANZ. (1998). *Safety MAP OH&S Auditing Standard*, Melbourne: VWA.
Vogel, C. (1992). Safety Surveys Build a Strong Foundation. *Safety and Health*, **145(1)**, 50–4.
WorkSafe WA. (1994). *Assessment of Occupational Health and Safety Management Systems*. Perth, WorkSafe WA, adapted from Workcover Corp (SA) Safety Achiever Bonus Scheme.

A number of websites provide health and safety checklists for particular applications.

Activities

1. Select a work environment and list the key areas you would inspect. Why have you chosen them?
2. How are standards and benchmarks arrived at in the inspection process? Identify some you would use for the key areas in Question 1.
3. Consider who would be the best people to have on an inspection team. Explain your selection.
4. Ask three workers to comment on their preferred type of inspection, i.e. planned or unplanned, formal or informal. Explain their answers.
5. Describe how work inspections are reported in a workplace.
6. In a chosen workplace, identify the action pathway that deals with inspection reports.
7. Consider two ways the inspection process could be improved in a work environment of the type you chose.

6

Accident prevention

WORKPLACE EXAMPLE

A contractor's employee was working at a food factory in the UK, and was electrocuted. The work being undertaken was the removal of redundant cables from a length of trunking following an upgrade in the electrical systems at the plant. As well as upgrading, the intention was to separate production and lighting circuits. The trunking contained a mixture of cables, many live, others not. Their source, what they supplied (whether production equipment, plant or lighting) and destination of many of the cables was not known.

The cable involved in the incident was a supply for lighting. The light fitting it served had long since been removed. A terminal block had been used to bridge a break in the cable, which was overlong. The cable was uninsulated for 40 mm where it entered the electrical connector and was energized to 253 volts AC to earth. The source of this was a 6 amp minature circuit breaker of a consumer unit located within the same plant room.

The contractor was fined GBP 25 000 for failing to ensure the safety of its employee. The food company was fined GBP 30 000 for failing to ensure the safety of a non-employee who was affected by the company's work. A Health and Safety Executive inspector commented that the food company could have avoided the death if it had ensured that its contractors followed, or had in place, a safe system of work, similar to that which the food company provided for its own employees. Isolation of the newly installed consumer unit would have been a valuable precaution and would in this case have prevented the incident. Isolation of any of the cables that were to remain if their sources could be traced would also have been an appropriate precaution. Live working methods throughout the whole course of the cable removal work would have been appropriate in the event that full isolation could not be achieved.

(Source: the Health and Safety Executive, UK.)

Accident causation factors

Accident causation

An important part of any safety strategy is the provision of an effective accident investigation methodology. For accident investigation to be meaningful, selecting the appropriate accident causation model is essential.

Any accident will be the result of a number of events which have occurred. These events can be divided into essential and contributory factors. These events can be identified as follows:

- essential – the accident would not have occurred without this, e.g. a person slips over because some fluid is on the floor
- contributory – if not present the accident could still occur, but the probability increases if this factor is present, e.g. when the person slipped over on the floor the lighting level in the area was very poor.

Once these events are clearly identified, appropriate intervention can be applied to prevent similar accidents occurring. In the USA a recent Liberty Mutual study gives the ten leading causes of workplace injury. In descending order they are: overexertion, fall on same level, bodily reaction, fall to lower level, struck by object, repetitive motion, highway accident, struck against object, caught in or compressed by equipment, and contact with temperature extremes.

Accident models

There are many models of accident causation by different authors from many disciplines such as engineering, psychology and science. These models relate events to outcomes and explain complex relationships between the employee and his or her work dynamics. A comprehensive list is given in the Appendix to this chapter.

Event tree analysis

This is a systematic approach to accident analysis and is particularly good for trying to determine where hazards exist in technologically complex operations. The space industry would be well suited to this type of analysis as there are many millions of operations which must come together to produce a successful mission launch.

Event tree analysis follows a process from inputs to outputs. Each situation or 'condition' is the result of previous events. They may have to happen together to produce the 'condition' (i.e. both 1 and 2) or either one may produce the 'condition' (i.e. either 1 or 2). As all possible situations are explored, an event tree begins to unfold. Mathematical probabilities can often be assigned to each condition and a quantitative analysis performed. Symbols, boxes and lines join the events and conditions to produce a visual representation of the event tree. Unwanted outcomes can be traced back in a reversal of the analysis above to determine which factors contributed to the unwanted outcomes. This is called 'fault tree analysis'. Faults differ from events in that faults are viewed as being a result of controllable human error. Events can include such faults.

Accident prevention

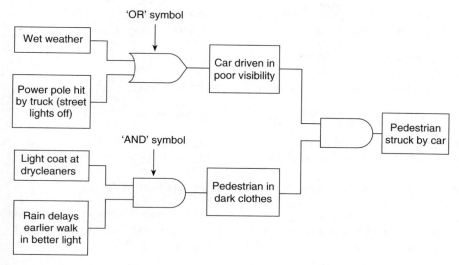

Figure 6.1 Event tree analysis

Figure 6.1 shows a possible event tree for a collision between a car and a pedestrian.

This can be taken further; for example, why was the car driver there? Why was the pedestrian not on a footpath?

Fault trees need not have a time line, whereas an event tree is drawn with a time line from left to right. If probabilities are shown on an event tree, these can derive from the fault tree because that can explain how the probability was arrived at. The causal factors of today's incident are tomorrow's risk factors. Errors can be viewed as consequences of acts or events, not as causes of acts or events. Once the consequence of an event is specified, and the causes of that consequence are identified, the likelihood of that event can be calculated. The analysis can be based on specified outcome, likely outcome or worst outcome.

Generally, human action is for *good* (explicable) reasons, because people are usually not careless and stupid, and their behaviour reflects their perception of risk.

Energy damage model

This model looks at the relationship between the source of the energy and the resultant effect of the energy exchange on the recipient (e.g. person, environment or equipment). The types of energy which cause the damage are numerous. Some of these are:

- kinetic energy
- thermal energy
- mechanical energy
- electrical energy
- chemical or biochemical energy
- acoustic/vibration energy.

If excessive, these types of energy become a problem when they are transferred to the body. This model, therefore, has three components:

- energy source, e.g. vibration
- transfer pathway, e.g. air
- individual's reception of the energy, e.g. damage to hearing.

Control measures are directed at each of the three components. That is:

- prevent, contain or limit the energy source, e.g. dampen the vibration
- remove the mechanism for pathway travel, e.g. use acoustic insulating material in the path of travel
- protect the individual from exposure, e.g. hearing protection devices.

Only two accident models have been examined. The models all represent different conceptual viewpoints. All are useful when considering control measures for accident prevention.

Identifying the factors

If we knew all of the causes of accidents, an assumption could be made that we would prevent accidents by eliminating the causes. This is unlikely to occur for two reasons. Firstly, accident causation is complex and accidents generally are caused by many factors. Rarely could you expect a single causative factor to result in an accident. It is most unlikely that two accidents would have the same causes, although the events may appear similar. There are currently several accident causation factors which are well known. Incorrect workplace layout, fatigue and lack of proper training are a few examples. The tendency is to control, rather than eliminate, most known accident causes. This is done for practical purposes.

Secondly, accident causation, like other complex issues, demands a great deal of understanding of the processes behind the occurrence. To improve our understanding, more information on accidents is required. Unfortunately, there are few reliable sources on accident causation. Most organizations collect information on injuries and seek single causative factors during accident investigation. Even where accidents are thoroughly analysed and reported, some organizations are often reluctant to release detailed information on the findings of accident investigations.

As a result of these and other limitations on our knowledge of accident causation, many of the current beliefs are based on what appears logical rather than what we know to be correct. The real influence alcohol and drugs have as accident causation factors in the workplace is assumed rather than known. It could be argued that all alcohol and drug exposure in the work environment should be eliminated. However, the present and future provision of resources available to target accident causation factors will dictate that we prioritize our expenditure on accident prevention in such a way as to achieve the best overall reduction in accidents for the effort and resources provided.

Accident prevention

The ultimate goal of targeting our efforts and resources at the real causes of accidents will only be achieved when our understanding of causation is improved.

What is an accident?

A simple but worthwhile definition for an accident is 'an unplanned event that may result in personal injury or property damage'.

This definition suggests, correctly, that an accident can occur without injury or damage resulting. Different terms are often used to describe an accident when there are no visible signs of injury or damage. Terms such as 'near miss', 'dangerous occurrence', and 'critical incident' are in common use; however, they are still accidents.

Accident causes

Traditional wisdom suggested that all accidents were preceded by:

- an unsafe act – actions or behaviour of people, usually the injured worker, and/or
- an unsafe condition – some environmental condition or hazard which caused the accident independent of the people.

It was considered by early pioneers in the accident prevention area that 90–95% of all accidents resulted from unsafe acts. This has proven to be a simplistic view of causation and, most importantly, has done little to advance accident prevention.

Predictably, this approach generated a ready supply of perceived causes that focused on the individual. These causes are still in wide use today. They are not causes but, rather, excuses; however, they are so widely accepted by both employers and employees that changing people's attitude to their use will prove difficult.

A quick look at supposed causes in accident investigation reports will reveal, amongst others:

- carelessness
- accident proneness
- inattention
- negligence
- not wearing safety equipment
- bad luck
- act of God
- not thinking
- laziness
- taking short cuts.

The use of these excuses seems to satisfy the management of many organizations. To accept an approach that does not go beyond finding fault in the individual indicates an acceptance that accidents are not preventable. This conclusion can be drawn from the fact that, in reality, people have limitations which are not necessarily blameworthy faults. Examples include skill level, experience level, age, level of training, level of motivation and general well-being, including the individual's mental state.

Elements of the work system influencing OHS

The broader view

The 'unsafe act–unsafe condition' approach to accident causation underscores what is a relatively complicated process. Accidents occur within systems, and the worker is part of a work system. Obviously if a person went to work, sat in a room and did nothing, that person would have little chance of being involved in an accident. The reality is that the worker is required to be part of a process, operate a machine or be part of a system.

In carrying out the work function, the person has demands placed on his or her physical and psychological make-up. It is in this broader concept of the individual being part of a system where the real causes of accidents are found. However, it must be understood and accepted that the individual imports into a sterile world of machines and work schedules, influences which occur outside the work environment. Family, financial or health problems are often examples of major concerns to individuals.

Given that our capacity to concentrate is limited and issues of a deep personal nature are unlikely to be set aside easily, these personal issues will be competing with factors in the person's workplace for attention. While it is difficult to find a solution to what is a perfectly natural occurrence (that is, bringing our problems to work), we cannot ignore the effect forces outside the workplace have on accident causation when designing work systems.

The work system

As previously discussed, the worker is part of an overall system when employed at the workplace.

The work system, as discussed in Chapter 1, is made up of five broad elements:

- people
- plant and equipment
- work methods, which includes procedures
- materials
- environment.

To these we could add management as a separate issue to people.

When combined, these elements allow for the production process to be carried out. Note that the fifth factor comprises both the natural environment – for example, a remote desert or Arctic minesite – and the work environment. The work environment is more than the physical aspects of the workplace (i.e. the lighting, heat, noise, air quality, etc.). It includes relationships between levels of employees, the communication system, attitudes of people, opportunities for employee participation and many other variables which are often difficult and sometimes impossible to measure accurately.

Given that work is a system rather than a collection of independent variables, there is sound reason to believe that an accident is a breakdown in a system rather than a fault in any variable – for example, the individual.

Let us consider such a system for the purposes of determining which parts of the system can be removed without affecting the overall work performance. Let us also

Accident prevention

Table 6.1 A welding system

• The welder	• Maintenance schedules
• Welding equipment	• Work environment
• Induction training	• Communication system
• Skills training	• Information access
• Supervision	• Personal factors
• Management – policy	• Feedback provisions
• Planning and costing	• Experience – self
• Time management	• Experience – others, etc.
• Work procedures	

consider whether there is a method of prioritizing the various parts of the work system to identify those aspects that may need greater attention in our attempt to determine accident causation.

Consider a system that requires a welder to work on a particular piece of plant during a planned maintenance shutdown. For some reason the welder has an accident and sustains an injury, requiring an investigation to identify causation. What might the system the welder operates in look like? Will understanding that system assist in identifying the cause of the accident? The system this welder works in will have certain elements common to systems that other welders work in. However, the nature of work environments is such that they may vary greatly between organizations and even within organizations.

As a consequence, there can be no standard operating system for a welder to work in. This result has a large bearing on identifying accident causation, as it implies that individual systems must be recognized and analysed to determine accident causation. The factors listed as part of the welding system in Table 6.1 are far from exhaustive.

Understanding the system

The work system is always complex even if the work itself appears to be simple. From the factors listed in Table 6.1, which areas could be discarded because they have no bearing on the accident? Only after a thorough investigation into the specific events that led to this particular accident would we be able to disregard any of these factors as part of accident causation.

It is possible to take any of the accident causation factors and break them down further in order to closer examine the influence that particular factor is having on the work system. This second tier analysis further increases knowledge about the work system through developing a better understanding of the individual processes that make up the system. It is in this area of investigation and analysis where we come closest to discovering true accident causation.

Use of a comprehensive accident model

The Synoptic Accident Model, see Fig. 6.2, looks at the accident process in two ways:

- Vertically through a series of transparent 'screens' where individual elements in a lower screen derive from 'macro' issues in a higher screen, and where the effects of macro issues are seen in a lower screen.

198 Enhancing occupational safety and health

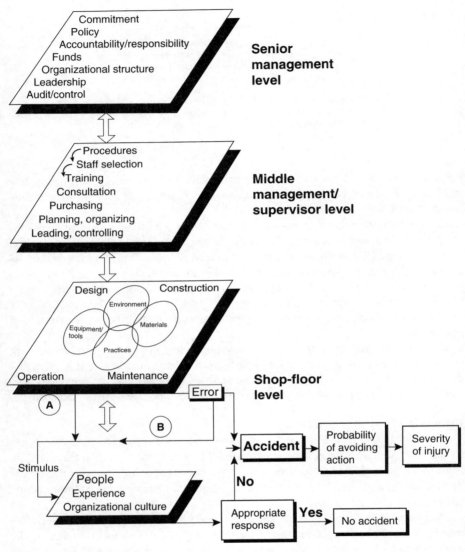

Figure 6.2 The synoptic accident model

- Horizontally at the 'shop-floor' level, where the interaction between people and four other elements of the work system occurs. The sixth element, management, is projected from above, even though participative management may be used.

The term 'shop floor' is used for the lower two levels rather than workplace, to indicate that all management is part of the workplace.

The connecting flows between procedures, staff selection and training are shown.

At the shop-floor level, people are deliberately separated from the other four elements in order to recognize human limitations and how the effects of them can be minimized. Culpability or blame as guiding concepts are deliberately avoided.

Explanation of the 'shop-floor' levels

1. In the upper shop-floor screens, issues which flow from management such as design, construction, operation and maintenance affect the four components of the system shown in the overlapping circles.
2. The third screen deals with design, construction, operation and maintenance of the machine or system. In a complex web of interactions such as this error can occur. However, there is a pathway of interaction between the machine or system and the personnel involved in its operation. Various signals or stimuli, both intentional (Path A – built into the system, e.g. warning light, audible signal, pressure gauge) or unintentional, (Path B – such as noise foretelling machine failure) alert personnel to a problem.
3. The training, selection, experience and organizational culture – key outcomes of good management practices – determine whether the person reacts appropriately to the stimulus or signal. Inattention, carelessness and negligence are not factors which should be given a significant weighting, because they are the outcomes, generally speaking, of poor management practices. Good selection will take account of human limitations such as strength, eyesight, size, hearing and information overload. In critical areas lack of experience, if unavoidable, must be offset by adequate training.
4. If the error meets with an appropriate response by the person, as the model shows, there is no accident or perhaps an accident of reduced severity.
5. If the error is not responded to appropriately, an accident occurs. At this point, personal injury can still be avoided or reduced in severity as the model shows. Appropriate avoiding techniques – for example, putting on a self-rescuer respirator and leaving an area with a chlorine leak – can still prevent injury. The probability of effectively avoiding action will come back to good management practices which ensure adequate emergency training, emergency equipment, and emergency systems and procedures, so it is important to note that upper level issues can also influence parts of the shop-floor accident chain once it starts.

Basic human behavioural aspects of accident and injury occurrence

Prevention or blame

While in a very few cases it can be concluded that the injured person's conduct had in no way contributed to his/her misfortune, in some instances that conduct undoubtedly arose from inadequate training or supervision. In other instances, enthusiasm for productivity, including zealous but ill-considered attempts to correct equipment malfunctions, played a part.

Without participating in the detail of accident investigation, it is very difficult to appreciate the role of human behaviour in accident causation and hence to recommend or implement the most effective responses. Similarly, sole reliance on numbers of accidents/incidents in the organization as a guide to safety performance can be misleading, if the behavioural component is not understood. They do not show whether individual managers/supervisors are effectively dealing with prevention or if further action is required to prevent future accidents and losses.

If an accident occurs, are the organization's means of identifying and correcting potential hazards at fault? Not necessarily. The incident may simply demonstrate the organization's limited experience.

The person–machine interface

The relationship and interaction between a person and the tool, equipment or machine they are using can be very important in either being a factor in accident or injury, or in preventing accident or injury.

Tools can have a poorly designed grip, causing fatigue if held for an extended period. Chain saws have caused the nerve and blood vessel injury known as 'vibration white finger' from the vibration transmitted over a period of time. The procedures which indicate how to use (interact with) the tool can also be important. For instance, starting a circular saw with the teeth already in contact with wood can cause kickback, so training is required in the correct procedure. The material used also plays a part in the safety of the person–tool interaction. A rotating cutter designed for wood may disintegrate if used on steel.

A person needs to be correctly seated or standing to make effective and safe use of the tool or equipment. On a vehicle, sudden movement may throw them off if this is not done. At a VDU work station, the wrong positions or too much time on the equipment may lead to occupational overuse syndrome affecting the tendons in the wrist.

Tools, equipment or machines must provide information in a way which is suited to the operator. Hard-to-read dials or ambiguous indicators are not suitable.

On some equipment other forms of feedback to the operator, or alternatively automatic cutouts, are important if the equipment is not to be overloaded.

As you can see, study of the person–machine interaction has many parts and can be a very important aspect of operator health and safety.

Principles behind the concept of non-culpable error

In looking at the factors contributing to accidents, a lot of attention has been paid not only to machine error but to human error, because ultimately most machine error can be traced back to a human decision, be it design, wrong use or lack of maintenance or repair, for example.

There are some accidents which occur through the deliberate intent of a person, for example, shooting a worker at a liquor store during a hold-up. However, most accidents occur through human error, not through conscious intent. This is the reason for moving away from culpability or blame and focusing on the type of error the person made and why they made it. James Reason in his book *Human Error* talks of error types and error forms. Error types include mistakes in planning an activity, lapses in storage of information, and slips in execution of an action. Error forms are types of fallibility which occur again and again in all kinds of activity involving thinking (cognition). To identify why accidents occurred, or to prevent them occurring we look at ways to reduce error. This can include these questions:

- Are the directions from a supervisor clear?
- Has the person been given the necessary authority to do the job?
- Are the materials supplied, e.g. building plans, flight plans, computer software, free of error?
- Does the person know the necessary procedure and have they been trained in the skills needed, i.e. to apply the procedure properly?
- Can the person transfer the learned skills to new situations?
- Can the person change the pattern of skill application properly if the unexpected occurs?
- Is the environment appropriate for the activity, e.g. correct lighting for fine work, good working temperature?
- Is the person subjected to information overload, so that ability to absorb, process and deal with it drops off, or underload, so that inattention creeps in?

Searching for the situational factors in a task which can lead to error, and addressing these to reduce accidents, is the principle behind the idea of non-culpable error. Where the original error occurred high up in the management tree and led to an accident lower down, the principle avoids making scapegoats of those who had no control over the original error.

Accident investigation

When is an investigation required?

The purpose of an accident investigation is to:

- Identify accident causes so that similar accidents can be prevented by mechanical improvements, better supervision or employee training.
- Determine the 'change' or 'factors' which produced an error that in turn resulted in the accident.
- Publicize the particular causes among employees and their supervisors, and to direct attention to accident prevention in general.

Basically, whatever type of investigation is undertaken, it must try to answer the following questions:

- Who was involved?
- What happened and what were the contributing factors?
- When did the accident occur?
- Where did the accident occur?

Finally, and most importantly:

- How can a similar accident be prevented from happening again?

An accident investigation should always be conducted as soon as possible after the event. It will be much easier if the investigator finds the situation at the scene of the accident exactly as it was when the accident took place. Consequently, after an accident, the site should be left undisturbed, unless changes have to be made to ensure the safety of persons or to prevent further damage.

Whether the site has been disturbed or not, it is desirable to try to reconstruct the sequence of events before, and during, the accident, possibly with the assistance of the injured person and with the cooperation of witnesses. It is extremely important that all accidents – serious, minor, or near miss – are investigated.

The difference between a 'fatal accident' and a 'near miss accident' may only be a fraction of a second in time or a fraction of a centimetre in space. The information gathered from the investigation of near miss accidents assists us in introducing remedial measures to reduce potential hazards in the workplace. If we gather information about all accident and incident causes, we are able to take positive preventative action.

Focusing on faults in the work system

Accidents are unwanted side effects of the production system. Therefore, a good accident investigation aims to establish the sequence of events which should have taken place. The discrepancies revealed by this comparison are areas where changes are needed to prevent a recurrence. The production system must then be monitored to see that corrective action, recommended on the basis of the investigation's findings, is being taken.

Causes not blame

The purpose of conducting an accident investigation is to establish causes. If attempts are made to apportion *blame*, people who might otherwise provide useful information (hence guidance on remedial action needed) will simply become defensive. The result could be:

- witnesses not revealing all of the circumstances and events surrounding the accident
- deliberate obstruction, or provision of false information
- removal of relevant information, documents or evidence.

The investigator(s) must remain impartial and objective if all of the causes are to be established.

Accident reporting and investigations are part of the management control loop because they tell us when something has gone wrong; that is, when there has been a deviation from the standard and work is not being done safely.

Systems of accident classification

Systems of accident classification are found in various national standards. They may include a numerical coding system which gives a code to each choice under the key factors relating to the accident. For example, the set of key factors from one country are:

- nature of injury
- nature of disease
- bodily location of injury
- mechanism of injury/disease
- breakdown agency and agency of injury/disease.

Steps in preparing and conducting an accident investigation

Accident investigation procedures need to be systematic. Good investigations:

- yield information needed to:
  - determine injury rates
  - identify trends and problem areas
  - permit comparisons
  - satisfy legal requirements (such as data required for personal injury involving workers' compensation payments)
- identify the basic causes that contributed directly, or indirectly, to each accident
- identify deficiencies in the production and management system that permitted the accident to occur
- suggest specific corrective action alternatives for the management system.

Under the self-regulation legislation, provision is made for health and safety representatives to be advised of such events and for the health and safety representatives to carry out 'any appropriate investigation'. While it is natural and necessary to show sympathy for the victim(s), future injuries will only be prevented if all the faults in the system are identified. Investigators should avoid preoccupation with injury outcomes, severity of injury or property damage.

Conducting an investigation

The leader of the accident investigation team should be chosen as soon as possible after the accident occurs. Management must designate or approve the leader and other members of the team. If a written investigation procedure has been established, this can be carried out automatically. Management is not required to approve the participation of the health and safety representative. The health and safety representative has a right to be involved with the investigation. The leader should have the authority to get the job done, and the experience to do it.

The leader's duties include:

- compilation of the report that will ultimately go to management
- convening and presiding over meetings

- controlling the scope of team activities by identifying the line of investigation to be pursued
- assigning tasks and establishing a schedule
- ensuring that no potentially useful data source is overlooked
- keeping interested parties advised of the investigation's progress
- overseeing the preparation of the final report
- arranging liaison with representatives, government agencies and news media.

Who should participate in an accident investigation?

The size and make-up of the investigation team is usually dictated by the accident's seriousness or complexity. The health and safety representative or committee members, where these exist, with the help of the employees involved, usually investigate minor injury or property damage.

The team for a major investigation involving serious injury, a fatality, or extensive property damage might include the employee(s) directly involved and their health and safety representative, the supervisor, safety personnel, engineers, designers, technical specialists, and employees familiar with the process or operation. The team must also include an appropriate member of the upper management team. It must be appreciated that government authorities will also investigate serious incidents, but with a view not just to establish contributing factors but often to assign culpability.

The important issue in setting up the investigation process is to minimize the stress on workers involved in cause identification. While certain amounts of stress are often unavoidable, stress added by multiple investigations of the same event must be avoided. To achieve a consensus approach to investigative procedures, there needs to be a planned agreed procedure on how accidents are to be investigated prior to their occurrence. Exceptions to the planned approach may result where outside authorities such as inspectors, police, etc., will be required to carry out independent inquiries.

Techniques for questioning witnesses

Types of witnesses
There are two main types of witnesses:

- eye witnesses – persons who actually saw the accident happen
- circumstantial witnesses – those who did not actually see the accident, but who can contribute valuable background information.

It is necessary to identify witnesses, including those who saw the events leading to the accident, those who saw the accident happen and those who came upon the scene immediately following the accident. Any others who may have useful information should not be overlooked. Witnesses and others should be interviewed as soon as possible to minimize the possibility that they will subconsciously adjust their stories to fit the interviewer's concept of what occurred, or to protect someone involved. Witnesses

Accident prevention

should be interviewed individually so that the comments of one do not influence the others.

Tactful, skilled investigators usually get cooperation from employees by eliminating any apprehension that they may have about incriminating themselves or others. Witnesses must be convinced that the investigators want to find all the factors that contributed to the accident and not to allocate blame.

Interview techniques

There are certain proven techniques for a successful interview. The following elements form the basic approach to 'investigation interviewing':

- Conduct the interview in private at the workplace.
- Put the interviewee at ease, don't hurry things.
- Ask for the interviewee's version of what happened.
- Only ask necessary questions.
- Repeat the interviewee's story as you (the interviewer) understand it.
- Close the interview on a positive note. Thank the witness.

Conducting the interview at the scene of the accident

The main advantage in interviewing at the scene is that it usually assists the memory of the person being interviewed if he/she can refer or point to physical conditions at the scene of the accident.

Putting the interviewee at ease

Remember that the accident victim or witness may be emotional. The best way to put the interviewee at ease is to remind him or her of the purposes of the interview, which are:

- You are solely interested in prevention, not affixing blame.
- You can only achieve prevention with his or her help in identifying all the factors.
- You are interested in fact not theory.

Be polite and reassuring, not aggressive. Ensure the interviewee understands that all statements are used to prevent future accidents.

The interviewee's version of the accident

Be sure the person being interviewed understands that it is his/her story of what happened which is wanted. Be careful not to interrupt or ask leading questions such as 'Surely you must have seen/heard this or that?'. If things are not too clear, make a note and wait until the end of the story. Above all, do not make judgements. This may defeat the whole purpose of the interview by putting the interviewee on the defensive.

Only ask necessary questions

Points to note include:

- The key word here is 'necessary'. Limit the questions to facts – for example, 'Where did the accident happen? What actually happened?'.

- Ask open-ended questions (ones that cannot be answered yes or no). They will prompt much more information.
- Make sure the questions are put in an objective and constructive manner.
- Wait until all factual information has been given before putting *why* questions to the interviewee. *Why* questions may prompt subjective answers. Remember that no one wants to incriminate himself/herself, or to point the finger of blame at a fellow worker.

Repeat the story as you understand it

Repetition is particularly important if a written statement has to be taken. It will ensure:

- correct understanding between the interviewer and the interviewee
- an opportunity for the interviewee to appreciate fully what has been said and, if necessary, provide a chance to make corrections.

Close the interview positively

Before closing the interview, check that everything has been covered by referring to the lists of questions above, or to any accident report pro forma.

The best way to close the interview is by reaffirming the purpose of the interview. This will promote further cooperation should a follow-up interview be required.

Listing relevant accident causation factors

For the accident investigation to be successful in identifying all of the causes, it will be necessary to establish:

a) Events leading up to the accident:
- the system of work being carried out
- the instructions given for the work
- variation from instruction or safe work system
- workplace conditions such as lighting, floor surfaces, stair treads and handrails, warning signs, temperature, weather if the incident occurred outside, etc.
- the exact location of the incident (with sufficient detail for the spot to be readily identified by somebody else reading the report)
- the materials in use or being handled
- the type of transport or equipment in use.

b) Facts of the accident itself:
- the state of the systems and the actions that occurred at the moment
- the persons directly involved, and those involved at a distance, if any
- the tools, equipment, materials and fixtures directly concerned
- the time
- the nature of any injury.

Persons who have knowledge of the work or conditions at the scene, whether or not they were there at the actual event, can also contribute to establishing accident causation.

Writing an accident report

Depending on the seriousness or severity of the accident or injury, an accident report may simply consist of a filled-in company accident/injury report form or it may be a full written report. Proformas may be available in national standards. A full written report would cover:

- who was on the team
- how the investigation was done
- what evidence was looked at
- who was interviewed and a summary of what they said
- evaluation of the information found
- conclusions about it, including how the team think the accident occurred and the contributing factors to it
- recommendations to prevent a recurrence.

It may be helpful to refer to an accident model which is appropriate to the type of accident to help in:

- analysing the accident
- not omitting important issues.

Summary

An effective accident investigation requires strong employer/employee commitment and involvement. Management must support the investigation process and act on the results. It must make sure that the investigators are capable and have sufficient resources for an adequate investigation. It is management's responsibility to evaluate the outcome of all accident investigations in consultation with the workforce; in particular, health and safety representatives and committees, should they exist.

Legal obligations in reporting of accidents

Most jurisdictions have requirements to report injuries above a certain level of severity, or diseases of certain types, to the occupational health and safety and/or workers' compensation authority. The types of injuries to be reported should be checked in the legislation relevant to you but may include:

- fractures
- injuries leading to more than 10 days off work
- loss of an eye, limb or part of a limb.

It will obviously include fatalities. Some jurisdictions also require reporting of serious incidents, near misses or dangerous occurrences even if there is no injury. Heavy fines generally apply to non-reporting. Reporting is also important in relation to workers' compensation payments.

A new way of viewing accidents

Introduction

The previous two sections dealt with accident causation factors and accident investigation. Here you are presented with a new way of looking at these issues.

Our ability to prevent accidents relies on our capacity to effectively manage error. The time has arrived for a return to basics. Some form of standardization is vital in the common terminologies, methods and structure used to record and analyse those events which cause accidents and injury to people.

At any given time the population is exposed to the likelihood of accidents that could result in injury or even death. Fortunately, the greater majority of accidents result in no detectable damage (NDD). While mortality and morbidity rates arising out of accidents affect our lives and create high financial burdens for people, families, employers and the community, we still know little, however, about the *causes* of accidents.

Central to the remedial actions required to prevent the repetition of accidents is the need to examine the past, model the future and above all manage the present. To achieve this, we must have the capacity to investigate the systems that have failed and try to establish causative factors. Currently, accident data is often lost or at best clouded by the variation in the way accidents are defined, reported and recorded. Essentially, most accident data is based on injury reports. Few, if any, organizations collect all accident data.

The use of standardized terminology is required across all industries and professions so that people working in prevention will have better access to relevant and meaningful data on causation factors. Effective risk communication expressed through clearly defined terms and simplified models is needed to assist in meaningful event analysis. The establishment of safe systems of work is not possible until our safety initiatives come to be based on valid and reliable data.

Currently, people responsible for the provision and development of safe systems of work are largely dependent on trial and error approaches to prevention. The strategies employed to prevent errors, accidents, hazards and injury at work generally focus on aspects of engineering, training, vigilance, self-awareness, behaviour, promotion, the use of personal protective equipment, reward and penalty. The application of these strategies, usually in combination, is commonly referred to as safety management, and the application of particular prevention initiatives is often *ad hoc*.

Some of these reasons include the following: the large amounts of energy available, or stored, in contemporary work systems; lack of control in many engineering systems; questionable integrity in computer decision making; the limitations of the individual's capacity to self manage; and growth in contracting out high risk activities. These variables are occurring in an environment where there are higher expectations of operational staff in regard to perceived non-core activities. Examples include: workplace reforms in social awareness; the increased influence of human resource management; occupational health and safety laws based around the principles of self-regulation; and issues linked to the environment and quality management.

While these are all factors worthy of consideration in their impact on workplace safety, here we consider the role of the accident.

Defining the problem

Obtaining credible and meaningful industry information on accidents and hazards is proving difficult, and hampering prevention work. Given the importance of the accident investigation process, there is surprising variation on what constitutes an accident. This variation occurs between different sectors of industry, within organizations, among the safety profession, in teaching institutions, in administrative authorities and in the wider community.

Occupational safety legislation and people in industry have traditionally recognized the adverse role of the accident in the workplace and have accepted the need to identify causation for the purpose of avoiding repetition. However, the statutes have generally resisted an attempt to provide a workable definition of an accident or have defined accident in terms of an outcome, i.e. injury, near miss, dangerous occurrence, incident, etc.

Reportable accidents referred to under most legislative frameworks are in fact reportable outcomes. This could be the case as a matter of practicability or as a result of the legislators not having a clear understanding of the OHS discipline.

Regardless, the approach is generally to define accident in terms of the outcome, and there are difficulties in accepting this definition. This has given rise to a plethora of descriptors including near miss, incident, serious accident, critical incident, and serious occurrence, for example. These all focus on the exposure outcome not the accident. It has now become a commonplace to recognize an accident only in the event of some reasonably dramatic outcome. This is creating a lost opportunity to capture the information leading to an accident which results in no detectable damage (NDD), even though the causes giving rise to the accident, that may result in death in the workplace, could be the same as those that result in NDD.

Accidents that result in NDD are accidents where there in fact is no damage or where any resultant immediate injury is so small as to go unidentified. Examples could include skin disorders, hearing loss, back damage, stress, lung disease, visual acuity, cancer, and asthma.

The list of event descriptors which avoid the use of the word accident will continue to expand. Industrial system failures have become spectacular, dramatic and newsworthy events – for example, Flixborough and Piper Alpha in the UK, Bhopal in India, Chernobyl in the Ukraine, Three Mile Island in the USA and the Longford gas plant in Australia. We seem to want to report events through the use of an ever-increasing glossary of terms while avoiding the simple use of the word accident. The need to create worker and public awareness of danger, which in some cases could cause catastrophic damage, is essential. However, the growing list of event descriptors is not contributing to the identification of causation factors. In fact the use of these descriptors is hampering the task of the epidemiologist, the researcher, prevention people, educators and legislators, as valuable data on causation is lost or not considered.

Accidents result from the demands of the work system exceeding the capacity of the controls employed to cope with the error factors in the system. Accidents can be defined as the point where control over a system is lost. A work system may contain several error factors both known and unknown; the system may well be chaotic, but still within control.

Error factors in a work system are known levels of risk, inherent in the work activity. If the individual considers the level of risk acceptable, then the system is considered safe. If the error factors are well managed the work system remains in control. Not all risk is known and we can only ever manage those risks which we have identified.

There are other possible definitions for accident; however, they generally carry a theme of randomness, some concept of ill-fate, bad luck or unplanned failure. Applying this type of traditional definition to accidents infers that an organization which has accidents is either unlucky or poorly managed.

This interpretation could explain the reluctance to use the word in certain industries. The use of terms such as 'incident' and 'near miss' sends a message of a less dramatic outcome than that conveyed through the use of the word 'accident'.

There is also an undeniable political correctness in the avoidance of the use of the word 'accident'. Because the word is often seen as reflecting a lack of management control over the system and its attendant risks, the term 'accident' is rarely found in safety dialogue affecting industry sectors engaged in sensitive activities, such as those found in the nuclear, aero-space, chemical, pharmaceutical and genetic engineering industries.

And who would want to fly with an airline that has accidents? It is currently far more commercially astute to experience an 'incident'.

Little will be achieved in the long term by the use of alarmist and dramatic words to sensationalize system failures. However, if we are to begin to understand the factors that give rise to the accidents, some common ground must be found on which to exchange information. Effective risk communication is essential in accident prevention, and in the development of the future directions in safety management.

We have become so successful at avoiding the use of the word 'accident' that many workers can not recognize that they are having or seeing accidents. They now only see accidents in terms of outcomes.

The real danger is that we will lose an aircraft, a tanker, a refinery or a mine to an accident cause that has occurred previously, but has gone unrecognized and unrecorded.

The pivotal role of accidents

The prevention of accidents, as has been defined above, is pivotal in injury causation for three reasons. Firstly, the accident is the point where the system actually fails. Secondly, the accident creates a hazard and, thirdly, the accident and hazard are essential for injury to occur. Consequently, to prevent injury, accidents must be prevented.

Notwithstanding that all accidents do not result in injury, the accident was identified in some of the earliest prevention work as a key factor in causing not just injury to, but disease in, workpeople. Even if we find that all accidents are not preventable, this should not diminish the need, or our resolve, to prevent those that are.

Work elements

Regardless of how much effort we put into the selection of the elements – person (people), machine, environment, materials, procedures – which make up the work system, their relationship can be expected to create some level of error. The best we can, and must do is to identify and minimize the effects of that error.

Accident prevention

We noted earlier that an accident is caused by the demands of the work system exceeding the capacity of the system to manage error. The elements that make up the work system mentioned earlier become unable to cope with the level of error in the system they create. There is a requirement on employers to provide a safe system of work. In determining if the system is safe, all five elements must be considered, not just individually, but in combination. The error occurs in the synergy, not in the individual elements.

Work is dynamic. Work systems react to meet existing stimuli. Consequently, any management strategy in place to monitor change and control the errors in the work system must also be dynamic. This would suggest that dependence on procedures to protect people from harm should be viewed carefully. While good procedures are vital to safety management, the actual work practices are critical to safe working. Further, to measure the capacity of a work system through the evaluation of each of the individual elements that make up the work is not possible. It is the elements of person, machine, materials, environment and procedures in combination, their synergy, that we must learn to measure. Work is a composite of the five elements but totally different. See Fig. 6.3.

Regrettably, at present, expressions such as operator error, lack of common sense, carelessness, act of God, bad luck, stupidity, etc., are part of the safety management mantra. Their use and inference brings an acceptance that the error or cause has been identified and explained. As a consequence, we stop looking for real causes. However, the model above assumes all work takes place in a system and consequently the focus on human error is diminished in favour of accidents resulting out of systems failure. A key requirement in prevention work is to believe that individuals are not at the core of accident causation, but rather that we design systems and place people in systems that fail to protect them.

Consequently, reference to human error or operator error is of little value other than to apportion blame. Isolating the people in a system failure is a convenience and will have only marginal significance in future system failures. The reality of the work system

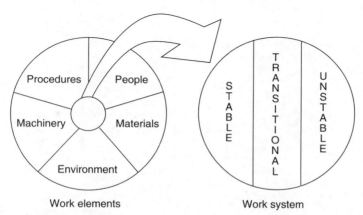

Figure 6.3 Creating a system of work

212 Enhancing occupational safety and health

is that it will always be in error and, as long as people are part of the system, they will have a role to play in the success and failure of the system. It is hardly earth-shattering news to hear, after an extensive inquiry into an accident, that it was 'operator' or 'pilot' error.

More importantly, the use of these causation factors creates an unhealthy acceptance in our perception. Because we have, from birth, been exposed and conditioned to personal fault finding, reinforcement of this behaviour by significant others could explain why we readily seek to blame individuals, including ourselves, when things go wrong.

The real tragedy is our immutable belief in nonsense causation factors, such as lack of common sense and carelessness. In every failed work system, the individual makes up only one component of the reason for failure. A more constructive mindset could be to always view causation as failure in a system. And a system failure will always result through human error in combination with other factors, never through the lone fault of the individual. Consequently, all accident investigation results must identify the role of the environment, the machinery, the materials and the procedures as other causation factors when, and if, human error is identified as a reason for the failure. This approach would largely ignore the adversarial ramifications of accident reporting in favour of a more productive outcome focused solely on prevention of system failure in the future.

The Hegney–Lawson system risk model

The traditional view of hazards in engineering activities includes chemicals, radiation, fire, earth moving, and construction, for example. However, identification of some other hazards is often complex if traditional energy models are used. People seeking to identify accidents and hazards linked to outcomes such as repetition and overuse injury, stress disorders, violence in the workplace and occupational asthma or cancer need a more general model. Notwithstanding that these outcomes can be fitted to energy exchange models (see Chapter 1), most people involved in the prevention process require simpler models.

The Hegney–Lawson System Risk model shown in Fig. 6.4 is intended to simplify an explanation of the injury causation sequence. However, it should be noted that the

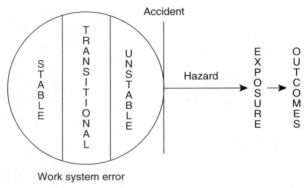

Figure 6.4 The Hegney–Lawson system risk model

model has application not only to injury but also to property and environmental damage. There is direct application of this model to risk management.

The main focus of the model is on the system created by the work elements from the commencement of work to the creation of a hazard. The model and the choice of terminology have been chosen to clearly identify the accident causation process. This clarity is necessary for the purpose of determining prevention strategies, accurate record keeping and risk communication, all necessary in decision making, training, education and the development of safe work systems.

The purpose of the model is to provide a mechanism which supports an analytical view of accident causation and the creation of hazards. Data collected is used to implement preventative strategies and provide information which will support decision making in risk management and general problem solving.

Here, the use of conventional terms is employed as model descriptors; however, the model does propose a quantum move away from some conventional thinking in that the model views the accident as the precursor to the creation of the hazard.

The model requires that an accident not only precede a hazard but that the accident actually creates the hazard. A hazard in this model is a system out of control.

Error factors

Simply stated, all work systems contain inherent stable error. When error factors in work systems are known and risk is accepted, the work system is often referred to as safe. This is notwithstanding the fact that the perception of those risks may result in diverse opinion over their acceptability. This variable in the acceptance of risk will always be present because safety is an individual's perception of risk.

However, when the error factors in the system exceed the controls available to manage them, the system becomes unstable and an accident is the likely result. When control over the work system is lost, the point at where control was lost is the accident. This loss of control has created a new system. This new system is outside our original safety parameters. As a result a hazard has been created.

The point of failure is the accident. The system out of control is the hazard. So it follows that, if this model is correct, then the accident has created the hazard.

Change and variation

Considerable effort is given to the identification and documentation of procedures for key tasks and activities. Procedures are important as a reference base for general guidance of the management and workforce. Additionally, there is an emerging legal and corporate imperative to have well documented procedures. However, as a tool in the prevention of accidents, the formal procedure approach to safety management fails to recognize and correct the subtle but critical unwanted variations in the work practices. Further, not all variation is unwanted, so how do you arrive at innovation without passing through variation? We need to embrace variation and control it, not avoid it, if we are to learn and grow. The need to embrace variation must be matched with the skill to manage it – 'there's the rub'. Organizations engaged in high risk activities have a commercial edge if they can manage variation effectively. 'Win their confidence then win their contracts': this appears to be the subtle commercial thinking behind the Du Pont approach to industrial safety. Good safety performance is a commercial edge.

It is control over the dynamic aspects of our activities, the work practices, which holds the key to accident prevention and good safety performance. Error control cannot be bought, copied or enforced – it must be intrinsic to the culture and inherent in every facet of organizational life. When a system fails, all employees, regardless of their corporate role, must feel and accept some responsibility for the failure. If any employee in any organization feels they can make no contribution to the well-being of every other employee, regardless of their role or position, then that employee is surplus to requirements.

Stable error

Stable error is the accumulated value of all the known and unknown error factors inherent in a system of work. Stable error is better regarded as the acceptable level of residual risk achieved through the reasonably practical approach necessary in establishing a safe system of work. Residual risk is the risk remaining after risk management strategies have been employed. Error factors are established in the work system either before the work commences, through a mismatch of person/machine/environment/materials/procedures, or as a result of unforeseen consequences of the activities involved. Additionally, error could result from unplanned or undetected changes in the work system during operation.

Stable error will always be present in the work system, so as a corollary a level of risk will always exist – there is no such place as a risk-free work environment. After accepting the risk inherent in the activity, we must establish the techniques needed to avoid losing our control over the system.

As an example, a common task such as driving a delivery vehicle in the city presents an array of acceptable stable errors in the work system. These become acceptable risks. If they were not acceptable the task or activity would be modified or eventually cease to exist. Examples of activities which are ceasing to exist are the mining of asbestos, smoking in the workplace or the use of Halon (BCF extinguishers). The risks they created were seen as unacceptable and the activities are being phased out.

Returning to the delivery truck, should we expand our city-based business to include country deliveries while using the same drivers, that is, same vehicles but a new environment? There would be a significant change in the work system and a change in the error factors. This would require a need to redefine the stable errors, and arrive at an acceptance of any change in risk levels prior to determining if a safe system of work still existed.

Even in work systems involving high risk activities there is stable error. Any task or activity can be performed as long as the stable errors are known, accepted, managed and monitored.

Transitional to unstable error

Once stable error has been identified, and the known risks assessed and accepted, the objective is to keep the errors stable. When a state of acceptance is reached, the system could be perceived by the individual as safe. However, as individuals are almost assured of having different perceptions, the opinions of what constitutes a safe system of work may vary greatly.

Should management of the work system fail to control error at this level, the error could go into a transitional phase. If not detected in this phase instability is imminent.

The nature of the error will determine the options to identify and redress the shift in the error phase. However, in some work systems the shift of error factors from stable, acceptable risk levels, through transition to unstable, unacceptable risk, is too rapid to afford detection and correction. The ability to redress the shift in error could be beyond the capacity of the operator or employer to control and correct.

Consequently, the safety management strategy used to govern the system of work will need to be carefully chosen to ensure a good match between the risk and the strategies employed to control the risk. The management techniques employed must be based primarily on the nature of the risks and have the capacity to identify and to cope with variations from the stable error.

This would suggest why traditional safety models have proven to provide ineffective prevention strategies in some work systems.

Applying the model

Using the delivery truck example, a proper maintenance plan should provide the opportunity to keep the braking system in good order. Regular service should identify any unusual wearing pattern and correct the brakes in the transitional phase, thereby preventing the braking system becoming unstable and possibly failing.

However, the delivery truck operates in an environment shared with other users. Should another vehicle enter the road space required by the delivery truck, either intentionally or unintentionally, the work system goes rapidly from stable through transitional into unstable. Importantly, the driver may not be aware of the impending danger. Let us imagine on this occasion our driver notices the vehicles are on a collision course, with each other, and our driver takes evasive action. The evasive action taken causes our truck to swerve violently to avoid contact with another vehicle and in so doing our driver loses control over direction. The truck slides sideways coming to rest one metre from a concrete bus shelter. No contact was made with any object, but there is a screech of brakes, some smoke from the tyres and the odd gasp from a group of people waiting for a bus. Assuming there is no detectable damage to any property or people (no one at the shelter had a heart attack), the outcome is likely to be presented in the following way:

- nobody injured
- no damage to property
- no damage to vehicle.

Has there been an accident?
If there was an accident where was it?
Was there a hazard?

The accident

By the definition given earlier there has been an accident. The accident occurred when control over the system was lost – the point when the driver could not control the direction of the vehicle. Up to the point where loss of control over direction occurred the system could have been redeemed.

The hazard

Following the accident we are able to identify the hazard. In this case the truck sliding sideways down the road suggests the system is well and truly out of control. The hazard exists for as long as the system is out of control.

It is not the intention at this time to elaborate on exposure or outcomes; however, for completeness and to provide an example let us assume a different ending to the delivery truck accident.

What if the truck had crashed into the bus shelter killing all the people, bounced off and struck a school bus which burst into flames.

- In such a catastrophic event, where is the accident?
- Where is the hazard?
- We know where the injured are and we can identify the damage.

The accident has not changed. It remains at the point where control over the system was lost. The only change is in the exposure to the hazard and the outcome. The hazard remains the system out of control; however, when the hazard met a new exposure the results are now death, injury and mayhem as opposed to the previous scenario which resulted in no detectable damage.

Summary

Regardless of the severity of an outcome, from death at one end of the scale to no detectable damage at the other, the causation factors which gave rise to the accident and created the hazard remain unchanged. It is the information on NDD accidents that we must collect, analyse and act on if hazards and injuries are to be avoided. This is based on the principle that not all accidents result in injury. We have known for years that there is a relationship between the total number of accidents a person experiences and the number that result in injury. It could be that continued exposure to accidents evokes the laws of probability and eventually the luck runs out. This in no way suggests that accidents are a function of luck but rather that the accident outcome may contain a component of ill fortune. The only way to ensure the prevention of injury is to prevent the accident.

As indicated in the delivery truck accident, all the factors which led to the tragedy in the second scenario were present in the accident that resulted in NDD.

Regardless of how time-consuming or how expensive and unpalatable the accurate reporting and recording of accidents is, there are no other practical proactive options for identifying methods of prevention. Even then we are relying on the laws of probability to protect us.

In some industries, conditions exist where one accident is too many and there is a need to focus totally on error management. In a perfect world, perhaps all workplaces could enjoy the capacity to allocate the resources needed to effectively manage error.

In the present industrial reality accident prevention seems to be a reasonable objective.

Principal elements in developing a workplace health and safety plan

Introduction

A health and safety plan is a series of activities within a safety management system (SMS) designed to implement a plan to manage hazards and prevent cases of occupational disease. In the USA, 29 CFR 1900.1 requires each employer to set up an OSH programme (=plan). The Canadian Centre for Occupational Health and Safety also refers to 'programme' rather than 'plan', but we use 'plan' to be consistent with ISO 9000/14 000, on which some current OHS management systems are modelled. Some form of plan is required to enable management to provide the framework for employee participation and orderly arrangements to deal with safety issues. The SMS provides a mechanism to review performance, identify the need for change and manage the change process. By having a proper identifiable system, and plan, both the employer and employee are better placed to meet their respective legal obligations. Because organizations differ, a plan developed for one cannot be expected to perfectly suit the needs of another.

Essential elements in a health and safety plan

Policy statement

An organization's occupational health and safety policy should be a clear statement of principles which serve as guides to action. Senior management must be totally committed to ensuring that the policy is carried out with no exceptions. Health and safety policy must be, and be seen to be, on a par with all other organizational policies.

As with health and safety plans, no one policy is suitable for all organizations. The policy statement can be brief, but it should mention:

- the objectives of the plan
- the organization's basic health and safety philosophy
- the general responsibilities of all employees
- the ways in which employees can participate in health and safety activities.

The policy should be:

- stated in clear and concise terms
- signed by the current Chief Executive Officer
- kept up to date
- communicated to each employee
- adhered to in all work activities.

Plan elements

While organizations will have different needs and scope for specific elements in their health and safety plan, certain basic items should be considered in each case. These are the building blocks in the development of a comprehensive plan. On a construction site, the plan is sometimes called the site safety plan.

Individual responsibility

All health and safety activities are based on specific individual responsibilities, most of which can be found in the pertinent legislation. It is somewhat surprising, but unfortunately true, that all these responsibilities are not well known in the average workplace. This situation can be improved by including details of specific responsibilities in the safety plan.

Responsibilities may be defined as an individual's obligation to carry out assigned duties. Authority implies the right to make decisions and the power to direct others. Responsibility and authority can be delegated to others, giving them the right to act for the employer. It is important to note that, while some responsibilities can be delegated, the management remains accountable for seeing that they are carried out.

The responsibilities of an employee apply to everybody in the workplace, including the Chief Executive Officer. First-line supervisors have additional responsibilities, and senior management have extra responsibilities. Where a health and safety coordinator/advisor/ manager has been appointed, it is best to spell out their special responsibilities. Defining responsibilities assists all employees to know exactly what is expected of each individual in health and safety terms. Examples of types of responsibility for each of these groups include:

Employee
- Know and comply with all safe working procedures.
- Report any injury or illness immediately.
- Report hazards.
- Participate in health and safety committees.

First-line supervisor
- Ensure all workers receive proper induction.
- Be familiar with relevant health and safety legislation.
- Encourage employee participation.
- Ensure that only authorized, properly trained workers operate equipment.
- Investigate all accidents/incidents.
- Inspect own area for hazards.
- Ensure equipment is properly maintained.
- Promote safety awareness in workers.

Management
- Set objectives and measure performance.
- Provide a safe and healthy workplace.
- Establish and maintain a health and safety plan.
- Ensure workers are trained.
- Report accidents/cases of occupational disease to the appropriate authority.
- Provide first aid facilities.

- Ensure personal protective equipment is available.
- Provide workers with health and safety information, and consult with them.
- Support supervisors in their health and safety activities.
- Evaluate health and safety performance of supervisors.
- Provide the climate for optimum participation.

Health and safety coordinator/advisor/manager
- Advise on health and safety matters.
- Coordinate interdepartmental health and safety activities.
- Collect and analyse health and safety statistics.
- Provide health and safety training support.
- Monitor compliance with the relevant legislation.
- Investigate and report on specific issues.

To fulfil their individual responsibilities, employees must:

- Know what these responsibilities are, (communication required).
- Have sufficient authority to carry them out, (organizational issue).
- Have the required ability, (training required).

Once all these criteria have been met, safety performances can be assessed on an equal basis with other key job elements. Health and safety are not an extra part of an employee's job, but an integral, full-time component of each individual's responsibilities.

Health and safety representatives

Legislation in many Robens-based jurisdictions provides for health and safety representatives. Functions of health and safety representatives may include:

- inspection of the workplace
- hazard identification
- accident investigation
- information source for other employees
- liaison with other employees on health and safety issues
- participation in the development of policy and procedural matters.

The health and safety representative is *not* the health and safety coordinator for the organization. The role of the health and safety representative is clearly defined. The duties and rights of the health and safety representative should be adhered to as closely as practical.

Health and safety representatives will be required to consider a wide range of health and safety issues. In order to participate actively and positively in the resolution of issues, and to gain the confidence of other employees, the representatives will be required to undertake regular and detailed training courses. The health and safety representative has the potential to be a key element in an organization's health and safety strategy.

Health and safety committee

Although all health and safety endeavours are included under individual responsibilities, an effective safety plan needs the cooperative involvement of all employees.

This joint effort can be fostered by the formation of a health and safety committee. For instance, in Alberta these are set up under S.31 of the OHS Act. Such a committee brings together labour's in-depth, practical knowledge of specific jobs and management's larger overview of job interrelationships, general company policies and procedures. This team can be more effective in solving general health and safety issues than a single individual.

To function properly, the committee needs an appropriate structure, a clear statement of purpose, duties and standard procedures for meetings.

Once the committee members have been chosen, the committee should participate in decisions on the details of its duties, and procedures. An early key decision which should be made is the question of reporting responsibility. The health and safety committee members should be active participants in the development, implementation and monitoring of all phases of the health and safety plan.

Health and safety procedures

Legislative health and safety regulations represent minimum requirements. In almost all cases, it is necessary for the organization to augment these regulations with specific procedures, which must be followed in order to manage employee well-being effectively.

While there should be little controversy over the need for procedures to protect the health and safety of workers, there are the dangers in having too few or too many. Too few may be interpreted as a sign that health and safety is not important, or that common sense is all that is required. Too many may be seen as not treating workers as adults and makes adherence less likely. Guidelines for establishing procedures include:

- the health and safety committee should participate in their formulation
- they should be stated in clearly understandable terms
- they are best stated in positive terms: 'thou shalt' not 'thou shalt not'
- the reasons for the procedure should be given
- they must be enforceable
- they should be available to all employees in written form.

Compliance with health and safety procedures should be considered a condition of employment. The question of how to deal with people who do not follow procedures must be addressed. Points which should be considered in establishing an approach to this issue are:

- all procedures are to be observed
- no breach will be disregarded
- the role of discipline is that of education not punishment
- action for breaches is taken promptly
- was the employee aware of the procedure or not?
- was the employee encouraged, coerced, or forced to disregard the procedure?
- while it may be desirable to have guidelines for penalties for the first offence, etc., some flexibility in their application is required, since each case will vary in its circumstances.
- action is taken in private, and recorded.

Employee induction

It is generally accepted that inexperienced workers are involved in accidents at a higher rate than others. While experience can only be gained through time, both health and safety education and job skills training can be used to improve this record. Health and safety education should start with employee induction when a worker joins the organization or is transferred to a new job. Induction sessions normally cover such items as explanation of the function of the work unit, organizational relationships, administrative arrangements, miscellaneous policies and rules.

Health and safety related items which should be included here are:

- health and safety representative and committee roles
- emergency procedures
- location of first aid stations
- health and safety responsibilities
- reporting of hazards
- use of personal protective equipment
- right to refuse hazardous work
- hazards, including those outside own work area
- reasons for each health and safety procedure
- where to obtain information.

A new employee can be expected to absorb only so much information in the first few days. A brochure outlining the points covered in the initial orientation sessions is useful as a handout to employees. It also serves as a checklist for the person conducting induction training. New employees should be encouraged to ask questions at any time when doubt exists as to correct procedures. Soon after the orientation sessions, they should be tested on their knowledge of the items discussed. In this way, both the quality of training and the level of understanding can be assessed. Any misunderstandings should be clarified.

Training

If general health and safety policies are to be incorporated into specific job practices and skill levels are to be raised to an acceptable standard, training is required. While all employees can benefit from health and safety training, special attention should be given to the training of supervisors and health and safety representatives.

For example, the following topics could be considered in supervisory training:

- know your workplace hazards
- know how people behave under stress
- consulting on safety
- maintaining interest in safety
- instructing for safety
- minimizing workplace air contaminants
- personal protective equipment
- workplace housekeeping

- material handling and storage
- guarding machines
- hand and portable power tools
- fire protection.

The supervisor is responsible to see workers receive appropriate training. This duty, however, is often delegated to somebody else.

Occasions when employee training is indicated are:

- commencement of employment
- reassignment to a new job
- introduction of new equipment, processes, or procedures
- inadequate performance.

Inspections

Regular inspections must be carried out to identify existing or potential hazards so that appropriate corrective action can be taken. It is expected that supervisors and workers report and take action on hazards as they are encountered. The frequency of planned formal inspections may be company policy. Past records of accidents and the potential for serious accidents and injuries are factors to be considered in determining if more frequent inspections are needed. Critical areas should receive extra attention.

Health and safety committee members are an obvious choice of personnel to carry out formal inspections. Other criteria for selecting those who will inspect are:

- knowledge of regulations and procedures
- knowledge of potential hazards
- experience with processes involved
- training in inspection.

It is important to spend time in pre-planning any inspection. Documents on previous inspections, accident investigation and maintenance reports, as well as health and safety committee minutes, should be referred to.

Checklists are useful aids in that they help to ensure that no items are overlooked in an inspection. One form of checklist is the critical parts inventory which details parts and items whose failure may result in a serious accident. While many ready-made checklists are available in the safety literature, it is best to adapt these to local conditions. The health and safety committee should participate in the preparation of these tailor-made checklists.

During the inspection, both *conditions* and *procedures* should be observed. If a hazard poses an immediate threat, preventative action must be taken right away, not after the inspection. Notes are made, specifying details of the hazard, including its exact location. When completing the inspection report, it is a good idea to classify each hazard by degree of possible consequences (for example: A = major, B = serious, C = minor). In this way, priorities for remedial action are established.

Accident prevention

Inspections serve a useful purpose only if remedial action is taken to correct shortcomings. Causes, not symptoms alone, must be rectified. Corrective action should be taken immediately with emphasis on engineering controls and work practices.

Reporting and investigating injuries and accidents

Occupational safety and health legislation requires that certain kinds of injury must be reported. The reportable injuries result from serious accidents. Put another way, serious accidents include those leading to reportable injuries. Thorough accident investigation policies and procedures need to be predetermined to ensure optimum benefit is realized from the investigation.

The accident investigation part of the health and safety plan should specify:

- what is to be reported
- to whom it will be reported
- how it is reported
- which accidents are investigated
- who will investigate them
- what forms are used
- what training investigators will receive
- what records are to be kept
- what summaries and statistics are to be developed.

Accidents and incidents are investigated so that measures can be taken to prevent a recurrence of similar events. Investigation represents an 'after-the-fact' response as far as any particular mishap is concerned. However, a thorough investigation may uncover hazards or problems which can be eliminated 'before-the-fact' for the future. After causes have been determined, prompt follow-up action is required to achieve the purpose of the investigation.

Other

Specific elements should be considered for inclusion in any basic health and safety plan. In some organizations, additional specific items may need to be incorporated. Examples of other elements which may be relevant in some plans are:

- lock-out procedures
- hot-work permits
- material handling
- plant maintenance
- fire safeguards
- vehicle safety
- off-the-job safety
- working alone
- heat stress
- chemical handling
- personal protective equipment
- engineering standards
- purchasing standards.

Promoting safety and health in the workplace

Once the various elements of the health and safety plan have been set in place and the plan appears to be running smoothly, effort is still required to maintain enthusiasm and interest.

The effectiveness of health and safety educational techniques depends to a large degree on how much importance management is seen to place on health and safety. Where they are sincerely concerned, interest in the plan can be maintained at a high level. Accountability for individual performance is a key motivator.

Safety awareness can be enhanced by:

- the setting of realistic goals and monitoring progress
- dissemination of all pertinent information
- recognition for superior performance
- maximum involvement of workers in health and safety activities
- incentive schemes
- short meetings at beginning of shift – toolbox meetings.

Incentive schemes are the most controversial. Most incentive schemes are based on the rationale that anything that raises safety awareness is worthwhile. However, there are those who do not share this viewpoint. They maintain that these programmes lead to under-reporting of accidents and promoting of the 'walking wounded' syndrome. Incentive programmes must not encourage workers to remain at work when it is unsafe for them to do so due to their physical condition. Therefore, when such a programme is launched, strict controls must be maintained to prevent this from happening.

While the health and safety committee, where it exists, can play a leading role in activities designed to promote the plan, participation of all employees must be encouraged if long-term success is to be achieved.

Implementing a health and safety plan

A good health and safety plan provides a clear set of guidelines for activities which, if rigorously followed, will reduce accidents and cases of occupational disease. The key to success is the manner in which the plan is implemented and maintained.

Senior management must visibly support the plan by:

- providing resources such as time, money and personnel
- ensuring that employees receive training as required
- making all applicable health and safety information available to all employees
- including health and safety performance as part of employee evaluations
- attending health and safety meetings.

The plan must be communicated to all employees. Special emphasis should be given to new workers, newly appointed supervisors, and new members of the health and safety committee. Revisions of policies and procedures should be clearly communicated to *all* employees.

Auditing health and safety plans and systems

Accident frequency and severity rates are an inadequate means to evaluate the effectiveness of a health and safety plan. Cases of occupational disease are under-reported in these statistics. The emphasis is on injury-producing accidents, not all accidents. Since accidents are a rare event, in small organizations the basis for comparison may be limited. Chance is a factor both in frequency and severity.

Rather than relying solely on these after-the-event measures, an audit provides a before-the-fact measure of control activities. A more valid comparison from year-to-year or site-to-site becomes possible. All elements of the plan, or the health and safety management system, can be critically reviewed so that priorities for future action may be determined.

The audit employs a checklist in which each element is given a weighting factor depending on its importance. Records, observations, interviews, and questionnaires are used to evaluate performance for each element. A number of audit systems are available. Many of these ready-made audit systems are based on indicators leading organizations have determined can be used to measure an acceptable level of performance in prevention.

A continuous audit programme ensures that critical areas and activities within the workplace are under constant review. The audit team, which should include representation from the health and safety committee, must receive appropriate training in audit procedures.

The audit identifies weaknesses in the health and safety management plan or system. Little is achieved unless a procedure is established to ensure prompt follow-up on deficiencies. This procedure should include provision for target dates for remedial action and checks to confirm completion.

Preparing a site emergency plan

A sub-plan on how to deal with emergencies should be included in the health and safety plan of every organization. Fires, explosions, major releases of hazardous materials, or natural hazards may present a threat. When such events occur, the urgent need for rapid decisions, shortage of time, lack of resources, and lack of trained personnel can lead to confusion that often results in unnecessary injury and damage. The objective of the sub-plan is to prevent or minimize fatalities or injuries, and control damage. The organization and procedures for handling these sudden and unexpected situations must be clearly defined and understood by *all* employees.

The development of the sub-plan follows a logical sequence. A list of potential hazards (e.g. fires, explosions, floods) is compiled. The possible major consequences of each is identified (e.g. casualties, damage). Required countermeasures are determined (e.g. evacuation, rescue, firefighting). Resources needed to carry out the planned actions are listed (e.g. medical supplies, rescue equipment, personnel). Based on these considerations, the emergency procedures can be established. Communication, training, and periodic drills are required to ensure adequate performance when the plan must be implemented.

Medical and first aid

There are generally minimum requirements for first aid facilities and the provision of medical aid prescribed under each jurisdiction's regulations. In many instances, these provisions would prove to be inadequate. Particular reference must always be made to the nature of the work and the availability of support for the injured.

The following information should be incorporated into the first aid part of the plan:

- location of first aid stations and other medical facilities
- identification of first aid attendants
- identification of other staff trained in first aid
- policy on pre-employment and follow-up medical examinations
- procedures for transporting injured employees to outside medical facilities
- provision of first aid training
- procedure for recording injuries and illnesses.

Other aspects of a plan

Rehabilitation

Since it has obvious health and safety implications, a policy on return to work after a lost-time accident might appropriately be included in this section of the plan. The fact that light duties are a controversial issue is all the more reason for the organization to have an agreed, clear policy made known to all employees.

If injured workers are offered alternative employment:

- the work must be suitable and productive
- the worker's physician must agree that such employment will not harm the worker or slow the recovery
- the worker will pose no threat to other workers
- consideration should be given to applying the policy equally to off-the-job injuries.

Under no circumstances should the reduction of severity ratings or lost-time injury statistics be a reason for returning an injured worker to employment.

Conclusion

The development, implementation and maintenance of an effective health and safety plan is not easy. It may be useful to develop a comprehensive series of programmes, allot priorities to the elements within the programmes, and implement these in stages. Success can be achieved if all segments of the workforce carry out their individual responsibilities and work together in a cooperative climate. Motivated and involved employees can be expected to do this.

(*Much of the material in this section on health and safety plans is taken from* A Basic OHS Program. *Reproduced with permission from the Canadian Centre for OHS.*)

Accident, injury, compensation and safety data

Assessing compensation data which may influence OHS decisions in a workplace

One useful step in making decisions about OHS in a particular workplace is to compare the data on work-related injury and disease in that workplace with national, state, province, territory figures for occupations in that workplace and for similar types of industry. This can give an idea of the types of workers most at risk, and the types of occurrences which need to be addressed first. It allows you to benchmark your workplace against other similar ones. For your particular workplace, it is possible to ask your compensation insurer for a breakdown of injury types, injury severity, injury frequency rates, injury duration, and claims costs, and relate these to particular sections or parts of your workplace.

Data to assess failure probability

Sources of data and groups of data can be found which will assist in the assessment of failure probability in engineering equipment and systems, and in human control/management systems.

Many possible sources of data of this type may be available and an in-depth look is not possible here.

You may be able to obtain manufacturer's information on expected risks of breakdown when purchasing equipment, and you can certainly find out recommended times for replacement or servicing of parts. Within a company, careful examination of maintenance records will give you information. Some electronically-based control systems store data on events which occur during operation so they can be analysed later.

Techniques such as THERP have been developed by, for example, Sandia Corp to predict human error within a control or management system, but these are beyond the scope of this book.

Careful analysis and recording of accidents will allow you to separate engineering failure or errors from human controls or management system errors. For example, a machine may not have failed but rather an operator may not have followed agreed procedures.

Collecting, sorting, accessing and validating data

Techniques for achieving a satisfactory level of incident, injury and illness reporting

As a general observation, perhaps the greatest assistance in achieving good reporting levels is through having good consultation arrangements and effective communication.

If employees know that the organization actively uses all accident data to reduce the risk of further injury, the motivation to report is improved.

Workers must also be made aware that compensation could be harder to obtain if even what appear to be minor injuries are not reported when they occur. Good training at all levels in collecting and recording accident data is important.

Organizations, teams or departments will soon make it known if they believe other organizations, teams or departments are *fudging* the figures.

Factors detracting from reporting safety performance

If competition in regard to safety between individual operating units is not carefully handled, under-reporting of accidents can occur, particularly if one unit feels another has an unfair advantage, e.g. newer equipment or an environment with fewer inherent high risks.

If workers feel that the organization only pays lip service to safety and doesn't really want to know about accidents this may detract from reporting.

Safety competitions with prizes can create difficulties. Opinions on these vary, but a lot of thought must go into them. Some organizations overcome the temptation to not report by having the prize go, not to the workers, but to a charity nominated by them.

Cross-checks to assess workplace reporting performance

Checks on reporting accident performance can include comparisons of the figures for similar areas within the workplace. While great care is needed to avoid appearing suspicious rather than pleased about an upturn in accident performance, sudden changes should be checked.

Other possibilities are that data on accidents is being wrongly classified or analysed. Going through some of the primary data can allow problems to be identified and fixed.

Sorting categories for workplace data

Sorting categories for accidents include non-injury-producing (dangerous occurrence or critical incident) and injury-producing. Injuries may be confined to one worker, affect a few, or many.

The injuries may be first aid cases – minor but medically treated; lost-time (did not complete at least one full shift) and lost-time and severe, (resulting in, or likely to result in, more than ten days off work, amputations, certain fractures or loss of sight).

This only covers levels of injury severity according to the occurrence categories, e.g. nature and bodily location of injury or disease. The other categories, e.g. type of accident, agency of accident, breakdown agency, can be used to classify accident rather than injury data. Obviously diseases which do not occur until many years after initial exposure are impossible to include in data for a workplace in a timely way.

Accident prevention

Gaining timely and reliable data on compensation costs

The best source of data on compensation costs, is your workers' compensation insurer.

Assessing data on reliability of engineering and human control systems

The data collected from accidents, injury-producing or not, can be examined to identify the contributing factors which led to the accident. It is possible to find out which part of the controls in the system broke down. For example, despite good and well documented maintenance, a part failed. This requires an engineering solution. Or perhaps a part has failed and it is clear from the documentation (or lack of it) on repairs and maintenance that the human control (management) system was sloppy. This may indicate poor policies, failure to develop procedures or failure to train people in procedures, and then audit observance of those procedures.

Accessing data from national, state or provincial, and broad industry sources

Government OHS, workers' compensation or statistics authorities publish data on accident and injury performance. So comparison of data across or within countries, provinces or states (in some cases) and across industry groupings is possible. It is also possible to compare male and female and age-based figures.

Factors affecting validity of data and its use for comparison purposes

Many things can affect the validity of the data. These include different cut-off dates for a year's data; poor reporting rates to the authorities (especially in some occupations and industries); poor classification skills; and poorly filled in primary forms or documents. There may only be small numbers of people in some groups in some jurisdictions. Workers' compensation laws may be national province, state or territory based and changes to those laws can affect classification and reporting of injuries or diseases. Some chest diseases may show up quite early, e.g. allergic reactions, whilst others may not occur for 40 years after initial exposure.

Data handling and analysis

Interpreting trends in injury types, and factors affecting trends

Reviewing accident data relating to types of injuries will allow you to spot trends in the types of injuries occurring. However, you should be aware that variations in

Internal

A number of internal factors can affect trends in injury types. They include process or equipment change; training changes; rapid turnover of staff; changes to procedures; and either achieving better reporting of particular injuries or, on the other hand, suffering a downturn in reporting.

External

External factors can also affect trends in injury types. Greater public awareness of a problem, e.g. occupational overuse syndrome, may encourage more workers to report it. An organization with many staff in vehicles may be affected by changes in roads and road layout, or by unusual weather patterns.

The economic situation could decrease workers' concentration or the ability to provide adequate staff levels or, on the other hand, it could lead people to pay more attention to commitment to working carefully.

Factors leading to unexpected trends after an intervention strategy

As mentioned above, greater awareness of a problem may lead to unexpected trends. A programme designed to reduce back injury may in the longer term reduce back injury. However, in the shorter term, the emphasis the programme may give to reporting back problems so that they can be dealt with, and not made worse through neglect, may have the effect of making the intervention strategy appear to lead to an increase in injuries.

Normal and skewed distributions and standard deviation

Distribution here refers to the *spread* of data which has been collected on a particular item.

For example, take hand injuries over 12 months. Let us say they are:

J	F	M	A	M	J	J	A	S	O	N	D
5	6	6	5	7	6	5	4	8	5	7	5

The mean number of hand injuries in a month is the total for the year divided by 12, i.e. 69/12 or 5.75. The range is from 4 to 8.

We can also look at the frequency with which a particular figure occurs. See Table 6.2 and Fig. 6.5.

The median value is 6. That is, the value that is *greater than or equal to* half the other values and *less than or equal to* half the other values.

The mode (most common value) is 5.

The distribution here is said to be skewed, i.e. more values one side of the mean than the other. In a so-called normal distribution the values are equally distributed each side of the mean. See Fig. 6.6.

Accident prevention

Table 6.2 Hand injuries

Number of hand injuries	Number of months this number occurred
1	
2	
3	
4	1
5	5
6	3
7	2
8	1

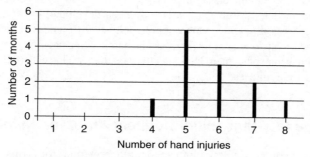

Figure 6.5 Hand injury occurrence by month

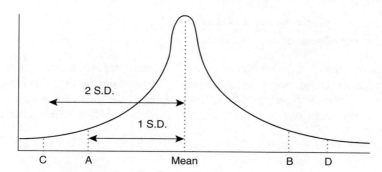

Figure 6.6 Normal distribution

In a normal distribution like that shown here, 68% of all values lie within what is called one standard deviation about the mean; that is, from A to B.

95% of all values lie within two standard deviations about the mean; that is, from C to D.

Using another example, since a person's height is important in designing healthy seating, then after collecting data on people's heights and graphing it as shown in Fig. 6.6, we might decide to design chairs which cater for all people within two standard deviations of the mean. These chairs would still not suit those either less tall than point C or taller than point D.

Factors affecting inclusion or comparability of data

The accident and injury data will depend on the definitions used and rules for reporting.

The culture in an organization may be such that people feel confident presenting for minor injuries and having them recorded. Or it may be the opposite. In some organizations people have been discouraged (for fear of dismissal) from reporting quite significant injuries.

Different reporting parameters at time of collection can also affect comparison of data.

As indicated earlier, various standards have been developed to try and adopt uniform definitions for reporting which should reduce differences in reporting parameters. However, while one organization may be happy to let an employee go home after treatment, another may have them back on alternate duties. This can affect the recorded number of lost-time injuries.

Establishing lost-time injury frequency rate for an industry

When doing this you need to take into account the potential for errors due to definition changes in reporting, and staff on varied hours.

The lost-time injury frequency (LTIFR) rate is calculated using:

$$\frac{\text{Number of lost-time injuries} \times 1\,000\,000}{\text{Number of person hours worked}}$$

As mentioned above, if the injuries counted as lost-time are influenced by a policy of getting people back on the job or the same shift after treatment, then this can reduce the LTIFR. Similarly, if standard hours rather than actual hours are used and there is considerable unofficial overtime, this may make the LTIFR look higher than it actually is.

Using compensation data to influence OHS decisions and compensation premiums

Gross dollar costs and severity of types of injuries expressed in dollar terms can be used to do this. An example will explain this:

If there are 30 claims with a gross cost of e.g. 4000 pounds, dollars, euros, rand, or ringgits, this 4000 currency units is made up of compensation costs (salaries) and other (medical, investigation). Of these claims, 25 may cost 400 currency units, while five cost currency units in total. We focus on that five. Out of the five, three are lower back claims worth 3200 currency units, two are finger and thumb claims totalling

400 currency units. These figures can be discussed with insurers to influence premiums. The lower back figure indicates more emphasis is needed on prevention of lower back injuries. However, note that one high cost injury of a particular type on its own may not be reason for a change in strategy. (*With acknowledgement to M. Gavin.*)

Implications of Heinrich's triangle and recent variations

The basis of the Heinrich triangle or the Bird–Heinrich triangle as it is known in modified form (see Chapter 1), is the careful collection of data over a range of industries on different types of accidents. The accidents can be classified as:

i) leading to disabling injuries, less serious injuries, or no injuries
ii) leading to serious injuries, minor injuries, property damage only, or incidents with no visible injury or damage (near misses, dangerous occurrences or critical incidents).

The figures behind the triangle are shown in Table 6.3.

The results vary in practice but look something like those in Table 6.3 (note, for example, the variation in the major injury to minor injury ratio). Some people say that the triangle tells us that if we want to control major injuries and fatalities, we must record and investigate all non-injury incidents. However, the common causes of minor injuries may not necessarily be very similar to the common causes of major injuries. Secondly, investigating every minor injury or near miss thoroughly could take up time better spent on other approaches to prevention.

Assessing the risk of types of failure

Failure may arise in engineering equipment units or systems, or in human control or management systems. A good and carefully analysed record of accident investigations may indicate the level of risk of failure that engineering equipment units or systems present, and the level of risk in failure of human control or management systems.

It may be possible to include factors such as weather conditions. It may also be possible, as mentioned earlier, to use computer processed control printouts, or maintenance or repair records to assess failure rates for equipment.

Table 6.3 Approximate ratio of types of incidents or injuries

(i) Heinrich	No. of injuries or incidents	(ii) Bird–Heinrich	No. of injuries or incidents
Disabling (including fatalities)	1	Serious injuries (including fatalities)	1
Less serious injuries	29	Minor injuries	10
Accidents without injuries	300	Property damage only	30
		Incidents with no visible injury or damage	600

Data reporting to management

Reporting channels to achieve action

This data can be used to set priorities for OHS tasks, so it is important to ensure that it follows the correct channels and ends up with those who can act on it. The reporting channels which are used to present occupational health and safety data for follow-up action will vary quite a lot depending on the organization you are concerned with. It will also depend on the type of data, who collects it, and where they fit into the organization.

You will need to make your own checks on the organization's arrangements for analysing, processing and handling the data. Some general comments apply, however. Firstly, check that the data has been collected correctly and properly analysed. You may, depending on where you sit in the organization, wish to comment on what the data tells you.

If you need to present the data, for example, to a meeting, prepare properly; make sure you understand the data, and you are clear about the action you are seeking. In an organization with good consultative processes, important occupational health and safety data will not be held back but will be provided quite deliberately to health and safety representatives and committees, where they exist.

Reporting methods, formats and report frequency for consolidated data

These methods and formats will vary quite a lot depending on the organization concerned. Some organizations will use electronic information technology quite extensively to process and pass on data, while others will still use paper. Reporting formats will include collated numeric data on, for example, first aid cases, accompanied by a written summary and proposed action, if any.

Graphical presentation of data to illustrate trends and comparisons is very useful, e.g. an audit profile graph which shows the scores achieved in prevention, heading by heading. The headings might include, for example, machine guarding and personal protective equipment.

How often reports are generated will depend on management requirements and the nature of the workplace. Workplaces with constant change (i.e. mines, construction sites) might use daily checklists. A furniture factory would not necessarily need such frequent reporting.

Setting logical priorities

Situations can occur where frequency and severity information and other influences are ignored with the result that there are 'flavour of the month' reactions. These may override more logical priorities.

Accident prevention

A good safety and health programme with appropriate objectives and goals requires a consistent and well-planned approach. This should be based on good research, known information, and existing statistics and workers' compensation data for the organization. If you are in a new organization (a 'greenfields' situation) you will need to obtain information on similar organizations to develop a clear view of potential hazards, the scale of risks they present and the control strategies which should be adopted. It is important to achieve a consensus between the parties in the workplace, so that everyone *owns* the strategy.

There is no doubt that a single unexpected serious injury or fatality cannot be ignored and the causes must be addressed, but sufficient balance must be kept in relation to the overall plan if you are not to find yourself simply putting out one 'bushfire' after another.

Influence on priorities may come from media publicity about a particular issue, e.g. the supposed effects of electric fields. However, the occupational health and safety budget must be used effectively, and clear-headedness kept about putting the effort and money into the biggest risks. You cannot ignore concerns workers have as a result of scare tactics, but you can address them logically and keep the safety boat on course. You should always, of course, be prepared to consider genuine new evidence which emerges about the risk from a particular workplace agent or stressor, such as occurred with keyboards and occupational overuse syndrome (OOS) some years ago.

The 'reasonable practicability' guideline used in much Robens-based legislation requires the cost of prevention to correspond with the size or magnitude of the risk. How to assess risk is addressed elsewhere in this text. For further guidance refer to G.L. McDonald's concept of Class I, Class II and Class III accidents in Chapter 1, and remember the Pareto principle – '80% of the cost comes from 20% of the accidents', and so 80% of the effort should go into reducing that 20% of accidents.

Violence in the workplace

Introduction

There have been some very serious incidents of violence in the workplace involving both staff and members of the public. These include the attack using aircraft on the World Trade Center in 2001, the attack on a nightclub in Bali in 2002 and the seizure of a Moscow theatre by terrorists, also in 2002. All resulted in a high loss of life.

Violence affecting workers, or workers and other members of the public, in a workplace, can take a variety of forms. The violence can result in death, serious or less serious physical injury, and psychological trauma.

This may include fear or apprehension about a repetition of the violence. The violence can take the form of actual physical assault, or threats of physical assault or other detrimental action by workplace groups (either other workers or patients or inmates) unless certain things are done. The violence can also take the form of verbal abuse, such as strong, loud, threatening, abusive or obscene language.

Reasons for violence

Violence in the workplace, or while working, can be the result of many factors. The most obvious, of course, is criminal activity, such as armed robbery of a store, garage, pharmacy or bank, or for example a desire by someone to get even with the person who sacked them. (This could occur away from work while the victim is off duty.) Police at work often face the possibility of coming into contact with violent people.

However, violence can also result from frustration at waiting in a queue, or from not being selected next for service where there is no queue, or from the poor mental health of a hospital patient or person in care, especially if they have little to occupy their time, or are in pain, and little slights are magnified in their mind.

Prison officers face the threat of violence, for example, because of the nature of the prisoner, or the way the prisoner perceives themself being handled. Substance abuse can lead to violence.

Anger developed by a member of the public at a long wait for service, is often the result of inadequate staffing levels, frustration at being told that 'the system', often inflexible software, doesn't allow something, or a misunderstanding because a staff member is fatigued from long hours and short staffing.

Intimidation, bullying or strong language may also be part of the culture in some workplaces, but certain staff may not find them acceptable.

Minority groups too can be the butt of violence. Schoolteachers may be the subject of violence from, for example, attempts to discipline students or because of marks awarded.

Working alone in jobs such as taxi and bus driving and fast food delivery may expose a worker to violence. Leaving a shop after hours with 'the takings' presents a risk.

Some areas of government operations not already mentioned, such as child and spouse protection, and family courts handling divorce and custody of children, can provoke strong feelings, which may be taken out on staff.

Effects of violence at work

Apart from the obvious effects such as death or serious physical injury, a violent incident can lead to effects such as a drop in morale, decisions by employees to leave, difficulty in staffing certain shifts, or sleep disturbance.

A person who constantly relives an incident, and thinks 'if only I had done that' is likely to find it difficult to concentrate and perform well. One incident may also result in an undesirable change of attitude to other inmates, patients, customers or members of the public.

What can be done to manage workplace violence?

The employer needs to consider the range of types of violence which might affect employees in their line of work, and the situations where it could occur. An assessment

Accident prevention

of which incidents are likely to be the more serious, and the more likely, needs to be made. The decisions which follow the assessments involve three steps:

- pre-incident
- during incident
- post-incident.

Pre-incident actions can include consultation with the workforce to gauge their feelings, opinions and ideas. It is important to develop procedures which aim to minimize the risk of violence. This is certainly feasible when the violence is not the result of planned criminal activity, but even in the latter case, certain procedures can reduce the risk of attack. One example is reduction of cash-based transactions, or minimization of cash held on premises with signs advertising this.

The procedures developed should then form the basis of staff training.

Actions during an incident must be aimed at minimizing the effect of the incident on staff, and other persons as the incident unfolds – calming and negotiation skills are important – and in certain cases staff must have a known line of escape.

Post-incident, these four matters need attention:

- providing medical care and hospitalization if anyone is injured
- providing counselling to employees and others who may have been traumatized by the incident
- resuming normal activity if possible once the incident has been brought to an end
- investigating the incident and making any changes needed to reduce the likelihood of a recurrence or to improve the outcomes through a better response.

It may be necessary to investigate and implement the improvements before normal activity can resume.

Once again procedures and training for the 'during incident' and post-incident phases are vital.

The approach to risk control for violence

The preferred order of control methods, as for other areas of health and safety at work, remains elimination, substitution, segregation, engineering, administrative and use of personal protective equipment.

An example of each is given here:

- elimination – a business operates during certain hours using card transactions only – no cash
- substitution – improve customer waiting areas to make them more welcoming, provide seating and a method to ensure that people retain their turn to be served
- segregation – prisoners or mental health patients known to be prone to violence are placed in another part of the facility with higher standards of security and surveillance and with full information flow between staff at shift handover
- engineering – closed circuit TV monitoring of areas, especially hidden areas like stairwells

- administrative – staff are rotated in and out of high stress customer contact tasks and support can be readily and rapidly summoned if needed
- personal protective equipment – bulletproof vests, or riot gear for police in some situations.

Violence between employees

It must be made very clear to employees what procedures are in place to deal with violent outbursts or incidents. There needs to be room to ascertain the reasons the incident has occurred, when not to proceed further against an employee with disciplinary action, and how to better manage relationships to reduce the chances of a repetition. Sometimes employees with a personality clash can be separated by assignment to different shifts or tasks.

Further reading

Annett, D. (2002). Why Cell Phones are Dangerous. *Accident Prevention (Canada)*, March/April, p. 9.

Butyn, S. (2002). Breaking the Code. (terrorism). *Accident Prevention (Canada)*, March/April, p. 23.

Di Pilla, S. and Vidal, K. (2002). State of the Art in Slip Resistance Measurements. A Review of Current Standards. *Professional Safety*, **47(6)**, 37–42.

Editorial. (2002). Liberty Mutual Study Reveals Leading Injury Causes. *Professional Safety*, **47(6)**.

Guastello, S.J. (1993). Do We Really Know How Well Our Occupational Accident Prevention Programs Work? *Safety Science*, **(16)**, 445–63.

Hopkins, A. (2003). Fault Trees, ICAM and Accimaps: A Methodological Analysis. *Safety in Australia*, **25(2)**, 13–21.

James, M. (2001). Learning the Ropes. *The Safety and Health Practitioner*, **19(3)**, 28–30.

Limb, R. (1999). That's Entertainment. *Health and Safety at Work*, **21(1)**, 14–15.

McEwen, S. (2000). Slips, Trips and Falls on the Same Level. *The Safety and Health Practitioner* **18(9)**, 40–1.

NIOSH. (2002). *Violence. Occupational Hazards in Hospitals*. Cincinnati: US DHHS-NIOSH.

NIOSH. (1996). *Violence in the Workplace. Risk Factors and Prevention Strategies*. Cincinnati: US DHHS–NIOSH.

NIOSH. (2000). *Worker Deaths by Falls. A Summary of Surveillance Findings and Investigative Case Reports*. Cincinnati: US DHHS NIOSH.

Pratt, S.G., Fosbroke, D.E. and Marsh, S.M. (2001). *Building Safer Highway Work Zones*. Cincinnati: US DHHS-NIOSH.

Quinley, K. and Schmidt, D. (2002). *Workplace Terrorism: Business at Risk*. Cincinnati: National Underwriter Co.

Schofield, D. et al. (2000). Recreating Reality: Using Computer Generated Forensic Animations to Reconstruct Accident Scenarios. *Proceedings of Minesafe International Conference*. Perth, Chamber of Minerals and Energy of Western Australia (Inc), 483–93.

Smith, S.J. (2002). Workplace Violence – 10 Tips for a Proactive Prevention Program. *Professional Safety*, **47(11)**, 34–43.

Standards Australia. (1990). AS1885.1 *Workplace Injury and Disease Recording Standard*. Sydney: Standards Australia.

Activities

1. In a workplace of your choice, interview five people at various levels and find out what they believe causes accidents and injuries at work. Report your results and comment on them based on your studies of the factors leading to accidents.
2. In the same workplace, describe the elements of the work system which are relevant to health and safety for one particular task. Include environment, tools/equipment/plant, people, materials, work methods and also organizational factors. Describe three interactions between two or more of these elements which you believe are critical to accident prevention.
3. Explain the method of workplace accident prevention (i.e. the key areas or coding used to describe the accident) under the standard applying in your jurisdiction.
4. Outline an action plan you would use for an accident investigation to establish these key factors.
5. Using the workplace you selected earlier, draw up a programme to promote health and safety in that workplace, using five approaches.
6. Explain how in your jurisdiction you would go about accessing a workplace accident and injury data from national, state, provincial or industry sources. Select an area of interest, e.g. women workers 18–25 years old, compare the data with that for male workers 18–25, and comment.
7. Explain some simpler methods of assessing the risk of engineering equipment unit (e.g. an elevating work platform) or engineering systems failure.
8. What steps would you take when reporting to management on health and safety so as to set logical priorities rather than trigger 'flavour of the month' reactions?

Appendix 6.1 – Accident models

The references can be found below this list.

Domino Theory (Heinrich, 1931)
Multicausal Model (Gordon, 1949)
Critical Incident Technique (Flanagan, 1954)
Combination of Factors Model (Schulzinger, 1956)
Goals Freedom Alertness Theory (Kerr, 1957)

Energy Exchange Model (Haddon et al., 1964)
Decision Model (Surry, 1969, in Viner 1991a)
Behavioural Methods (Hale and Hale, 1970, Anderson et al., 1978)
Fault Tree Analysis II (Meister 1971, Hoyos and Zimolong, 1988)
Error Model (Wigglesworth, 1972)
Life Change Unit Model (Alkov, 1972)
Hazard Carrier Model (Skiba, 1973, Hoyos and Zimolong, 1988)
Task-Demand Model (Waller and Klein, 1973)
Multilinear Events Sequencing Model (Benner 1975)
Management Oversight and Risk Tree (MORT) (Johnson, 1975)
Multilinear Events Sequencing (Benner, 1975)
Systems Safety Analysis (Smillie and Ayoub, 1976)
Risk Estimation Model (Rowe, 1977)
Danger Response Model (Hale and Perusse, 1977)
Incidental Factor Analysis Model (Leplat, 1978)
Accident Sequence Model (Ramsey, 1978, quoted in Sanders and McCormick, 1987, and Ramsey, 1985)
Psychological Model (Corlett and Gilbank, 1978)
Domino/Energy Release (Zabetakis, quoted in Heinrich et al., 1980)
Stair Step Model (Douglas, quoted in Heinrich et al., 1980)
Motivation Reward Satisfaction Model (Petersen, quoted in Heinrich et al., 1980)
Energy Model (Ball, quoted in Heinrich et al., 1980)
Systems Model (Firenze, quoted in Heinrich et al., 1980)
Epidemiological Model (Suchman, quoted in Heinrich et al., 1980)
Updated Domino Model (Bird Jr, quoted in Heinrich et al., 1980)
Updated Domino Model (Adams, quoted in Heinrich et al., 1980)
Updated Domino Model II (Weaver, quoted in Heinrich et al., 1980)
Task Ability Model (Drury and Brill, 1980)
OARU Model (Kjellen and Hovden, 1981, Kjellen and Larsson, 1981).
Traffic Conflicts Technique (Zimolong, 1982)
Signals Passed at Danger Decision Tree Model (Taylor, R.K. and Lucas, D.A. in Ch. 8 of Van der Schaaf, Lucas and Hale, 1991)
Ergonomic and Behavioural Methods (Kjellen, 1984)
Human Causation Model (Mager and Pipe, 1984)
Near Accidents and Incidents (Swain, 1985)
Behaviour Model (Rasmussen, 1986)
Contributing Factors Model (Sanders and Shaw, 1987)
Hazard Carrier Model (Hoyos and Zimolong, 1988)
'Comet' Model (Boylston, 1990)
Comprehensive Human Factors Model (DeJoy, 1990)
View of Worker on Safety Decisions Model (Saari, 1990)
Epidemiological Model (Kriebel, quoted in Cone et al., 1990)
Universal Model (McClay, 1990)
Federation of Accident Insurance Institutions (Finland) Model (Seppanen, 1997)
Question Tree Model (Hale et al. in Van Der Schaaf, Lucas and Hale, 1991)

Occurrence Consequence Process Model (Viner, 1991b)
Onward Mappings Model based on Resident Pathogens Metaphor (Reason, 1991)
Functional Levels Model (Hurst et al., 1992)
Tripod Tree (Wheelahan, 1994)
Attribution Theory Model (DeJoy, 1994)
Cindynic Hyperspace (Kervern, 1995)

In addition Laflamme (1990) cites decisional models by Saari, Singleton, Hale and Hale, Lagerlof, Andersson, Hale, and Hale and Glendon, sequential models by Cuny and Leplat with insights by Winsemius, Saari and Faverge, sequential models using the INRS method by Monteau, Leplat, Moyen et al., and Quinot and Moyen, and the organizational models of Arsenault and Cloutier and Laflamme. Some of the above models relate to information processing or information flow (e.g. Surry), and Saari (1984) cites more such models (apart from those already mentioned) by Goeller, Fell, Hakkinen, Feggetter, Hoyos et al., WHO, and Lawrence. There is also a Perception Response Mismatch model similar to Surry model which could not be sourced.

References to accident models

Adams, E.E. (1991). The Quality Revolution: A Challenge to Safety. *Professional Safety*, **36(8)**, 22–8.

Adams, N. (1988). Education and Training for OHS: The Need to Integrate Two System Approaches. *Journal of Occupational Health and Safety – Aust. NZ*, **4(4)**, 301–5.

AFCC (Australian Federation of Construction Contractors). (1987). *Safety – A Matter of Management*. Melbourne: AFCC.

Alkov, R.A. (1972). The Life Change Unit and Accident Behaviour. *Lifeline* Sep.–Oct., Norfolk, Va., US Naval Safety Center. See also Heinrich et al., 1980.

Bensiali, K., Booth, A.T. and Glendon, A.I. (1992). Models for Problem Solving in Health and Safety. *Safety Sci.*, **15**, 183–205.

Benner, L. (1975). Accidents Investigations – Multilinear Events Sequencing Methods. *Jounral of Safety Research*, **7**, 67–73.

Benner, L. Jr. (1985). Rating Accident Models and Investigation Methodologies. *Journal of Safety Research*, **16(3)**, 105–26.

Boylston, R.P. (1990). *Managing Safety and Health Programs*. p.100, New York: Van Nostrand Reinhold.

Bradley, G.L. (1989). The Forgotten Role of Environmental Control: Some Thoughts on the Psychology of Safety. *Journal of Occupational Health and Safety – Aust. NZ*, **5(6)**, 501–8.

Burgoyne, J.H. (1993). Reflections on Accident Investigation. *Safety Sci.*, **16**, 401–6.

Cone, J.E., Makofsky, D. and Daponte, A. (1990). Fatal Occupational Injury and Energy Exchange: A Replication of Kriebel's Model. *Journal of Occupational Accidents*, **12**, 187.

Corlett, E.N. and Gilbank, G. (1978). A Systematic Technique for Accident Analysis. *Journal of Occupational Accidents*, **2**, 25–38.

Davies, E. (1993). Safe Production – Performance, Principles and Practice. *Journal of Occupational Health and Safety – Aust. NZ*, **9(6)**, 565–76.

Dejoy, D.M. (1990). Towards a Comprehensive Human Factors Model of Workplace Accident Causation. *Professional Safety*, **35(5)**, 11–16.

DeJoy, D.M. (1994). Managing Safety in the Workplace: An Attribution Theory Analysis and Model. *Journal of Safety Research*, **25(1)**, 3–17.

Drury, C.G. and Brill, M. (1980). *New Methods of Consumer Product Accident Investigation in Proceedings of the Symposium – Human Factors and Industrial Design in Consumer Products*. Medford Mass., Tufts University, May, 196–211.

Ferry, T.S. (1988). *Modern Accident Investigation and Analysis*. (2nd edn.). New York: Wiley.

Ferry, T.S. (1990). *Bridging Engineering Gaps and Vacuums from Design to Disposal*. Journal of Occupational Accidents, **13**, 17–31.

Fine, B.J. (1963). Introversion – Extroversion and the Motor Vehicle Driver Behaviour. *Perceptual and Motor Skill*, **16**, 95–100.

Flanagan, J.C. (1954). The Critical Incident Technique. *Psychological Bulletin*, **51**, 327–58.

Gibson, J.J. (1961). The Contribution of Experimental Psychology to the Formulation of the Problem of Safety. In *Behavioural Approaches to Accident Research*, pp. 77–89. New York: Association for the Aid of Crippled Children.

Gordon, J.E. (1949). The Epidemiology of Accidents. *American Journal of Public Health*, **9**, 504–15.

Guarnieri, M. (1992). Landmarks in the History of Safety. *Journal of Safety Research*, **23(3)**, 151–8.

Haddon, W., Suchman, E.A. and Klein, D. (1964). *Accident Research Methods and Approaches*. New York: Harper and Row.

Haddon, W. (Jr.), (1973). Energy Damage and the Ten Countermeasure Strategies. *Journal of Trauma*, **13(4)**, 321–31.

Hale, A.R. and Perusse, M. (1977). Attitudes to Safety: Facts and Assumptions, in Phillips, J. (ed). *Safety at Work*, SSRC, Conference Paper No.1, Centre of Socio-legal Studies, Wolfson College, Oxford, reported in Kjellen q.v.

Harms-Ringdahl, L. (1993). *Safety Analysis. Principles and Practice in Occupational Safety*. London: Elsevier.

Heinrich, H.W. (1931). *Industrial Accident Prevention*. New York: McGraw-Hill.

Heinrich, H.W., Petersen, D. and Roos, N.R. (1980). *Industrial Accident Prevention*. New York: McGraw-Hill.

Hocking, B. and Thompson, C. (1992). Chaos Theory of Occupational Accidents. *Journal of Occupational Health and Safety – ANZ*, **8(2)**, 99–108.

Hoyes, T.W. and Glendon, A.I. (1993). Risk Homeostasis: Issues for Future Research. *Safety Sci.*, **16(1)**, 19–34.

Hoyos, C.G. and Zimolong, B. (1988). Advances in Human Factors/Ergonomics 11. *Occupational Safety and Accident Prevention*. Amsterdam: Elsevier.

Hudson, P.T.W. and Verschuur, W.L.G. (1994). *A Review of the Necessity for Procedures and the Problems Associated With Them*. (draft). University of Leiden.

Hurst, N., Bellamy, L.J. and Wright, M.S. (1992). *Research Models of Safety Management of Onshore Major Hazards and their Possible Application to Offshore Safety in Major Hazards Offshore and Onshore*. Rugby: Institute of Chemical Engineers, pp. 129–48.

Illywhacker, (1994). Opinion, in The Health and Safety Professional, Newsletter of the Safety Institute of Australia Victorian Division, **3(5)**, 8.

Johnson, W.G. (1980). *MORT Safety Assurance Systems*. New York: M. Dekker Inc.

Kerr, W. (1957). Complementary Theories of Safety Psychology. *Journal of Social Psychology*, **45**, 3–9.

Kervern, G.-Y. (1995). Cindynics: The Science of Danger. *Risk Management*, **42(3)**, 34–42.

Kjellen, U. and Horden, J. (1993). Reducing Risks by Deviation Control. *Safety Science*. **16**, 417–38.

Kjellen, U. and Larsson, T.J. (1981). Investigating Accidents and Reducing Risks – A Dynamic Approach. *Journal of Occupational Accidents*, 3, 129–40.

Kjellen, U. (ed.). (1984). *Occupational Accident Research*. Amsterdam: Elsevier.

Kletz, T.A. (1983). *HAZOP and HAZAN. Notes on the Identification and Assessment of Hazards*. Rugby, England: The Institute of Chemical Engineers.

Komaki, J. et al. (1978). Effect of Training and Feedback: Component Analysis of Behavioural Safety Program. *Journal of Applied Psychology*, **65(3)**, 434–45.

Krause, T.R. and Hidley, J.H. (1992). On Their Best Behaviour. *Industrial Safety and Loss Control*, 29–33.

Laflamme, L. (1990). A Better Understanding of Occupational Accident Genesis to Improve Safety in the Workplace. *Journal of Occupational Accidents*, 12, 155–65.

Lepisto, J. (1990). *Journal of Occupational Accidents*, **12(1–3)**, 99.

Leplat, J. (1984). *Occupational Accident Research and Systems Approach* in Kjellen, U. (ed.) supra.

Mager, R.F. and Pipe, P. (1984). *Analyzing Human Performance Problems*. Belmont Cal., David S. Lake, quoted in Thygerson, p. 72, q.v.

McClay, R.E. (1990). Finding the Proximal Cause of Loss Incidents Using the Universal Model. *Journal of Occupational Accidents*, 12, 187–8.

McKenna, F.P. (1983). Accident Proneness: A Conceptual Analysis. *Accident Analysis and Prevention*, **15(1)**, 65–71.

McKenna, F.P. (1985). Do Safety Measures Really Work? An Examination of Risk Homeostasis Theory. *Ergonomics*, **28(2)**, 489–98.

McKenna, F.P. (1988). What Role Should the Concept of Risk Play in Theories of Accident Involvement? *Ergonomics*, **4**, 469–84.

McLean, C. (1994). Accident Investigation – An Opportunity for Prevention. *Australian Journal of Workplace Health and Safety*, **12(1)**, 9–13.

Meister, D. (1971). *Human Factors: Theory and Practice*. New York: Wiley.

Petersen, D. (1978). *Techniques of Safety Management*, 2nd edn. 16–19, USA: McGraw-Hill Kogakusha.

Pitzer, C. (1993). *Safety Psychology – Managing Safety Attitudes in the Real World*. Proceedings of the Minesafe Conference, Perth, Western Australia, March, pp. 543–58. Perth: Chamber of Mines and Energy of WA (Inc.)

Purswell, J.L. and Rumar, K. *Occupational Accident Research: Where Have We Been and Where Are We Going?* in Kjellen, U. (ed.), supra.

Ramsey, J.D. (1985). Ergonomic Factors in Task Analysis for Consumer Product Safety. *Journal of Occupational Accidents*, **7**, 113–23.

Rasmussen, J. (1986). *Information Processing and Human Machine Interaction*. Amsterdam: North Holland.

Reason, J. (1991). *Too Little and Too Late; A Commentary on Accident and Incident Reporting Systems*, in Van der Schaaf et al. (eds), pp. 9–26, q.v.

Rowe, W.D. (1977). *An Anatomy of Risk*. New York: John Wiley and Sons, quoted in Viner 1991a, p. 39, q.v.

Saari, J. (1990). On Strategies and Methods in Safety Work. *Journal of Occupational Accidents*, **12**, 107–17.

Salminen, S. (1994). Risk Taking and Serious Occupational Accidents. *Journal of Occupational Health and Safety; Aust. NZ*, **10(3)**, 267–74.

Sanders, M.S. and McCormick, E.J. (1987). *Human Factors in Engineering and Design* 6th edn, p. 622. New York: McGraw Hill.

Sanders, M.S. and Shaw, B.E. (1987). *Research to Determine the Frequency and Cause of Injury Accidents in Underground Mining*. Proceedings of the Human Factors Society, 31st Annual Meeting New York, 19–23.10.87. Vol. 2, 926–30. Santa Monica, Human Factors Society.

Schulzinger, J.S. (1956). The Accident Syndrome. *Springfield*, **11**. Charles C. Thomas.

Seppanen, S. (1997). *The Investigation Procedure of Fatal Employment Accidents*. Work Safety Health, Helsinki, Finnish Institute of Occupational Health.

Simard, M. and Marchand, A. (1994). The Behaviour of First Line Supervisors in Accident Prevention and Effectiveness in Occupational Safety. *Safety Sci.*, **17**, 169–85.

Smillie, R.J. and Ayoub, M.A. (1976). Accident Causation Theories: A Simulation Approach. *Journal of Occupational Accidents*, **1**, 47–68.

Swain, A.D. and Guttman, H.E. (1983). *Handbook of Human Reliability Analysis with Emphasis on Nuclear Power Plant Applications*. Washington: Sandia National Laboratories, NUREG/CR-1278, US Nuclear Regulatory Commission.

Thompson, C., Thompson, B. and Hocking, B. (1992). Application of Chaos Theory to the Management of OHS. *Journal of Occupational Health and Safety – ANZ*, **8(2)**, 109–99.

Thygerson, A.L. (1986). *Essentials of Safety, 3rd edn*, p. 85. New Jersey: Prentice-Hall.

Van Der Schaaf, T.W., Lucas, D.A. and Hale, A.R. (eds) (1991). *Near Miss Reporting as a Safety Tool*. Oxford: Butterworth-Heinemann.

Viner, D. (1991a). *Accident Analysis and Risk Control*. Carlton South, Victoria: VRJ Delphi.

Viner, D. (1991b). Quantitative Risk Management. *Journal of Occupational Health Safety – Aust. NZ*, **(1)**, 59–67.

Waller, J. and Klein, D. (1973). Society, Energy and Injury, Inevitable Triad? in *Research Directions Towards the Prevention of Injury*, pp. 1–37, Bethesda Maryland, US DHEW.

Wheelahan, B. (1994). Accident Investigations Using the Tripod Method. *Australian Petroleum Exploration Association Journal*, **34**, Pt. 1, 155–9.

Wigglesworth, E.C. (1972). A Teaching Model of Injury Causation and a Guide for Selecting Countermeasures. *Occupational Psychology*, **46**, 69–78.

Worksafe Australia. (1994). *National Standard for Plant*. Canberra: AGPS.

Zimolong, B. (1982). *Verkehrskonflikttechnik–Grundlagen und Anwendungsbeispiele*. Koln, Bundesanstalt fur Strassenwesen, Unfall-und Sicherheitsforschung Strassenverkehr, Heft 35.

Included in these references are some which discuss accident causation and prevention generally. *Purswell and Rumar* and *Hoyos and Zimolong* discuss types of accident models while *Benner* evaluates the usefulness of different models and the use made of them by accident investigation agencies.

7

Risk engineering

WORKPLACE EXAMPLE

In the UK, an employee went into the pit of a carding machine in a yarn factory. The machine was in operation and he went in to retrieve a fallen 'end'. When his hand became trapped in the condenser, he was unable to be rescued until the time delay released the door lock. This was because he had shut the safety-interlocked door after him when entering.

The employee broke his finger in the accident only a week after an inspector had visited and referred the company to guidance on the safety of carding machines. In court the company managing director was quoted as saying: 'We have to admit that we have been woefully negligent but genuinely thought at the time we had a good system.' The fine was GBP 4000.

(Source: the Health and Safety Executive, UK.)

Risk and reliability

Safety technology

In 1854 Elisha Otis was at the Crystal Palace exposition in New York. After being hauled up in an elevator in an open-sided shaft, halfway up he had the rope cut with an axe, calling 'All safe, gentlemen'. The shaft conveyance safety arrest mechanism (safety dogs) then took over.

Safety technology has a long history and is an important part of risk engineering. In the eighteenth century a competition was run for a chronometer which would operate accurately at sea so that longitude could be measured – essential for the safe operation of seagoing vessels. It was won by a John Harrison.

The growth in risk engineering since those far off days has been huge. One way of dividing the field is to decide whether the engineering contributes to active safety – the factors which aim to prevent an accident, or passive safety – the factors which limit injury if there is an accident. Note that in fire protection the reverse terminology is in use – passive protection being a mixture of features which both prevent and limit fire and injury, and active protection being fire detection and extinguishment.

The range of safety technology is vast and includes guarding; presence-sensing and interlocks on machinery; pressure relief on boilers; explosion dampers in ductwork carrying flammable vapours or fine solids; double insulation; earth leakage (ground fault) devices for electrical safety; temperature pressure and flow control software in chemical process industries; fall restraints for roof work; and rockbolts and shotcrete for underground mining.

Risk engineering

The concept of risk engineering is not as daunting as it sounds. It involves balanced judgement and quantifying or measuring the risks faced. The frequency of loss events depends on both those resulting from failure in operation and those from failure of equipment. The first of these in turn depends on how complex a task is, as well as the quality of operational management. The latter partly reflects training and supervisory standards. In some cases, the desire to quantify risk leads people to make assumptions, yet those assumptions, in part, have a wide margin of error. Risk-based regulations and codes may have flexibility leading to economic benefits for an organization. However, if the previous administrative controls (e.g. inspection frequency) are to be relaxed, the assessment of the risk must be precise enough not to lessen safety standards. Otherwise new safety technology or improved safety management systems may be needed to offset any potential lessening of safety resulting from the flexible approach. This improvement in the techniques of achieving safety is, of course, a linchpin of the Robens-based idea of a flexible approach which encourages such initiatives.

(*With acknowledgement for some of the above ideas to Tweeddale, M. Quantify Exposures*, Corporate Risk, *September 1994, p. 35.*)

Reliability

The standard IEC 61508 deals with electrical, electronic, or electronic programmable systems, but many of the principles it applies to the system life cycle extend well beyond those. System errors can result in failure to safety with a revealed fault, a neutral fault with no effect or failure to danger, which can be revealed or unrevealed. Clearly, the key is to focus on what must be the unrevealed failures to danger. The standard also considers reliability. Here there are two main considerations – low demand mode of operation and high demand, or continuous mode, of operation. The safety integrity level (SIL) for the first is the average probability of failure to perform the design function on demand. For the second, the SIL is set by the probability per hour of a dangerous failure. It is essential to control random failure in the type of equipment IEC 61508 deals with, due to hardware, and also systematic failure due to hardware, software and operations.

Random failure comes from equipment degrading due to the stresses on it. The timing of such failures is not predictable but they uniformly degrade throughout the service life.

248 Enhancing occupational safety and health

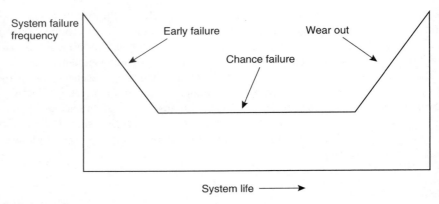

Figure 7.1 Bathtub or reliability curve

The causes of systematic failures can include modification errors; maintenance error; software error; systematic failure of hardware; and specifications error.

(*With acknowledgement for some of the above ideas to Walters, K.* IEC 61508 Using Electrical/Electronic/Programmable Electronic Systems to Manage Risk. *Engineers Australia paper 4.11.99, Perth.*)

One long established concept in risk engineering is the bathtub curve. See Fig. 7.1.

The slope downwards on the left represents the decrease in faults (increase in reliability) during the commissioning, ramp-up, run-in or teething period. The flat base represents the period of sustained low fault operation (chance failure). The upward curve on the right represents the steady increase in faults (decrease in reliability) as equipment or a system reaches the end of its useful life. Obviously design, planned inspection, maintenance or parts replacement can have an impact on various parts of the curve.

Causes of early failure include poor policy, planning, training, and system design as well as resistance to change. Factors in chance failure may include drop in employee skill level; lack of maintenance; overload; environment, e.g. dust; no performance monitoring; lack of effective controls; or no planned reviews. Wearout failure is more likely where financial planning is poor and resources are switched, there is failure to retrain employees, only parts of a system are corrected, outdated methods of work persist, or staff are not consulted.

Technological risk

The relationship between people and the technology used to build the world we live in has changed since the New York Exposition of 1854 mentioned earlier, and the example given of the elevator arrest mechanism.

We still marvel at the often apparent wizardry of new technology and its capacity to enhance productivity, improve efficiency and generally improve our quality of life. However, there is a dark side and that is that we have reached a point in our history where new technologies are championed without a clear understanding in the minds of designers and engineers as to all the risks created by that technology.

It could be argued that the introduction of the motor car and the use of asbestos and lead introduced the dark side of technology well before now.

Although the technologies and activities associated with these three items have served us well there has been a price to pay in terms of our exposures. All three have taken many lives and created much suffering and misery. Their value to the development of our society will need to be measured in the long term.

Safety technology strategies

Safety technology should be seen in the context of the four main strategies used to manage danger and prevent harm:

- injury control
- hazard management
- accident prevention
- error control.

Progress in all four areas makes a contribution to prevention of injury.

The most common advances in safety technology are generally found in 'injury control' where the application of personal protective equipment (PPE) dominates the technology.

The market is awash with safety equipment aimed at injury prevention and control. Safety footwear, glasses, ear protection and respiratory protection are now being designed to offer this control and there is a recognition of the need to consider comfort, fashion and better compatibility where multiple pieces of equipment are required.

'Hazard management' is a four-step process – identification, evaluation, assessment and control.

Safety technology has improved in the application of each step in this safety process. The use of mobile analytical equipment has made the identification and evaluation of some chemical hazards faster, and reduced the associated costs. The developing areas of robotics and video application have enabled the increasing automation of high energy work. Examples are found in wood chipping, milling, steel rolling mills and remote mining equipment. These ongoing work changes are increasing the opportunities for hazard control.

'Accident prevention' is enhanced through the use of techniques, particularly electronically-based equipment, to monitor the risks or shifts in the risk exposure in the work process. Equipment is available to detect systems and environments in the unstable state. Critical error management occurs in many industries where the use and application of warning devices is employed to determine variations in everything from ground stability in mines to chemical exposure levels. All such devices are subject to regular calibration and maintenance in order to maintain high levels of reliability.

'Error control' is the optimum level of control. Most organizations have some form of error control. However, the type of work activity and the work culture, along with the associated start-up costs of data collection and the need to employ well trained and competent staff, have limited the potential of error control. Some industries, such as the aerospace industry, are typical error control users. There are several software programmes available to assist in error identification through the development of logic- and event-tree analysis applications.

Systems engineering

Technological change

All technological change will have an impact on the risk levels in the system. Some risks will be reduced, even eliminated, while new risks are created. Proper identification and evaluation are essential elements of change management.

For example, people working from ladders faced several risks when working at heights. These included poor working posture, restricted reach, slipping of ladder, slipping on rungs, splinters from stiles, ladder struck by mobile equipment, twisting damage, and lifting and falls injuries.

Introducing the elevating work platform (EWP, e.g. cherrypicker or scissorlift) would offer a reduction in the injuries and accidents associated with working from ladders. Clearly, this is the case; however, we have seen new risks and injuries arise out of the use of EWPs. The errors that now result in 'system yielding' result from misuse of the technology, poor machine control design, prime mover instability, overextension of the apparatus, and working outside restraining limits. Yielding systems result from the loss of integrity of the system controls. Yielding systems may occur at the wearout stage of the equipment life cycle, when the system starts to decay – as yielding systems are generally the result of an ageing process, lack of resources and loss of skill (e.g. cracking of a dam wall, arcing in electrical equipment, tension and stress in an office, or rusting). See Fig. 7.1.

This lack of risk identification appears common in many areas of human progress. As engineers and entrepreneurs seek to reap the benefits of technological change, openly announcing the emergence of new risk exposures could be seen as having a negative benefit. The attention deficit while using a mobile (cell) phone in a vehicle is a current example.

Managing the risks of technology

Risk analysis in the application of technology has generally focused on the failure of technology in either the functioning of the technology or the immediate environment where the technology is used. Road vehicle usage plays a critical role in many work activities and in getting to and from work, as well as, of course, in off-the-job activities, so it provides a relevant example, given here.

A tyre blow-out on a car may lead to loss of control over travel direction. The immediate concern is the driver, the passengers and the vehicle. The controls we use to deal with loss of directional control and to avoid adverse outcomes and consequences may include car design features and enhanced driving skills. We cannot expect the other road users to cope effectively with this type of sudden yielding of part of the system. However, they are exposed to the hazard created through the loss of control over directional travel as a result of the tyre failure.

The following are typical ways we have traditionally managed safety and risk from technology.

Hazard management

The hazard created as a result of the accident is that the car is out of control.

If the operator has the skill and the time opportunity to bring the car back under control, the hazard will subside. Control means not merely to rest, but to a state where danger has also subsided.

Accident prevention

This option is not available once control over directional travel is lost.

The principal strategy to prevent an accident after the blow-out occurs becomes the handling skill of the operator.

Error control

This involves improving the design of the technology to account for sudden failure, so control over the direction of travel is not lost (i.e. not operator dependent).

Exposures

Once a hazard is created there are immediate direct and indirect exposures. Direct exposures include other road users, pedestrians, people in their homes, animals, plants, property, etc. Indirect exposures also result in the form of family, friends, social involvement, rescue workers, and costs to society.

Outcomes

The outcome of such an event is unlikely to have a nil result. The concept of No Detectable Damage (NDD) is preferred in that if the outcome does not result in immediate direct loss, the event outcome is often quickly forgotten. However, the event may have latent effects that at the time of the exposure go undetected.

(See Chapter 6 for a detailed explanation of NDD in the Hegney–Lawson System Risk Model.)

Consequences

The outcome will govern our perception of the risk (safety) and our acceptance of continuing our relationship with the technology and the environment in which we employ its use.

Given that tyre failure is not a rare event (a recent example is a particular brand used on a recreational use vehicle – RUV or US, SUV), some tyre manufacturers will promote the idea that their tyres are less prone to failure. No manufacturer could reasonably be expected to produce a tyre that is certain not to fail. Tyre research and consumer demand will lead to an improvement in tyre reliability. However, failure rates will remain. Regardless of to what level that rate falls, tyre failure will continue.

The principal action taken to control the tyre blow-out experience will ultimately depend on the effect this event has on our society. To manage this type of technological failure we consider controls at each level of the event. This suggests a systems approach to the engineering associated with the risk.

The method of dealing with loss events should cover all the engineering capabilities of the system. Only after all the engineering improvements have been considered and made, should attention be turned to the role of the operator. Engineering change will always be impacted upon by cost factors and available knowledge. To this end any change to the status quo will be governed by practicability. Practicability in this context must be seen as a real barrier against progressive change, not as an excuse for inaction.

Hazard management provides a range of options:

- strengthen cars against impact
- separate cars travelling in opposite directions
- reduce speed limits
- protect assets with roadside barriers
- separate road users and pedestrians by physical barriers.

Accident prevention includes the following options:

- car design improvements to cope with tyre failure
- formal inspections by approved inspectors every six months
- monitoring by traffic police for tyre maintenance performance.

Error control for tyres can include:

- improve tyre design to reduce wall failures
- develop methods of field testing for tyre faults
- install incorrect-inflation-warning device
- introduce written maintenance activity book
- introduce ongoing performance standards criteria
- capacity for non-destructive tests.

Exposure and outcome

A system failure such as a blow-out may result in an exposure and with an outcome so rapid that there is no time to avoid the damage. The outcome can still range in severity from NDD to death and even catastrophe.

Any damage will be determined by the amount of energy available to cause harm. The energy exchange potential is governed by factors such as the speed, mass and shape of the car. Physical damage will be determined, if contact is made, by the capacity of the contact object to absorb the energy exchange. Psychological damage is far more difficult to readily identify. However, post traumatic stress syndrome is finally receiving the attention it rightly deserves. Horrible events such as the September 11 World Trade Center terrorist attack in New York and the Bali Bombing in Denpasar Indonesia have highlighted the need to consider the psychological damage, both personal and societal, as well as the physical trauma that result from such events.

Errors and reasons for errors

The difficulty faced in risk engineering and by those responsible for safety management is the huge variation in event outcomes. This variation affects decision making and may lead to over compensation or a delay in making the appropriate changes. Changes in the direction of laws governing occupational safety and health will continue to favour self-regulation ahead of prescriptive laws. Prescriptive laws through regulations, etc. will be unable to keep pace with new technologies and the environments they operate in.

Errors, accidents and hazards are reasonably predictable, however injury experience is far more difficult to predict. It may well be that, presently, exposure outcomes are, in the main, a function of luck. There would be some clear examples where that is not the case. If an aircraft loses power at 30 000 feet and control is lost and unable to be recovered, the outcome is able to be predicted with a high degree of certainty. The same cannot be said of the tyre blow-out.

Technology is also creating a challenge in areas where we need greater vigilance, such as the spread of terrorism and disease. The rapid spread of SARS would not be possible without the opportunities afforded to the virus through aircraft and the economic implications of global trade and travel. The speed of electronic transmission of a software virus, with its potential to affect information technology systems and hence safety, is even faster.

The mechanisms set up in countries and organizations to deal with threats, such as the spread of disease and injury, will find it difficult to cope with the rapid onset of them. The hazard-exposure-outcome experience is too fast. Traditional agencies for detection such as public health functions, general medical practice, disease control centres and workplace OHS functions are unable to respond fast enough to counter dangers arising from the effects of some exposures.

Another example was the onset of what became known as RSI, *repetitive strain injury* and later as *overuse injury*. The disorder resulted from our lack of understanding and management of the opportunities to be had from computing.

People employed in data entry and typing activities were worst affected. Although the injury had been well documented and understood for centuries, the sheer scale of the damage now linked to the new computer technology in the 1980s caused massive disruption in some communities and industries.

The exposure-outcome results often form the basis of the application of safety technology to injury prevention and control. Industrial deafness has drawn the attention of safety engineering experts in the application of hearing protection devices. Respiratory protection has required the development of a range of controls to address a wide range of risks, from simple dust masks to the more elaborate use of positive pressure air-supplied respirators.

Deaths and injury resulting from falls are always difficult to address. However over the years there has been the development of restraining and capture devices to minimize accident outcomes. Regrettably, the back remains a vulnerable part of the worker's anatomy in terms of exposure to medium- and long-term damage. The application of safety technology to reduce the lifting and twisting tasks in industry, through mechanization and automation, has been ongoing for decades. However, the nature of work and the activities of people will see the back continue to be exposed to damage. It will be in the management of this part of the body by methods other than our current concept of safety technology that the solution to back damage will be found.

Electrical safety

Introduction

Electricity is one of the most common power sources in our home and workplace environment. Electrical equipment, which uses electrical energy, forms a system which needs to be treated with respect. In other words, users and those in the vicinity of electricity are required to be safety aware. Electrical equipment used in workplace environments and tasks needs to be considered for safety management. This includes such areas as the type of equipment used, handling of equipment, and maintenance of systems in place to reduce the risks from electrical faults.

The risks with using electrical equipment will be explored in this section, along with practices which enhance electrical equipment safety.

How does shock affect the body?

An electric shock affects the body when a current is applied through an entry point. The entry point is where a live conductor exposes the body to current.

Once current passes through the body on its way to earth, nerves and muscles experience a disturbance of function. The path that the current takes will depend on the location of the entry site. The exit site is where the current leaves the body to earth. At the entry and exit points, a person may experience severe stabbing and burning pain due to muscle contraction. The degree of shock an individual may receive is dependent on several factors, including skin resistance. Skin resistance is greatly reduced when the skin is wet or moist, and so the degree of shock will be greater.

Other factors include the:

- magnitude of electrical current
- length of time the current is applied
- length of current path through the body – for example, if a current has an entry point through the hand and an exit point through the foot, a greater systemic effect is likely
- medical and health status of the individual.

Terminology

Throughout the electrical safety section, it is important that you are familiar with some common terms that are associated with electricity.

Current
: Refers to the actual flow of electricity and is measured in amperes.

$$\text{Current} = \frac{\text{Voltage}}{\text{Resistance}}$$

Voltage
: Is the driving force or pressure of electrical energy and is measured in volts.

Resistance
: Refers to the characteristics of the circuit path which offers resistance or retards current flow. Resistance is measured in ohms (Ω).

Power
: Is the current multiplied by voltage and is measured in watts (W).

Alternating current frequency
: Is measured in Hertz (Hz) (i.e. cycles per second).

The transfer of electrical energy

Electrical power supply is predominantly in the form of alternating current (a.c.). Individual national systems may operate on 100 or 110 v, 240 or 250 v, and 50 or 60 Hz.

Direct current (d.c.) is supplied in some regions, notably where there are mining communities, through generator power supplies. The hazard with using direct current as compared with alternating current, is that it is much more difficult to switch off. It may also cause more severe shock injuries.

The majority of electrical appliances used today convert a.c. to d.c. through built-in rectifiers. Therefore, equipment servicing should strictly follow regulations, codes and applicable standards.

Acts, regulations, standards and codes on electrical safety

Specific legislation

Specific requirements under the occupational health and safety and electrical safety acts, regulations, and codes of practice in your country, province, state, or territory need to be followed.

Regulations, codes and standards

Employers must ensure that electrical wiring of plant and equipment complies with standards for electrical wiring in buildings, structures and premises, and that the use of portable electrical equipment and heaters in hazardous areas is as described in appropriate standards. Electrical cords or leads should carry a current tag indicating that they have recently been tested for safety.

Recognizing electrical hazards on a worksite

Electrical dangers

When electrical equipment is being used, if system faults occur electrical energy can severely injure or kill a person who is in contact with a live conductor. The danger is not apparent in the size of equipment, as a small hand-held electrical device can be just as damaging as industrial equipment, if faulty.

It has been found through experience that the following current magnitudes are likely to affect a recipient in the way shown if contact is made with a live conductor:

1–3 mA	generally cannot be felt
10–15 mA	difficult to let go or be pushed away
25–30 mA	contraction of chest muscles is affected, resulting in heart fluttering and an inability of the heart to pump blood effectively.

These levels must not be taken as more than indicative. Other possible effects from shock include burns (contact, arc, radiation, vaporized metal), arc eye and other injuries resulting from a fall.

Electrical hazards in the workplace

Faulty electrical equipment can result in fires and explosions. When a short circuit occurs on any system, the resulting current is limited only by the electrical constraints of the system (including the fault) and can reach a value as high as 20 times the normal load current for the plant item concerned. Thus, in many cases, conditions at the fault are rather like the explosion of a bomb. Metals melt and very hot gases are liberated, often in a small confined space, so that there is a great risk of damage and injury to persons.

Electrical equipment which has the possibility of generating heat, or sparking with use, requires special consideration. See the section below on flammable atmospheres.

Eliminating fire and explosion risks from arcing, sparking and electrical installations

Electrical safety management

The prime objective of electrical safety is to protect people from electric shock, and also from fire and burns arising from contact with electricity. Two basic preventative measures against electric shock are:

1. Protection against direct contact, for example by providing proper insulation for parts of equipment liable to be charged with electricity.
2. Protection against indirect contact, for example by providing effective earthing for metallic enclosures which are liable to be charged with electricity if the basic insulation fails for any reason.

(*Preventative measures reproduced with permission of Pearson Education from Stranks, J. (1991). The Handbook of Health and Safety Practice, 2nd edn., London, Pitman.*)

Control measures
Control measures which reduce the risk of fire and explosions involve the use of fuses, current circuit breakers and earthing devices. Equipment insulation is also recognized as an essential control measure.

The purpose of the fuse in an electrical circuit
A fuse is a device for protecting a circuit against damage from an excessive current flowing in it, by opening the circuit when the fuse-element is melted by such excessive electric current.

Semi-enclosed rewirable fuse
The semi-enclosed rewirable fuse consists of a tinned copper wire placed in the circuit conductor path. The fuse will blow when it is approximately 160% overloaded. In the case of a short circuit fault, which can allow thousands of amps to flow, the fuse will blow within 10 milliseconds (half a cycle). The tinned copper is ionized and can be thrown from the fuse into the surrounding area and, in some cases, can cause secondary fires and explosions. The advantages of the semi-enclosed fuse are few except for their low cost; the fact that a blown fuse is easily seen; and the ease with which a user can replace them. The disadvantages are that they may be open to abuse with the wrong size wire being used on rewiring; to risk of fire; to deterioration of fuse element with age and use; and to slow operation when compared with HRCs (see below).

High rupturing capacity or HRC fuse
The HRC fuse consists of a silver fuse element enclosed in a cartridge which is filled with silica/quartz/sand. On blowing the fuse, the arc is contained within the cartridge. Due to the quality of the element, the fault-clearing time is limited to 0.003 seconds (3 milli-seconds) which is less than one quarter of a cycle. The advantages of this type of fuse are containment of the arc; quick fault disconnection; little if any deterioration of fuse with age; less likelihood of abuse; and virtually no risk of secondary fire or explosion. The HRC works more quickly to operate than an air circuit breaker. The disadvantages of the HRC are mainly cost and availability of spares at the time of the fuse blowing.

Advantages of earthing the metal bodywork
It is considered an advantage to earth any metal framework of a building, in order that any bare conductors of stray electrical current can be diverted to earth. If a metal frame was not earthed and became active, any person in direct contact with the frame and earth could be electrocuted. The earthed frame means that the fault to the frame would immediately be earthed.

The above system does not, however, guarantee that the supply fuse will blow in the event of a fault to earth. If the fault current path is of a high resistance, e.g. paint, rust, poor contact, then the fuse protecting the circuit may not blow. If the frame is not of sound construction but has loose joints, etc., then part of the frame could still be alive.

Advantages and limitations of earth leakage circuit breakers
The installation of earth leakage circuit breakers (ELCBs, or GFCIs – ground fault circuit interrupters in the USA), or residual current devices (RCDs), can be seen as an

additional protective device. In the event of an out-of-balance supply and return route for the current, the ELCB will trip and disconnect the supply. For example, domestic ELCBs are set to trip with a 30 mA cut-off balance. This 30 mA can be best understood as the equivalent of approximately 1/200th of the current that an electric kettle uses. If 30 mA of current flow leaks from a circuit, then the ELCB will disconnect the supply.

The advantages are the quick response time, low current sensing and protection of the individual. The limitations include the fact that although 30 mA is a small amount of current, the human being can be killed by smaller levels of current. It should be realised that the ELCB is not an overcurrent protector like the fuse; therefore, the ELCB should be used in addition to fuses. Further disadvantages of the ELCB are the initial cost, and nuisance tripping, such as when starting motors, fluorescent lights, refrigerators and stoves. The ELCB can, in some cases, be too sensitive.

Another hazard existing within an electrical system includes arc burns during the welding process. Exposure to flash or arcing is evidenced in cases where there is equipment short-circuiting, use of incorrect tools, dropping things onto the apparatus, and penetrating cables or conduits.

Intrinsically safe equipment

Flammable atmospheres
Safety concerns over ignition sources in flammable atmospheres exist in many work environments. Failure to provide intrinsically safe equipment has led to disasters in a wide range of industries and environments, such as mining, space, hospitals and mills to name a few. The prime considerations with equipment used in flammable or explosive atmospheres is the nature of the risk and the level of exposure.

This area of safety and risk management requires specialized skills and knowledge in the selection and installation of appropriate equipment. Standards and up-to-date advice from technically qualified people must be sought before carrying out any activities in flammable atmospheres. Various recognized testing agencies certify equipment for different classes of flammable environments.

Preparing an electrical safety checklist

General electrical safety within the workplace
Electrical safety within the workplace can be enhanced through:

- correct installation of equipment and power source
- correct insulation of equipment conductors and terminals
- earthing apparatus frames to earth so that any current leakage is diverted to earth
- earth leakage fault protection, i.e. the use of fuse and earth leakage devices which divert current leakage, tripping off power supply to a unit

- isolation – correct procedures should be used during maintenance and repair of electrical items, and this includes danger tagging and isolation procedures
- ensuring equipment users are trained in the correct operational procedures and that safe working procedures are followed
- regular and effective maintenance and prompt attention to any faults or damage.

Fixed machinery hazards

Relationship between person, machine, environment, work methods and materials

The relationship between a person, machine and the working environment forms an important consideration in a work system, as we noted earlier in this book.

The work process and system are linked together by procedures that are adopted to actually perform tasks. Materials used may also be important. The relationship of the work process to the work system is represented in Fig. 7.2, as first seen in Chapter 1.

As can be seen, the work process is reliant on the five parts of this relationship being effective and efficient. The interface, or linking, between these factors can be disrupted if compromises in safety occur. Poor linking of the factors can, on the other side of the coin, compromise safety.

Need for machine guarding

Contact between people and the moving parts of machinery often results in some of the most serious forms of industrial injury. The earliest advances in injury prevention dealt with the hazards associated with overhead transmission shafts. The transmission shaft involved that part of the power source which provided motion to the actual operational

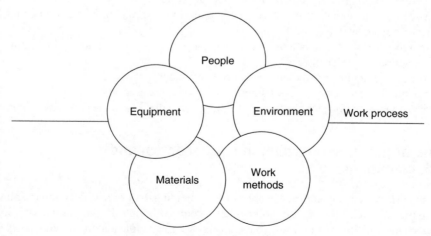

Figure 7.2 The work process and the work system

parts of the machine. The transfer of energy from the overhead transmission shaft to the machine usually occurred through a series of belts. Initially, the use of guards was introduced to prevent contact and so 'nip' and other injuries from the belts. Ultimately, the overhead transmission shaft was replaced by stand alone machines. This change in engineering led to an immediate reduction in belt-related injuries; however, the new machines still presented a significant hazard to operators. Today, there is a requirement to guard machines so that contact between the moving parts and the operator, or passer-by, cannot occur.

Principal hazards associated with moving machinery

When we view the working relationship between person and machinery, primary hazards of machinery need to be examined.

Machinery is an assembly of linked parts or components (at least one of which moves), with the appropriate actuators, control and power circuits joined together for a specific application in the processing, treatment, moving or packaging of materials. This includes an assembly of machines which in order to achieve the same end are arranged and controlled so that they function as an integral unit.

Examples of hazards associated with the use of machinery include:

- contact with or being trapped between parts of the machinery and other materials
- trapped in the workings or motion of the machinery
- struck by objects thrown out of moving machinery
- poor or inadequate maintenance procedures.

Design faults in machinery are often responsible for operating equipment hazards. If you consider that machinery is composed of power-driven moving parts, to omit guarding from moving parts would place the operator at a great risk if he or she is working in close proximity to the machinery.

Areas on machinery which can present a hazard to operators include:

- a reciprocating trap – a horizontal trap occurs when there is a horizontal motion of a part on a machine; at the extreme of travel this part may pin a person against a fixed object, such as a wall
- shearing traps – where one or two moving parts cut across each other; for example, the action of garden shears
- in-running nips where moving belts or chains meet
- ejection of machinery parts or supplies used on the equipment
- direct contact with machines which can expose an individual to the potential for burns.

Role of machine guarding in injury control and risk management

Should there be contact between people, or their clothing, hair or jewellery and the moving parts of machinery, serious injury may occur. Although design engineering has reduced the incidence of injuries from contact between people and the moving parts of machines in more modern equipment, accidents continue to occur with aged machines which may not have adequate safety guards in place. Injuries may also occur from maintenance oversights where work has not been attended to on machinery, safety systems are not replaced, and/or there has been lack of maintenance on machines. It should also be noted that guarding alone may not be the only answer to managing machinery hazards. If the guard can be removed while the machinery continues to operate, operator hazards still remain.

Often machine designers fit guards for the purpose of satisfying some legal requirement or standard without considering the effects of the guard on the operational or production performance. The purchaser, usually an employer, may modify the equipment in order to achieve both safety and efficiency. Countries around the world have different standards and requirements for machine guarding.

Machine guarding requirements

Machine guarding is covered by the various parts of national and international standards. In general, design and construction requirements involved in machinery safety guarding can be summarized to include the following:

- The owner of machinery shall provide, and at all times maintain, safety devices, fences, barriers or guards in accordance with the minimum requirements.
- Safety devices, fences, barriers and guards shall be provided on every dangerous part of machinery.
- Safeguarding of machinery should be planned into the design of machinery. Safety devices, fences, barriers, and guards should not in themselves create a hazard.
- Safeguarding shall be designed to prevent persons reaching into the danger area of machinery. Human capabilities have to be considered when determining minimum clearances.

Further criteria for guarding are:

- provide positive protection
- prevent all access to the danger area during operation
- cause no discomfort or inconvenience to the operator
- be compatible with the production process
- operate with minimum effort
- preferably be built in
- provide for maintenance, inspection and servicing
- be durable enough to withstand extreme conditions
- not constitute a hazard in themselves
- protect against unforeseen operational conditions
- comply with relevant standards.

Types of machine guards

There are several techniques available in guard design using electronic systems; however, the vast majority of guarding on machines is still mechanical. Wherever machinery is used, there is likelihood of a hazard resulting from transverse, rotating and reciprocating motions; in-running nips; or shearing, bending, punching or cutting actions. The type of guard necessary to be fitted to a particular part of the machine depends on the design purpose of that part.

Where protection against machine parts is necessary, guards can be considered under two broad categories.

1. Non-operational – for use on transmission gears, belt drives, shafts and parts that usually do not require frequent adjustments or approach by the operator.
2. Operational – where the part performs the function or purpose of the machine, such as cutters, blades and chucks.

Safety of machinery may be provided through several options. These include specific techniques, or where necessary, combinations of different methods, such as:

- tool or machine design
- fixed guards
- adjustable fixed guards
- interlocked guards
- automatic guards
- trip or quick-stop devices, including presence-sensing systems.

Application of relevant standards to machine guarding

Tool or machine design

Tool or machine design can eliminate the need for separate guards. Using remote control, automatic feeding or enclosure, the operator is isolated from the dangerous area. Two-handed operating devices, requiring simultaneous action by the operator, are used to prevent the operator's hand entering the danger area. However, this technique does not account for a second person becoming involved in the process, thereby reducing the reliability of the guarding system. Hence this method of guarding may require the addition of a mechanical guard.

Specially designed jigs and tools may be used for manipulating work in conjunction with fixed guards, thus allowing the operator's hands to be kept clear of the danger points. Automatic and semi-automatic feeding mechanisms are also used in combination with a fixed guard. Whatever technique is used, it is important that it does not create additional danger points.

Fixed guards

Fixed guard enclosures provide a high level of protection. The advantages include: no moving parts, enclosure of the dangerous area at all times, allowing the feeding and ejecting of material but preventing hands entering the danger zone, enclosing power transmission, and restraining bursting parts from flying about. The essential requirements of a fixed guard system include:

- effective prevention of access of any part of the body to the danger area from any direction

- strong rigid construction
- effective attachment.

Failure to observe the fundamentals leaves the guard partially ineffective and likely to fail in its main purpose.

Adjustable fixed guards

The positioning of fixed guarding around the operations point on a machine may not allow sufficient flexibility during production, severely limiting the use of the machine. An option is to provide adjustable sections to permit the machine to accept different-sized materials. The method, while increasing flexibility, increases the potential for error.

Interlocked guards

When fixed or adjustable guarding is impractical, an interlocked device should be considered. The interlocked guard provides access to the danger area when the moving parts of the machinery come to rest. The guard is required to:

- shut off or isolate the power source to prevent operation when the guard is open
- remain closed until the moving parts have stopped
- stop the machine immediately the guard is opened
- completely enclose the moving parts in the danger area.

Both mechanical and electrical interlocking is used in this technique, depending on the type of machine to be guarded. The system must be designed such that should a fault occur, the guard will fail safely, making the machine inoperable.

Automatic guards

The automatic guard is activated by the machine itself through connection to the operating system. The guard is linked to the working cycle of the machine and is designed to protect the operator even in the event of a machine fault. Speed conditions and stroke are critical to the effectiveness of automatic guards. Considerable skill and judgement are called for in setting and testing the device, and attention is needed to ensure the guard itself does not create a shear point during operation.

Trip or quick-stop guards

In special circumstances, where physical guarding is not fully possible, the application of quick-stop devices can be incorporated. In particular, use is made of photo-electric beams or other presence-sensing systems. As is the case with automatic guards, these types of devices present technical difficulties and reliability concerns if not properly installed and integrated into the operations of the machine.

Checklist for machine guarding

Machine safety checklists can be used to identify potential and actual hazards in the workplace. Checklists provide a baseline for hazard identification and action planning that may be required to correct a hazard.

Examples of machine safety checklists are available from various sources, such as those on internet sites. However, it is important to remember that checklist items will vary in their requirements depending upon the type of industry and workplace in which they will be applied.

Here are some issues to consider:

Maintenance of guards

A guard should be as close as possible to the part which it protects. This allows safe approach to the machine's controls, lubricating points and other functions which require regular monitoring. All guards should remain in position whenever the machinery is in motion or use. If a guard is removed when a machine is at rest, this must only be done by an authorized person and must be replaced prior to start-up. Accidents have occurred during maintenance when machines have been started with people still working on moving parts. A danger tag and lock-out system must always be used when guards have been removed from machinery to allow work to be carried out.

Fences around machines

Fences should be designed to prevent people crawling under or reaching over them and coming in contact with dangerous parts of the machinery. Where gates are required, normal interlocks may be inadequate. If it is possible to enter the fenced area, special provision must be made to ensure the equipment does not operate if the gate is accidentally closed.

Planning in advance

Guards and safety devices should be considered when the machinery is being planned, fabricated or purchased. The guard should, as far as possible, appear and function as part of the total machine.

Special requirements for specified plant, tools or equipment

Your local OHS legislation may include specific controls to be applied to certain types of plant, tools or equipment. These controls may specifically include the need for a person operating plant to be competent. There may be national certification standards for competency to operate, for example, cranes and hoists, and pressure vessels. Scaffolding and rigging may also be covered. Certain types of plant may need to be registered with a government authority.

Safe use of pressure vessels and lifting equipment

Risks in using pressure vessels

What are pressure vessels?

A pressure vessel is a vessel which is subjected to either internal or external pressure. This includes all parts of the vessel up to the point of connection. Because such vessels contain compressed gas in some form, safety is maximized through correct handling, maintenance, storage, use and work practices within the workplace.

Common pressure vessels in the workplace

The most common pressure vessels in the workplace are boilers and cylinders containing oxygen or fuel gas. A boiler is a closed vessel in which water is heated by combustion of fuel or heat from other sources to form steam, hot water, or high temperature water under pressure.

Gas cylinders may vary depending on the type of gas which is under pressure and may include fuels such as those in:

- LPG gas cylinders
- acetylene cylinders

or other gases such as those in

- helium cylinders
- nitrous oxide cylinders
- oxygen cylinders.

Risks associated with pressure systems can be summarized to include:

1. The release of stored energy which is uncontrolled. This may result from structural or corrosion damage to the cylinder or boiler. When stored energy is released from these systems, in the case of a gas cylinder, it may act as a missile until all energy has been released.

    The release of volatile gases may also place an environment at risk of vapour contamination and, if flammable, of potential fire or explosion; therefore, regular maintenance and structural checks should be carried out.
2. Unsafe operating and handling practices. Regulations which outline storage and handling of compressed gas cylinders may vary between jurisdictions, however, the principles are set out below.

Handling and storage of compressed gases

Handle high pressure cylinders carefully and do not drop or jar them. They must be stored in the areas specifically provided.

- Secure gas cylinders to prevent cylinders from falling and accumulating damage.
- Store cylinders in a cool area to prevent gas expansion.
- Store cylinders in an upright position to prevent safety valves from being blocked by liquid.
- Store oxygen cylinders separately from cylinders containing acetylene or other combustible gases.
- Separate full cylinders from empty ones.
- Plainly mark empty cylinders and close the valves.
- Chock or tie cylinders securely while being transported to prevent them from rolling or being knocked over.
- Don't fill damaged, corroded or out-of-test cylinders.

- Do not attempt to repair or destroy defective cylinders.
- Do not overfill gas cylinders as this may result in the safety valve opening or a weak cylinder splitting.
- Do not hoist cylinders by the valve assembly.
- Keep grease away from cylinder connections.

Boilers

Boilers should be regularly maintained and inspected to ensure that the integrity of the system is maintained.

Standards on pressure vessels will provide valuable further detail. The definition of specified (including registered) plant, which includes boilers and pressure vessels, may vary between jurisdictions. Generally, however, this classification includes any designated plant, with exclusions for particular pressure vessels and boilers with specific capacities.

Critical factors in safe use of lifting equipment

Equipment within the workplace which is used for lifting will vary considerably depending on the type of industry and the load which is to be lifted. For example, in the construction industry, cranes and slings are predominantly used to lift loads. Other industries often use fork-lift trucks to lift and move loads. Lifting devices should be selected to suit the job or task required. For this reason, lifting equipment has limitations on the load weight it may safely lift and the conditions in which it can be used. Environmental factors may play an important part in the selection of equipment.

An example of some environmental factors which should be considered include:

- whether equipment is to be operated internal or external to a building
- the ground surface – slope/stability/surface type
- radius limits for safe manoeuvring.

Once lifting equipment has been selected for the task at hand, equipment should be operated in accordance with the manufacturer's recommendations and statutory requirements. A workplace may also introduce safe operating procedures for the use of equipment, to ensure that safety rules are followed during operation.

Cranes

Only those people who hold relevant certification are permitted to operate cranes. Safety measures include the following:

- Cranes must not be loaded beyond their rated capacity, and crane operators should not move a loaded crane over other workers. Employees should not walk under a suspended load. On overhead cranes, the alarm must be sounded before and during crane travel.
- Before putting the crane into service each day, the various motions of the unit, as well as the braking systems and all safety devices, should be checked to ensure

Figure 7.3 Sheave block fall

that they are effective, and where applicable, outriggers on mobile cranes must be extended before the boom is used. The outrigger must be on a surface which can take the load. Care should be exercised in lowering or raising outriggers.
- Crane booms must not be allowed to move in the vicinity of live electrical conductors unless effective clearance is maintained, as required, and adequate safety precautions are taken for the voltage concerned.
- People must not ride on cables, slings or material being moved by a crane.

Figure 7.3 shows the result of failure of a crane braking system.

Crane design

Cranes operate safely, but can also fall over or fail structurally. Both are based around the principles of physics. Central to this are the laws that govern gravity and Newton's third law of motion 'for every action there is an equal and opposite reaction'.

A crane's rated lifting capacity is based on three key factors:

- stability – the ability not to tip under load
- weight of the crane – size and structure
- strength – ability of structure and components not to yield under load.

The safe lifting capacity of a crane is often considered in the context of the working radius. Structural strength is a major factor in a small working radius, and stability becomes more critical as the working radius increases.

Working radius is measured from the centre of rotation and the distance through the load after the drawdown as the crane takes up the load.

Lifting capacity charts will always be based on optimum conditions, and workplace conditions which are below optimum will affect safe working.

Wind, visibility, machine condition and stability are key performance factors in safe operation. Stability of the crane ensures optimum safe performance. When outriggers are used in a lift, they must be firmly positioned on level ground. Ground can vary from hard solid surfaces to soft – even ground fill. Bearing capacity is affected by the nature of the ground and as such is a prime consideration in the set-up process. Outriggers must not be used where safety would be compromised, such as over loose, shifting or uncertain ground or surfaces, e.g. a concrete floor with unknown cavities.

Chains

Attention to the points below is important:

- only tested chains should be used for handling, binding and hoisting
- chains must not be overloaded, and must be kept free from twists, knots and kinks
- ensure that chains are not trapped under a load
- take up any slack in the chain, and apply the load slowly
- chains must be inspected regularly for wear, stretch, cracking and deformation.

Wire ropes

Care and use of these is also critical to safety:

- Gloves must be worn at all times when handling wire rope.
- Defective wire rope must not be used. Defects such as corrosion or pitted surface of wire; excessive wear; kinks; 'bird-caging'; fatigue indicated by broken wires; or mechanical abuse, such as pinched or partially cut strands, should be considered sufficient reasons for removing wire rope from service. Newer wire ropes have defect indicators.
- Wire ropes must not be overloaded, and short radius bends around loads or small hooks should be avoided.
- Avoid exposing wire rope to corrosive fumes or liquids.
- Every precaution must be taken to keep clear of winches and winch cables when they are under load.
- Dragging wire rope from under loads or over obstacles should be avoided.
- Store wire ropes where they cannot pick up dust and grit.

Natural fibre and synthetic ropes

These features are critical to safe operation:

- Only approved fibre ropes should be used. People must be trained in how to select the correct type and size of rope to perform the job safely.
- Ropes must be carefully inspected before use for defects such as cuts, broken strands, wear or abrasion.
- Ropes must not be overloaded, or exposed to acids or acid fumes during use or storage.
- Sharp bending or kinking of ropes should be avoided, and square-edged objects must be padded.
- Knots must be avoided wherever possible because they can reduce the load-carrying capacity of a rope by as much as 50%.

Natural fibre ropes (manilla, sisal)
In addition to the general rules above, the following steps apply to these ropes:

- When inspecting natural fibre ropes, discoloration and rotting must also be looked for. This includes internal as well as external fibres.
- All practical precautions must be taken to keep these ropes clean and dry; if a rope becomes wet, it should be thoroughly dried before storing. When drying ropes, care must be exercised to avoid exposure to high temperatures which may dry out the oil in the fibres.

Synthetic fibre ropes
In addition to the general rules above, the following precautions apply to these ropes.

- When inspecting synthetic fibre ropes, also look for melting of the fibres due to heat.
- Because synthetic fibre is a thermo-plastic material which melts at relatively low temperatures and which is degraded by ultraviolet light, care must be taken to avoid exposure to prolonged sunlight or high temperatures in either storage or use.
- Synthetic fibre ropes must not be used in applications where friction could cause the rope to melt or fuse.
- Extra care must be taken when knotting these ropes because of their somewhat slippery surface, particularly when new.
- When using these ropes, remember that they stretch more than manilla rope.

Slings and spreaders
All people concerned with hoisting equipment must familiarize themselves with the use of slings and spreader bars with respect to size and safe loads for various sling angles. And so:

- All slings must be carefully inspected before use. Defective slings must be destroyed.
- Slings and spreaders must not be overloaded. The optimum angle between two ropes of a sling suspended from a hook is 45 degrees.
- Slings must not be placed around sharp edges of loads unless adequate padding is used to avoid damaging the sling.
- Kinks must be removed before loading slings.
- Spreaders must be marked with their rated load and their own weight.
- Load must be distributed evenly on all legs of the sling.
- People must not stand under any suspended loads.
- Slings must be stored in a dry place.
- Synthetic slings are banned in some areas because fret abrasion of the fibres is not always obvious.

Jacks and stands
Every part of the system making up an operating crane is important, including the jacks and stands, so:

- Only jacks and stands which are in good working condition must be used.
- The rated capacity of a jack should be stamped on the body of the jack and this capacity must not be exceeded.

- Jacks and stands must be set on an adequate base to prevent sinking, tipping or slipping.
- Adequate chocks must be used to support the load, as well as jacks, if people are to work underneath. Fatalities during house transport have occurred through failure to observe this precaution.

Handling steel

Severe hand, foot and other injury can result without proper precautions including these:

- gloves must be worn by employees loading and unloading steel
- loads should not be lifted until people are clear
- people should avoid having their hands on a load while it is being lifted, moved or lowered
- steel should be stored so that there is no danger of tipping or falling.

For further safety steps refer also to standards for cranes.

Other powered materials handling equipment

Hand trucks

There are various types of hand trucks in use. Although they may appear to be easy to handle, there are handling practices which should be observed:

- The load should not be top heavy. Heavy objects should be placed at the bottom of the load.
- The load should be placed well forward so that the weight is carried by the axle and not by the handle of the truck.
- The load should be placed so that it will not slip, shift or fall off, or obstruct the operator's vision.
- The truck must carry the load. The operator should only maintain its balance and provide the motive power.
- When proceeding down an incline, the truck should be ahead of the operator. When proceeding up an incline, the truck should be behind the operator.
- The truck must be kept under control and moved at a safe speed.
- When lifting heavy loads, the wheels should be chocked to prevent the truck from moving back while balance is being achieved.

Powered trucks, fork-lifts and similar vehicles

Only employees who have been tested and certified should be permitted to operate fork-lift trucks, and should operate these vehicles only at speeds which are safe for the existing conditions. Steps include:

- Traffic rules must be observed. Operators should keep to the left wherever possible, slow down or stop at intersections or blind corners, and sound the horn.

When a truck is left unattended the power should be shut off, the controls neutralized and the brakes set.
- Fork-lift trucks should be operated with the forks or the pallet about 100 mm off the floor. On down grades a loaded truck should proceed with the load last, and on upward grades, with the load first. The truck must be so controlled that an emergency stop can be made within the clear distance ahead.
- The operator's legs and feet must be kept inside the guard or operating station of the truck at all times, and a seat belt worn.
- When operating in confined spaces, the operator's hands must be so positioned that they cannot be pinched between the steering control and projecting stationary objects.
- Trucks should never be parked in an aisle or doorway or where they obstruct material or equipment that someone else may need.
- People must not be lifted from one elevation to another by a fork-lift truck, or ride on the forks of fork-lift trucks.
- A passenger may be carried only when seating is specifically provided for that purpose.
- Fork-lifts must not be overloaded, and the load must be kept as close to the vehicle as possible and never suspended by slings around the vehicle forks.

Refer also to standards on fork-lifts.

Conveyors

Conveyors have been responsible for some shocking injuries, so safe use includes the following:

- Only employees who are delegated the responsibility should start, load and unload conveyor systems.
- Conveyors must be inspected before use to ensure that they are in sound working condition.
- All moving parts of conveyors must be effectively guarded.
- People must not ride on a conveyor.
- People working near moving conveyors must be instructed in the emergency stop procedures.
- Conveyors must not be overloaded and loads must not project outside the limits of the conveyor.
- An emergency stop lanyard should be provided.

Refer also to standards on conveyors.

Care and maintenance requirements for lifting equipment

The care and maintenance of lifting equipment should be based on the designer's, manufacturer's, installer's or erector's recommendations. Owners who hire out equipment may also be consulted.

Maintenance schedules are required to ensure that equipment continues to perform in the safest possible way. Careful inspection is also required for cracking, corrosion and signs of overloading. Some of the gear associated with lifting requires careful storage, labelling and records, e.g., slings, wire and natural fibre ropes. Some wire ropes now have built-in overload indicators.

Storage requirements include:

- low dust
- dry
- off the ground.

Particular care is needed in marine or acid mist environments.

Jack supports require constant attention.

Inspection of equipment should be followed as per regulations and recognized national standards. Maintenance schedules are required to ensure that equipment continues to perform structurally in the safest possible way.

Carrying out an inspection on selected plant

Using national standards, codes of practice, and suppliers' manuals, such as those on, for example, fork-lift operation, you would need to prepare a checklist, then carry out such an inspection and report on it. The inspection should include features of the plant itself, as well as observations on its safe operation. Include in your report those items you found satisfactory and any matters which need attention, together with your recommendations.

Explosives

Only those employees who have been authorized as being competent to handle explosives should be permitted to use them. The transporting, storing and handling of explosives is governed by legislation.

Fire hazard identification and extinguisher use

Principles and conditions of combustion

Fire results from a fast chemical reaction between a combustible substance and oxygen, accompanied by the generation of heat. There are four requirements for fire to occur:

1. Oxygen – except in special circumstances, the chief source is air
2. A fuel or combustible material
3. A source of energy, usually in the form of heat
4. A chemical chain reaction.

These four requirements are often represented as the fire tetrahedron, see Fig. 7.4.

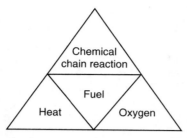

Figure 7.4 The fire tetrahedron

The chemical reaction

Fuel plus oxygen produces combustion and heat. If the fuel is natural gas (methane) the reaction is written as:

$$CH_4 + 2O_2 \rightarrow CO_2 + 2H_2O + HEAT\Uparrow$$
Methane + Oxygen → Carbon dioxide + Water

If the fuel was hydrogen, the equation would be:

$$2H_2 + O_2 \rightarrow 2H_2O + HEAT\Uparrow$$
Hydrogen + Oxygen → Water

Both methane (CH_4) and hydrogen (H_2) are combustible substances. Methane is an organic substance (i.e. it has carbon in its molecule), hydrogen is inorganic (i.e. it does not have carbon in its molecule) and both need a supply of oxygen for combustion to occur. However, there are substances that contain their own source of oxygen, e.g. ammonium nitrate NH_4NO_3.

$$NH_4NO_3 \rightarrow N_2O + 2H_2O + HEAT\Uparrow$$
Ammonium nitrate → Nitrous oxide + Water

Spontaneous combustion

Fire is generally associated with rapid oxidation; however, slow oxidation is the cause of spontaneous combustion. Sometimes, material bursts into flames without an obvious means of ignition. As a substance oxidizes slowly, a small amount of heat is generated. If this heat is not permitted to escape, it gradually raises the temperature of the substance until the ignition temperature is reached. At this point the substance begins to burn.

Examples are:

- oily rags in poorly ventilated cupboards
- coal dumps
- 'green' hay stored before being thoroughly dried.

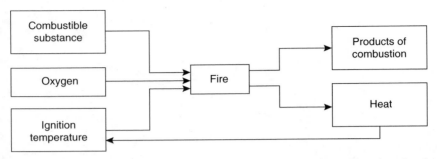

Figure 7.5 The fire process (Figure reproduced with permission of Pearson Education from Stranks, J. (1991). *The Handbook of Health and Safety Practice*, 2nd edn., London, Pitman)

The fire process

A fire may be described as a mixture in gaseous form of a combustible substance and oxygen, which occurs after sufficient energy is put into the mixture by a source of ignition to start this process and lead to burning. Once started, the energy output from the fire provides a continuous source of combustion with excess energy given off as heat. See Fig. 7.5. Unless the fire triangle is broken the fire will continue.

Measures relating to the initiation of a fire

A number of measures are used:

1. *Flash point*. The flash point is the lowest temperature at which a flash results when a mixture of flammable liquid vapour and air comes in contact with an ignition source.
2. *Fire point*. Fire point is the lowest temperature at which a flammable liquid vapour can be ignited and combustion is sustained.
3. *Auto ignition temperature or spontaneous ignition temperature* is the lowest temperature at which the substance will ignite spontaneously.
4. *Lower flammable limit* (lower explosive limit – LEL) is the smallest concentration of flammable gas or vapour which, when mixed with air, is capable of ignition and sustained combustion.
5. *Upper flammable limit* (upper explosive limit – UEL) is the highest concentration of flammable gas or vapour which, when mixed with air, is capable of ignition and sustained combustion. See Table 7.1.
6. *Ignition energy*. This is the minimum energy (usually measured in a spark) needed to cause ignition.

Table 7.1 Explosive limits

Fuel	LEL %	UEL %
Methane	5	15
Acetylene	2.5	100
Hydrogen	4	75
Petrol (normal)	1.3	6.0

Most common cause of fires

Using figures compiled recently for metropolitan Melbourne, Australia the most common causes of fire (in order) are:

	Per cent
Misuse of ignition source (e.g. thrown away cigarette)	27.8
Suspicious (suspected of being deliberately caused)	20.4
Mechanical failure (e.g. a part failure or short circuit)	16.6
Operation deficiency (e.g. overload)	12.6
Cause not known	9.0
Misuse of material which was ignited (e.g. something combustible too close to heat)	4.9
Deliberately lit	4.3

Design deficiencies, nature (e.g. lightning) and exposure to another fire are other identified causes. You may wish to access similar data for your area.

The outskirts of some cities in hotter climates are fairly frequently the site of summer bushfires, some of which are deliberately lit.

Inspecting to identify poor housekeeping practices constituting fire hazards

Fire is one of the most dangerous events which can occur in the work environment. It is much easier to prevent a fire than to extinguish one. The observance of safe practices and good housekeeping at all times is the most important factor associated with fire prevention.

The following are some of the practices which can help ensure that fire hazards are reduced to their absolute minimum. As with any inspection, it is efficiently carried out if a checklist is prepared first. You may wish to consider the following in the checklist you prepare:

- Usage of paper and cardboard in the area; proper storage of it; disposal of waste paper
- Restrictions on smoking; defined smoking areas
- Suitability and lack of overloading of electrical appliances
- Can water heaters boil dry?
- Storage and use of materials with flammable solvents, e.g. thinners, glues, inks, varnishes
- Use of floor and wall bar or strip heaters (Can they tip? Do they have a tip switch? Are they too close to flammables, e.g. curtains, chairs, paper?)
- Is required emergency fire signage in place and operative?
- Are extinguishers freely accessible, with current markings? Are hose reels and valves ready for easy use?
- Are all fire doors as they should be and not locked against egress or jammed open?

- Are all emergency exits free of stored materials?
- Storage of wooden materials
- Clean up of flammable dust, e.g. wood dust
- Proper shielding for welding
- Correct storage and segregation of flammable liquids and oxidizing agents
- Flammable liquids must not be carried or stored in open containers. They must be kept in approved containers and identified by the proper class labels for dangerous goods. The contents of the containers must also be identified by legible labels.
- Open flames and smoking are prohibited in all areas where flammable liquids or gases are stored or dispensed. Such areas must have appropriate warning signs displayed and these must be strictly observed.
- Whenever there is danger of static electrical build-up and discharge in the presence of flammable liquids or gases, all equipment must be bonded together and earthed.
- Rubbish should be burned in an approved incinerator. If this is not available, a safe distance must be maintained between a fire and any flammable material. In any event, a fire must not be lit or allowed to remain alight in contravention of acts and regulations, fire bans or local government by-laws.
- When welding or cutting operations are performed at elevated positions, precautions must be taken to prevent hot metal from falling on people or onto combustible materials.
- Welding or cutting must not be done near exposed flammable liquids or gases. Flammable dust in the work area should be removed or shielded to prevent ignition.
- When flammable material cannot be removed from exposure to welding or cutting sparks, it should be protected by a shield of non-combustible, or fire-resistant, material. Suitable fire extinguishing equipment must be at hand.
- All electrical wiring should be installed to the appropriate standard with suitable fusing, circuit breaking and leak detection.
- Flammable liquids must not be used to start or maintain fires except in, or with, equipment specifically designed to use flammable liquids in this way.
- Open fires should not be left unattended unless adequate precautions are taken.

This only covers housekeeping. There are of course other precautions to take against fire, and to minimize its effects if it does get started. Details of building design requirements can be found in the appropriate building codes for your area.

Classes of fires and methods of extinguishment

Fires are commonly classified under broad categories according to the fuel type and means of extinction. See the relevant standard for further details on portable fire extinguishers.

Class A
Fires involving solid materials, normally of an organic nature, e.g. wood, paper, coal and natural fibres. Water applied as a jet or spray is the most effective way of achieving

extinction. Water must *not* be used to extinguish fires in energized electrical equipment, or where flammable liquids are involved.

Class B

Fires involving flammable liquids, e.g. solvents, greases, petrol. Generally, these types of fires require a dry powder, or foam extinguisher. There may be some restrictions on the type of foam used because some foams break down on contact with alcohols. In the case of both dry powder and foam, extinction is achieved by smothering (see below). Foam is best for putting out burning liquids in containers when the liquid is likely to be hot enough to reignite on contact with oxygen. Foam must *not* be used to extinguish fires in live electrical equipment, as the foam is a conductor of electricity.

Class C

Fires involving gases or liquefied gases, e.g. methane, propane and natural gas. Both foam and dry chemicals can be used on small fires, backed up by water to cool leaking containers. To control a fire from a major gas leak may require isolating the fuel supply. Direct flame extinguishment is difficult and may be counter-productive. If the leak continues there may be a secondary ignition.

Class D

Fires involving metals, e.g. magnesium and aluminium. These fires are extinguished using special types of dry powder. Extinguishment is achieved by smothering.

Class E

Fires involving electrical equipment, e.g. electronic equipment, switch rooms and power boards. This type of fire should be tackled firstly by isolating the electricity and minimizing the risk of injury. The fire can then be treated under one of the other classifications. Where this is not possible or practical, the use of carbon dioxide (CO_2), or dry powder is suggested. Carbon dioxide is suitable for extinguishing most types of fires, including fires involving live electrical and delicate laboratory equipment, and small fires in flammable liquid escaping over both vertical and horizontal surfaces. A direct discharge from this extinguisher at close range may be harmful to the skin, as frostbite is possible due to the extremely low temperature of the discharge.

A word of caution! Exposure for some time to carbon dioxide in a confined space could cause suffocation. Vacate the area immediately after use. Ventilate the space to disperse the gas as soon as possible.

Class F

Fires involving cooking oils and fats.

Extinction

Extinction means putting out a fire. This can be achieved by:

- smothering – reducing the oxygen supply
- cooling – reducing the rate of energy input
- starvation – reducing the fuel supply.

It is a reduction of one or more of the elements above that is necessary to extinguish a fire, not the complete removal of any element. For fire to occur and be sustained, the oxygen/fuel mix must be within the lower and upper flammability limits, and it must receive a minimum amount of energy.

Smothering
Smothering is achieved by allowing the fire to consume the oxygen while preventing the inflow of more oxygen, or through the addition of an inert gas to the burning mixture.

Cooling
Cooling is the most common, and cheapest form of extinguishing medium. When water is added to a fire, the heat output vaporizes the water, which provides an alternative heat sink. Eventually, insufficient heat is added to the fuel and continuous ignition ceases. In order to assist rapid absorption of heat through rapid vaporization, water is best applied to the fire as a spray rather than as a direct stream or jet.

Starvation
Starvation is achieved by taking the fuel away from the fire, the fire away from the fuel, or reducing the amount of the fuel available. In a fire involving supply pipelines, fuel is often removed from a fire by means of isolation valves. Fire is removed from the fuel by breaking down and separating burning materials. Reducing the amount or bulk of a fire can be achieved by breaking the fire into smaller and more easily controlled units.

(*'Fire types' and 'Extinction' reproduced with permission of Pearson Education from Stranks, J. (1991). The Handbook of Health and Safety Practice, 2nd edn., London, Pitman.*)

Identification, selection and use of correct fire extinguishers

Portable fire extinguishers
These appliances are designed to be carried and operated by hand. They contain various extinguishing media as noted above, which can be expelled by the action of internal pressure and directed onto a fire.

A summary is shown in Table 7.2.

X indicates 'Do not use'.

For Class D fires, involving combustible metals, only special purpose extinguishers are to be used. Older water, foam and wet chemical extinguishers may simply be one colour – red, blue or oatmeal respectively.

(*With acknowledgement to FESA, Western Australia.*)

Two newer extinguishing gases are available: Inergen is a gas mixture designed to flood an area but be breathable for a short time; FM 200 is an extinguisher gas which is also friendly to the ozone layer. There is also a Finnish water micromist system which can replace inert gas in some applications.

Table 7.2 Extinguishers and their applications

Class of fire	Extinguisher type: Water	Alcohol-resistant AFFF foam	Wet chemical	Vaporizing liquid NAFP 3 – commercial use only	Carbon dioxide	Dry chemical powder AB(E) and B(E) types
Colour:	All red or unpainted stainless steel	Red with blue band	Red with oatmeal band	Red with yellow band	Red with black band	Red with white band
A Ordinary combustibles (wood, paper cotton)	√	√	√	√	√ limited effect	√ – AB(E) type X – B(E) type
B Flammable and combustible liquids	X	X – non AFFF √ – AFFF except for alcohol	X	√	√	√
C Flammable gases	X	X	X	X	X	√
E Fire in or near energized electrical equipment	X	X	X	√	√	√
F Fire involving cooking oils and fats	X	√	√	X	√ limited effect	X – AB(E) type √ – B(E) type

Interpreting fire emergency response from HAZCHEM signs

An explanation of the Hazchem code will be found in the section '*Main factors in transport of hazardous chemicals*', in Chapter 9.

Fire safety managers

Roles and responsibilities

A manager with sole responsibility for fire safety will only exist in some larger organizations, or perhaps also be employed by building managers for a large high-rise. In other cases, the role may be just one of several held by the one person. That person may also be responsible for security, first aid and health and safety. So, in thinking about the role and responsibilities of a fire safety manager, remember that these must still exist to a degree in any organization, and be assigned to a person or persons, even if their main role is something different.

The fire safety management role includes:

- ensuring proper building and process design and alteration, bearing in mind requirements such as those in the relevant building code
- ensuring materials purchasing is in line with policy on fire-risk minimization
- ensuring appropriate extinguishment is available, currently dated and operative
- ensuring automatic fire detection and control systems are operable
- developing and maintaining an emergency response system including selection of emergency roles for personnel, training of key emergency response personnel and development of an effective emergency communication system
- training staff in fire prevention measures and fire emergency response and, by extension, bomb threat response
- running emergency drills
- liaising with key outside parties in relation to fire prevention and emergency response, e.g. government authorities and fire equipment suppliers
- ongoing audit of housekeeping and fire control and response issues
- considering the special needs of non-employees, e.g. shoppers, patients. (Prisoners are a special problem. Petrol-powered vehicles should not be used for carrying prisoners if this involves locking the prisoners in the rear.)
- investigation of fire incidents.

Building regulations and fire safety compliance

Performance-based and prescriptive fire safety provisions

Performance-based fire safety provisions look at the outcomes of a successful system for fire minimization and control in a workplace or organization, without describing the detail of how this is to be achieved. For example, in Australia the *Building Code* has a series of performance requirements for the structural elements of a building, one of which is the FRL or fire-resistance level. This has up to three parts: structural adequacy; integrity; and insulation (in that order). Each is expressed in terms of minutes of required satisfactory performance; e.g. an FRL of 240/240/240 means that the structural elements would perform for four hours on all three criteria. There are performance requirements for, for example, fire doors, fire shutters and fire-stopping material (material for plugging 'penetrations' in floors, etc.).

On the other hand, an example of prescriptive requirements is that for penetration of a floor, wall or ceiling by a cable or cluster of wires. This requires penetrations to have a specified maximum cross-section. Refer also to 'Alternative methods' further on in this chapter for a note on the issue of performance versus prescription.

Legal requirements for fire safety

Specific reference to fire safety is not generally included in the actual OHS act, but there will be reference to fire safety in most of the accompanying regulations. The regulations vary in adequacy.

Building regulations and minimum standards of fire safety

Obviously, building design alone can only be partly successful in preventing fire. Other important factors are:

- good housekeeping, separation and segregation of materials and proper storage, e.g. fire cabinets for solvents
- the transport and handling of flammable materials
- the behaviour of personnel, e.g. smoking, disposal of paper, cigarettes and matches
- adhering only to uses for which the building was designed, e.g. no solvent-based printing if not designed for that
- selection of wall and floor coverings and furniture which will not spread flame from an ignition source, and will minimize the speed of travel
- extreme care with radiant heaters, and preferably use of tilt-switched radiant heaters, or wall-mounted radiant heaters
- great care with high temperature sources used in maintenance such as welding equipment
- electrical design, installation, repair and maintenance to meet national or accepted standards.

The approach to fire in buildings falls into two key areas:

- design to prevent fire
- design to limit the spread of a fire if it starts, and to limit the effects of that fire.

Before building any new buildings, an application for approval must generally be submitted to the appropriate level of government.

Enhancing occupational safety and health

Fire design of buildings is based on the observed characteristics of fires, some of which have been carefully reproduced in places such as the UK Fire Research Station. These characteristics are:

- in most cases the fire will only develop if there is fuel above the initial ignition source
- combustible materials in the path of the flames increase the size and intensity of the fire
- hot gases and flames, as they are lighter than air, travel upwards
- fire tends to follow vertical paths of travel – i.e. chimneys, flues, stairways, and the interior of stud walls – but drafts or forced airflows, e.g. in underground mines, can change this characteristic
- ceilings can be a barrier to this upward spread
- sometimes the lateral and downward spread is increased by the presence of highly flammable coatings or bondings, glues, varnishes and lacquers
- the likelihood of flammable surfaces catching fire from radiant heat energy, rather than direct spread, depends to some extent on the rate of burning of the fire in materials providing the radiant heat.

Building compliance with fire safety regulations

Building codes may include:

- fire-resisting construction
- compartmentation and separation
- protection of openings
- structural tests for lightweight construction
- early fire hazard indices
- fire doors, smoke doors, fire windows and shutters
- penetrations
- emergency lighting, exit signs and warning systems
- smoke hazard management
- fire-fighting equipment, and when sprinklers are required
- lift installations
- fire isolated exits
- number of exits
- distances to exits and second exits.

Fire-resisting construction (fire resistance and stability)

This includes constructing a building to protect it from fire in another, and using materials which minimize the spread of fire and generation of smoke and toxic gases. 'Stability' must be enough to allow escape and firefighter safety, and to minimize collapse onto nearby property. Standard fire tests are used to decide if different parts of a building will perform satisfactorily. 'Structural adequacy' looks at the ability of a structural part to continue to support a load. The 'integrity' aspect looks at how well a structural member prevents fire, gases and flames getting through it. Insulation is designed to limit heat transmission so that something on the other side of a wall or floor does not receive enough heat flow to ignite.

Compartmentation and separation

Compartmentation refers to the division of a building into 'compartments' separated by structural material of specified fire resistance, such as a fire wall, to prevent spread of fire and smoke, and facilitate access by firefighters. 'Separation' limits the opportunity for the fire to spread to other buildings. (There was recently a fatal blaze in a park home (semiportable home) which threatened to ignite others nearby.) It also refers to separation of certain key equipment such as sprinkler valve equipment. Separation for dangerous goods also separates the goods from fire sources such as roadways.

Protection of openings

Certain types and layouts of doorways, windows, infill panels and fixed or openable glazed areas may be covered by a building code. It may include, for example, distances between windows on either side of a fire wall, and where protection of the opening is required, internal or external wall-wetting sprinklers, or automatic fire doors for doorways, automatic fire shutters for windows or construction of approved FRL for other openings.

Structural tests for lightweight construction

These may cover the requirements for materials such as sheet or board, plaster, sprayed insulation, and concrete mixed with soft products such as pumice, which can be damaged by impact, pressure or abrasion, and thinner forms of masonry.

Early fire hazard indices (EFHIs)

These deal with materials, linings and surface finishes in buildings, particularly fire-isolated exits. Three indices are used – flammability, spread-of-flame and smoke-developed, which are measured by standard tests. In addition, protection of sides and edges from exposure to air may be a requirement.

Fire doors, smoke doors, fire windows and shutters

Fire doors are required to meet certain specifications and glazed parts must meet the integrity requirement in the FRL. Smoke doors must of course prevent smoke passing and, if glazed, minimize the risk of injury if a person accidentally walks into them.

Penetrations

Penetrations refer to services which penetrate walls, floors and ceilings required to have a FRL. Metal and UPVC pipes, wires and cables, electrical switches and outlets, and the fire-stopping material are all considered.

Emergency lighting, exit signs, and warning systems

This part of a building code may cover the requirements for provision and design of lighting and emergency signage which will retain illumination in occupied areas and in egress ways, such as fire-isolated stairways, ramps or passageways independent of the normal power supply.

Smoke hazard management

Various classifications of buildings may be required by a code to have smoke control systems. This includes particular requirements such as natural smoke venting, smoke exhaust systems, air handling systems of a particular design, and smoke doors. In particular, smoke must be excluded from fire-isolated exits. For natural venting, openings must be either openable or – if on a ground floor – shatterable.

Lift installations

A number of issues arise with lifts. These include restrictions on use in a fire, fitting out as an emergency lift for mobility-challenged people and fire service personnel, and prevention of spread-of-fire by way of lift shafts and doors.

Fire-isolated exits

Fire-isolated exits, and the provision for reaching them via fire-isolated stairways and ramps, are important features in some buildings.

Distances to exits and second exits

The distance to an exit is clearly an important issue. Requirements vary with the class of building.

Fire-fighting equipment and when sprinklers are required

Fire-fighting equipment includes fire hydrants, hose reels, sprinklers and fire control centres. Special care is needed in a building being constructed, in which sources of ignition may be more likely and fire systems are yet to be installed.

Alternative methods of satisfying the intent of the regulations on building fire safety

Individual variations will need to be discussed with the relevant government and fire authorities. For example, in some cases a roof with a lower FRL may be approved if an automatic sprinkler system is installed in a situation where it is not a requirement under the building code. The relevant building code should be consulted for further information on performance requirements; any 'deemed-to-satisfy' provisions; and alternative solutions, as part of a flexible approach.

Determining building classifications, fire load and fire resistance

Basic building classes

Classifications of buildings in building codes

These will vary with the country, type of terrain, weather patterns, and available materials of construction. However, generally there will be divisions covering:

- single and grouped dwellings (terraces, villas, town houses, units, apartment blocks)
- boarding and guest houses and hostels
- hotels or motels
- school residential accommodation
- accommodation for children, the elderly, and disabled
- residential accommodation in a health care institution
- health care facilities
- office buildings
- shops and retail facilities
- an eating room, cafe, restaurant, milk or soft-drink bar
- a dining room, bar, shop or kiosk part of a hotel or motel
- a hairdresser's or barber's shop, public laundry, or undertaker's establishment
- a market or sale room, showroom, or service station
- a car park
- a manufacturing building
- a laboratory
- a workshop
- non-habitable buildings such as garages.

The codes generally deal with classifying buildings with multiple use.

Required fire resistance for a building

Codes may set standards for fire resistance of a building, and hence the required type of construction. To do this they may use tables which consider the rise in storeys, the floor area, and whether it is a multitenanted residential complex, for example.

Calculating fire load of a compartment

The fire load of a compartment within a building is based on the calorific or heat value of the contents of the compartment. Doing this accurately is a specialized task. The fire load is usually given in megajoules (MJ) per square metre of floor area. Calculating the fire load requires grouping together materials of similar calorific value (e.g. chipboard furniture) and estimating the weight in kilograms. The results are then added, knowing how many MJ each kilogram of a particular type of material can produce. The total heat output figure is then divided by the floor area of that compartment in the building.

An approximate estimate can be made with a trained eye, grouping fire load into low, moderate or high status.

Preparing a technical brief of fire safety requirements for a building

Design features of a building affecting structural integrity

Type of construction
As mentioned earlier the particular method of construction of a building, and its layout, affect its structural integrity in a fire. For a building of multiple classification, the most fire-resisting type of construction may be required, applying the most classification for storeys to all storey, with perhaps certain exceptions. Sports spectator venues need special attention. A soccer ground fire at the Bradford Football Stadium in the United Kingdom in 1985 indicates how important this is.

Compartmentation and separation
Compartmentation and separation affect the spread of fire and smoke, and codes may cover maximum sizes of fire compartments and atria. Atria (atriums) in buildings potentially could allow easy travel of fire and smoke from level to level.

Vertical separation and fire walls
Vertical separation of openings such as windows is generally covered in codes, and fire walls which create compartment. (In a recent two-storey school fire, the fire spread rapidly along the ceiling because fire walls did not break up the ceiling space.)

Lift (elevator) shafts, sprinkler valves and openings
Lift shafts could easily transmit fire and smoke and these are covered in codes. Certain equipment such as sprinkler valves must be fire-isolated. Openings in external walls and fire walls require protection.

Services and penetrations
Services passing through a floor are covered by codes. Penetrations of floors, walls or ceilings between compartments for cables, etc., have a limit on size and must be fire-stopped, e.g. caulked with suitable material. Materials, linings and surface finishes are covered, and may deal with what are called 'early fire hazard indices', EFHIs.

Sprinklers
Codes generally set out sprinkler requirements.

Active and passive methods of improving fire safety of buildings

'Passive protection' refers to protecting a building by attempting to confine a fire to the area in which it started. Automatic venting or blocking of the products of combustion is involved.

Passive protection includes:

- fire-retarding treatments and material
- air conditioning which can switch to fire mode, i.e. no return air recycling
- stairway, passageway and liftshaft pressurization if a fire occurs
- the correct fire rating of all walls, roofs, doors, floors, ceilings, windows and structural members
- fire doors, fire windows and fire shutters.

Smoke and heat management is important to allow occupants to escape and to reduce the temperature build-up.

'Active protection' is aimed at detecting and extinguishing a fire once it has started. It includes:

- fire and smoke detectors and alarms
- fire hydrants, hose reels and extinguishers, including water pressure booster pumps
- automatic sprinkler systems
- gas flooding systems
- evacuation systems.

The detection system can be wired to a fire panel and also linked to the fire services.

Requirements for egress from a building

Building codes cover access and egress, and emergency lighting, exit signs and warning systems. It may involve fire-isolated passageways, ramps and stairways. An 'exit' has a wide range of meanings; generally it takes the form of an internal or external stairway, a doorway opening to a street or open space, a fire-isolated passageway or a doorway situated within a fire wall. The door has to open in the direction of travel to the outside.

The best building design in the world is no use if the exit is locked or blocked with goods. Exit doors must be readily openable while the building is occupied. Exit doors must not be chained, bolted, fastened, or obstructed in any way. They should be openable by a single-handed action from the inside by a single device. They must not be blocked by traffic or parked vehicles on the outside. Special attention is required for invalids, people in bed due to illness, elderly people, people whose movement is impaired, very young children, and people in captivity.

Number and dimensions of exits

Exits must be adequate and must always include an alternative means of escape. This is probably the most critical factor in fire safety provision. In the event of a fire, or emergency, the occupants must be able to get out. The building code will give more accurate specifications on the number of exits required. The dimensions of exits can vary depending on the number, or estimated number, of occupants. The size can also vary according to the class of building, e.g. hospitals clearly need special specifications.

Defining and specifying travel distance for a given occupancy

In regard to exit travel distance, it is important that the occupant does not have to travel an excessive distance to reach an exit. These distances vary according to the class of building. Typically, the distance is 20 m to 40 m, with more than one choice. Check with the building code for the specifications.

(*For most of the last three sections, acknowledgement to ACTRAC (Australia), OH&S Fire Safety, used with permission.*)

Features controlling fire and smoke spread

As indicated above, the spread of fire and smoke can be controlled by both passive and active means. On the one hand, the measures include separation of buildings; compartmentation; fire walls; fire doors; fire and smoke shutters; controls on air conditioning; and pressurization of critical areas such as stairways; and, on the other, detectors and alarms; extinguishers; hose reels and sprinkler systems. Smoke detectors generally detect fire more quickly than heat detectors.

Communications requirements for emergency control functions within a building

Building codes deal with the requirements for emergency warning and intercommunication systems. For buildings over a certain number of storeys and for process plants this is essential. The system must detect the fire, initiate an alarm and alert emergency services. Many buildings have PA (public address or Tannoy) systems which are used by designated fire control staff to ensure orderly evacuation. It is also used to sound the all-clear. Many mines and process plants also have a designated internal phone number for direct communication to a command centre if an emergency arises – fire or otherwise.

Preparing a fire safety survey checklist

You will need to make use of the material covered above to put together this checklist. It should cover likely fire sources; correct building design and construction for a building's current usage; passive and active protection factors; and communications and emergency planning for fires.

The human element in fire causation and behaviour during fire emergencies

Building design, emergency procedures and human behaviour

In situations where people have died in a fire, generally it is smoke and toxic gas inhalation which kills – or renders people unconscious so they cannot escape – rather than heat itself. While it is necessary to sound an alarm and alert people so that they respond ('Fire!'), there is the risk that panic can arise. This is more likely if there is a delay in giving a warning, and hence time is short to evacuate a large number of people. If exits to the outside are blocked or have been locked, and a crush develops, people can panic further. In some cases, such as the Bradford soccer stadium fire in the UK, many people died from crushing, not from fire. (Multiple deaths also resulted from exits with illegal locking or from inadequate egress in the Whisky A Go Go nightclub fire in Brisbane, the Stardust nightclub fire in Dublin and a more recent nightclub fire in Shanghai.)

A prompt and clear response by well-trained staff will ensure that panic is minimized and evacuation is effective. It is important that physically challenged people are properly briefed in advance, because if they set off down an emergency stair at the wrong time they can obstruct the exit of many others.

Problems can arise in buildings where large numbers of people inside are casual visitors, such as in major shopping centres. Tenants may change and appointment of new wardens overlooked. Exit routes should be well marked and signposting good. Panic can cause people to overlook obvious escape routes, and can lead to individual competitive responses rather than orderly behaviour where members of a group look after each other.

Provisions for special-needs groups

Obviously this needs careful consideration. Unpredictable behaviour in emergency situations is undesirable. As an example, some years ago a computer installation was planned with CO_2 gas flooding for fire control. Exit times after warning before the flood activated were calculated. However, this did not allow for an employee in a wheelchair. As mentioned earlier, a new flooding gas called Inergen allows breathing for a limited time. A water micromist system is another solution, even in computer facilities.

Hospitals, nursing homes, special care facilities, sheltered workshops, prisons and child-care facilities all require special attention. The steps to take are:

- Ensure data on people with special needs and any special information relating to them are kept at a designated control point. Where there is a turnover, data must be kept current.
- Communicate the procedures for dealing with the people concerned.
- A safe holding area, constantly supervised, must be set up during an evacuation; it is important that appointed staff remain there to give reassurance and confidence.

- Where people are not going to be able to use stairs, the alternative means of escape must have been planned in advance and must be used.
- The most able must be evacuated before those with most impaired mobility. This is not an easy concept to have accepted. The evacuation can take place in parallel if a non-stair route is being used for those with impaired mobility.

A fire in the early hours of the morning at a nursing home some years ago resulted in fatalities as staff rostering did not provide sufficient staff to respond to a fire.

The reader may wish to note once again what has been said about priorities for evacuation of occupants with various levels of mobility in fire emergency situations, and what has been said about planning for otherwise healthy but physically challenged people. It is important that physically challenged people are not allowed to enter stairwells unsupervised, to avoid blocking exits, which would be contrary to the evacuation plan.

Motivation and characteristics of people committing arson

A variety of reasons for arson exist. They include:

- Covering up detectable traces of a crime, such as burglary or murder.
- Intimidating someone such as a business competitor, someone from a racial group disliked by someone else, or someone who refuses to pay 'protection' money. (Perth, Australia, had a spate of fires in Chinese restaurants in the late 1980s, lit by a neo-nazi group.)
- Setting fire to premises to claim insurance, as a means of overcoming a poor business situation.
- Getting back at people because of perceived poor treatment, such as when students or ex-students set fire to a school for revenge.
- Wanting to play a heroic role. Some bushfires are set by volunteer firefighters who relish the idea of being an important part of the action.
- Pyromania. A person enjoys seeing what results from fire, secretly knowing they started it. A variety of types of social, sexual, marital or psychological inadequacy may explain such behaviour.

Measures against arson

It is very difficult to prevent or limit the opportunity for people to deliberately light fires if they are so-minded. Fortunately, most people are aware of the risks and the devastation – often much greater than such a person intended – which can result. For instance, deliberately lit bushfires have led to loss of life; put large numbers of firefighters and others at risk; and caused severe disruption and huge economic losses.

Education campaigns may be of some effect, especially among young children still learning about fire. They can be effective in reducing the risk of accidentally initiated fires such as bushfires from picnickers, and office fires from smokers. Possibly the only other measure which will result in some success is to employ effective perimeter security and entry detection systems on premises.

Once a fire is started, good building design and automatic control systems can limit its effect.

The effectiveness of locking systems in providing building security and safe egress in fire emergencies is described in the next section.

Locking systems – security versus safe egress

Deficiencies in locking systems

Unfortunately building codes may allow for release mechanisms which are not immediately obvious and which may not be on the door. A lock should be easy to find, and its operation obvious, even in darkness. A lever action handle, for example, is fairly easy to use. Key-boxes and door releases requiring breaking a glass shield, alarm-triggered release doors and push-button electronic releases can all have problems in operation. If the computer system which controls release doesn't allow locks to fail safe, this is a major concern. An explosion in the 110 storey World Trade Center in New York in 1998 in which 50 000 people were working, disabled most of the electronic systems. In the 1981 Stardust Disco fire in Dublin mentioned earlier, 68 people died. The security exits were locked. A very good video, the *Anatomy of Fire,* was made by the UK Fire Research Establishment showing a recreation of that fire, to establish how fittings and linings could have caught fire so rapidly.

It is worth remembering too that in some countries, even hotels belonging to established chains may not match the fire safety standards in the country of the parent company. Other accommodation may likewise not meet best practice standards. For example, beginning in the mid-70s, the United Kingdom adopted new fire safety standards for all homes offering bed and breakfast stays. In any hotel, it is always useful to study the map and instructions on fire which are placed on the back of the door to your room and to check the emergency exit. It is much harder to study it in the early hours of the morning by emergency light (if any) with a fire in progress. The Childers backpackers hostel fire in Queensland, Australia in 2000 killed fifteen young people as a result of arson.

Fire safety needs and conflict with building security

There have been significant moves to upgrade the security of many premises because of a perceived increase in crime. In addition, premises such as nightclubs can be concerned about entry other than by way of the front door. The result of these issues can be a conflict between security and the need for easy egress in the event of an emergency. Security breaches, being more common than fire, may also lead to people in charge of security emphasizing security over egress.

Acceptable locking systems

Any system must provide adequate security while allowing for adequate egress as required by building regulations. The requirements usually are set out in the relevant building code.

A recent catalogue from a lock manufacturer identifies twenty-three different types of locking devices. A latch is easy to open from the inside but can be forced from the outside. The deadlatch is an improvement and can be locked from the inside. Locks approved for exit should allow easy exit without a key. Commercial exits can now be more fire-safe while still secure, because of electronic checks on merchandise leaving a store, sensors which unlock doors as people approach, and the use of security video cameras.

The increase in break-ins has led to pressure for better domestic security, not least from insurance companies. People in high crime areas will have extra locking devices fitted to doors and windows. However, these can restrict ability to escape in a fire. It is essential to ensure ready escape in a fire. In poorer areas, fires can result from not being able to leave children supervised, or from using unsafe low cost heating. Not having a smoke detector is an additional risk. Better domestic fire safety will result if:

- every room has an alternate means of escape
- windows can open far enough to allow escape
- barred windows have a quick release mechanism which people know how to use
- deadlocks are off while people are indoors
- practice fire drills are held
- there are fire extinguishers
- electrical gear is in good condition
- curtains, furniture and clothing are kept away from heaters
- there is a guard in front of an open fireplace
- flammable liquids are stored outside the house
- there is no smoking in bed
- matches and lighters are kept away from children
- power points are used only for their rated load.

Fire prevention and emergency training programmes

Identifying groups or individuals in the organization requiring training

The response needed for a fire will vary with the type of workplace involved. Workplaces include small office premises and factory units; large sites with process industries; schools, colleges and universities; multi-storey office buildings; hotels; shopping centres; hospitals and nursing homes; warehouses; offshore rigs; and underground mines. An orderly response without panic results when people have had proper training, but this is more difficult when many occupants are only casual and non-employees.

Risk engineering

The need for staff training in these situations is consequently higher. Universities, colleges and, to a lesser extent, schools have the problem that staff come and go during the day, and there is no record of who is in attendance for particular classes.

Before the training needs can be addressed, in workplaces which warrant it, a suitable group should meet to work on emergency planning. This will include, for example, any in-house engineers.

Particular issues to address are:

- minimization of fire risks
- ensuring first response extinguishers are available and people know how to use them
- planning escape routes, without using lifts
- planning muster points which are safe and allow people to move further away
- a system for accounting for everyone, including those in back rooms, toilets, store rooms, etc.
- the role of the switchboard
- marshalling points and marshals or wardens on each floor or in each area
- keeping new entrants out, e.g. the public, during an emergency
- safeguarding money and other valuables
- contacting emergency services
- giving particular attention to people whose mobility is impaired
- paying particular attention to night response and the role of night staff.

After addressing these issues a written plan should be drawn up. A short form of this, which is easy to use, should be widely distributed and read. The plan will identify the training needs of different people or groups of people; this training should then be carried out.

The effectiveness of the emergency plan will depend on it being tested – firstly after forewarning occupants – and then followed up by a debriefing. Once appropriate corrections are made after the test, a further unannounced drill must be held. Once again, after a debriefing make corrections, if any, based on the experience of the drill.

Particularly in a large complex workplace, correct selection of key personnel is crucial. They will need to be the right type of person – cool in a crisis; likely to continue to be employed in the building or workplace for some time; and likely to work principally in the building rather than in outside work. In a dispersed site it may be satisfactory for the selected individuals to generally work anywhere on the site, providing they have portable communications equipment, such as a pager or mobile phone with them at all times. Deputies must be appointed so that absence of key personnel still allows effective response.

Roles and responsibilities of employees and managers in fire emergencies

Generally speaking, in a workplace large enough to warrant it, the roles will include:

- chief warden
- deputy chief warden

- communications officer who knows the communications equipment and speaks confidently
- floor or area wardens – people who are generally available in their area
- wardens, with a minimum of two. (Extra wardens assist the floor or area warden on the basis of one for each 20 people.)
- deputies to chief warden and communications officer in the event that they are not there.

If there is an emergency warning and intercommunication system (EWIS), the communications officer is to use it. Otherwise, the telephonist on the switchboard may need to operate any system such as a PA (public address, or Tannoy).

Different levels of emergency staff are identified by different-coloured helmets, marked with the title of the job and the floor or area number. The helmets are:

Chief or deputy chief warden	– white
Floor or area warden	– yellow
Warden	– red

The roles of people with emergency responsibilities are set out below.

Chief warden

On becoming aware of an emergency the chief warden should take the following actions:

- ascertain the nature of the emergency and determine appropriate action
- ensure that the appropriate emergency service has been notified – ensuring that communications officer, floor or area wardens are advised of the situation
- if necessary, initiate evacuation and controlled entry procedure
- brief the emergency services personnel upon arrival, on type, scope, and location of the emergency and the status of the evacuation, and thereafter act on the senior officer's instructions.

Deputy chief warden

The deputy chief warden shall be required to assume the responsibilities normally carried out by the chief warden if the chief warden is unavailable, and otherwise assist as required.

Communications officer

On becoming aware of the emergency, the communications officer should be responsible to the chief warden for the following:

- ascertaining the nature and location of the emergency
- notifying appropriate ECO (emergency control operations) personnel either by the EWIS or other means
- transmitting and recording instructions and information between the chief warden and the floor or area wardens and occupants
- recording the progress of the evacuation and any action taken by the wardens.

Floor or area wardens

On hearing an alarm, discovering or becoming aware of an emergency, the floor or area wardens should take the following actions:

- commence evacuation if the circumstances on their floor warrant this
- implement the emergency procedures for their floors
- communicate with the chief warden by whatever means available and act on instructions
- direct wardens to check the floor or area for any abnormal situation
- advise the chief warden as soon as possible of the circumstances and action taken
- co-opt persons as required to perform warden duties.

Wardens

Wardens may be required to carry out a number of activities, including the following:

- acting as floor or area wardens, if they are unavailable
- calling fire brigade or other appropriate emergency service by operating the manual alarm point and telephone (Note: always back up the activation of a fire alarm with a telephone call. It guarantees that the alarm has gone through and it gives the emergency service further information with which they can pre-plan their course of action.)
- operating the intercommunication system
- checking to ensure fire doors and smoke doors are properly closed
- searching the floor or area to ensure all persons are accounted for
- ensuring orderly flow of persons into protected areas, e.g. stairwell
- assisting mobility-impaired persons
- acting as leader of groups moving to nominated assembly areas
- operating first attack firefighting equipment, e.g. portable fire extinguishers, hose reels and fire blankets, when suitably trained.

Identifying skills and knowledge required by groups and individuals

Once the roles, duties, and tasks required have been established, the training needs to meet these requirements. For example, a very important training need is the experience of fire, heat and smoke in a confined area similar to a fire environment. It is important that key ECO personnel are not taken by surprise by the fire environment. In this way they will be familiar with it, they will be more effective in their duties, and be less likely to panic or act irrationally. ECO personnel should be familiar with – and, therefore, trained where necessary – in the following:

- practical experience of fire, heat and smoke in a confined area similar to a fire environment (see below)
- operation of, and any procedures for using, the communications equipment installed
- operation of all fire detection and suppression systems that may be installed on their floor or their area

- operation of basic fire-fighting equipment such as: portable fire extinguishers, hose reels, fire blankets, and fire alarms. (This could include the use of breathing apparatus.)
- any special procedures that may exist to protect strategically significant items located on their floor or in their area
- escape routes and safe holding areas
- any dangerous goods which may need special attention or isolation.

Note: fire and emergency services need to train their own staff in the very real experience of confined fire, heat and smoke. They have access to facilities that enable this training, which is considered a vital part of fire safety training. Your local fire and emergency service should be able to offer or assist in any fire safety training required.

The person in charge of fire safety or emergency procedures in a building should ensure that every occupant and every new occupant of the building is advised of the procedure to be taken in the event of an emergency, and each occupant should be given the name, location and telephone number of the warden of the area in which they work. Occupants should be encouraged to approach their warden for information and clarification of procedures.

Arrangements should be made for regular short demonstrations explaining the various types of first-attack fire-fighting equipment, its uses and limitations, and the correct methods by which it is operated. All ECO personnel should be familiar with the EWIS, and it should be tested weekly.

ECO personnel should meet together at intervals of not greater than six months. These meetings should also be used as short training sessions to maintain the interest of personnel and improve knowledge and skills. In addition to general fire safety, bomb threat, and evacuation procedures, training sessions could include: methods of assisting mobility-impaired persons; the study of behaviour of people during emergencies, and resuscitation.

(*With acknowledgement to ACTRAC (Australia), OH&S Fire Safety for material in the two sections above, reprinted by permission.*)

Preparing a short training programme

It is clear that there are a number of defined areas of training required. For each group, the learning outcomes must be included in the course of training so that the person has the skills to carry out the role and responsibilities of their emergency position. The skills are best defined by writing them down as assessment criteria for each learning outcome. The assessment criteria are the basis for ensuring that the person has acquired the necessary skills. This is important to give key personnel experience in operating in smoke, and this can be done with simulation facilities possessed by some training organizations.

It is suggested that the reader selects one of the emergency groups above, e.g. floor warden, and writes out the learning outcomes and assessment criteria for a short course for floor wardens. Use active verbs, e.g. identify, discuss, transmit.

Delivery of a short training programme in an aspect of fire safety

The reader can choose which training programme to give, but generally the training for the key fire safety positions will be given by someone with considerable experience. Unless you have that experience yourself, you could therefore decide to limit yourself to presenting an induction programme suitable for any new staff member, including the opportunity, if possible, to operate and handle extinguishing equipment. Refer to Chapters 2 and 13 or other sources for information on giving effective presentations and training.

Further reading

Australian Building Codes Board. (1996). *Building Code of Australia*. Sydney: CCH Australia. (Also available from Standards Australia with associated standards on CD-ROM.)

Bailey, A. (1990). Industrial Hazards Due to Electrostatics (and other electrical issues). *Health and Safety at Work*, **12(5)**, 29–44.

Canter, D. (1990). *Fires and Human Behaviour*, 2nd edn. David Fulton Publishers Ltd.

DHHS-CDCP-NIOSH (US). (2002). *Electrical Safety Student Manual*. Cincinnati, NIOSH.

Editorial. (2002). Negotiated Rulemaking. OSHA Proposes Crane and Derrick Standard Changes. *Professional Safety*, **47(9)**, 15.

Ford, J. (1993). Machine Safety – A Formal Education. *Accident Prevention*, Jan/Feb. 1993, 20–5.

Hamilton, M. (1990). Facing Up to Welding Safety. *Health and Safety at Work*, **4(2)**, 24–8.

IEC 61508. Electrical, *Electronic and Programmable Electronic Systems*. Geneva, IEC.

Krieger, G.R. (ed.). (1994). *Accident Prevention Manual For Business And Industry: Engineering & Technology*. 11th edn. Itasca Ill: National Safety Council.

Laing, P.M. (ed.). (1997). *Accident Prevention Manual For Business And Industry: Administration & Programs*. 11th edn. Itasca Ill: National Safety Council.

Pearson, M. (1992). Earthquake Safety Planning – Are You Prepared?, *Safeguard (New Zealand)*, July, pp. 12–15.

Sheppard, J. (2001). The Future of High Visibility Clothing in Canada. *Accident Prevention (Canada)*, September/October, p. 18.

Stranks, J. (1999). *The Handbook of Health and Safety Practice*. 5th edn. London: Financial Times/Prentice-Hall.

Thomas, G. (1995). Safety at Sports Grounds–Part 2. Fire Safety at Football Stadia. *Health and Safety at Work*, **17(5)**, 14–16.

Worksafe Australia. (1994). *National Standard for Plant*. NOHSC 1010, Canberra: AGPS.

Worksafe Australia. (1992). *National Occupational Health and Safety Certification Standard for Users and Operators of Industrial Equipment*. NOHSC:1006 Canberra: AGPS.

Worksafe Australia. (1996). *National Standard and National Code of Practice for the Control of Major Hazard Facilities*. NOHSC:1014, 2016 Canberra: AGPS.

Legislation, codes, rules, standards, and guidance notes on machinery, conveyors, cranes, electrical equipment, gas cylinders, mobile equipment, fire and fire services.

Activities

1. Select a workplace and run a check on the electrical safety aspects. Include, for example, fuses, insulation, tools, cords, power outlets, switchboxes, and safety devices.
2. Write down the requirements for the licensing of cranes and pressure vessels, and their operators, in your OHS jurisdiction.
3. Prepare a list of the different types of machine guards in a selected workplace. Comment on their suitability.
4. Access a copy of the building code applicable in your jurisdiction and outline the key requirements in relation to fire protection and control.
5. Draw up a fire protection and control checklist for a workplace you select.
6. Access a copy of the fire response plan for an office building, and comment on its adequacy. If you can, find out the frequency with which fire drills are conducted.
7. Draw up a plan for a one hour presentation to new workers on fire including, for example, extinguishers, access, egress and the locking/security system.

8

Health at work

WORKPLACE EXAMPLE

'In a year or two (the company) will produce the richest and most lethal crop of cases of asbestosis in the world's literature... Naturally I think some of the chests should be looked into.'
(letter to Commissioner of Public Health, Western Australia, 1948)
and

I said, 'Look, you've got problems'... They pooh-poohed me and in fact they labelled me a trouble-maker... They didn't listen. I was just a young man, a flying doctor... I'd only just come out from England and they thought, "who's this jumped-up fellow?" That was the attitude.
(comment from Emeritus Professor Eric Saint, Department of Medicine, University of Western Australia, District Medical Officer in the Pilbara region of Australia in the late 1940s/early 1950s, when he saw blue asbestos being handled at the Wittenoom mine and the nearby port).

Development history of occupational health

Contribution of key historic figures

The early Egyptians recognized the value of veils as a form of respiratory protection when mining cinnabar (red mercury oxide) for cosmetics. There are also records of the effects of the sun felt by workers in King Solomon's mines in Arabia.

Prior to the nineteenth century the key historic figures who contributed to the development of occupational health were Georgius Agricola (1494–1555) and Theophrastus Bombastus van Hohenheim Paracelsus (1493–1541).

Both observed miners and their diseases. Agricola wrote a book *Of Things Metallic* and Paracelsus made the basic observation of toxicology – the study of poisons – which was that 'the dose makes the poison'. This also is the basis of drug therapy. The third figure was Bernardino Ramazzini (1633–1714), known as the father of occupational health. He wrote a book on diseases of trades – *A Diatribe on Diseases of Workers*, and it was he who suggested a doctor should ask a patient what his or her work was.

Percivall Pott in 1776 made the connection between an occupational disease and its cause, linking scrotal cancer in chimney sweeps with soot, which we now know contains the cancer-causing compounds PAHs (polynuclear aromatic hydrocarbons).

Further contributions in the English-speaking world in the nineteenth century were made by:

- Anthony Ashley Cooper, 7th Earl of Shaftesbury (1801–1885: reduced hours and improved conditions of young persons and women in mines, factories and other workplaces)
- Dr Thomas Percival (1740–1804: report on young persons at work in textile mills)
- Robert Owen (1771–1858: good conditions in his textile mills)
- Sir Robert Peel Sr. (1788–1850: early factory legislation)
- Michael Sadler, M.P. (1780–1835: supported changes in Parliament)
- Dr Charles Thackrah (1795–1833: book on occupational health 1832)
- Dr William Farr (1807–83: used birth and death statistics for occupational health in mid-nineteenth century)
- Dr Edward Headlam Greenhow (1814–88: study of effects of lead, dust and fumes)
- Dr John Arlidge (1822–99: statistics on potters' diseases, especially from silica)
- Dr Thomas Legge (1863–1932: first UK national medical inspector of factories, and writer of *Industrial Maladies*, 1934)
- Dr Alice Hamilton (1869–1970: worked in the USA, especially on lead poisoning).

In Russia, F.F. Erisman (1842–1915) played a major role.

Other key people deserving of mention include Donald Hunter (1898–1978) who wrote successive editions of *Diseases of Occupations*, and Luigi Parmeggiani who edited the ILO *Encyclopaedia of Occupational Health and Safety* for many years.

(Adapted from Waldron, H.A. [ed.]. Occupational Health Practice. Butterworth-Heinemann. Reprinted by permission. See Further Reading at the end of this chapter.)

Trends in the development of occupational health

Over the years a number of trends have emerged in the development of occupational health:

- the development of the idea of dose: an acceptable dose of one substance of a certain toxicity may be different from an acceptable dose of another substance of another toxicity level
- a recognition of the connection between type of work and health status or even type of disease
- linking certain agents in the workplace with their effects on health, e.g. soot with cancer of the scrotum

Health at work

- the gradual development of government intervention in the workplace to protect the health of workers
- the establishment of a base of data from the births, marriages and deaths register
- the start of epidemiology or the statistical study of the links between disease and possible causes
- reporting of occupational diseases and injuries to the authorities
- compensation for some health problems developed as a result of work.

Effect of the World Wars on occupational health

Occupational health was paradoxically given added impetus by the two World Wars. A key effect of the First World War was the need to keep skilled munitions workers (mainly women) healthy while working with toxic and explosive materials. Also, aircraft workers then used solvents in the dope (glue). In the Second World War there were two key developments:

- Bedford's work on thermal comfort (heat stress) as a result of the need for men to operate submarines while retaining their mental and physical effectiveness. This followed up work after the First World War by Yaglou on men in US army training.
- US work on ergonomics as a result of the need to improve the comfort and ease of operation of tanks.

Cumpston (see Further Reading at the end of this chapter) described the development of occupational health in Australia. There were two key driving forces:

- silicosis (scarring or fibrosis of the lung) from underground hardrock mining
- lead poisoning from mining silver, lead, and zinc.

Legal developments

Until the introduction of Robens legislation in many jurisdictions, they had separate Acts primarily on safety for areas such as construction, factories, shops and warehouses, and machinery. These were often administered by a labour department while occupational health issues were separately administered by a health department.

Current developments in occupational health

International conventions on OHS

The beginnings of the International Labour Organisation (ILO) were set up in the early part of the twentieth century. One of the reasons was to ensure that nations who set high labour standards were not undercut competitively by those who did not. The ILO is based in Geneva and one of its key activities is the development of international conventions. Countries which are ILO members are encouraged to adopt or *ratify* these Conventions, which obliges them to transform them into law in that country.

The European Union, the United States and some non-government organizations have been working to try to incorporate seven so-called *core* ILO conventions into international trading agreements such as the World Trade Agreement. Within the Asia-Pacific regional trade forum, APEC, some countries do not support this. There are currently 175 ILO conventions (some supersede others). These include Convention 155 on occupational health and safety, and Convention 161 on the provision of occupational health services to all employees. (It is generally understood that the words 'occupational health' in the convention include occupational safety.) When a country ratifies a convention it agrees to ensure that it is backed up by the law of the country.

Convention 161 and Recommendation 171 pose the question – 'how do we provide occupational health services to all employees, and to what extent do these services overlap with primary health care, that is, general health care provided to the population as a whole?'.

To provide financial incentives, in Finland, for example, the Finnish Social Security Institution was paying 55% of a firm's cost of providing the service. The service provider firm had to be an approved one.

A number of models for providing occupational health and safety services exist. They include:

- Provision of salaried people within a company or organization. The professions and services included depend on the size and complexity of the company's or organization's operations.
- The occupational health service of a large organization is also made available to other (usually smaller) organizations nearby.
- An organization contracts for the services it needs from a provider of a range of services or several providers of different services.
- Government provides the services to organizations.
- Government provides the services but only to small employers.
- A group of organizations jointly finance and develop long-term contracts for the provision of occupational health services for their operations.

Disciplines in occupational health

A range of disciplines are involved in providing expert advice and assistance within occupational health services. These include:

- occupational physicians – doctors with qualifications and/or experience in workplace health
- occupational health nurses – nurses with qualifications and/or experience in workplace health
- health and safety practitioners – people with qualifications and/or experience in general recognition, evaluation and control of hazards at work

- occupational hygienists – experts in recognition, evaluation and control of workplace health hazards
- engineers, especially those with qualifications and/or experience in safety and acoustic engineering
- ergonomists – who look at how the demands and design of the workplace and workplace tasks fit the mental, psychological and physical abilities of women and men
- psychologists – who look at the 'management climate' of the workplace and its 'safety culture'; play a role as counsellors in the rehabilitation of injured workers; and examine behavioural aspects of accidents, such as error.

Workplace stressors, processes, diseases and disabilities

Introduction

Toxins are chemical substances which interfere with the normal operation of a particular part, or parts, of the body. They can do this in a variety of ways. The chemistry of the toxin decides how, where and how badly it will interfere with the cells, tissues, organs or systems of the body. In some cases the chemistry of the surface of particles may be important, e.g. in asbestos fibres and quartz dust. However, we cannot just look at the chemical make-up of a substance and know in advance what it will do to the body. Over the years, as knowledge has built up, we can sometimes make some reasonable predictions but, as new substances come along, it is a process of trial and error. Medical specialists, for example, get permission to run human trials on new drugs after the drugs have been tested on animals, colonies of human cells grown in laboratories and bacteria.

Routes of absorption of toxins, target organs and detoxification

Routes of absorption
Toxins are absorbed into the body by the following routes:

- inhalation – breathing
- ingestion – through the mouth and partly from swallowing toxins in nasal secretions which run into the throat
- through the skin – dermal absorption
- injection – by a needle-stick injury or high pressure jet.

It is important to remember that people other than the worker exposed to a toxin can be affected. Close contact with a child can result in the child breathing dust from hair or clothes, or the child's skin contacting a liquid toxin on overalls. The sexual partner of a worker risks genital contamination (e.g. asbestos fibres have been found in ovaries) if the worker's hand, fingernail, and other hygiene is not good. Ingestion of toxins by

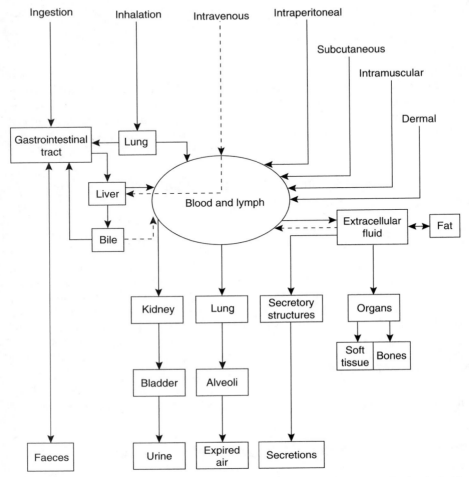

Figure 8.1 Routes of absorption, distribution and excretion of toxins in the body (with acknowledgement to Amdur, M.O. and Doull, J. (eds) (1991). *Casarett and Doull's Toxicology*. (4th edn.) New York, Pergamon, reproduced by permission of the McGraw-Hill Companies.)

a worker can result from poor personal hygiene such as eating or smoking without washing, or biting fingernails.

Target organs in the body

Toxins are generally linked with particular conditions in specific parts of the body, but the toxin may have effects in several parts of the body. For example, lead affects:

- production of haemoglobin (the red blood cell pigment) in bone marrow
- the liver
- brain function.

Detoxification and elimination of toxins

Substances which are not naturally a part of the body but can potentially interact with it are sometimes called 'xenobiotics' ('xeno' means foreign). Many foodstuffs contain substances which are potentially damaging to the body if eaten or drunk in excess – alcohol being the best example.

Over time the survival of the species has meant that the body has developed ways to deal with some of these substances by reducing their toxicity, quite often making them more water soluble where necessary and, if possible, passing them out in the urine. The liver, the body's biochemical processing factory, does a lot of this work. However, the body today is exposed to many synthetic substances. The human body copes with some of these because its existing systems work on them, but it has not had enough evolutionary time to adapt to dealing with others.

When chemical 'X' enters the body and is changed into chemical 'Y' before being passed out in the urine, chemical 'Y' is called a 'metabolite' of chemical 'X'. However, between the liver and the kidney, the metabolite, which may be more or less toxic than the original chemical, circulates throughout the body in the bloodstream and may enter other organs, tissues or systems of the body – for example, the brain.

A diagram of the routes which toxins can take between absorption and elimination is shown in Fig. 8.1.

Note that some swallowed chemicals are not very well absorbed into the body from the intestines, while others are. For instance, only around 10% of swallowed lead is absorbed from the intestines, so a chemical may be partly excreted in faeces, partly in urine, or if it is a volatile (gives off vapour) or gaseous compound in the blood, some may be eliminated in exhaled breath or through sweating. Arsenic finds its way into hair, fingernails and toenails and is lost that way.

Dose-response, synergism and additivity

Dose-response

An important principle of toxicology (the study of toxins and their effects) was set out by Paracelsus over 400 years ago: '*The dose makes the poison.*' We have found a few exceptions since, such as blue asbestos, but generally speaking, if we graph the effect of the toxin (the body's response to the toxin) against the amount of toxin absorbed we get a picture like the following (see Fig. 8.2).

The point at which such a curve turns upwards is generally called the 'threshold'. There is quite a lot of debate about whether doses below the threshold are acceptable when setting standards for exposure to workplace or environmental toxins, but that is what is done for most of the 700 or so workplace chemicals for which standards have been set and, also, for ionizing radiation standards.

It is important to note that *dose = concentration × time* – for example, a low concentration over a long time or a high concentration over a short time. In both cases, with some toxins the body may recover between exposures – that is, when the worker is not exposed or is not in the workplace – so that there is no progressive effect. However, some toxins are cumulative (*chronic* poisons) and the effect is progressive. With some

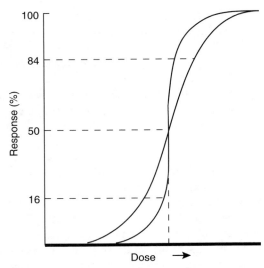

Figure 8.2 Dose-response curves (with acknowledgement to Amdur, M.O. and Doull, J., see Fig. 8.1 for publication details)

cancers, there is a latency period; that is, many years pass between first exposure and obvious disease. Toxins which have an immediate effect are called 'acute' toxins.

Toxins may act locally at the point of contact or systemically; that is, spread through the system to other target tissues or organs.

Additivity
When the body is exposed to two or more chemicals at once, the effects of one may simply add to the effects of the other if both affect the same organ, tissue or system. This is called 'additivity'.

Synergism
This term describes the situation when a combined effect is more than you might expect from simply adding both effects together; an example is alcohol and some antihistamine drugs. The same can happen with certain workplace chemicals – for example, exposure to chlorinated degreasing solvent followed by drinks at the pub.

Considerable care and judgement are required to set standards for workplace exposure where several toxins are encountered at once, because of the possibility of synergism occurring.

For example, an underground nickel miner's lungs may be exposed to:

- nickel
- quartz
- asbestos fibres
- radon (a radioactive gas)
- diesel exhaust fumes (PAHs, aldehydes and nitrogen oxides).

Links between workplace stressors, processes and diseases

Introduction

Diseases and conditions of the lung are important when considering occupational health. The lung can be affected by fumes, smokes, mists, dusts, fibres, gases and biological agents such as spores (for example, from fungi) and enzymes (for example, from raw cotton).

Some dusts may be radioactive. Disease of the lung caused by dust is called pneumoconiosis, Greek *pneumon* (lung), *koni* (dust). These have been practically eliminated in underground mining in many countries by using effective mine ventilation.

Particle behaviour in the lung

The very small particles which make up dust can be broken into three groups. These are:

- 'non inspirable' – particles which settle out too fast to be breathed in through the nose
- 'inspirable' – particles whose settling speed is slow enough to allow them to be breathed in through the nose
- 'respirable' – particles whose settling speed is even lower. They will remain suspended long enough to reach the deeper vessels in the lung – the bronchioles and alveoli.

Dust particles have very irregular shapes. For example, a fibre is a particle with a length of at least three times its width. A respirable fibre has a diameter of approximately 3 μm or less, depending on its density.

Actual diameters are difficult to establish; therefore, an 'equivalent aerodynamic diameter (e.a.d.)' is used. The e.a.d. is the diameter of a small sphere with a density of 1 g/ml which settles at the same speed as a particle.

The vast majority of particles over 30 μm e.a.d. are not inspirable. Virtually 100% of particles 1 μm e.a.d. and below are respirable; 50% of particles 5 μm e.a.d. are respirable; and almost no particles greater than 7 μm e.a.d. are respirable.

Generally for dust diseases, the greater the weight of dust breathed, the higher the chance of disease. While there may be more particles 1 μm and below, a 0.5 μm particle, for example, is only 1/1000 the weight of a 5 μm particle. So the larger particles have a much greater effect per particle than the smaller, even though there are generally more smaller particles. For fibrous particles, numbers not weight are more important.

Protective mechanisms of the lung

The respiratory tract (nose, throat, windpipe, lungs) has a number of protective mechanisms against dust and fibre particles. Only a very small fraction of the dust and fibres we breathe actually remains in the lung permanently, otherwise lung function would be obstructed significantly.

The protective mechanisms are:

- The mucus-covered hairs in the nose.
- The turbinates or conchae in the nose (large wet surfaces which trap particles).
- The reduction in airflow rate as the lung vessels become smaller. This allows dust to 'sediment' out on the walls of the vessels (sedimentation).
- The forking of the vessels. Particles are stopped by 'impaction' as they hit the inside of the fork.
- As the lung vessels get smaller, the larger particles, especially fibres, jam across the cavity (lumen) of the vessel. This is called 'interception'.
- As the air speed slows, very small particles are also deposited by 'diffusion' – they are knocked against the walls by the air molecules.
- The *mucociliary escalator*. Fine hairs, with a coating of mucus, on the walls of the lung vessels carry particles up to the throat with a spiralling action.
- *Macrophages*. These are mobile 'scavenger' cells which 'eat' particles and attempt to break them down with enzymes. Unfortunately dusts such as quartz and asbestos kill the macrophages.

Processes which can produce dust-caused diseases

These include:

- foundry work – shaking out the sand moulds
- brick manufacture – from any sand used
- hard-rock mining and coal mining in rock strata containing quartz, e.g. sandstone
- construction work involving tunnelling and excavations in sandstone
- processes involving raw vegetable fibres such as hemp, cotton, jute, sugar cane fibres (bagasse)
- pottery – moulding, smoothing, dust on floors
- abrasive grinding
- abrasive blasting with dry sand
- concrete and brick cutting with powered tools
- quarrying, e.g. slate
- removal of asbestos insulation, mining in rock containing asbestos fibres, manufacture of products containing asbestos
- furnace lining and refractory brick maintenance.

Different health effects of fibres and silica

The different characteristics of these types of dust particles are described here:

- Silicosis is caused by crystalline silica in three forms – quartz, tridymite and cristobalite. The last two are a result of the heating of materials containing quartz, such as bricks in furnace linings. Crystalline silica causes thickening of the walls of the alveoli and fibrous scar tissue develops. On X-rays this shows as isolated spots which often gradually spread. It is probable that crystalline silica may also cause lung cancer in some cases.

- Asbestos and possibly some synthetic mineral fibres also cause scarring. They may also cause cancer of the lung, particularly in smokers. Asbestos can also cause cancer of the pleura (lung lining), called 'mesothelioma'.
- Some vegetable fibres such as raw cotton can cause constrictive lung disease. Breathing becomes difficult when workers return to work after a weekend off. There is swelling of the cells of the alveoli, and with long-term exposure scar tissue (fibrosis) develops.

The four principal types of synthetic mineral fibres are rockwool, slagwool, glasswool and ceramic fibre. There is some evidence of excess lung cancer in early workers in the rockwool, slagwool and glass fibre industry when conditions were poor. Lung cancer is related to fibre fineness (less than 0.5 μm) and length (greater than 7 μm). Some filter grades of glasswool are in this category. A lot of rockwool, slagwool and glasswool fibre is too coarse to be respirable, i.e. reach the alveoli. Ceramic fibre is newer; is not dissolved over time in lung fluid like the other synthetic mineral fibres (SMF); and is fine enough to be 'respirable'. However, all these fibres break across, not along, the fibre. Asbestos breaks along the fibre to produce finer fibrils and it is the very fine fibres which are most likely to cause cancer of the lung.

Toxic gases and asphyxiants

Asphyxiation refers to lack of breathing due to a lack of oxygen. There are two basic types of asphyxiation – simple and chemical.

Simple asphyxiants

Simple asphyxiants are those which work simply by displacing oxygen. These include nitrogen (79% of air), argon (used in welding) and, to a degree, carbon dioxide, although it also has a biochemical effect in the body. In confined spaces where oxygen has been used up by rusting, oxidation of minerals in the soil or walls of a tunnel, or by biological growth, only nitrogen and argon remain in appreciable quantities.

Wineries and breweries produce carbon dioxide during fermentation and, being heavier than air, it sits in vats and pits and can be generated by stirring up the 'lees' (residue) in vats when mopping out. Deaths have resulted.

Chemical asphyxiants

Chemical asphyxiants work by interfering with the lung's gas exchange mechanisms, blood oxygen transport or the chemistry of body cells.

Some examples are shown in Table 8.1.

Carbon dioxide is also partly a chemical asphyxiant as it affects the acidity/alkalinity of the blood and carbaminohaemoglobin levels.

Effects of the chemical asphyxiants

Chlorine, nitrogen dioxide, ammonia, sulfur dioxide and sulfur trioxide are primarily irritants which lead to the lungs filling with fluid (pulmonary oedema). Hydrogen sulfide

Enhancing occupational safety and health

Table 8.1 Examples of chemical asphyxiants

Gas	Where found
Ammonia	Some refrigeration units, plan printers, nickel/cobalt extraction
Carbon monoxide	Petrol (gasoline) motor exhausts, poorly adjusted fuel burners, iron smelting
Chlorine	Water treatment, pool disinfection
Hydrogen cyanide	Fumigation, carbon-in-pulp gold plants
Hydrogen sulfide	Sewers, rotting estuary marine vegetation
Nitrogen dioxide	Acid manufacture, diesel engine exhaust, welding
Ozone	Odour control, photocopiers, electric motors, welding
Phosphine	Fumigation
Sulfur dioxide	Acid manufacture, sulfide ore roasting, plant fumigation, power generation (coal or oil), black shale oxidation in iron ore mines
Sulfur trioxide	Acid manufacture

at higher concentrations anaesthetizes the sense of smell, so the person is not aware of its presence; it is more toxic than hydrogen cyanide. The effects of nitrogen dioxide may be delayed. Hydrogen cyanide interferes with the use of oxygen by the cells of the body (interference with cytochrome oxidase). Phosphine causes the break-up of the oxygen carrying pigment in the blood (haemolysis). Carbon monoxide is absorbed by the haemoglobin (red pigment) in the blood in preference to oxygen. It is tasteless and odourless, and is present in cigarette smoke. It has caused many deaths.

Toxic metal exposure

Many of the worst effects of metals in industry have been recognized and controlled in a range of countries.

The metals found in workplaces which can cause disease or disability include those in Table 8.2.

This does not include metals which threaten health due to radioactivity, e.g. uranium, plutonium and thorium. Thorium is found in the mineral sands industry in the mineral monazite.

Metal-like elements (metalloids) which can cause disease and where they are found include:

Metalloid	Where found
arsenic	insecticides, arsine gas as a by-product of moisture in ferro silicon
antimony	some alloys, formerly in some pigments.

Effects of toxic metals
Some of the effects of the toxic metals are set out in Table 8.3.

Biology of mutation and toxins causing occupational cancer

Cancer related to work activity is estimated to make up to 4% of all work-related disease. Cancer is the uncontrolled abnormal growth of cells which may lead to impairment of

Table 8.2 Sources of metals in workplaces

Metal	Where found
Beryllium	Aircraft alloys, electrical controls, fluorescent lights or tubing for 'neon' sign manufacture (no longer generally in these last two)
Cadmium	'Silver' solder brazing, marine galvanizing, pigments, pottery glazes
Chromium	Metal plating, welding of zinc chromate painted steel
Cobalt	Animal feed preparation, abrasive wheel use
Lead	Pigments, leaded petrol (gasoline), batteries, ore-fluxing, glass manufacture, pottery glazes, paints
Manganese	Hard face welding, fertilizer manufacture
Mercury	Laboratory instrumentation, electrical switches, dentistry, formerly in some fungicides
Nickel	Mining and processing, steel manufacture, hard face welding, metal plating, coinage
Osmium	Tissue fixing in histological laboratories
Tin	Heating of tinned steel, sanding marine antifouling paint
Zinc	Welding and cutting of galvanized steel
Vanadium	Vanadium ore mining, by-product found in flues from burning oil in furnaces, acid plant catalyst

Table 8.3 Effects of some metals

Metal	Effect
Cadmium	Interference with kidney function, fume produces acute irritation of the lung.
Chromium	Can cause ulceration of skin on hands, and of the septum in the nose leading to nasal cancer. Can also cause lung cancer.
Lead	Interference with nervous system and brain function, interference in the production of red blood cells.
Manganese	Interference with nerve and muscle function, in severe cases causing *chicken walk* (sometimes misinterpreted as a problem with the semicircular canals – the organs of balance). Manganese in the right quantities and forms, however, is an essential element.
Vanadium	Tremor, eczema, chronic bronchitis.

body function and, if not corrected, death. Cancer is commonly in the form of a malignant tumour. Not all tumours are malignant; some are benign. In the workplace, cancer can result from radiation (solar UV; radon gas in uranium mining; poor practices in medical radiology); possibly from stress; and from some chemicals.

The chemical causes of cancer work either by altering the genetic material (genes, DNA) in our body (genotoxic), or in a variety of other ways (epigenetic). Alteration of DNA is described as mutation; however, not all mutations cause cancer.

DNA is composed of sugar molecules joined together with phosphate molecules and molecules of four nucleotide bases; that is, cytosine, guanine, adenine and thymine (C, G, A, T). The C, G, A, T are arranged in two lines, or strands, which spiral around each other (see Fig. 8.3).

In DNA, C in one strand is always linked to G in the other and vice versa. A always links to T and vice versa, so one strand is a *mirror image* or template for the other. When our body cells divide, the DNA strands separate and each one is the template for a new strand in a new cell. The order and pattern of C, G, A and T are what make every plant, animal and person different, however slightly. Sections of the pattern are 'instructions'

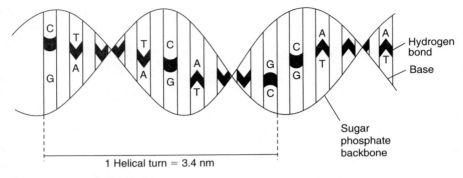

Figure 8.3 DNA double helix

for making the different proteins we need from amino acids (some we make from food, some we must get from food).

In a mutation, something goes wrong with the process of matching the template. Perhaps a C, G, A or T is simply missed out. When the abnormal strand becomes a template for the next cell, the new strand is wrong, and so on. One way in which a chemical can cause mutation is shown here. This chemical is bromouracil, and it can be mistaken by A for T, and by G for C. What happens is shown in Fig. 8.4.

This illustrates one way in which a chemical can cause mutation of DNA. Mutation has a greater effect in rapidly dividing cells. Cells divide more rapidly in children and in the developing foetus of a pregnant worker, for example. However, the body has repair mechanisms, the effectiveness of which decreases with age or impaired immune system. There is often a latent period with cancer where the disease does not appear until many years after exposure to a chemical or radiation.

The chemicals and processes which can or may cause occupational cancer are quite extensive, and are listed by the International Agency for Research on Cancer (IARC) in groups:

Group 1   The agent (mixture) is carcinogenic to humans. The exposure circumstance entails exposures that are carcinogenic to humans.

Group 2

    Group 2A   The agent (mixture) is probably carcinogenic to humans. The exposure circumstances entail exposures that are probably carcinogenic to humans.

    Group 2B   The agent (mixture) is possibly carcinogenic to humans. The exposure circumstances entail exposure that are possibly carcinogenic to humans.

Group 3   The agent (mixture) or exposure circumstance is not classifiable as to its carcinogenicity to humans.

Group 4   The agent (mixture) is probably not carcinogenic to humans.

Table 8.4 shows some workplace chemicals definitely linked to cancer.

Partially burnt carbon compounds from combustion, such as diesel exhaust, contain cancer-causing substances known as polynuclear aromatic hydrocarbons (PAHs).

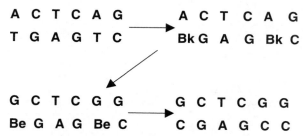

Figure 8.4 This shows three replications of part of a DNA strand containing six base pairs. The mutagen bromouracil (B) exists in two forms, k and e, and can switch from one to the other. Bk can pair with A, Be can pair with G. In the first replication of the DNA, bromouracil in the k form has replaced T in the lower part-strand of the daughter shown. It has then switched to the e form so that in the second replication the top part-strand of the daughter shown now contains two Gs. In the third replication the part-strand of the daughter shown is that which no longer contains B. GC pairs have replaced the AT pairs. The DNA has mutated.

Table 8.4 Cancer-causing workplace chemicals

Substance	Cancer
Asbestos	Lung and lung lining (pleura)
Benzene	Leukaemia (blood cancer)
Chromium compounds	Nasal cancer
Soots, tars, oils	Skin, scrotum

Industrial sources of skin disorders

Skin disorders (dermatoses) are a common form of work-caused problems. The skin can be exposed to dusts, fibres, plant materials, metal compounds, glues, solvents, oils, tars, resins, acids, alkalis e.g. soap or detergent, viruses, and UV and ionizing radiation. Irritant dermatitis is one form of dermatosis and consists of redness, swelling and cracking.

Allergic dermatitis consists of redness, itching, blistering and weeping of the skin.

Both can take a long time to heal after exposure ceases and gloves cannot usually be worn until recovery is complete.

Table 8.5 shows some sources of dermatitis.

Mechanism of allergy and workplace sources of allergy

Allergy comes from the operation of the body's immune system which is set up to fight foreign protein such as that in bacteria. An 'allergen' (an allergy-causing substance) causes the immune system to produce an 'antibody'. Some non-protein substances,

Table 8.5 Some sources of dermatitis

Irritant	Allergic
Acids	Some plants (horticulture) e.g. grevillea, primula
Alkalis, e.g. soaps, cleaning agents	Epoxy resins
Solvents which dissolve fats	Nickel
Styrene (fibreglass)	Some dyes, e.g. used in hairdressing

Table 8.6 Sources of workplace allergens

Allergen	Where found
Amoebae	Air conditioning (humidifier fever)
Animal protein	Hair cutting, veterinary work
Enzymes	Raw cotton
Isocyanates	Used in belt-joining glues, two-pack paints, manufacture of foam and polyurethane rubbers
Moulds	Hay, grain, cheese

e.g. nickel, may cause allergy. These 'haptens' link with a body protein and the immune system then develops an antibody. The allergen-antibody 'fight' produces 'histamine', which causes swelling and itchiness. It is generally treated with antihistamines. The problem then is that once the body has developed an antibody for a specific allergen, it will react to smaller and smaller quantities on each successive exposure.

Lung allergens

Certain workplace substances are potent lung allergens, which cause asthma (constriction of lung vessels).

Examples are given in Table 8.6.

Some types of wood dust are potent irritants, e.g. African iroko, Victorian blackwood.

Effects of workplace solvents

Solvents in industry are used for a wide variety of purposes.

Some tasks, or products used in tasks, and solvents used are shown in Table 8.7.

The key effect of many of these solvents is that they dissolve fat which protects the surface of the skin and so provide easier access to the skin by bacteria. This may cause dermatitis.

A second effect results from inhalation. For example, petroleum products are simple narcotics; toluene affects blood cell production (the effect can be reversed), and also lowers sperm count; chlorinated hydrocarbons can cause liver and kidney damage and disturbance of heart rhythm which is sometimes fatal.

Table 8.7 Sources of workplace solvents

Task or product used in task	Type of solvent
Cleaning electrical parts	Fluorinated hydrocarbons
Decarbonizing	Orthodichlorobenzene, cresol (cresylic acid)
Degreasing	Trichloroethylene
Glues	Toluene, methylethyl ketone
Liquid paper	1,1,1-trichloroethane
Mastics	Dichloromethane (methylene chloride)
Paints	Xylene, various petroleum fractions, e.g. mineral turps
PVC pipe cement	Tetrahydrofuran, cyclohexanone
Printing	Variety of solvents in inks and press cleaners
Spray painting	Toluene, acetone
Sterilization	Alcohol

Dichloromethane can cause the same effects as carbon monoxide (production of carboxyhaemoglobin) but more slowly.

The fluorinated hydrocarbon cleaners are relatively safe, although skin protection is required, but some have the potential to affect the ozone layer.

Some solvents are absorbed through the skin into the blood stream in significant quantities. Consult exposure standards for those with a 'skin' notation. Solvent vapours can also be eye irritants.

At all times it should be also remembered that many industrial solvents are *flammable*.

Vibration, solvents and peripheral nervous system disease

Some ketone solvents (not acetone or methylethyl ketone) can produce disease of the peripheral nerves in the body leading to shakes, twitching, or loss of nerve control or muscle action. Vibration, particularly from hand tools, can also produce this effect in the hand and lower arm.

Health hazards from welding and cutting

Welding with an electric arc, and brazing and cutting with a gas torch require appropriate controls to prevent harm to health. And with many solvents, there is a significant risk of fire from welding sparks. The high temperature vaporizes the metals involved and produces a fine fume containing metals, metal oxides, products from metal coatings and any flux in the rod or wire.

The problems are summarized in Table 8.8.

UV radiation (see radiation below) especially with stainless steel and aluminium welding causes burns of the skin and the retina of the eye.

Precautions are also needed against burns, electrocution, manual handling injuries and, with oxy-acetylene gas, leakage (e.g. in vehicles and from damage to oxygen or acetylene hoses, especially in confined spaces). Bystanders are at risk from welding operations. Arc-air gouging, a cutting operation, is particularly dirty and noisy. Laser cutting can also give rise to fumes which may be a hazard depending on what is cut.

Table 8.8 Toxins from welding

Aspect of welding or cutting	Problem
Arc or oxy-acetylene flame	Nitrogen dioxide and ozone (see toxic gases)
Hard face welding	Manganese, nickel, chromium fume (see metal health hazards above)
Low hydrogen electrodes	Metal fluorides – bone and teeth mottling
Rust-resisting coatings on steel	Zinc, chromium, lead fume; carbon monoxide
Other coatings on steel, e.g. tar-based or polyurethane paint	Variety of toxic products – cyanide if a polyurethane paint
Welding galvanized metal	Zinc fume fever

Decompression sickness

Decompression sickness can arise when workers work in 'hyperbaric' (higher than at atmospheric pressure) environments such as in tunnelling, construction work, or in diving. Nitrogen dissolves in the bloodstream to a greater degree than normal due to the pressure. If workers are not 'staged' as they exit from the pressurized area, or rise to the surface of the water, i.e. don't take time to depressurize, nitrogen bubbles form in the blood. These can cut off blood supply in the smaller vessels of the brain and cause 'the bends'. A longer-term effect is death (necrosis) of bone tissue in the ends of long bones such as the femur. Helium does not cause the same problem, and is used in deep-diving gas mixtures. Nitrogen breathed at depth can also cause narcosis including altered perceptions.

Biological hazards and zoonoses

Hazards from biological agents arise in a number of workplaces. Some were described earlier under 'Mechanisms of allergy and workplace sources of allergy'. Table 8.9 outlines some of the possibilities.

The serious and debilitating effects of the zoonoses (animal-borne diseases) are not always realized among those who have contact with animals (vets or vet nurses), or are in animal processing – for example, abattoir workers or meat inspectors.

Effects of workplace climate on health and performance

Workplace climate can be divided into two basic areas:

1. Climate-controlled environments with light or seated activity.
2. Other environments where the climate cannot be controlled and activity is light, moderate or heavy.

The 'organizational climate' is a separate issue and will not be considered here.

Table 8.9 Biological hazards

Workplace or activity	Possible problem
Air conditioned buildings	Humidifier fever (allergy), Legionnaire's disease (a pneumonia), transmission of disease, e.g. influenza
Animal or animal products handling: veterinary; meat processing; meat inspection	Erysipelas, Q-fever, brucellosis, leptospirosis, hydatid
Farming; cheese-making	Moulds – lung allergies
Ceiling spaces with bird droppings; certain eucalyptus trees	Cryptococcus neoformans – potentially fatal lung infection
Nursing; medicine	Hepatitis B, HIV-AIDS from needle-stick injuries
Medical pathology	Viral and bacterial infections
Cotton-processing	Enzyme produced inflammation of alveoli
Police; community service workers; corective services staff; ambulance staff	Hepatitis B, HIV-AIDS from skin abrasion
Prostitution; professional escort services	Hepatitis B, HIV-AIDS, gonorrhoea, syphilis, genital warts

Controlled environments

Indoor controlled climate needs to meet certain specifications if people are to work comfortably. Relative humidity should be around 55%. Lower levels can lead to dry eyes, throats and skin (this can happen in air conditioned buildings drawing in subzero air, without adequate humidification). Higher levels can produce feelings of discomfort. Temperature generally should be between 20.5°C and 23°C. Poor air distribution will lead to discomfort from lack of air movement (stuffiness) and people complaining of feeling too cold or too warm. Solar load can also produce discomfort on the sunny side of air conditioned buildings. Biological control procedures in air handling systems will minimize problems such as humidifier fever and Legionnaire's disease. Careful selection of indoor materials and instrumentation will reduce 'sick' or 'tight building' syndrome. Some plastics, carpet glues, the formaldehyde in chipboard, printing presses in areas not designed for them, and ozone from photocopiers are part of the problem. Carbon dioxide build-up from a high ratio of recycled air also contributes to the syndrome.

Natural and uncontrolled environments

Heat strain can result from the heat stress imposed on a worker. The environmental factors involved in heat stress are:

- air movement
- air temperature
- humidity
- radiant heat.

Factors within the heat balance equation

The body's heat balance involves the following factors:

M – the heat produced from metabolism by the body, for example, the oxidation of glucose to fuel the muscles

R — radiant heat energy, either received by the body from hotter surrounding surfaces or given off by the body to cooler surrounding surfaces
C — heat gained or lost by convection; that is, the flow of air over the surface of the body
K — heat lost or gained by conduction; that is, skin contact with a hot or cold surface
S — the heat stored in the body if any, due to inability to lose sufficient heat through radiation, convection or evaporation
E — heat lost from the body due to evaporation of sweat.

For core body temperature to remain constant (i.e. S = O),

E = M ± R ± C ± K

Normally, K is quite small and can be ignored.

Heat strain

Undesirable heat strain results from a combination of factors – clothing, activity levels, age, health and fitness status. The body can pick up or lose heat from conduction (very little); convection (movement of air over the body); or radiation. It can also lose heat by evaporation of sweat; however, it is not desirable for even a normal fit healthy young man to sweat more than 1 litre/hour over 8 hours (women's tolerances have not been as well studied). Also, in conditions where heat retention is at the rate of 73 W for an hour, the body core temperature will rise approximately 1.2°C. A rise above this can lead to heat syncope (fainting); heat exhaustion; anhidrosis (sweating stops); and heat stroke.

Figures 8.5 and 8.6 show the full range of effects.

Heat strain also affects the ability to think and concentrate. Short cuts can be tempting. Sweating can also affect grip. As a result safety can suffer.

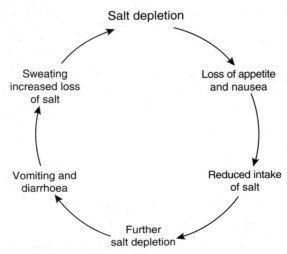

Figure 8.5 Vicious circle of salt depletion

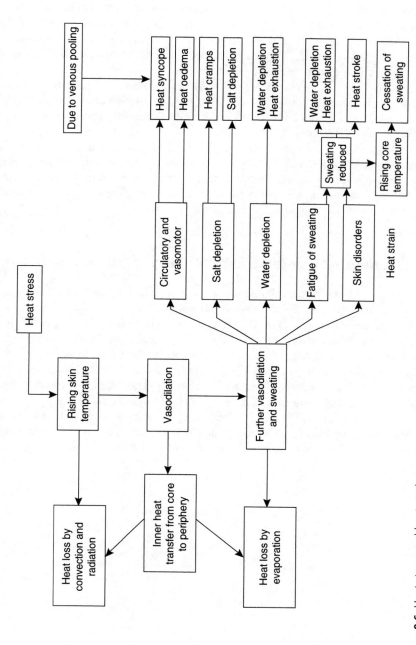

Figure 8.6 Heat stress and heat strain

Physiological and psychological strain from heat stress

Sweating is an active process which consumes energy. In addition, dilation of the peripheral blood vessels, which is required to conduct heat to the skin surface, and the work rate both increase the demand on the heart. Excessive sweating alters the electrolyte balance in the body and this must be restored through intake of adequate fluid and, if necessary, salt or electrolyte replacement drinks. One simple check on dehydration is that urine is dark, not pale yellow or colourless.

At air temperatures above 35°C, the body actually takes in heat from the air. However, providing humidity is not very high, good circulation of air over the body is an effective means of cooling the body, so spot air movement by means of, for example, a fan, can be quite effective.

Psychological strain can also be a product of heat stress, a familiar situation. The energy expended by the body to keep cool, and the greater volume of blood in the peripheral circulatory system (so reducing that available to the brain) produce feelings of sleepiness and fatigue, and cause difficulty in concentrating on things requiring a high level of precision and attention to detail.

Relative merits of clothing in relation to thermoregulation

There is a unit of insulation value for clothing known as the 'clo', which is actually $0.155°C\ m^2/W$. Some values are shown in Table 8.10.

In the sun, there is a need to balance SPF (solar protection factor) values for clothing against the ease with which it permits air circulation and hence evaporation of sweat. Heat exchange through the scalp is important and headwear for solar protection should allow for this. Moist clothing in cold conditions assists loss of heat, i.e. the insulation value is reduced. Open weave clothing is not good for retaining layers of warm air in cold conditions, but does allow air circulation in the heat, although it may not have a good SPF value. Colour of clothing affects radiant heat uptake. Clearly, for male workers a bare torso aids heat control, but poses problems when solar UV radiation is considered. Sweating will tend to remove any solar protection cream applied.

Table 8.10 Insulation values of clothing

Clothing ensemble	Clo
Naked	0
Shorts	0.1
Light summer clothing	0.5
Typical indoor clothing	1.0
Heavy suit and underclothes	1.5
Polar	3–4
Practical maximum	5

People tend to make logical adjustments to clothing as needed, but these must be balanced against site requirements on dress to meet other safety concerns, e.g. no loose clothing (shirt flaps) near moving machinery.

Metabolic rates vary from: Sitting = 95 W; Standing – Heavy Hand Work = 210–260 W; to Heavy Work With the Whole Body = 560 W.

Moderate and heavy work in uncontrolled hot environments with natural and/or process heat sources (indoor and outdoor) requires careful attention to the following:

- fluid intake
- suitable rest breaks
- provision of shade if possible
- choice of clothing – to allow sweating and airflow but not allow through too much UV
- headgear which 'breathes'
- time to acclimatize.

This is necessary for much of the year in the tropics. Also, 14 day on/14 day off 'fly-in, fly-out' operations in the tropics need particular attention if time-off is in a temperate area, because the time for acclimatization is about 10 days.

Cold stress

The effects of cold stress should also be considered where appropriate. At a critical level it can lead to increasing lethargy and apathy, and poor control of movement. There are obviously areas and jobs where cold stress is a problem. This is a product of air temperature and airflow (the wind chill factor). Jobs where this is a concern include:

- outdoor work in alpine and high southern or northern latitudes, especially in winter, such as telecommunications tower maintenance
- freezer and chiller work.

It is important to recognize that a considerable proportion of body heat is lost through the scalp, so head covering is important along with other properly designed clothing.

Causes of musculoskeletal injury

Damage to the muscles, skeleton and joints of the body makes up a very significant proportion of total workplace injuries and is responsible for large compensation costs. Although actual fractures are serious, they are considered separately from musculoskeletal injury. Musculoskeletal injuries can result from a single damaging incident, a second damaging incident which aggravates existing injury or from a series of consecutive small incidents (cumulative trauma disorder – CTD). An example of the first type is a lifting job which puts excessive pressure on a disc in the spine. Fluid leaks from the ruptured disc, there is swelling and a nerve is pinched, resulting in pain. See Fig. 8.7.

An example of the third type is lower back injury resulting from poor disc nourishment while sitting in an off-road vehicle (skid-steer loader, earthmover, scraper, etc.) for long periods, with the spine vibrated by movement over rough terrain. This may lead to permanent damage to tissues in the spine.

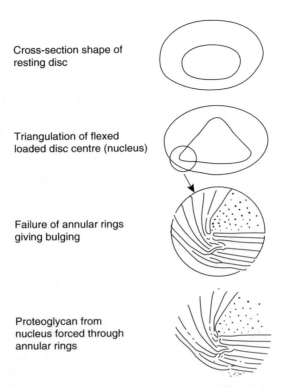

Figure 8.7 Internal change of shape and failure of loaded flexed disc (by permission of Lippincott, Williams and Wilkins, from Adams, M.A. and Hutton, W.C. (1985). *Gradual Disc Prolapse*, Spine, **10(6)**, 524–31)

Musculoskeletal damage can particularly affect the lower back, neck and shoulders. Occupational overuse syndrome, OOS (repetition strain injury), is generally a result of fine repetitive work involving the hand and wrists. It can involve inflammation of the tendons of the wrists and hands (tenosynovitis), or squeezing (due to inflammation) of the tendons and nerves entering the hand (carpal tunnel syndrome). Other types of OOS exist and can also involve the nervous system. Slips, trips and falls are other obvious sources of sprains and strains of muscles and joints.

The causes and prevention of musculoskeletal injury from handling of loads and materials by hand are dealt with in Chapter 11 and in guidance on manual handling from OHS authorities.

Effects of IR and UV radiation

Infra-red radiation

Work around furnaces such as in foundries, steel making, glass manufacture and metal casting can produce strong infra-red (heat) radiation. Infra-red radiation can lead to

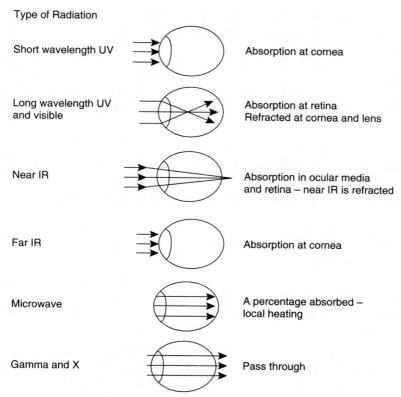

Figure 8.8 Effects on eye of radiation

cataracts (growth of white tissue) on the cornea (outer surface of the eye) and burns of the retina (the tissue at the back of the eye which detects the intensity and colour of light).

Ultraviolet radiation

Ultraviolet radiation can be split into three types: UV-A, UV-B and UV-C. UV-A is the least damaging to the body and can be found in sunlamps, etc.; however, eye protection is still required. UV-C carries the most energy and is most damaging. The eye can be exposed to such radiation from looking directly at the sun (as in early morning and late afternoon driving, although the extra atmosphere the rays pass through at those times of day reduces their intensity); from arc and oxy-acetylene welding processes; and from reflected or direct UV laser radiation.

Because the lens of the eye focuses radiation, the UV rays can be concentrated by 100 000 times on a section of the retina, therefore damaging the retina permanently. The UV rays can also produce burns of the cornea, with a feeling like sand in the eyes (arc eye), even from reflections while welding in steel sheds, tanks or railway tank-cars.

Effects of electromagnetic radiation on the eye

These are summarized in Fig. 8.8.

Linking examples of body systems, organs and tissue with the workplace stressors affecting them

What you have read above has dealt with many of the workplace sources of disease, disability and ill health. These sources as we noted earlier are called 'stressors'. Each stressor affects particular target cells, organs, tissues or systems in the body. For information on noise, vibration and ionizing radiation as stressors, see Chapter 10.

Worker health monitoring

Airborne sampling and medical monitoring of stressors

There are a number of approaches to assessing the risks presented to workers by physical and chemical stressors in the workplace.

Two key approaches are:

1. To measure by representative sampling the amount of the stressor to which the worker can be exposed.
2. To use various measures to assess how much of the stressor is actually absorbed by the body or how much effect the stressor has had on the body.

The first approach is dealt with in more detail in Chapter 10. The second approach is basically explained here, and can be described as 'medical monitoring'.

Some of the approaches to medical monitoring are set out in Table 8.11.

Table 8.11 Medical monitoring

Test	Used for
Audiometry (hearing)	Exposure to noise – done with an audiometer
Biological monitoring:	
– blood	Exposure to metals, organic chemicals including pesticides, e.g. blood cholinesterase tests
– fat biopsy	Exposure to some pesticides
– liver	Rather non-specific – exposure to certain chemicals
– urine	Exposure to metals, solvents, organic chemicals
– urinary specific gravity	Dehydration
Blood pressure	Signs of cardiovascular disorder caused, for example, by stress – done with a sphygmomanometer
Lung condition	Stressors which damage lung tissue – this uses, for example, X-rays
Lung function	Stressors which damage, irritate or cause obstruction of the lung – uses spirometer, vitalograph and other techniques
Nerve conduction	Stressors such as lead and some solvents which can affect nervous activity – uses skin electrodes and an oscilloscope
Whole body test of ionizing radiation	Absorption of radioactive substance

An example of an intermediate technique lying between medical monitoring and measurement of the amount of stressor in the workplace is the use of film badges and TLD dosimeters to measure exposure to ionizing radiation.

Various psychological tests may also be used to determine the effect of stressors on the individual, particularly effects from psychosocial stressors. However, some toxic agents also cause a reduction, or change, in intellectual capacity.

Medical monitoring usually starts at, or before, employment. It can include fitness and strength tests. Some types of medical monitoring may be required by law – for example, initial and periodic checks on the lungs of underground miners; an initial audiogram; and audiograms at specified intervals in workplaces with noise over the action level. For organophosphorus and carbamate pesticides, the normal cholinesterase level must be measured for each person likely to be exposed, so as to assess later exposure. Fitness tests play an important role where a workplace is pursuing a worksite health improvement programme.

Sources of standards for assessing screening results

Once health monitoring is carried out, it is necessary to look at standard values. With most measures of health and fitness, a normal range of values will be given; that is, there is not generally one exact value which applies to all. For lung X-rays, the examining doctor will refer to the International Labour Organisation U/C classification of pneumoconiosis (dust disease of the lungs).

For a number of toxic agents or their metabolites (i.e. what is produced from the toxin by the liver and excreted in the urine), normal values are given in the American Conference of Governmental Industrial Hygienists (ACGIH) handbook which contains Biological Exposure Indices. (See Further Reading at the end of this chapter.) The German MAK Commission also produces BAT values.

While there is general international agreement on acceptable levels for lead in blood, for example, acceptable values for some toxins may vary between countries. They can be obtained by reference to your OHS authority.

The normal levels of hearing at various ages are also available, so that noise-induced hearing loss can be calculated.

Values for toxins in urine are sometimes given as per gram of creatinine. This is a way of allowing for variations in urinary output (1–2.5 litres/day). Thus a 'spot' (one-off specimen) sample of urine can be used, rather than a whole-day sample which is unreliable because of the difficulty of collection.

Note that while trends in the records of workplace groups can be discussed in health and safety committee and management meetings, the records of an individual are confidential and cannot be disclosed without her or his written agreement. It is important to ensure that environmental and medical monitoring test results are properly recorded, collated and stored securely. In some cases, e.g. companies working with specified carcinogenic substances, such records must by law be kept for long periods of time. However, even without a legal requirement to keep records, such records are valuable to any organization in reviewing health and safety performance.

Stressors inside and outside the workplace

Introduction

In this section we take a brief look at the links between health in the workplace and health outside work. It is sometimes difficult to decide the extent to which health problems are due to the workplace or to life outside work (lifestyle) factors.

The words 'work-related' are used in two ways. Some people use them to describe conditions which are caused almost entirely by factors at work. However, others use the words 'work-related' to mean conditions such as the following:

(i) where a condition which results from causes outside the workplace is made worse by factors in the workplace – for example, noise from machinery is added to the effects of loud music heard away from work
(ii) where a factor in the workplace makes the effects of factors outside the workplace (including lifestyle factors) worse – for example, stress leading to substance abuse.

We will use 'work-caused' for those conditions which are almost completely caused by workplace factors.

Medical factors conducive to work-related ill health and disease

Some individual medical factors may lead to increased work-related ill health, disability and disease. These include:

- eye-sight problems
- diabetes
- proneness to asthma (screening tests are available)
- proneness to skin allergy (screening tests are available)
- physical and mental impairment
- certain conditions which may result from ageing, e.g. spinal degeneration, loss of skeletal and muscular strength.

Workplace standards are set to ensure the health of most individuals, but some people are *hypersensitive* to a particular workplace stressor. A developing foetus, for example, is likely to be hypersensitive to chemical intake. Certain activities are undesirable for a woman in advanced pregnancy – for example, prolonged standing. These needs must be balanced against anti-discrimination legislation which prohibits preventing women from undertaking certain work.

Under anti-discrimination laws, as far as reasonably practicable, workplaces and jobs should be designed to allow employment of physically and mentally impaired people.

Health at work

Other workplace and external factors affecting health of people at work

There are some other factors, both inside and outside the workplace, which when combined affect the health of people at work. It is important to recognize that people bring concerns and preoccupations to work. The supervisor must foster good interpersonal relations by knowing his or her people. Supervisors have been known to pressure someone over poor performance, not knowing that a member of the worker's family had a serious illness. Safety can be affected when outside concerns mean a worker's mind is not on the job.

Workplace factors affecting the worker and others outside the workplace

There are some obvious factors at work which affect the health of a person, both at work and outside. These include dust as a cause of lung disease; noise resulting in poor hearing and so social isolation; and ill health due to chemical poisoning. Note that poor personal hygiene and habits may result in other persons, such as family members, being affected by chemicals. Examples would be asbestos on clothing, and dirty fingers possibly affecting sexual partners.

A number of other factors in the workplace can affect the health outside work of the person and others.

Stress at work

Stress can be viewed from a number of angles. Figure 8.9 relates to information processing.

Excessive task demand, or task demand for which the person lacks training and so must spend extra time working on responses, reduces effective processing of information,

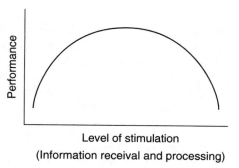

Figure 8.9 Stress vs performance

328 Enhancing occupational safety and health

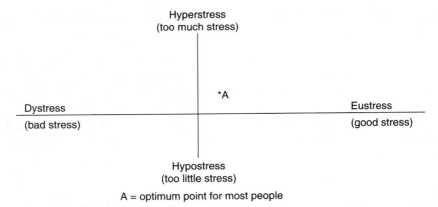

Figure 8.10 Task demand and stress

and ability to cope. It can cause accidents and ill health. Task demands which are too low (in quantity and quality) may also cause illness or accidents, because the person becomes bored or 'switches off'.

The results of quality and quantity of task demand are best shown in the following diagram, Fig. 8.10.

Note that some stress is good, some not, and that for many people occasional over-stress is challenging, although it is not healthy over a long period.

Contributing factors to stress

Five main factors contributing to undesirable stress at work have been identified by Karasek (see Further Reading at the end of this chapter).

These are:

- lack of meaning
- underwork or overwork (understimulation or overstimulation)
- lack of predictability
- lack of control
- conflict.

While some people would like to confine safety and health issues to a neat little separate compartment, isolated from industrial relations, this is not always possible. General hours and general conditions of labour have a significant effect on health and safety, an example being extended shopping hours. Fatigue can threaten safety. The common law duty of care to provide adequate staff sometimes results in 'staffing level' issues. Staffing levels can have a significant impact on workers' health and safety.

Lack of meaning
Lack of meaning in a job can lead to a poor image of the organization, a loss of self-esteem and possible social problems, although a person may also find a healthy balance

through more stimulating activities outside work. Multi-skilling; job enrichment; a 'bottom to top' approach to quality management; and a caring management (who care about the product, the company, the person, and the customer) are all approaches to overcome lack of meaning. Providing pay is reasonably adequate, surveys have shown that job satisfaction is more important in the long run than pay.

Under and overwork

Underwork can refer to quantity and quality of work – that is, not enough work and work which provides little mental or, for some people, physical challenge. Moves to greater efficiency tend to reduce underwork, but where people are underworked, sky-larking and practical jokes can lead to accidents, and there is a lot of time to find and create problems.

Overwork, or 'karoshi' as the Japanese call it, is an obvious source of stress and health breakdown, if prolonged. It can be organizationally or individually driven, or a combination of both. It is not new. It is described in a story about an overworked accountant written by Sir Arthur Conan-Doyle a century ago, *The Silver Mirror*.

Examples of the first case are not hard to find. In many performance-driven organizations, such as small businesses, the hours needed to achieve what is expected can be very long. This can have an obvious impact on personal relationships.

In some cases, the overwork primarily comes from the individual – the 'workaholic'. It can be an escape from unsatisfying personal relationships. In some cases it is a genuine love of the activities in the job, and in some cases it is because the person is not good at planning, organizing, scheduling and prioritizing work, or just cannot politely but firmly say no to people.

In the third case, in some jobs there is an immediate direct relationship between effort and earnings from piecework payment, production bonuses, or sales commissions. And yet, for some people, immediate rewards for more work suit them; for example, sales people in a survey in Western Australian were shown to be a relatively healthy group. This is not necessarily a criticism of the varying approaches to work described here. But it does emphasize the need for promotion of effective time management; a balanced lifestyle with time for personal relationships and physical activity; and careful monitoring of workers to ensure that overwork does not get to the point of reduced performance, ill health, poor interpersonal relationships, substance abuse or accidents.

Lack of predictability

Lack of predictability is a two-edged sword. Personality is important – some people prefer unpredictability which throws up new challenges. In many areas change is something which we cope with. Perceptions can make it either a threat or an opportunity. Some people in the construction industry like the challenge of a series of different contracts, even though there is not always any certainty as one winds down where the next one will be.

Political changes can affect public sector workers because of decisions on privatization. Contracting-out, downsizing and e-business affect workers in private and public sectors. Many businesses operate in an unpredictable environment – changes in the economy, in tariffs, government regulations, wage changes, new competitors, and change in customer preferences.

What is important in making unpredictability less stressful is to maximize the involvement of workers at all levels in understanding the organization's problems and opportunities, and having a hand in deciding how to deal with them – that is, to share control.

Under Robens-based health and safety acts, there is a requirement for employers to consult with employees on planned changes which may affect their health, safety and welfare. If the necessary changes cannot be made without reducing staff, then every effort should be made to redeploy people or provide equitable retrenchment packages.

What is unfortunate is to see poor performance in a company through lack of consultative management, and then to see shopfloor workers laid off rather than poor managers.

Lack of control

Lack of control on the day-to-day, or week-to-week level is also a problem. Part of job enrichment to lift the quality of the job for the employee is to give her or him greater say in how it is done and, within a work area, who should do what, and who should share duties. Both the quality improvement approach to product or service quality and the consultative approach to health and safety emphasize the need for workers to contribute ideas towards working safely and productively. The participative style also works because it does not require someone to be there cracking the whip. Such joint decision making does not prevent work teams agreeing on a command structure to deal with critical incidents or emergencies. One experiment in a participative approach was to use production nodes rather than a production line at Volvo's Kalmar plant, with each node's work team jointly deciding how to deal with assembly of a section of the car. The Norwegian telecom company has designed a building and building services to suit a management style based on a high degree of trust of employees.

Conflict

Conflict can be a serious cause of stress. At the group level it is often seen in poor labour relations, with unions and management constantly at loggerheads. However, it can also occur between supervisor and employee, between departments, and between individuals. Solutions vary and may include:

- steps to improve safety culture and job pride
- redeploying a person into another position where they are more suited to the type of work
- counselling
- training in conflict resolution, negotiation and problem-solving skills
- changing a supervisor or moving an individual to another supervisor
- clear definition of roles and expectations (that is, where possible, not answering to two or more bosses) and a job description form.

Care must be taken that the steps do not infringe the requirements of anti-discrimination legislation.

Management styles, shiftwork, piecework, production methods, ages of man (or woman) and worker health

Management styles

Management styles, as they might affect the health of people at work, have already been dealt with to some degree in the section above. They are generally broken down into McGregor's Theory X, autocratic, and Theory Y, participative/democratic.

There is a third style – 'laissez faire' (let go, or free-for-all). The organizational structures are also important. An organization which requires rapid response to opportunities cannot work effectively in a rigid hierarchical mode. Many organizations only work well in a matrix style with many direct horizontal links between functional areas. Frustration at poor organizational structure can lead to poor health.

Shiftwork

Shiftwork, particularly that involving night shifts, affects the human body. The body is designed to be active during the day and asleep at night. Night shift work does interfere with the body's circadian rhythm (response to light), and with digestive and brain activity patterns. No shift system can totally overcome this, but some are much better than others. Shifts should be forward-rotating.

Twelve hour shifts in particular need careful fatigue management.

Piecework

Piecework, or pay by production, can be a positive stimulus and incentive, up to a point. But unless production quality (including lack of accidents) is an integral part of the production process, it can lead to short cuts, poor quality and accidents. The aim should be to give a reasonable overall rate of pay where reasonable hours are put in. The piecework 'rag trade' has often seen fierce competition for work and hence a job with excessive hours, low pay per piece, and a work environment outside official control (often the home, so that home-life and work-life merge). In Melbourne one fashion house stocks traceable 'no sweat' clothing produced by outworkers with approved conditions of work.

A second example is the cottage-industry side of automobile parts production in Japan; the positive side being that such people can maintain a partly rural lifestyle. A third example of piecework is found in parts of the owner/operator trucking industry.

As fair wages for piecework are based on the norm, there will always be people below the norm, for various reasons. They are not all lazy – some are just more meticulous; more careful. Therefore, group production bonuses consistent with quality, safety and health make more sense. Individual excellence can be recognized in other ways, such as promotion.

Taylorism

Taylor production methods are named after Frederick W. Taylor, an early methods study expert. Gilbreth held similar views. These methods consist of breaking a job down into simple steps requiring relatively little training, and having each worker on a production line carry out one step. It was used extensively, for example, in the Australian vehicle industry with migrant labour, and it was an immediate solution to poor education and poor English skills. However, each worker must work to the 'beat' of the production line – no faster, no slower – and there is no real call on intelligence. For the reasons described in the sections above, attempts are being made to design the manufacture of goods to provide a reasonably stimulating, varied and challenging work environment for people. The more repetitive, mundane tasks are being performed by industrial robots or mechanized processes.

The ages of man (and now woman)

We have described the interaction of work and organization, management, outside lifestyle and personal relationships on the health of women and men at work. However, there is another factor to consider and that is what used to be described as 'the ages of man' but now needs to be called 'the ages of man and woman'. Because men have been in the workforce from teenager to their sixties for a long time, we know more about 'the ages of man'. As more and more women spend an entire lifetime at work (broken only by maternity leave) 'the ages of women' in relation to work will be better documented. One additional factor for example is menopause, but others which potentially affect women at work are calcium depletion; pre-menstrual tension; iron deficiency from menstruation; child-care; and running a home.

The statistics show that young workers, lacking experience and training, have more accidents. But age also affects the perception men have of work, and this affects the way they interact with the organization, and their mental health. Until now, a man's whole life has been based upon realizing his ambitions and largely realizing his success by work. Retirement as such, not just age, is a major cause of premature death in men. Women have in the past realized their life's ambitions and found their satisfaction in the bringing up of a family. For women, return to work after a break for child bearing or to bring up a family, generally has meant that fewer women have been promoted to top positions.

This, in turn, has resulted in less women in the power structure and the so-called 'glass ceiling' – hidden barriers to promotion. This can affect women's mental health. The same power structure has in the past played down sexual harassment; again a source of undesirable mental stress for some people, mainly women, at work.

Emerging work-related diseases

In the past there have been some major types of work-caused disease. These include skin disease; immediate traumatic musculoskeletal injury; silicosis; and lead poisoning. Heart disease was probably also a problem, even though we see it now as an emerging work-related disease. In times past it may have been regarded as more just the norm. There was more heavy lifting in some jobs but also many jobs involved more good physical exercise. The control of these diseases has improved, sometimes due to a reduction in exposure to the agent causing the problem as demand by technology for it has lessened. An example would be lead in paint and petrol. However, it is worth noting that in less developed countries the problem may still exist.

As statistics for the 'classical' work-caused diseases have reduced, other problems have come to the fore or emerged. In many cases they can be described as 'work-related' because only part of their cause lies inside the workplace. For example, musculo-skeletal injuries at work, or at home, are made more likely by a sedentary job and sedentary lifestyle, resulting in less fit spinal discs, tendons, ligaments and muscles. Passive smoking (or sidestream smoking) – that is, inhaling smoke from someone else's cigarette – can be a cause of lung conditions and may occur at home or work.

Community, environmental and workplace links to cancer

Issues
There are a great many causes of the various types of cancer. The causes of some types are well understood, others not so well understood. It is possible that stress may cause or contribute to cancer because the body repair and disease fighting mechanisms in the immune system become less effective as a result of long-term, undesirable stress. Some of the specific workplace toxins which can cause cancer were described or referred to in a previous section.

Just as *offsite* safety programmes are now run by some companies (a worker injured painting the gutter on the weekend is still a worker off work on Monday), many work organizations now encourage a healthy lifestyle programme which includes nutrition, exercise, controlled alcohol intake and quitting or reducing smoking.

Nutrition
In workplaces such as offshore gas and oil rigs, and at remote minesites, or in fact any worksite where there is an on-site food and drink facility or canteen, nutrition becomes an aspect of concern, because it has a bearing on cancer and on other diseases, including those emerging above, such as diseases of the heart and arteries (cardiovascular disease – CVD). Undesirable stress and smoking also contribute to CVD.

Why is nutrition important in relation to cancer? Doll and Peto in the UK produced some data showing striking correlations between colon cancer and meat consumption and between breast cancer mortality and fat consumption. The first graph compared daily consumption of meat per head against colon cancer incidence in women in twenty-three countries ranging from Nigeria at the bottom to New Zealand at the top. The second graph compared total daily dietary fat intake with the age-adjusted death rate from breast cancer in thirty-eight countries ranging from Thailand at the bottom to Netherlands at the top. Fat intake until the relatively recent introduction of polyunsaturated margarine has been largely animal fat or saturated vegetable fat, although olive oil was common in the Mediterranean. The meat consumption in the first graph was a reflection of saturated fat intake.

While all we can say is that as the fat or meat consumption increases, so does the occurrence of disease, with no established link between one as the cause and the other as the effect, the correlations are striking. The exact way in which fat might cause cancer in general is not known, although there are ideas about how it causes colon cancer.

Smoking

Smoking is a known cause of cancer, while the intake of certain spirits, which may be a part of certain jobs which involve a lot of socializing, can lead to cancer of the stomach and oesophagus, and cirrhosis of the liver.

General environment

The general environment has an important effect on health both in and outside work. Air and water quality generally are well controlled in some countries; in rapidly industrializing or heavily industrialized cities, however, certain industrial effluents finding their way into air or drinking water supplies may be a cause of cancer or other types of ill health. The combination of chlorination of water and certain substances arising from soil and vegetation in water can result in substances which are suspected carcinogens (cancer-causing agents).

An unexpected event such as an explosion in a chemical plant can sprinkle a chemical over the countryside including fruit and vegetable growing, sheep grazing and dairy cattle grazing areas. An example was Seveso in Italy, where dioxin (TCDD), which can cause malformed babies in pregnant women, spread over a radius of a number of kilometres. The same chemical was in Agent Orange, the defoliant (plant and tree killer) used in Vietnam. A similar defoliant was used by Australian farmers. Similarly, the nuclear plant explosion at Chernobyl, and the nuclear testing at Semipalatinsk in Kazakhstan, Los Alamos in the USA, and Woomera and Monte Bello in Australia, sprinkled the countryside with radioactive iodine 131 and strontium 90. The nuclear plant at Mayak in the former USSR made the Techa River and Lake Karachay highly radioactive. Radioactive iodine 131 and strontium 90 are taken up by cattle. Where dairy cattle are present, the milk picks up the iodine and strontium. The thyroid gland, the body's main activity gland, uses iodine, radioactive or not. The bones take in strontium like calcium, and the radiation from the strontium can affect red cell production in bone marrow.

Another type of unexpected event is the overflow of a tailings dam due to heavy rain. If it is uranium mine tailings, radioactivity can affect fish and wildlife, and people who eat these. If it is gold tailings, the cyanide can pollute local water supplies, though if the fish survive they are not a hazard to man. An Australian company's tailings overflow at the Esmeralda mine in Hungary in 2000 is a good example. The cyanide killed large numbers of fish in a major river.

Environmental and community factors can also affect the health of workers and hence performance and absenteeism in other ways. These include food-borne infections (good food hygiene is needed in workplace food facilities, and places such as cafes, mobile food vans, lunch bars, restaurants and hotels where workers buy food); infections by parasites; and mosquito-borne infections such as malaria, yellow fever, dengue and Ross River virus.

Physical, mental and spiritual fitness and worker health

Physical fitness
In addition, it is clear that controlling the factors which lead to reduced health status is important. However, being physically fit as distinct from merely being free from obvious disease, injury or disability is important. Physical fitness increases ability to cope with heat and cold. Exercise removes stress and fatigue metabolites from the body; improves ability to deal with undesirable stress; improves physical and mental endurance; improves alertness; reduces the likelihood of musculoskeletal injury; and improves one's outlook on life. The type of exercise can vary to suit the individual, and their age. The important thing is to select an exercise programme which you enjoy (because it is part of social or family activities) and stick to it. Physical fitness is linked to mental health because it reduces the effects of stress and tends to improve people's ability to perceive problems and challenges positively. Through a nutrition and exercise programme, an Australian colliery has achieved a huge improvement in the percentage of workers with a blood cholesterol level of less than 5.5 mmol/litre. High blood cholesterol levels increase the risk of heart disease.

Mental fitness
The effects of mental fitness on the health of people at work have been partly dealt with in a previous section. Mental fitness can be affected adversely by the use of some mood-altering drugs, including overuse or prolonged use of some prescription items. Equally, prescribed drugs can help people to deal effectively with some mental problems such as anxiety and depression.

In some work settings, particularly isolated work environments where social life is lacking, illicit drug taking may present a problem, both in relation to safe behaviour at work and in relation to general health. In less physically isolated situations, lifestyle may lead to drug taking. It is a concern for the health and safety practitioner because of its effects on the behaviour of the person at work, including issues such as theft from other workers and embezzlement. Behaviour outside work may include violence.

Spiritual fitness
Spiritual fitness is also important. It does not necessarily involve the practice of religion, although many people seek spiritual fitness through religious practices. What is

important is to recognize that there is within every one of us something which goes beyond body and mind, and is the basis for our desire to relate to other people, to seek companionship, to love and be loved, to respect and be respected. Spiritual fitness involves developing an attitude to life or a philosophy of life and working on the development of satisfying personal relationships.

Where people do develop unsatisfactory psychological responses to problems at work or at home, which result in substance abuse or deterioration in performance, they should be counselled by a supervisor who should try to address unsatisfactory organizational factors which may have led to the responses. However, it may be necessary to refer the person to an 'employee assistance programme' (EAP), if counselling by a supervisor is unsuccessful.

Alcohol and drugs in the workplace

Introduction

Alcohol, illicit and medicinal drugs, and other substances which affect the brain and central nervous system, such as solvents, can all affect a person's ability to work safely. In some cases mining safety legislation specifically prohibits being at a mine if adversely affected by alcohol or drugs. It also prohibits a person having, without the mine manager's permission, any liquor or drug in their possession at the mine, or consuming it at the mine without that permission. So it does not preclude a social function at the mine if the manager has agreed.

Stimulants are often used to combat the effects of fatigue from shiftwork or from driving long distance haulage vehicles. These can range from caffeine-containing drinks to wake-up pills containing other drugs. A more serious problem can arise if the effect of the stimulant carries over into a time set aside for sleeping, and alcohol or other drugs are then used in an attempt to offset this. Combining different drugs (including alcohol) can raise serious further problems.

Fatigue from upwards of fifteen hours non-stop work can produce the inattention, altered perception, inappropriate reaction to events, and delayed reaction times which alcohol does. The effects of so-called recreational drug use, and this includes alcohol, can carry over into the workplace. In the case of alcohol, there may be a zero blood alcohol level by the time the employee starts work, but an increased level of irritability. This can lead to industrial relations disputes, or accusations of workplace bullying.

On the other hand, tetrahydrocannabinol (THC) can be detected in the bloodstream for weeks after using cannabis, and yet there are no obvious remaining effects on behaviour or perception, although prolonged cannabis use is said to lead to a progressive slowing of thought processes. This long detection period has led to misunderstandings where staff are stood down after traces of THC have been found in a routine test.

Impaired judgement can affect high-level business decisions if these are taken at a business meal which involves substance abuse.

Policy

It is important for an employer, in conjunction with workplace safety representatives, to develop a policy on substance usage. In a society where legal drug use is permitted (alcohol, tobacco, medicinal), and illicit drug use is widespread (cannabis, heroin, amphetamines, ecstasy), employees need to have a clear idea of what is permitted and what is not. The return to work by management employees after a 'long lunch' must be included just as much as wages employees having too much alcohol at a bar at lunchtime, if the policy is to have genuine credibility.

Issues like employees driving home after an on-the-premises 'sundowner' or 'happy hour' need to be considered. The policy must not be seen to be an undue interference in an employee's private life, but that is not to say that workplace wellness campaigns shouldn't target alcohol, drugs, smoking, and fitness and eating habits. The problem arising from workplace social functions can be reduced if low or non-alcoholic drinks are available and adequate food is provided.

Special provisions may be needed in relation to alcohol usage in offwork time at the camp or mine village on a remote minesite.

Procedures

The procedures developed to implement the alcohol and drug policy need to be clear so as to avoid any misunderstandings. If someone is stood down, it needs to be done in a way which is clearly outlined in advance and fair to the individual. The procedures should deal with what is to occur if the policy is breached by an employee, and describe the sanctions which apply, which should increase in severity for repeated breaches.

The procedures need to make clear what criteria are to be used to decide if a person is affected by alcohol or other drugs. It is important not to mistake the effects of medicinal drug use or fatigue for substance abuse. Employees using medicinal drugs which may affect motor coordination, alertness or judgement should advise their employer if they are involved in driving or such tasks as using machinery.

In the interests of the safety of fellow workers, those fellow workers should be able to find, in the workplace procedures, the way in which affected persons are to be identified and managed. Where employees have a concern about a fellow worker, they should know who to go to, to deal with the matter. Procedures must include all who enter the workplace – employees, regulars or not.

Searches for alcohol and drugs being brought into a site also need to be dealt with in the procedures.

Addiction

Where employees know they have developed a substance use problem, it is important that they are able to access an employee assistance programme. They should be able to access this voluntarily without the knowledge of the employer if they wish. Or a supervisor may offer this as an alternative to disciplinary action or dismissal. Confidentiality and privacy must be respected. Employee assistance programmes offer counselling and rehabilitation.

Induction and training

Induction and training are important facets of substance use management. They provide opportunities to explain the policy, and let people know what is expected of them. They also give people the skills to implement the procedures appropriate to their level of employment. Periodic follow-up is important.

Drug testing

Written procedures on drug testing are important if random compulsory testing is to be used. What constitutes a suitable valid test method, and a positive test, needs to be clear. The actions flowing from a positive test must also be clearly spelt out, and included in the workplace training.

Workplace wellness

Personal and corporate health status

The nature of a corporate health culture

The first substantial evidence of workplace health promotion, seen in the United States in the 1970s and 1980s, resulted from a combination of cultural, corporate and health factors. The absence in the USA of a national health insurance scheme meant that corporations paid for a large portion of the national health bill. Corporate health costs in the late 1970s were rising as much as 20–30% a year, and thus became an issue of immense concern to corporate executives. In the 1980s, corporate health policies emphasized the need to reduce medical care costs by improving the health of workers. During the last three decades, the idea also of 'cultural wellness' boomed in the USA (and then other countries), as seen, for example, in the health food industry, health clubs and gyms, and escalating interest in jogging, exercise and aerobics, particularly with the middle class. These interests appeared to result from changes in cultural and social values rather than any actual scientific evidence which may have shown that this type of activity was in fact beneficial to one's health. This development of a health culture proved a timely opportunity for those corporations seeking solutions to their escalating health costs. Although the majority of organizations aim to maximize profit and success, a fundamental concern is the welfare of their employees. In recent times it has become more evident that many companies are increasingly recognizing the influence of employee health on worker satisfaction and morale, and the subsequent benefits on the entire corporate culture measured by various factors. These include company image; decreased turnover; increased performance and productivity; lower health insurance premiums; reduced workers' compensation costs; and less employee absenteeism.

The most obvious benefits that result from health programmes at work are improvements in health for programme participants. Most of the potential benefits of a workplace health programme can be categorized as:

- physiological, i.e. improving physical health
- psychological, e.g. improving social opportunities, communication, morale
- organizational, e.g. improving worker relations
- economic – reducing costs as a result of less sickness and absenteeism and a corresponding increase in productivity.

These categories represent both personal and corporate health status (i.e. condition). Obviously, regardless of the category, there is going to be a desirable impact on the overall corporate structure due to the associated changes in employee attitude, morale, well-being and productivity. Thus there is a definite relationship between personal health status and corporate health status. The health goals that we set as individuals will have an impact on the organization as a whole and, conversely, the goals set by the organization will be dependent on the status and goals of each individual within the organization.

Health and lifestyle issues

By far the largest proportion of the resources of health care systems is used in the treatment and cure of illness. The system is designed to treat the consequences of disease rather than to prevent ill health from occurring. Prior to the 1960s most people took little responsibility for their own health, and relied on the health system for treating illness. Acute infectious (communicable) diseases were largely under control at this time; however, chronic disease (particularly heart disease, cancer, respiratory disease and stroke) had begun to be a major threat. These chronic diseases have become known as 'lifestyle' diseases, due to the fact that they are related to lifestyle and as such are mostly preventable. Too much alcohol, smoking, poor nutrition, sun exposure and inactivity are lifestyle behaviours which are responsible for the majority of morbidity (sickness) and mortality (death), particularly in developed countries today; yet preventative health funding still remains low on the health budget in comparison to the costs for health treatment. The implications that these behaviours have on the health status of individuals are extremely diverse, affecting their entire lifestyle, whether it be in the capacity of occupation, education, family or leisure. In addition to these lifestyle behaviours, injury is also of immense concern. As with the lifestyle diseases, the majority of the injury statistics are acknowledged as preventable.

Health status in the workplace

In order to be able to better influence the health of individuals and the overall organization, it is important to understand the type of factors which combine to determine health status of both individuals, the organizations to which they belong and the countries in which they reside. In addition to the types of factors outlined above, it is also important to consider the multi-faceted concept of health and realize that we can also examine health from the point of view of:

- social health (including the ability to make and maintain relationships with other people)
- emotional health (including the ability to recognize and express emotions)
- spiritual health (including religious beliefs or an appreciation of nature)
- societal health (including understanding the individual in the context of the environment).

Relationship between work and health

Work occupies approximately half an employed adult's *waking* life and must be expected to influence his or her health, primarily through industrial hazards, injury and stress-related illnesses. These direct risks encountered in the work environment are the traditional province of the occupational health and safety services. However, it is also clear that the health people bring to work influences their ability to function. The relationship between work and health indicates that many factors contribute to the health of an individual.

Health issues in the workplace do not just involve lifestyle. It is essential to consider the workplace issues of hours of duties; work practices; job design; ergonomic factors, such as lighting; and disability and ill health issues such as noise-induced hearing loss, respiratory sensitization and musculoskeletal disorders in the workplace. Every time someone takes action to prevent an employee becoming injured, or work-related ill health occurring, they are involved in workplace health promotion.

Rationale for health promotion in the workplace

A major problem for health promotion is how to reach the population which is targeted for intervention, with minimum cost and maximum benefit. Using specific settings which enable greater access to *captive* groups is one way of being able to do this. Work settings provide greater access to adults than any other community programme.

There are three main reasons which justify health promotion in the workplace:

- the workplace provides a stable and favourable framework for implementing health promotion action, including organizational structures and management to support programmes
- within individual worksites, specific needs of particular social classes or cultural backgrounds can be addressed

- the mutual benefits which can result (workers and families, employers and governments all stand to benefit from the resulting improvements in health) offer tangible evidence of the multi-dimensional nature of health promotion at work.

(*With acknowledgement for these reasons to McMichael, A. in Egger, G. et al. – see Further Reading at the end of this chapter, reproduced by permission of the McGraw-Hill companies.*)

Within its broad organizational framework, business and industry can provide a full spectrum of services, from prevention and early detection to referral and treatment.

For example, in the case of alcohol and other substance abuse problems, employers can use deterioration of job performance to detect problems and to encourage the employee to seek early intervention. It is difficult for traditional medical care and public health institutions to provide such comprehensive preventive services. Thus, the work setting represents the single most important channel through which a large proportion of the adult population can be reached. (*With acknowledgement for this para to Lovato and colleagues – see Opatz (ed.) in Further Reading at the end of this chapter, reproduced by permission of Human Kinetics Publishers.*)

Much of the workplace health promotion literature offers a range of reasons justifying why worksites should offer health promotion programmes. Unfortunately, up until a few years ago there was little empirical evidence to support many of the claimed benefits, thus resulting in the failure of some programmes to achieve the high success rates that were unrealistically set and consequently expected by programme coordinators and management. Much has been learned from the mistakes of the past, however, and there have been many recent programmes which have demonstrated impressive results with a variety of health issues.

Programme planning must emphasize the importance of precise goal and objective setting when considering desired or expected programme outcomes. It is essential that sufficient consultation (particularly between management and employees) occurs during this planning process, so as to clarify various aspects of the programme. It is critical to set goals and objectives that are not only desired by both employer and employees, but also realistic and achievable in terms of targets set by local, state or provincial and national bodies, and in terms of available resources (e.g. time, money and personnel).

Rather than present here a rationale which justifies the inclusion of workplace health promotion in all worksites, it is more appropriate and useful to develop your own personal rationale and philosophy based on your own reading, experiences and associated attitudes and beliefs. This statement is also more relevant if you can relate it to a specific worksite in which you are working (either past, present or future).

Benefits of health promotion improvements in the workplace

Health care cost reduction
Reports in the literature indicate that health promotion can be cost-effective and provide a positive cost–benefit ratio.

The benefits are not always evident in the short term, so it is important to set realistic time frames for the assessment of any potential programme effects. Some of the intangible benefits described in the workplace health promotion literature have been summarized as:

- improvement in the individual's self-esteem and self-image
- improvement of employee health with a relatively captive audience
- the ability of programmes to address deficits in knowledge
- improved relationships between individuals as there are open channels of communication
- employee loyalty and morality as people are working together for the benefit of all
- lifting of team spirit as people are able to express their needs and wants
- the reduction in health problems experienced by the wider community
- the ability of the work environment to provide social support in a range of healthy behaviours
- people having the opportunity to utilize professional assistance they would not necessarily have access to otherwise
- such programmes help to ensure that legislative requirements are met, as health and safety issues are being addressed in the workplace.

(With acknowledgement for this summary to Lukowski – see Further Reading at the end of this chapter.)

There are case studies which highlight a number of tangible benefits. These benefits are obviously extremely attractive to employers, and include increased productivity, increased profitability through an overall cut in expenses, and lowering of absenteeism rates. Further benefits are a reduction in staff turnover, and a reduction in workers compensation claims.

These tangible benefits are also described as corporate health indicators. Similarly, there are a number of tangible benefits which are particularly important for employees, and these will produce associated benefits throughout an organization. It is useful to categorize these employee benefits as either physical, emotional or social, although it is the physical benefits which are more easily recognizable. Benefits can be grouped according to the type of personal health behaviour improvement – in nutrition, exercise, and in the reduction of smoking, alcohol intake and stress.

Further reading

Amdur, M.O. and Doull, J. (eds). (1991). *Casarett and Doull's Toxicology*. 4th edn. New York: Pergamon.

American Conference of Governmental Industrial Hygienists ACGIH (latest year) *Threshold Limit Values for Chemical Substances and Physical Agents and Biological Exposure Indices*. Cincinnati: ACGIH.

Benison, S. (1985). *History of Occupational Disease*. Cincinnati: University of Cincinnati Press.

Blaze-Temple, D. (1992). The Evaluation of Alcohol and other Drug Interventions in the Workplace. *Health Promotion Journal of Australia*, **2(2)**: 22–7.

Brookbanks, K. (2002). A Sense Of Balance (work/life), *Safeguard (New Zealand)*, September/October, pp. 44–9.

Brown-Haysom, J. (2000). Salon Safety Only Skin Deep (hairdressing), *Safeguard (New Zealand)*, November/December, pp. 44–6.

Charlton, R. (1993). *Should Lifestyle and Health Promotion of the Workforce be Employer Responsibilities?* Proceedings of the Minesafe International Conference, Perth, March. Perth: Chamber of Mines and Energy of Western Australia (Inc.).

Cumpston, A. (1979). Health and the Worker. *Journal of Occupational Health – ANZ*, Feb., 32–4 and Aug., 18–25.

DHHS–US. (2001). *Updated US Public Health Service Guidelines for the Management of Occupational Exposures to HBV, HCV and HIV and Recommendations for Postexposure Prophylaxis.* Atlanta, US DHHS–CDC.

Doll, R. and Peto, R. (1981). *The Causes of Cancer*. Oxford: OUP.

Editorial. (2002). Canada Human Rights Commission Workplace Testing Policy (drugs). *Accident Prevention (Canada)*, September/October, p. 5.

Egger, G., Spark, R., Lawson, J. and Donovan, R. (1999). *Health Promotion Strategies and Methods*. Sydney: McGraw Hill.

Everley, M. (1995). Safety Screening and Ethics. *Health and Safety at Work*, **17(5)**, 10–12.

Glass, W. (2003). Absent Friends (absenteeism and culture), *Safeguard (New Zealand)*, January/February, pp. 28–31.

Griffin, M. (1997). What Shall We Do With the Drunken Sailor, and Miner, and Bank Worker? *Australian Safety News*, **67(11)**, 24–7.

Ide, C.W. (1995). Time Gentlemen Please. *The Safety and Health Practitioner.* **13(5)**, 23–6.

International Labour Organisation. (1985). *Convention No. 161 and Recommendation No. 171. Occupational Health Services.* Geneva: ILO.

Karasek, R.A. (1979). Job Demands, Job Decision Latitude and Mental Strain. Implications for Redesign. *Admin. Sci. Q.* **(24)** 285–308.

Koh, D., Chia, K.S. and Jeyaratnam, J. (eds). (2001). *Textbook of Occupational Medicine Practice*. 2nd edn. Singapore: World Scientific.

Lloyd, P. Health Concepts, in Pantry, S. (ed.) (1995). *Occupational Health*. New York: Chapman and Hall, pp. 22–35.

Lukowski, S. (1997). Safety and Health Promotion at Work: How do we Create a Health Promotion Culture at Work? *Safety Institute of (Western) Australia Journal*; Autumn, 16–20.

Macfie, R. (1996). The Juggling Act: Balancing Family And Work. *Safeguard (New Zealand)*, September/October, pp. 20–2.

MARCSTA and Circadian Technologies. (2001). *Managing a Shiftwork Lifestyle – A Personal Approach*. Perth, MARCSTA.

Midford, R. et al. (1997). Workforce Alcohol Consumption Patterns at Two Pilbara Mining-Related Worksites. *Journal of Occupational Health and Safety – Aust. NZ*, **13(3)**, 267–74.

Murphy, A. and Smith, D. (2001). Being Cool (And Coping) (extremes of temperature), *Safeguard (New Zealand)*, July/August, pp. 44–7.

NIOSH. (2002). *Health Effects of Occupational Exposure to Respirable Crystalline Silica*. Cincinnati: US DHHS-NIOSH.

Opatz, J.P. (ed.) (1994). *Economic Impact of Worksite Health Promotion*. Illinois: Human Kinetics Publishers, Illinois.

Pantry, S. (ed.) (1995). *Occupational Health*. New York: Chapman and Hall.

Perry, G.F. (Jr.) (1994). Occupational Medicine Forum. *Journal of Occupational Medicine*, **36(10)**, 1061–63.

Pinto, M.A and Janke, D. (2002). Mold 101. An Overview for SH&E Professionals (indoor air). *Professional Safety*, **47(8)**, 34–8.

Raffle, P.B. (1987). *Hunter's Diseases of Occupations*. London: Hodder and Stoughton.

Sherman, M.T. (1990). *Wellness in the Workplace: How To Plan, Implement and Evaluate a Wellness Program*. California: Crisp Publications Inc., pp. 32–3; 56–61.

Stellman, J. (ed.) (1998). *Encyclopedia of Occupational Health and Safety* (5th edn.) ILO: Geneva. (4 vols.) Also on CD-ROM.

Suruda, A. and Agnew, J. (1989). Deaths from Asphyxiation and Poisoning at Work in the United States 1984–86. *British Journal of Industrial Medicine*, **46**, 541–6.

Tripartite Working Party on Labour Standards. (1996). *Report on Labour Standards in the Asia-Pacific Region*. Canberra, AGPS.

Waldron, H.A. (ed.) (1997). *Occupational Health Practice*. 4th edn. London: Butterworth-Heinemann.

Wilton, A. (2003). Try Before You Buy (pre-employment health checks), *Safeguard (New Zealand)*, May/June, pp. 32–7.

World Health Organisation (WHO), (1997). *The Jakarta Declaration on Health Promotion into the 21st Century*. Fourth International Conference on Health Promotion, Jakarta. Jakarta: WHO.

Zenz, C., Dickerson, O.B., Horvath, E.P and Horvath, E.P. (Jr.) (1994). *Occupational Medicine. Principles and Practical Applications*. 2nd edn. St Louis, Mosby – Yearbook.

Activities

1. Describe the historic development of OHS in the jurisdiction you live in.
2. Identify two OHS problems which were identified in the nineteenth century, and are still problems in the twenty-first century. Explain the reasons for this.
3. Explain the role and key activities of your national OHS authority.
4. In a suitable selected workplace, identify toxins present and explain the target organs in the body which they can affect.
5. Write a short presentation to give to a workplace group on a particular chemical, physical or biological stressor. Include its effects, how bodily exposure can be assessed, and the control measures.

6. Arrange to visit a workplace occupational health centre, and describe its activities.
7. Draw up the key points in a stress reduction programme to be included in a workplace health and safety plan.
8. Obtain a suitable health risk appraisal form or a sample health checklist. Select an appraisal which you feel would best suit individuals who work in a workplace with which you are, or have been, associated. Complete the appraisal in order to establish an overview of your own personal health. From this appraisal, identify two specific aspects of your own health which you feel you either need to, or would like to, modify.

9

Hazardous substance management

> ### WORKPLACE EXAMPLE
>
> A man was working alone in an Australian geological fossil laboratory at the back of a former residential premises which had a swimming pool. While seated, he spilt approximately 100 ml of hydrofluoric acid on his trousers. Lacking an emergency shower or calcium gluconate gel, he jumped into the pool. However his legs became extensively necrosed and he died several days later in hospital. The premises were not approved for such activities by the local government authority. The manager of the laboratory was prosecuted under OHS law.

Chemical elements, compounds, classes and physical state

A basic knowledge of hazardous substances will help you to understand the way they look and behave, making the names more meaningful.

Nature of atoms and molecules, elements and compounds

Atoms

Atoms have a nucleus (centre) surrounded by a cloud of electrons (negatively charged particles) in orbit around the nucleus. The word electron comes from the Greek word for amber, because electricity was first discovered by rubbing amber (fossilized tree gum).

Chemicals consist of atoms. The word atom means something that cannot be cut. An atom is the smallest particle of a chemical element which still has the chemical properties or chemical behaviour of that element. Cut the atom further and it is no longer the same chemical element.

Elements

The nucleus contains protons (positively charged particles) and neutrons (neutral particles). The number of protons in an element equals the number of electrons. For a size comparison, if the nucleus is a football, the outer electrons would be the outer rim of a sports stadium. Elements are listed in order of increasing size in the 'periodic table'. See Fig. 9.1. Each element in the periodic table has one more proton and one more electron than the one before. This is seen in the atomic number, which is the total number of protons in an element. The number of neutrons in an element also increases the further it is down the table and this together with the number of electrons is reflected in the atomic weight.

Everything in the natural world is made up of some of the 91 naturally occurring elements. Symbols are used to name the elements. Some of the symbols are easy to follow, e.g. Ar for argon. Some come from Latin words, e.g. Fe for iron (*ferrum*), Na for sodium (*natrium*), K for kalium, from Arabic *kali*, ash, hence the word *alkali*.

Use of the periodic table to identify classes of elements

The periodic table of elements is divided into horizontal rows and vertical columns. Elements in a particular column have similar chemical behaviour. Looking at the periodic table, the metals are in Row 2 (lithium, beryllium), Row 3 (sodium, magnesium, aluminium), Row 4 (potassium, K through to gallium, Ga), Row 5 (rubidium through to tin), Row 6 (caesium to bismuth) and Row 7 (francium to actinium). There are two special series of metals from atomic number 58–71 and 89–103. The first are the rare earth metals and the second the radioactive metals (those beyond 92 do not occur naturally). Nos 90 and 92 occur naturally and are used for atomic power. The rest of the elements in the table are non-metals. Some have some metal-like properties and are called metalloids, e.g. nos 5, 14, 32, 33, 51, 52, 84 and 85.

Other non-metals are gases, i.e. nos 1, 2, 7, 8, 9, 10, 17, 18, 36, 54 and 86. Number 86, radon, is given off by uranium in uranium mining and milling operations. In column 1 of the table shown in Fig. 9.1, the elements running down from lithium are called alkali metals, because when they combine with oxygen, and are then mixed with water, a strong alkaline or caustic solution results. Sodium, for instance, gives caustic soda. The column 2 metals, e.g. calcium, are called alkaline earth metals. Their oxides also form alkalis when mixed with water, e.g. lime. Elements 1, 6, 7, 8, 15, 16 and 17 are some of the important elements composing the basic structure of all life. Other elements used by living organisms include numbers 11, 12, 19, 20, 23, 24, 25, 26, 27, 29, 30, 34, 42 and 53.

In column 7, the elements from fluorine down are called the halogens. They are all potent oxidizers. Chlorine, Cl, is used to purify water, and iodine is used to kill bacteria in wounds.

On the very far right of the table are elements which normally don't combine at all – the inert or noble gases. Argon, for instance, is used as shielding gas in welding because it does not combine with metals even at high temperature. Helium is used in deep-diving air.

Enhancing occupational safety and health

I	II											III	IV	V	VI	VII	VIII
1 H 1.0																	2 He 4.0
3 Li 6.9	4 Be 9.0											5 B 10.8	6 C 12.0	7 N 14.0	8 O 16.0	9 F 19.0	10 Ne 20.2
11 Na 23.0	12 Mg 24.3	Transition metals										13 Al 27.0	14 Si 28.1	15 P 31.0	16 S 32.1	17 Cl 35.5	18 Ar 39.9
19 K 39.1	20 Ca 40.1	21 Sc 45.0	22 Ti 47.9	23 V 50.9	24 Cr 52.0	25 Mn 54.9	26 Fe 55.8	27 Co 58.9	28 Ni 58.7	29 Cu 63.5	30 Zn 65.4	31 Ga 69.7	32 Ge 72.6	33 As 74.9	34 Se 79.0	35 Br 79.9	36 Kr 83.8
37 Rb 85.5	38 Sr 87.6	39 Y 88.9	40 Zr 91.2	41 Nb 92.9	42 Mo 95.9	43 Tc (98)	44 Ru 101.1	45 Rh 102.9	46 Pd 106.4	47 Ag 107.9	48 Cd 112.4	49 In 114.8	50 Sn 118.7	51 Sb 121.8	52 Te 127.6	53 I 126.9	54 Xe 131.3
55 Cs 132.9	56 Ba 137.3	57 *La 138.9	72 Hf 178.5	73 Ta 180.9	74 W 183.8	75 Re 186.2	76 Os 190.2	77 Ir 192.2	78 Pt 195.1	79 Au 197.0	80 Hg 200.6	81 Tl 204.4	82 Pb 207.2	83 Bi 209.0	84 Po (209)	85 At (210)	86 Rn (222)
87 Fr (223)	88 Ra (226)	89 §Ac 227.0															

*Lanthanide Series	58 Ce 140.1	59 Pr 140.9	60 Nd 144.2	61 Pm (145)	62 Sm 150.4	63 Eu 152.0	64 Gd 157.2	65 Tb 158.9	66 Dy 162.5	67 Ho 164.9	68 Er 167.3	69 Tm 168.9	70 Yb 173.0	71 Lu 175.0
§Actinide Series	90 Th 232.0	91 Pa 231.0	92 U 238.0	93 Np 237.0	94 Pu (244)	95 Am (243)	96 Cm (247)	97 Bk (247)	98 Cf (251)	99 Es (252)	100 Fm (257)	101 Md (258)	102 No (259)	103 Lr (260)

6 → Atomic number
C → Symbol
12.0 → Relative atomic mass (atomic weight)

A relative atomic mass in brackets is the mass number of the isotope with the longest half-life.

Figure 9.1 Periodic table of the elements

Isotopes

The number of protons identifies an element. Some elements may have atoms that contain different numbers of neutrons. The different atoms of the same element are then called isotopes. Because of the different number of neutrons in one isotope of an element, it can be quite radioactive; that is, the nucleus is progressively breaking apart and giving off radiation. The type and amount of radiation determines the degree of risk. The radiation from isotopes is used in controlled situations – for example, cobalt therapy, for treating cancer. Cobalt 60 therapy uses a cobalt isotope with 33 neutrons. Normal cobalt in vitamin B12 or soil fertilizer almost always contains only 32 neutrons. Your home smoke detector may contain a 37 kBq source of americium 241. Some natural elements, e.g. chlorine, are a mixture of two long-lived isotopes of the element.

Chemical compounds and their formation

Ions

Ions occur when elements or combinations of elements have either acquired more electrons than protons or lost some electrons, leaving them with less electrons than protons. Protons are electrically positive, and electrons electrically negative. Ions are electrically charged particles and can be either positive or negative. They can be attracted by charged electrical poles such as cathodes or anodes.

Chemical compounds

Elements combine with each other to form chemical compounds. The smallest particle of a compound which still keeps the chemical properties or behaviour of the compound is called a 'molecule'.

Chemical compounds are divided up into two main groups – 'inorganic' and 'organic'. Organic compounds, originally named because they were found in living things, now cover most compounds containing carbon. Organic compounds may also contain hydrogen, oxygen and varying amounts of other elements, e.g. sulfur, nitrogen or phosphorus. They can include metals, e.g. iron in haemoglobin.

Inorganic compounds

Inorganic compounds are those which are not organic. Generally, they are named with the positive ion first, followed by a variation of the negative ion next. Positive ions generally come from elements to the left of the periodic table; negative ions from those to the right. For example, the compound made of sodium and chlorine is called sodium chloride – this is common salt. The compound made of sodium and the bicarbonate ion is called sodium bicarbonate – this is carb soda. The bicarbonate ion in turn is made up of three elements – hydrogen, carbon and oxygen.

Organic compounds

Organic compounds made from carbon are based on the fact that carbon shares electrons, even with other carbon atoms. This allows long strings of carbon atoms to join together to make the polymers we call plastics. The properties of compounds can be very different from the elements which make them up. Sugar (sucrose) is a compound that is sweet, non-toxic and comes in white crystals; however, it is made of carbon, which is usually a black powder, hydrogen, a colourless flammable gas, and oxygen, another colourless gas in which substances burn and which the body needs. Sugar crystals are clear because the spaces between the atoms in the crystals allow light waves of all colours through.

Calculating weight of a 'mole' of pure chemical compound

Atomic weight

The weight of an atom is compared with one sixteenth the weight of an oxygen atom. The result is called the atomic weight. The atomic weight of each element is given in the periodic table. The water (H_2O) molecule is made up of two hydrogen atoms and one oxygen atom. To determine the molecular weight of water, add the weight of both hydrogen atoms to the weight of the oxygen atom.

$$\text{Molecular weight } (H_2O) = 2 \times 1(H) + 16(O)$$
$$= 18 \text{ (the exact atomic weight of hydrogen is 1.008).}$$

A mole of a compound is the molecular weight in grams, so for water this is 18 grams.

The amount of a chemical element, ion or compound in the body is often measured by doctors and biochemists in millimoles or micromoles per litre of blood or urine.

Reactions

The chemical combination of elements such as sodium and oxygen is called a reaction. This particular reaction is called oxidation, which is a very common reaction; for example, rusting is one type of oxidation. In many reactions, energy, in the form of heat, light or noise, is released. Some reactions actually absorb heat – for example, urea or ammonium nitrate dissolving in water (useful for cooling a sports injury cold pack without ice). In explosions, energy is released in very large amounts, while in the glowing bracelets children buy at fairs only a small amount of energy, in the form of light, is released.

Elements combine or bond together by giving away electrons, taking electrons from another element, or by sharing them. Only outer electrons normally get involved in chemical reactions. When an element loses electrons, it is oxidized and becomes a positive ion. When an element takes electrons it is reduced, and becomes a negative ion. Oxidation by chlorine, for example, may occur even though oxygen is not involved. An oxidation reaction which involves chlorine and is a safety hazard, for example, is the reaction between pool chlorine (calcium hypochlorite) and brake fluid.

What's inside an atom?

From the word 'nucleus' comes the word 'nuclear'. Nuclear reactions are different from chemical reactions as they involve the breaking up or the fusing together of the nuclei (nucleuses) of elements.

Common chemical hazard classes and descriptors

Meaning of some descriptors

Chemicals vary in how toxic (poisonous) they are. Toxicity is a chemical property which causes damage to a tissue, organ or system in the body. A chemical can be a hazard because of its toxicity. How much of a risk the hazard presents depends on the circumstances, e.g. is anyone nearby? is the chemical in powder, liquid, mist, vapour or gas form? We also talk of risk – how likely is it that the hazard will result in an accident – of something unplanned and unwanted occurring. A chemical can also be a hazard in other ways. For example, it may be explosive on its own, or it may be explosive when mixed with other chemicals and detonated, for example ANFO, a mixture of ammonium nitrate (a fertilizer) and fuel oil.

On chemicals labelled under the EU scheme, the word 'harmful' is used for the lower end of the toxicity scale; the word 'toxic' for the middle; and the words 'very toxic' for the higher end (in English).

'Flammable' and 'highly flammable' are words which describe how easily a substance catches fire and burns. 'Flash point' measures how easily a liquid gives off vapour and how easily that vapour catches fire from a spark or flame. Liquids with flash points up to 61°C are called 'flammable', and those with flash points between 61° and 150°C, 'combustible'. Under the current UN Recommendations, the classifications for flammable liquids are Class 3 (Packing Groups I – great danger, and II – medium danger), and Class 3 (combustible liquids in bulk).

Spontaneously combustible substances are those which catch fire without any spark or flame being applied.

Corrosive substances are those which burn or eat away skin, the eyes, or some metals – for example, alkalis, like caustic soda, and some acids, like hydrochloric acid (spirits of salts).

Oxidizing substances may supply oxygen which causes another substance to burn, or they may not supply oxygen but still pull electrons out of other substances and change them. For example, chlorine will bleach the dye in a cloth – chlorine is an oxidizer.

Infectious substances are those containing a biological agent hazardous to health – for example some fungi, bacteria and viruses.

Irritant substances are those which either directly irritate the skin or eyes, causing pain or itching, or trigger off an allergic response with the same symptoms.

Radioactive substances are those which give off harmful radiation, such as alpha, beta or gamma radiation, or neutrons.

Carcinogenic substances are those which produce cancer by interfering with the genetic blueprint, DNA, in body cells, or by other mechanisms.

Mutagenic substances are those which cause changes to the genetic material in particular cells of the body. The cell then does not grow or divide normally.

Teratogenic substances cause abnormalities in the foetus developing in the womb, resulting in birth defects. The word teratogenic means 'causing monsters' but not all birth defects are as serious as that.

Terms relating to fire hazards

There are a number of words or terms which you need to understand when looking at the Material Safety Data Sheet (MSDS) for a chemical. This is a document which the supplier of a chemical must provide to a chemical user.

When the word 'volatile' is used to describe a substance, it means that it gives off vapour. It usually refers to a liquid which vaporizes fairly easily. 'Vapour pressure' is the term used to give an idea of how easily a substance vaporizes. Today it is measured in Pascals, but you will still see vapour pressure given in mmHg (mercury) (760 mm = 1 atmosphere or 101.3 kilopascals), or millibar. (1016 millibars = 1 atmosphere). A common industrial solvent like toluene, found in paint thinners, has a vapour pressure at 30°C of 340 Pa. Vapour pressure approximately doubles for every 10°C (18°F) rise; therefore, a volatile chemical packed in Germany in midwinter can end up causing a swollen drum if stored outside in summer desert heat at an oilfield or mine, for example, or can squirt a worker in the face if not opened carefully.

'Vapour density' is a term which compares the weight of 1 litre of the vapour of a chemical with the weight of 1 litre of air. So a vapour density of four means the vapour is four times heavier than air and likely to find its way into channels and pits in a workplace. Some degreasing solvents fit into this category.

'Flash point' is a measure of how likely a liquid is to catch fire if a spark or flame is applied – the lower the flash point, the greater the hazard.

'Auto-ignition' temperature is the temperature at which the vapour will ignite without any spark or flame. For some chemicals this can be quite low. Carbon disulfide, an industrial solvent, auto-ignites at around the temperature of boiling water.

LEL (lower explosive limit) is the lowest percentage of a flammable vapour or gas in air which will burn with a spark or flame. UEL (upper explosive limit) is the highest percentage of such a mixture which will catch fire. Below the LEL the mixture is too lean; above the UEL the mixture is too rich. In practice, the percentage of a flammable vapour in a confined space, like a workroom or in an accidentally released gas cloud, is not constant or uniform. It varies, especially near the edge of the cloud, or near doors or draughts in the confined space. So a good safety margin must be allowed when using LEL and UEL. An explosimeter can be used to test confined spaces to see if flammable gases or vapours are near or above the LEL. The LEL–UEL range for methane (natural gas) is 5.3–15%, while that for hydrogen is 4.1–74.2%. Methane is considered safe to use in kitchens. Hydrogen would not be, for this reason and because of the higher rate of energy release when it catches fire.

Distinguishing other terms relating to chemicals

The following words are some words we use which are important when it comes to understanding and controlling chemical hazards.

Pure and impure

A pure chemical contains molecules from only one chemical compound. In practice, this is never completely the case and chemicals are made to a degree of purity. Many industrial chemicals of technical grade might only be 80% pure, for example. Many other industrial chemicals are a formulation or mixture of a number of different chemical substances. Labelling and classifying mixtures is based on a look at the most hazardous substances in the mixture, and how much there is of them. Some mixtures may only have a small amount of a certain chemical, but that small amount may also be the most toxic or hazardous in some other way – for example, the chemical in minor amounts could be a cancer-causing agent such as dioxin in the weedkiller 2, 4, 5-T, the Agent Orange defoliant used in the Vietnam war.

Solid, liquid and gas

If you could look inside them, you would see that in a 'solid', if it is crystalline, the atoms are linked into a regular three dimensional lattice. The links are there even in non-crystalline solids like glass, but the atoms are not arranged regularly. Diamond is a form of the element carbon and can be cut with very smooth flat faces. This is because it is crystalline. In 'liquids' there is still an attraction between the individual molecules, but there is hardly any attraction between them in gases. That is why when a gas is released it diffuses or spreads quite rapidly. A strong-smelling vapour is added to natural gas so that you can detect a leak, and it has been smelt at up to 27 km (17 miles) from the gas mixing station after a leak. The molecules of the vapour have travelled 27 km.

One important thing to note when a solid changes to a liquid, and a liquid to a gas, is that heat energy must go in, but there is no rise in the temperature of the substance until all of it has made the change. When the substance returns from gas to liquid, for example, that heat energy is released. So steam returning to water, while condensing on the skin, can cause scalding because of the amount of this hidden or latent heat energy released.

Homogeneous and heterogeneous

These simply mean a uniform or a non-uniform mixture of a substance respectively.

Other terms

A 'vapour' is a gas given off by a liquid at a temperature below the boiling point of the liquid version of the substance. There is water vapour in the atmosphere but the atmosphere is not at 100°C, the boiling point of water.

Some gases are given off directly from solid substances – like the gas carbon dioxide from dry ice (solid carbon dioxide). They 'sublime'.

'Dust' is finely divided solid particles suspended in air produced by a mechanical process such as crushing, or from deliberate or natural grinding, abrading or blasting.

A 'fume' is the result of heating of a solid (such as molten metal, welding) until a gas is released. The gas is recondensed into tiny solid or liquid particles. This happens when steel is welded. The fume is a mixture of very fine iron and other metals, iron oxide and other metal oxides and flux.

'Aerosol' means fine liquid or solid particles suspended in air. The spray cans we call aerosols actually produce aerosols.

'Smoke' is a mixture of solid and liquid particles suspended in air, resulting from a substance burning.

'Mists' and 'fogs' are fine liquid particles suspended in air. In mists the particles are slightly larger and so do not remain suspended as long.

'Natural' and 'synthetic' are words often used. Synthetic simply means made by humans.

The following are a few more words which are commonly used to describe chemicals.

An 'acid' can be either strong (and easily eat away some metals – for example, hydrochloric acid or spirits of salts), or weak, for instance citric acid, which we eat in lemons. Strong acids are a threat to human tissue and metal structures, and even some weak acids like hydrofluoric acid, used for cleaning aluminium and building exteriors, are very toxic to body tissue. Acids are acids because they release so-called ions of the element hydrogen.

'Bases' are substances which can neutralize the effect of acids. They do so because in water (they are then called 'alkalis') they release so-called hydroxide ions which mop up the hydrogen ions to make more water. Bases come from the oxides of metals in columns 1 and 2 of the periodic table, and also from a range of chemical compounds based on nitrogen, for instance. The commonest of the second type is a solution in water of the gas ammonia. Ammonia is used in some refrigeration plants and if it is released, dissolves rapidly in the liquid of the nose, mouth, lungs, eyes and skin causing irritation, or if it is strong enough, burns, because the liquid has become very alkaline. Acidity and alkalinity are measured on the pH scale, with 1, very acid, to 7, neutral, to 14 very alkaline (caustic). Caustic means burning. The more hydrogen ions the substance releases, the lower the pH; the more hydroxide ions released, the higher the pH.

Organic solvents

These are liquids which dissolve other substances. They can be more or less pure organic chemical compounds or mixtures. They can be synthetic or if not natural, obtained from natural sources; for example, many of the oil refinery 'cuts' sold as industrial solvents are mixtures of organic chemical compounds which are distilled from petroleum or distilled from it after 'cracking' (chemical conversion). Many are flammable, but synthetic compounds where atoms of chlorine or fluorine are built-in (so-called safety solvents) may be non-flammable. The substance 1,1,1 – trichloroethane used in fabric protector sprays and fat-dissolving drain cleaners, is not flammable. However, beware of the words 'safety solvent'. The solvents with chlorine in them, and to a lesser extent fluorine, still pose toxicity hazards, or a hazard to the ozone layer. Some classes of organic solvents are shown in Table 9.1.

Gas and vapour pressure

Earlier we talked about vapour pressure as one reason why solvents can cause problems, but there is another thing which applies to all gases and vapours which you need to know. It is based on the Universal Gas Law. This says that, as the temperature of a gas rises, so does the pressure if you do not change the volume. So pressure will rise as temperature does in a closed drum.

The law also says that volume must increase as temperature increases if the pressure is to stay the same, and that if the volume is reduced while the temperature stays the

Table 9.1 Organic solvents

Class of organic solvent	Examples
Aliphatic hydrocarbons	n-hexane, iso octane
Cyclic aliphatic hydrocarbons	Cyclohexane
Nitro hydrocarbons	Nitromethane
Aromatic hydrocarbons	Benzene, toluene, xylene
Halogenated hydrocarbons	Trichloroethane
Esters	Ethyl acetate, amyl acetate
Ketones	Methyl ethyl ketone, acetone
Aldehydes	Acetaldehyde
Alcohols	Methanol, ethanol, propanol
Ethers	Ethyl ether
Glycols	Ethylene glycol
Glycol ethers	Ethyleneglycol monomethyl ether (EGMME)

same, then the pressure must go up. Think of a compressor. As the volume is reduced and the pressure is increased, the temperature rises.

So, mathematically:

$$\frac{P_1 V_1}{T_1} = \frac{P_2 V_2}{T_2}$$

P_1 = original pressure
V_1 = original volume
T_1 = original temperature (°K)
(°K = °C + 273)

P_2 = final pressure
V_2 = final volume
T_2 = final temperature (°K)

Chemical reactions and structures

What happens during a chemical reaction

Chemical reactions of the oxidation type (where an element is oxidized) involve electrons being exchanged or shared. It is part and parcel of such a reaction that another process called 'reduction' takes place at the same time. Reduction generally describes a number of processes used industrially to recover metal from oxide and sulfide ores, such as blast furnace operation and aluminium smelting.

Other chemical reactions can involve one element exchanging for another. The thing which drives a reaction is that the total amount of energy in the system of chemicals involved is less after than before a reaction. However, to start a reaction often involves putting a small amount of energy in – for instance tinder or a lighter for wood, and a detonator for explosives. You can think of a chemical reaction like a river running downhill. There is less energy to do useful work left in the river at the end, but on the way it can be used to drive turbines and produce electrical energy or, in the old days, to drive a water wheel and grind wheat. Equally, a chemical reaction like burning diesel fuel in a motor can be used to do useful work.

Exothermic and endothermic reactions

Exothermic reactions give out heat. Endothermic reactions, like the one described earlier for cooling an injury pack, take in heat. Reactions can be fast or slow, and concentration, temperature and catalysts affect the speed. The concentration is the strength of the chemicals involved, whether they are in a weak or strong solution. Concentration is important because reactions involve molecules getting together, and the more there are in a space the more likely they are to meet.

Temperature speeds up a reaction because as the temperature rises, molecules move faster, so again they are more likely to meet. If a solid is involved, the molecules will vibrate more as the temperature increases until they break the bonds that bind them.

A catalyst is involved in the reaction but is not changed itself. For example, the platinum-rhodium catalyst in your vehicle exhaust catalytic converter causes reactions to occur which reduce pollution, but the catalyst is good for probably 150 000 km (93 000 miles).

Reactivity of metals

You have already noticed that some metals are more reactive than others. For example, iron waterpipes rust more readily than copper pipes corrode, though both of these reactions involve combination with oxygen. Some acids will attack iron and zinc but barely touch lead.

We hardly ever see some of the metals like lithium, sodium and potassium in metallic form because they react rapidly and catch fire in the presence of air and moisture. A table of oxidation potentials shows which metals are most reactive and which least so. The table also allows a user to work out the voltage to be obtained from various chemical cells. Sometimes you will find the table printed as a table of reduction potentials with the positive and negative signs of the voltages reversed. However, in that form, it serves the same purpose.

The following table shows the standard oxidation potentials for half-reactions.

Table 9.2 Standard oxidation potentials for half-reactions

	Half-reaction	$E°$(volts)	
Very strong reducing agents	$Zn \rightarrow 2e^- + Zn^{+2}$	0.76	Very weak oxidizing agents
↑	$Fe \rightarrow 2e^- + Fe^{+2}$	0.44	↓
↑	$Sn \rightarrow 2e^- + Sn^{+2}$	0.14	↓
Reducing strength increases	$Pb \rightarrow 2e^- + Pb^{+2}$	0.13	Oxidizing strength increases
↑	$H_2(g) \rightarrow 2e^- + 2H^+$	0.00	↓
↑			↓
Very weak reducing agents	$Cu \rightarrow 2e^- + Cu^{+2}$	−0.34	Very strong oxidizing agents
	$2Cl^- \rightarrow 2e^- + Cl_2(g)$	−1.36	

IONIC CONCENTRATIONS, 1 M (mole/litre) IN WATER AT 25°C
All ions are aquated, that is the ion is weakly bonded to water molecules.
The reaction which gives the higher voltage is more likely to succeed.

Explanation: As zinc is more easily oxidized (loses electrons more easily) than iron, zinc can be used as a sacrificial anode to protect iron.

Tin and lead give voltages little different from that of hydrogen gas (which is generated when many metals contact water or acid); they do not corrode easily. In addition, copper gives a voltage lower than hydrogen gas, so it is used in water pipes for its lack of corrosion.

At the bottom end of the scale, it can be seen that chlorine has a strong tendency to pull electrons out of other materials; that is, it oxidizes them. Hence its use to kill bacteria, etc., in swimming pools by oxidation.

At any time there must be an oxidation and reduction reaction together. An oxidation reaction (towards top of table) pushes a reaction (lower down), in the opposite direction to the arrow. So chlorine gas can be reduced to chloride ion in water, but in the process other materials in the parallel reaction are oxidized.

Types of explosions and explosion arrest

In safety and health, reactions involved in explosions are obviously of particular importance. Explosions are basically of two types – 'detonation' and 'deflagration'. Explosions produce a flame front and a pressure wave as a result of gases expanding due to the heat produced at the flame front. In a deflagration explosion the flame front can range from metres per second to hundreds of metres per second, but it is less than the speed of sound. In a detonation explosion the flame speed is supersonic and there is a shock wave with it. Explosions of flammable dusts such as flour, wood and coal are deflagrations. The pressure wave from the heat generated at the flame front travels ahead of the flame front at, or above, the speed of sound. This allows pressure detectors to pick up the explosion and operate explosion relief vents or explosion suppressors. This is not possible with detonation explosions.

Fig. 9.2 shows the result of an ethylene explosion in a banana ripening room due to faulty metering equipment. It shattered all the glass windows in a mezzanine office but occurred before office hours, so there was no serious injury.

Bonds, crystalline structures and properties of materials

It is useful to have some understanding of the chemical background to industrial materials.

Iron as it is made is normally in the form of cast iron, which has a relatively high carbon content, up to 6 per cent, as a result of using coke in the reduction of the iron ore. Cast iron is brittle and cannot be beaten into a variety of shapes, unlike, for example some forms of steel.

The molecules of iron and carbon arrange themselves in different three dimensional grids as the proportion of carbon and the temperature changes. Reduction in the

Figure 9.2 Ripening chamber blast

percentage of carbon occurs when cast iron is converted into steel. It is also possible to get desirable crystalline forms which exist only at higher temperature normally, by heating the metal and then cooling it rapidly (quenching). The crystalline structure can also be altered for particular purposes by adding a certain proportion of atoms from metals such as molybdenum, chromium, nickel, manganese and tungsten; for example, stainless steel contains nickel. Use of such metals and metallurgical processes can give high tensile steel; that is, it can take design stresses and not crack.

Carbon is also interesting. Carbon comes in three different forms – amorphous carbon (the common form); graphite, a soft waxy form; and diamond, a hard, clear crystalline form. In graphite the attractions or bonds between the carbon atoms are strong in a series of layers, but weak between layers. So graphite – used, for example, in pencil leads – rolls or peels easily from the 'lead'. Carbon can also conduct electricity, unusual for a non-metal. Silicon is a semi-conductor.

Plastics are also based primarily on carbon. Plastics or polymers are based on linking small molecules chemically to form long chains. The main bonds are carbon to carbon. It is also possible to make plastics in which the bonds are not just along the chains, but across the chains. This crosslinking, as it is called, produces firmer, harder plastics. Polyethylene is a softer plastic. Some grades of PVC, such as in irrigation pipe, are harder.

Chemical basis of corrosion

Corrosion of metal parts and structures may seriously affect their strength and hence their safety. Rusting of iron basically involves three steps. Iron, especially in the presence of moisture, loses electrons, or to put it another way, is oxidized. The reduction reaction which goes with this is that water and oxygen take up electrons to form hydroxide ions, the same ions found in alkaline solutions. Step three involves the positively charged iron ions and the negatively charged hydroxide ions being attracted towards each other (because they are electrically opposite) to form so-called hydrated iron oxide, rust.

The oxidation and reduction reactions do not always occur at the same place in the metal, but the ions from the iron flow through the metal or through the thin film of moisture on the surface to be taken up by the oxygen atoms. Prevention of rusting takes a number of forms, e.g. phosphoric acid treatment followed by painting, electroplating with say chromium, or galvanizing by coating with zinc. The zinc oxidizes in preference to the iron.

Another method used involves attaching a metal which is more readily oxidized to the iron; that is, a metal which appears higher up the table of oxidation potentials. Zinc anodes (an anode gives off electrons into a solution) are used, for example, and the zinc gradually dissolves instead of the iron. The anode is replaced when eaten away.

Corrosion and failure of metals due to incorrect choice of type can be important sources of equipment and structural accidents, especially if the wrong metals are in contact.

Law on solution of gas in water

A number of gases dissolve in water to a greater or lesser extent – oxygen is a little bit soluble in water (which is just as well for fish, who extract it with gills), while a gas like ammonia is very soluble in water. The higher the pressure of the gas, the more gas dissolves, and the lower the temperature of the water, the more gas dissolves. So colder sea water contains more oxygen, and generally contains more fish as a result.

As you know, if you want enough carbon dioxide in water to make good soda you must use a container which will allow you to hold the carbon dioxide from the 'sparklet' under pressure. You also used chilled water. Once you pour the soda, the pressure is released, and some of the gas is released which results in bubbles. Or, while you watch a cold beer warm up, the warming drives more of the gas out of the beer.

This law explaining the behaviour of gases is mentioned here, because it allows you to understand what happens to nitrogen gas in the air we breathe when people are breathing that gas under pressure during a dive. It also explains why tiny bubbles of nitrogen gas form in the body's blood vessels if the diver comes up too fast, causing 'the bends'. Pressure has made the nitrogen more soluble in the blood. Releasing the pressure makes it less soluble. The same can happen in tunnelling work if the people there are working at greater than normal atmospheric pressure.

For this reason, some deep diving involves breathing a helium/oxygen mix. Helium does not form bubbles in the blood (it is less soluble in blood).

Classifying chemicals

Introduction

Dangerous goods

Substances can be divided into non-hazardous and hazardous. Many hazardous substances are also listed as so-called dangerous goods, and so there are requirements set for their storage and transport.

The UN Recommendations for air, land and sea transport (see Further Reading at the end of this chapter), are the basis for a uniform system of control on transport of dangerous goods internationally and within countries, with flow-on effects to storage requirements.

The Recommendations cover those 'hazardous substances' which are listed or otherwise covered in the Recommendations as 'dangerous goods'. The IMDG (maritime) and IATA/ICAO (air) codes are consistent with the Recommendations.

There are nine classes of dangerous goods, with divisions of some classes. The classes are: explosive; flammable and non-flammable non-toxic gases; flammable liquids; flammable solids and spontaneously combustible materials; oxidizing substances and organic peroxides; toxic and infectious substances; radioactive and fissile materials; corrosive substances; and miscellaneous.

For hazardous substances used in the workplace and by the public, there are EU and US classification systems, or other national systems derived from them.

Markings and labels under the UN Recommendations

Dangerous goods labelling provides essential information to those working with them or who may come across them or have to deal with them in an emergency.

In dangerous goods language, the 'label' is only the diamond-shaped class label (which is called a 'placard' during transport), and what people generally call the label on a container is called the 'marking'. The word 'placard' is also used for signage on various parts of dangerous goods premises. The label is of a specified colour and contains, in all but two cases, a distinctive pictogram, e.g. a flame.

The UN Recommendations set out in detail what should appear on dangerous goods markings including the class (in the lower half of the label), and the United Nations (UN) number. A substance with a subsidiary risk requires a second label.

Check your relevant dangerous goods legislation (in the USA, DOT Hazmat legislation) for the requirements on marking and labelling of packaging. Combustible liquids as noted earlier are those with flash points and fire points below their boiling point, and a flash point from 61–150°C. They only become dangerous goods when transported in bulk with bulk flammable liquids (flash point less than 61°C), or 1000 or more litres of flammable liquid.

The fire point is tested by a US test and is similar to flash point, but the flame on ignition must last five seconds.

Further information on the application of the labelling system to substances in transit or being transported, and in relation to imports, will also be needed by those responsible for such matters.

Under the UN Recommendations, differences apply to bulk containers and packages. Bulk refers to:

- Class 2 dangerous goods (gases) in containers greater than 500 L
- goods of other classes in a container of not less than 450 L and a container with a net mass of not less than 400 kg.

Workplace and consumer labelling

There are a number of systems for labelling hazardous chemicals for workplace and normal consumer use, rather than as dangerous goods (if they are those too), including those

Hazardous substance management

of the USA and the European Union. The classification and standardized risk and safety phrase system for workplace hazardous chemicals in Australia, for example, is now based on the European Union system. This labelling may include the dangerous goods labelling requirements if the hazardous substance is also a dangerous good. Canada uses the WHMIS (Workplace Hazardous Materials Information System), the USA has labelling requirements under the Hazard Communication Standard, and the UK covers them in the COSHH (Control of Substances Hazardous to Health) Regulations.

Selecting correct dangerous goods classification for a workplace hazardous substance

The UN Recommendations cover not only substances made up of one chemical compound but also many mixtures.

Hazardous substances which are listed as dangerous goods are, as noted earlier, put into one of nine classes. Substances can be given a class on the basis of the main or primary hazard, and a subsidiary class on the basis of another class of hazard. For example, a chemical like red fuming nitric acid may be both corrosive and oxidizing, Class 8 and Class 5.

The national version of the UN Recommendations may make use of standards and other publications, some of which are listed in the Recommendations. The Recommendations and standards are not law unless they are included in, or given legal force by, the legislation covering dangerous goods in your jurisdiction. This is usually after adoption by your country in international agreements such as those on shipping and aviation.

Dangerous goods are listed by number and alpha order in the UN Recommendations. Many of those listed are not pure chemical compounds but mixtures or groups of chemicals.

The Recommendations cover all dangerous goods, but explosives (Class 1) are also covered by a separate code. While the UN Recommendations refer to transport, much of what they contain has, as noted, been used as a basis for the regulations on storage.

The Recommendations provide the following information (using dipropylamine as an example – see Table 9.3).

1. **UN number:** 2383
2. **Name and description:** Dipropylamine
3. **Class or division:** 3 (flammable)
4. **Subsidiary risk:** 8 (corrosive)
5. **UN packing group:** II (medium danger)
6. **Special provisions:** Nil
7. **Limited quantities:**
8. **and 9 Packagings and intermediate bulk containers:** Column 8 gives the packing instruction and column 9 special provisions.
9. **and 11 Portable tanks:** There are also two further columns giving coded instructions for portable tanks and portable tank special provisions (omitted from Table 9.3).

Table 9.3 Sample dangerous goods entry

Col. 1 UN No.	Col. 2 Name and description	Col. 3 Class or Division	Col. 4 Subsidiary risk	Col. 5 UN packing group	Col. 6 Special provisions	Col. 7 Limited quantities	Packagings and IBCs	
							Col. 8 Packing instruction	Col. 9 Special provisions
2383	Dipropylamine	3	8	II		1 L	P001 IBC02	

The meaning of the coding such as P 001 and IBC 02 is given in the Recommendations.

Correct workplace placards for dangerous goods storage

Placards are placed on dangerous goods storages to provide information to emergency services, people at the workplace and the general public about the location and type of chemicals stored, and information on emergency procedures for spills or fire. Different requirements apply for:

- package stores
- tanks and bulk stores
- underground tanks and stores.

What are called 'exemption limits' tell you the quantities above which placards are required. For national exemption limits check your relevant local legislation. Individual dangerous goods authorities may also issue guidance notes. There are three types of placards: the one to be placed at the main entrance to the premises; the placard or placards needed for specific package storage areas within the premises; and the placard for tanks or bulk stores. Check your local legislation for the formats required.

Designing appropriate labels for workplace hazardous substances

The requirements for a workplace hazardous substance label, as distinct from a dangerous goods label, are set out in the OHS legislation in your jurisdiction. Some national examples were noted earlier. In some federations, the requirements are set nationally. For example in Canada, WHMIS is given legal authority by the Hazardous Products Act. The risk phrases, safety phrases and first aid phrases for a label are also prescribed. In the USA the Hazard Communication Standard 29 CFR 1910.1200 gives the details. Note that US requirements on chemicals are found in 29 CFR for OSHA, 30 CFR for MSHA, 40 CFR for the Environmental Protection Agency and in 49 CFR for the Department of Transportation. The Toxic Substances Control Act, the Hazardous Materials Transportation Act and the Clean Air Act constitute important US legislation on hazardous substances control. The US Acts known as CERCLA, CWA, SARA, and EPCRA are also relevant.

The Australian National Occupational Health and Safety Commission (NOHSC) system, as noted, closely follows the European Union system. The standardized risk phrases and safety phrases to use for hazardous substances labels can be found in NOHSC's *List of Designated Hazardous Substances*. This derives from EEC Directive 67/548/EEC as updated by 96/54/EEC Annexes 1 and 6. However the Australian state and territory legislation is what gives the labelling actual legal force.

On US labels, you may find the US National Fire Protection Association (NFPA) diamond-shaped label, with four divisions which give a concise warning of the hazards associated with a particular material. Clockwise from the top they are: flammability; reactivity; special; and health. The figures indicate degree of risk and range from 0 – insignificant to 4 – extreme. A star indicates a chronic health hazard. The US National Paint and Coating Association's HMIS is similar but contains a PPE code.

Classification of single substances and chemical mixtures as hazardous or not hazardous

Substances need to be classified as hazardous or not hazardous to allow for appropriate labelling and to decide which substances need to be assessed for health risks, and hence control measures. For a large number of substances used either alone or in mixtures, in the European Union system, there is as noted above a 'List' with rules for determining whether a workplace chemical is hazardous or not. A separate document on classifying substances is required for those not found in the list. Most of this work will be done by manufacturers of chemicals. For the EU, an online database (see Further Reading at the end of this chapter) can assist.

Main factors in transport of hazardous chemicals

Relevant regulations and Recommendations

If the hazardous chemicals are dangerous goods, i.e. listed in or otherwise covered by the UN Recommendations, the Recommendations include requirements for:

- classification of the chemical
- assignment of packing group
- marking of packages
- vehicles and transport containers documentation
- packing
- transport in bulk
- stowage and compatibility of different dangerous goods requirements,
- procedures for transport
- construction and testing of packages, intermediate bulk containers, large packages and portable tanks.

Some countries include the Hazchem emergency action code (see pages 366 and 377) in their local version of Table 9.3. Mixtures and unlisted substances are dealt with in technical appendices to the Recommendations.

The dangerous goods regulations relevant to your jurisdiction should also be consulted in relation to transport. These may include:

- special licensing (road taxing) of vehicles
- reciprocal recognition of dangerous goods licences
- vehicle requirements
- transport procedures including responsibilities of consignor, prime contractor, owner and driver
- licensing of drivers.

Transport by pipelines may also be included.

Basic vehicle requirements

Generally speaking, the relevant dangerous goods legislation may require vehicles to have a current licence (UK – be currently road taxed) and be approved by the authority in relation to compliance with the UN Recommendations. There are exceptions in some cases where bulk containers are not carried. Check your local requirements.

Specifications may be given for various aspects of vehicle usage.

Some of the issues covered may include:

- vehicle safety standards
- segregation of different classes of dangerous goods and stowage
- transfer of bulk goods
- documentation
- emergency information
- safety equipment
- transport procedures
- carrying of documents and emergency procedures guides
- packaging of goods
- testing of containers
- emergencies
- placarding of vehicles
- vehicle licensing (UK – road tax) and driver licences
- insurance
- provision of safety equipment
- fire extinguishers
- breakdowns, parking and stopping
- health screening of drivers.

The duties of consignor, prime contractor, driver, owner, occupier, loader, or transferor may also be set out in legislation or codes implementing the Recommendations. Rail transport will generally have some additional requirements and air transport has a separate set of controls.

General checks on vehicles may be set out. Some substances may be subject to transport only over designated routes.

Training requirements

These may be set out in legislation implementing the Recommendations. In addition to an appropriate vehicle driver's licence, the driver may be required to satisfy the competent authority and this may be done by completion of an approved course of training. Obviously, emergency services must also have a coordinated plan to respond to emergencies and this requires appropriate training.

Placarding of transport vehicles

On vehicles transporting packaged dangerous goods making up a full load of a single dangerous good, or in tank transport units containing solids, liquids or gases which are dangerous goods, in general the front and rear must carry marking made up of the placard, the class number (also on the placard) and the UN number on the placard, or next to it as described earlier. Placards are not required on transport units carrying any quantity of Division 1.4 explosives Compatibility Group S; dangerous goods packed in limited quantities (as defined); and excepted packages of radioactive materials. A second placard must also be displayed if the Dangerous Goods List specifies such a risk, except if other goods in the same transport unit already require such a placard due to their primary risk.

Packaging requirements

Packing groups

Packing Group I (PG I) contains goods of greatest danger, while PG III are those of minor danger. The Recommendations distinguish between outer packaging and inner packaging. The Recommendations specify the minimum quantities for which proper labelling is required. Package specifications are described in detail, which include types of containers, package markings and tests for strength. Packing groups apply to all dangerous goods classes except 1, 2, and 7.

Planning segregation for a set load of dangerous goods

Segregation is required when dangerous goods of different classes are transported (or stored) if those classes are incompatible. If placed incorrectly, they increase the risk to people, property and/or the environment. The rules for segregation are set out in the local legislation or code, mirroring the requirements in the UN Recommendations.

Emergency equipment and emergency procedures

Emergency equipment to be carried with transported dangerous goods is to be provided in the cabin by the owner of the vehicle and the prime contractor. What is to be provided may be set down in legislation or a code. Suitable fire extinguishers are required. In addition to the documentation required by the Recommendations to be carried by the carrier, emergency procedures may be set out in emergency procedure guides (EPGs). The driver and prime contractor would then ensure that the driver has a copy of the EPGs, that they are in the required holder, and that the driver is able to produce them to the authority administering dangerous goods legislation, to the police, or to an emergency response authority, e.g. fire and rescue service.

The consignor must provide the driver with shipping documents. General steps in all emergencies may be set out in dangerous goods codes or legislation. These may include, for the driver, where able:

- notifying police or fire and rescue service
- notifying the prime contractor
- providing reasonable assistance to an authorized officer
- carrying out emergency procedures in the emergency information, and in any emergency plan
- if there has been an escape of a dangerous good, preventing other vehicles, dangerous goods or sources of ignition coming within a specified distance
- warning people in the vicinity
- preventing or minimizing escape of a dangerous good into drains, sewers or watercourses.

An additional feature which is very important is the local emergency action plan (LEAP). This allows for appropriate forewarning of communities likely to be affected by a hazardous substances escape. A LEAP includes the appropriate response, together with opportunities for training local response groups, and alerting the public.

Interpreting the Hazchem Code

The Hazchem Code, adopted in some countries as part of dangerous goods placarding of vehicles and stores, is important in case of spill or fire. For example, in the Hazchem Code 3YE, the number refers to the type of extinguisher to use (foam); the first letter refers to the protective gear to be worn (breathing apparatus) and the second letter refers to evacuation. See Fig. 9.3.

Note that if the white on black notation is missed, the safest choice of protective equipment results.

1. **JETS**
2. **FOG**
3. **FOAM**
4. **DRY**

Hazardous substance management

P	V	FULL	DILUTE
R			
S	V	BA	
S		BA for FIRE only	
T		BA	
T		BA for FIRE only	
W	V	FULL	
X			
Y	V	BA	CONTAIN
Y		BA for FIRE only	
Z		BA	
Z		BA for FIRE only	
E		CONSIDER EVACUATION	

Figure 9.3 Hazchem code

FOG — In the absence of fog equipment a fine spray may be used.
DRY AGENT — Water **must not** be allowed to come into contact with the substance at risk.
V — Can be violently or even explosively reactive.
FULL — Full body protective clothing with breathing apparatus.
DILUTE — May be washed to drain with large quantities of water.
CONTAIN — Prevent, by any means available, spillage from entering drains or watercourses.

Dangerous goods vehicle and driver checks

Checking a vehicle and driver for compliance with the regulations on transport of dangerous goods

Such a check, taken to the limit, could be quite complex, and a task for specialists. However, on an everyday working basis, a suitable checklist based on the information in the last section will allow routine checks. Different requirements apply to rail vehicles.

Checking packaging and segregation for a particular load

This check will require identifying the goods, and ensuring that packaging is suitable and appropriately marked. You will then need to check segregation following the legislation or code implementing the UN Recommendations. Note that such a code may ban 'fire risk substances', e.g. waste paper, hay, sawdust or wood chips, being carried with dangerous goods. Combustible liquids are also affected.

Storage of hazardous substances

Checking premises for storage of hazardous substances

This section can only give an overview of the main items to look for in a storage area for hazardous substances, or in planning such an area. You will find the appropriate criteria in your local dangerous goods regulations, if the substance is a dangerous good and quantities are sufficient to come within the regulations. While the UN Recommendations refer to transport, many of them form, as noted, the basis for legislation, guides and standards on storage and handling of dangerous goods.

However, for hazardous substances in general, the principal points to consider are:

1. **Location** in relation to other parts of a plant, mine site, or premises. When deciding on a location, you must consider the following:
   - that it is appropriate for drainage from washing down the storage area or collecting spills
   - that it is close to the delivery or receival, and dispatch points (if dispatch applies) of hazardous substances
   - the overall location of plants with stores of particular chemicals is also controlled by environmental planning and major hazard legislation, and county or local government by-laws
   - separation distances from other buildings and roadways.

   These need careful checking in the planning stage. The Coode Island fire in Melbourne, the more recent firework factory fire at Entschede in The Netherlands, and the ammonium nitrate store explosion in Toulouse show how important this is.

2. The **type of building.** This includes the type of construction; fire rating; fire rating of dividing walls; ventilation (natural and mechanical); flooring materials; and floor design (drains and bunds). Bunds are low walls for containing possible spills.

3. **Lighting, electrical fittings and electrical equipment** require attention. While this is fairly obvious for stores of flammable liquids and gases, attention is also required for the many chemicals which give off corrosive vapours or gases. Depending on the situation, certain mobile and other equipment used in stores for flammables needs to be appropriately designed; for example, electrical sparks within equipment must not be allowed to come in contact with flammable vapours.

4. **Ventilation** is required to prevent the build-up of gases and vapours. Cost savings can be achieved if natural ventilation is used where possible.

5. **Sunlight.** Some chemical stores are outdoors, and in summer conditions in some countries this can be intense. Goods, labels and containers may require protection from the effects of ultraviolet rays (UV) in sunlight and at least some protection from heat by being shaded.

6. **Cold.** Very cold conditions may lead to embrittlement of containers or freezing of the contents, both of which may lead to escape of the contents.
7. **Bunding, drainage, spill recovery and spill neutralization.** For example, floor and drain design must ensure no build-up of toxic or flammable vapours. Stores of sodium cyanide may require that ferrous sulphate is kept on hand for neutralization.
8. **Signage.** This includes placards mentioned earlier, but will also show areas where personal protective equipment (PPE) is required and the type. It will indicate exits, first aid facilities and eyewashes. It will also indicate the nature of certain chemical hazards. Different colours indicate the different classes of signs – black on yellow for hazards; white on green for emergency facilities; red, black and white for danger; blue on white for PPE; a barred red circle for prohibition; and white on red for fire safety.
9. **Multilevel stacking.** Vertical segregation is ineffective if liquids or powders, for example, stored high are incompatible with chemicals lower down (see the segregation chart in the UN Recommendations or dangerous goods regulations). Access-ways must be adequate for equipment used, provide pedestrian safety and have sufficient floor strength. The materials from which shelving is made must be suitable for the chemicals – for example, no metal shelves for strong acids.
10. **Segregation** of different classes of goods based on the segregation chart in the dangerous goods regulations for your jurisdiction is essential, and screen walls must meet approved or required standards.

    Separation distances between buildings and from boundaries and 'protected works' must also be observed. Protected works includes certain types of buildings such as dwellings; places where people congregate; hospitals; other adjacent places where people are employed; a ship at berth; another dangerous goods storage or another potential source of radiant heat from combustible materials.
11. **Fire precautions** are essential particularly with certain types of stored chemicals. Apart from those which are flammable, fire can cause many others to give off very toxic smoke and gases. The precautions centre around planning, design and location described above, and include fire and smoke detection. They also include: doors which slow the spread of fire; compartmentation using fire-rated walls; fire detection and arresting equipment; smoke control; and training of personnel to respond effectively.
12. **Emergency plans.** Both fire and spillage require proper emergency plans, emergency response equipment and staff training. Where outside assistance will be needed, this should be thought of in advance. There must be prior discussions and agreement with emergency services such as police, fire and rescue service, dangerous goods inspectorate and environmental authorities. A group scheme may exist where a number of companies in an area support each other in a major emergency. Some jurisdictions will require the preparation of a total hazard control plan (UK – major accident prevention plan) for large stores or major hazard facilities.
13. **Documentation.** A chemical register which clearly lists the chemicals held, and their quantities and location, is essential and must be kept up to date. An Emergency Services Manifest should be prepared and stored at an agreed location where a fire, explosion or spill would not make it inaccessible. This manifest describes where, what and how much chemical there is for the emergency services called in

to tackle a problem. As mentioned earlier, placarding will also assist emergency combat teams. Emergency procedure guides should be obtained, as well as MSDS, for the chemicals stored.
14. **Training.** Apart from training for emergencies, a key feature of preventing accidents and injuries is adequate training of staff in procedures for the operation of stores, and proper record keeping of those procedures and that training. Training goes beyond the chemicals themselves and will, where necessary, involve training in other areas – for example, safe lift truck operation.

Static discharge

Static discharge is often not considered when electrical safety is planned or discussed. There are links between the lack of electrical conductivity of some liquids and powders and the dangers of static electricity discharge. Finely divided flammable powders (some plastics) and flammable industrial solvents of low electrical conductivity (petroleum-based solvents, for example) can generate considerable static charges while flowing. Synthetic fibre clothing and devices involving plastic sheeting for wrapping can generate spark discharges. Stepping down from rubber-tyred equipment may cause static discharge.

The hazard from transfer of liquids can be minimized by attaching earthing clips at each end of a conductor to the discharge and receival containers, if they are made of conducting metal. Rubber-tyred vehicles and shoes require special treatment in some flammable areas. Anti-static clothing is required.

Information sources on chemical hazards

Key features

Key features of national requirements may include:

- provision of information by suppliers
- provision of information by employers
- induction and training
- assessment
- control
- monitoring
- health surveillance
- record keeping
- employees' duties
- role of relevant public authorities and emergency services.

You need to refer to your local occupational health and safety regulations for the exact legal requirements.

Using data sheets, labels and other information to identify hazards

This involves firstly recognizing chemical hazards and identifying them. This will mean surveying the use of chemicals in the workplace. The next step is to use data sources to identify hazards associated with individual chemical compounds (where they are known) or using MSDS to identify hazards associated with trade name chemicals. Other documentation which may help includes stock records; the chemical register; emergency services manifest; placards; and container labels.

Information sources for safe use of chemicals

The key information sources to use in controlling chemical hazards in the workplace are:

- your national and local OHS and dangerous goods authorities
- materials safety data sheets (MSDS)
- substance labels
- chemical registers
- emergency procedure guides
- UN/ILO/EU/NIOSH International Chemical Safety Cards (on CD-ROM, and Internet, www.ilo.org)
- the UN Recommendations
- dangerous goods and occupational safety and health acts and regulations in your jurisdiction.

The above sources will allow supervisors, health and safety officers, and health and safety committees to have a basis for controlling hazards.

Other information sources which will assist are:

- Lewis, R.J. Sr. (1995). *Sax's Dangerous Properties of Industrial Materials.* 9th edn. New York: Van Nostrand Reinhold
- Bretherick, L. (1999). *Bretherick's Handbook of Reactive Chemical Hazards.* 6th edn. London: Butterworths
- referenced journal articles on tackling particular hazards in particular situations (These may be looked up on CCINFO-DISC CD-ROM and selected internet sites.)
- relevant national standards
- NIOSH *Pocket Guide to Chemical Hazards and Other Databases* (on CD-ROM)
- chemical supplier information lines
- other databases available on Internet, e.g. www.cas.org.

Complex situations and major processes will obviously require access to in-house expertise or expert consultant assistance.

Evaluation of chemical hazards

Once the identification or recognition phase is complete, the next step is to assess or evaluate the hazards.

Assessment involves these questions:

1. Is anyone exposed to the chemical hazard? If so, who? How often? How much? The last question may require expert assistance with air sampling, but in many cases this will not be necessary. For instance, many substances may not result in airborne exposure from dust, vapour, gas, fume or mist. Instead, any potential contact may be with the skin. The substance may or may not be absorbed via the skin. Does the substance act on contact or is it absorbed to go on and affect other target organs? Poor personal hygiene could result in ingestion (swallowing) if the chemical is on hands and the worker bites nails or eats or smokes with unwashed hands.
2. What form is the chemical in? This affects the degree of risk. Is it solid, liquid, paste, gas, vapour, fume, mist, dust or fibrous dust? Will it be heated and give off vapour or fume? For instance, low hydrogen electrodes in welding give off fluoride fume, a toxic hazard, during use. The possibility of exposure may depend on process control, for example maintaining the correct pH of cyanide solutions in a gold recovery plant. If the pH is too low, hydrogen cyanide gas is given off.
3. Are workers close to or distant from the source of the hazard? Is the work area open, closed, ventilated? Are there heavier-than-air vapours which can flow elsewhere? Will airflow carry vapour or dust away from the source towards workers?
4. Is process control a critical factor? For example, brick kiln vapours may normally be led away to proper filtering equipment and high level exhausts, but emergency vents may open to assist process control and so expose workers.
5. Is unexpected contact from spillage of containers due to poor handling practices (climbing ladders, pouring instead of pumping) a possibility?
6. How toxic is the chemical? Which parts of the body are affected? Is the danger acute (immediate), or chronic (where the effects can result from repeated small exposures)?

In the Australian National Occupational Health and Safety Commission *Guidance Note for the Assessment of Health Risks Arising From the Use of Hazardous Substances in the Workplace*, for example, the key steps in assessment are:

1. Decide who will do it.
2. Divide the work into units.
3. Identify substances in the work.
4. Determine which substances are hazardous.
5. Obtain information about hazardous substances.
6. Inspect the workplace and evaluate exposure.
7. Evaluate the risk.
8. Identify actions resulting from conclusions about risks.
9. Record the assessment.
10. Review the assessment.

Producing the records of that assessment to an inspector is now required under some Australian OHS legislation.

Hazardous substance management

Preferred order or hierarchy of controls

After assessing and evaluating the hazard against accepted standards, the basic approach to control of chemical hazards can be summed up in what is called the preferred order (or hierarchy of control measures). It is shown here:

MANAGEMENT T
                                   R   Elimination
                                   A   Substitution
                                   I   Segregation
                                   N   Engineering controls
                                   I   Work practices
                                   N   Personal protective equipment
                                   G

The idea is to use control measures at the top in preference to those further down, although a mixture of measures may be used.

Recommending appropriate measures to minimize risk

Control measures

Consider the following:

- Can the chemical be eliminated entirely? Do we need to use it? Do we need to carry out this process?
- If the chemical must be used, does it have to be exactly this one? Is there another one which will do the job just as well and is less of a hazard?
- A third control measure, even if a less hazardous choice is made, is to isolate workers from the chemical (by a sealed process) or to segregate the chemical from workers (another room, lids on mixers, 'glove box', etc.)
- If isolation or segregation cannot be used, or only partially used, then consider additional engineering controls. Typically with chemicals posing an inhalation hazard this means improving ventilation. General dilution ventilation using mechanical means may be replaced by capture at source with a captor hood or slot, carrying the contaminant away in ducting to a suitable filter or to the outside.
- A fifth option, which can be used in conjunction with the other options, is administrative controls such as changing work practices using revised procedures and training. This can involve rotating workers in and out of hazardous areas, or changing the procedures used. As this control is dependent on human reliability, the higher options are preferred.
- A last choice, but least preferred as a choice on its own, is to use personal protective equipment – chemical goggles; respirator or breathing apparatus; head covering; apron or suitable overalls; suitable footwear and gloves for a particular chemical.

All of these control measures require active **management** and **training** to make them successful.

Monitoring and health surveillance

Once control measures are introduced it is essential to follow them up to see if they are successful. This may involve measuring airborne contaminants, analysis of urine or blood, or other medical screening of workers – for example, flow and volume testing of lung function by a trained nurse.

It is also important to run checks on ventilation equipment to ensure it continues to operate at planned effectiveness levels. In some jurisdictions, mining law requires trained officers to regularly conduct air sampling for airborne contaminants.

Record keeping

The control measures and the procedures which go with them must be recorded; so too must the results of monitoring. In some cases, for example with cancer-causing substances, these records must, by law, be kept for many years, and handed to the responsible government authority if the company closes. Check with your occupational health and safety authority on the requirements of any specific legislation on carcinogenic substances.

Employees' duties

Another issue to remember is that the general duties imposed on employees under occupational health and safety legislation apply to chemical hazard control. Employees, having received appropriate training and proper protective gear, protective procedures, and protective equipment or plant, must:

- work as trained
- follow supervisor's directions
- wear protective equipment supplied
- not endanger other workers
- report promptly any defects or problems which may cause a change in the degree of risk.

Emergencies

These will not be dealt with in detail here as they are covered in other sections and chapters. However, it is essential that procedures to deal with possible emergencies are developed, people are trained in what to do and where to go, it is clear who takes control, and there is liaison at the planning stage with emergency authorities – police, fire and emergency services, environment, energy supply and medical providers.

Requirements for chemical waste disposal

These procedures include containment, treatment or neutralization, sediment collection, filtering and disposal in an approved manner. Disposal could involve dumping at a suitable toxic waste site, incineration, running to an evaporation pan, or approved disposal into mains sewerage or a sewerage outfall.

Advice must be sought from bodies such as government authorities which administer environmental protection, mines regulation, and occupational health and safety matters. These may include waterway and national parks authorities, and county, borough or other local government authorities.

Checking premises using hazardous substances

Safety equipment in premises

Premises where hazardous chemicals are manufactured, used, handled or stored will require a variety of safety equipment. This includes:

- Safety showers and eyewash stations which must be kept free of obstacles. The water must be run often, so that the pH of the water is correct, and the water in the pipes cannot be allowed to become too hot for the eyes or body. Nor must it be allowed to freeze.
- Adequate facilities to wash and shower.
- Waste disposal which is suitable for the purpose; for example, rags and tissues soaked in flammable solvents need metal containers with well-fitting lids. Plastic bins may 'melt' if in contact with some solvents.
- Fume cupboards, slot captor or receptor hood ventilation. These must be driven by a suitable and adequate fan, and will often need water wash, dust filtration or precipitators built in. Local air pollution control requirements may require capture of vapours, which cannot simply be discharged from an exhaust vent. The ventilation system pressures should be checked regularly to spot obstructions or reductions in fan performance, and the pressure figures recorded.

Permitted exposure levels for airborne chemicals

You will find the maximum exposure levels to a range of over 700 dusts, vapours and gases in the exposure standards for atmospheric contaminants in the occupational environment adopted in your country. In the USA the Occupational Safety and Health Administration call them 'PELs'. In Germany they are called 'MAKs', in the UK 'OESs' and 'MELs', in Australia 'ESs', for example. In some jurisdictions recommended standards, e.g. the threshold limit values (TLVs) from the US ACGIH, have been included in regulations.

There may be three types of standards:

1. Time weighted average exposure standard for an eight hour day
2. Short-term exposure standard
3. Peak or ceiling exposure standard.

The first is designed for contaminants where average concentration over time is the important factor – for example, some industrial solvents such as toluene. The second is designed for 'cyclic' exposures, such as where an employee might remove a mixer lid in batch process operations four times a day. The third is designed for substances where even very short-term exposure is harmful – for example, hydrogen cyanide and acute irritants.

Each of these standards refer to exposure in the operator's breathing zone, not to background 'ambient' levels. The time weighted average standards will need to be varied downward for shifts of longer than eight hours.

When sampling is undertaken with portable air sampling equipment, the results are compared with the permissible or recommended exposure standards. Control measures are required if the samples are representative of normal activity in a work area and exceed the exposure standard. Exposure should be reduced to below the standard in order to improve conditions relating to health and safety in the work environment and to meet legal obligations.

Procedures for receipt and dispatch of hazardous substances

Documentation

Some of this documentation follows from what has been discussed earlier.

Incoming

Incoming hazardous substances must be accompanied by a delivery docket that adequately identifies them; by an MSDS if one has not been previously supplied; and by any additional documentation on toxicological testing of the substance. If the purchaser requests it, the OECD new chemical summary assessment report – if one has been prepared – must be supplied, in some jurisdictions.

The dangerous goods marking (including the label) or the placard, or the label (hazardous substances), is also in a sense 'documentation'. A dangerous goods marking indicates what the substance is by the description and UN Number and the class of dangerous good. In some countries the placard (augmented to be an emergency information panel) contains an emergency contact and the Hazchem Code for dealing with chemical accidents. Your local version of the UN Recommendations may also require the marking to fit in with the national OHS labelling requirements, such as those of WHMIS in Canada. So there may be information on recommended mode of use; there may be risk symbols; and there may be risk and safety phrases and first aid information.

The requirements cover both inner and outer packaging. The receiver of hazardous substances should ensure that the Emergency Services Manifest (ESM) is updated to reflect the goods newly arrived on site.

Outgoing

Outgoing documentation will vary to some extent with the status of the consignor. However, as seen above, a manufacturer, importer or supplier must provide:

- an MSDS – this may require appropriate testing of the substance/s
- any additional health and safety information on the substance they possess if requested (subject to trade secrecy provisions)
- properly labelled (hazardous substances) or marked (dangerous goods) packages

Hazardous substance management 377

- where the hazardous substances are dangerous goods, the documentation described in greater detail below – the UN Recommendations give an example of a suitable form as part of this documentation
- in some jurisdictions, an OECD new chemicals assessment summary report if one exists.

There are requirements for freight containers, intermediate bulk containers and bulk containers used for storage. A transport unit used for transport of a dangerous good must carry the correct dangerous goods placard. The UN Recommendations have details. A transport unit includes road transport tank and freight vehicles; railway transport tank and freight wagons; and multimodal containers and portable tanks. A consignment of dangerous goods must have the correct marking of the type the UN Recommendations require. This marking consists of adding the UN number to the bottom of the placard or adjacent to it on an orange background with black border. Consignments include solids, liquids or gases in tank transport units; packaged dangerous goods of a single commodity making up a full load; and certain radioactive materials. Markings may be required for both the primary risk of a substance and the subsidiary risk. There are certain exceptions to this given in the Recommendations. Generally, local dangerous goods regulations require consignors to keep records of the name and address of consignees and the date dangerous goods were consigned to them, together with the description and amount of goods consigned. This doesn't apply if the amount supplied is less than that requiring placarding (under the 'exemption limit') or the delivery is into a fuel tank of a vehicle or vessel. Records in some jurisdictions must be held for a set period in safe custody and the supplier must ensure that premises to which deliveries go, if they are required to have a dangerous goods licence, are licensed.

The driver of the vehicle carrying dangerous goods must ensure he or she carries a dangerous goods transport document . This is to show:

- consignor, consignee and date
- dangerous goods description – UN number preceded by 'UN'
- proper shipping name
- class or division and for Class 1 explosives the compatibility group
- the packing group if one is assigned
- generic descriptions require in addition the technical name/s
- for empty uncleaned containers, with 'Residue last contained' before the shipping name
- for containers of waste, 'Waste' before the proper shipping name
- for substances transported at elevated temperature (liquid greater than 100°C, solid greater than 240°C), the word 'Hot' if other words don't indicate that
- total quantity
- where the exceptions for limited quantities are being applied, the words 'Limited quantity', and similar appropriate words for goods in salvage packaging
- a statement of actions by the carrier for loading, stowage, transport, handling and unloading including precautions for heat dissipation (or a statement saying that they are not necessary)
- instructions on transport mode or route

- emergency arrangements
- certification of acceptability for transport, and that the goods are properly packaged marked and labelled and in proper condition for transport
- for sea transport (except for tanks) a container/vehicle packing certificate certifying that the operation has been done according to nine set conditions.

There are further specialized requirements for certain chemicals, infectious material and radioactive substances.

The emergency information may be in the shipping document, provided in an attached MSDS or in a separate document, but must be kept away from the dangerous goods packages.

Some jurisdictions require relevant 'emergency procedure guide' (EPG) cards for the goods carried. The marking including placarding previously mentioned must be on the vehicle. There is a need to ensure that emergency service manifests (ESMs) are updated to reflect the quantities remaining at the consignor's premises after despatch of goods.

Authorization procedures for internal hazardous substances movement

Clearly the exact way in which an organization arranges its documentation for movement of hazardous substances within a site will vary. It must be remembered too that the substances may move in bulk or package, or in pipelines. You will need to check with your own company to see what is done. However, some guidelines will help your assessment:

1. Is the documentation at the stores and purchasing end currently adequate? For example:
   - stock control records
   - checks that hazardous substances meet purchasing specifications and packaging requirements
   - chemical register
   - if hazardous substances are delivered by an external supplier to points within the site, the checks and documentation which exist before the substances are offloaded or discharged to bulk containers, e.g. underground tanks
   - approved filling station
   - documentary mechanism to update the emergency services manifest as receivals occur
   - documentation that any product received is labelled in accordance with OHS regulations
   - documentation that specified materials to deal with spills are available
   - documentation that substances are correctly assigned for storage as per segregation requirements
   - when packaged hazardous substances are checked out of the store, adequate authorization from the department concerned for the type, and quantity and for the person receiving them

Hazardous substance management

- if a new product, documentation checking the supply of an MSDS and the existence of an assessment report for that substance as per OHS regulations
- documentation of emergency procedures applying to movement on site, e.g. spills
- documentation of labelling and suitability of the secondary container if the substance is decanted
- procedures for opening packages at the use point or storage depot, closing packages, disposal of empty packages, and repacking of contents of damaged packaging
- procedures to ensure the correct form of discharge from delivery vehicle to bulk container (e.g. air padding) is used.

You are also referred to your relevant OHS regulations.

2. Is there documentation within the user department or at the point of use? For example:
   - authorization to obtain type and quantity of hazardous substances from stores (check against existing departmental stocks)
   - documentation from the department to the person maintaining the currency of the emergency services manifest
   - if a new product, documentation that a copy of MSDS has been received with the product
   - documentation of consultation on use of a new hazardous substance as per OHS regulations
   - if a substance is decanted, documentation of labelling of a secondary container if not for immediate consumption
   - documentation that users in the department have been trained to use the substance concerned, e.g. hydrofluoric acid
   - procedures for appropriate storage and use, and emergencies, within the department, including segregation.
3. Appropriate procedures for communication between security on the gate, purchasing staff and staff at the point of receival of chemicals delivered from outside, e.g. bulk delivery of anhydrous hydrofluoric acid.
4. Delivery by pipeline from storage point to use point, especially after maintenance of the line, requires:
   - documentation of compliance with finalization of lock-out/tag-out procedures, including removal of isolation disc if any, replacement and resetting of, e.g., pressure/flow control/relief valves, surge/ballast tanks, and temperature controls
   - notification to the point of use and an answer from them indicating they know that supply to the pipe is to be restored, or that substance is to flow
   - documentation that the correct pipeline colouring has been used on the replaced sections
   - that the altered pipeline has been given regulatory authority approval if required by dangerous goods regulations
   - documentation of examinations of the pipeline as per the relevant standard and checks based on an approved code of practice
   - documentation that pipelines do not contravene segregation requirements.

5. Delivery by pipeline from outside to a point on site, requires documentation of appropriate procedures to deal with restoration of supply after cut-off, and with disruptions to supply, e.g. natural gas to process equipment.

You are referred to relevant parts of the UN Recommendations, your dangerous goods regulations, and any guidance notes which accompany them.

Authorization procedures for external hazardous substances movement

The authorization procedures for movement from outside your site to your site are the responsibility of the firm consigning the hazardous substances. The same procedures will apply to your organization if it is sending materials offsite.

You will need to consider the following if you are consigning:

- Documentation of name and address of the person to whom dangerous goods are consigned and the date, as well as a description of the goods and the amount. Also to be included are the UN number, packaging group, mode of packaging, name of manufacturer or consignor or local agent. These must be kept in a safe place for two years.
- Documentation of types, quantities and dates to be consigned, drawn up in advance.
- Scheduling of correct transport, transport capacity and trained drivers, either in-house or on contract.
- Checks that the consignee holds the appropriate dangerous goods licence, if required.
- Checks that the vehicle carries the correct placard (or emergency information panel if required), a licensed/trained driver, personal protective equipment, emergency procedure guide, and manifest for goods carried.
- Checks that the driver and vehicle from an outside company are licensed for carrying dangerous goods.
- Checks that the packaging for transport conforms to dangerous goods requirements including suitability and marking.
- Checks on identification of separate compartments in a bulk container.
- Checks on written approval to use a road train (in some jurisdictions).
- Checks on the driver's knowledge of a designated or required route if one has been laid down.
- Checks that there is the correct distance between electrical fittings and loading and discharge points on the vehicle for flammable dangerous goods.
- Checks that the vehicle carries the correct fire extinguishers (number and type), and three (or another specified number of) double-sided reflective triangles in case of breakdown.

The above points are not all-embracing. Your legislation may have special requirements for packaging; for licensing bulk dangerous goods vehicles; and for vehicles carrying, e.g. flammable dangerous goods or explosives. You should check on these.

You are also advised to research the matter further. For example, unit loads (such as shrink-wrap pallets) and freight containers, and the quantities of packaged goods requiring vehicle marking, need to be addressed.

Compliance of workplace labelling and MSDS with accepted or mandated formats

Labelling (workplace, not dangerous goods as such)

The key issues to look for here are firstly a durable label which will not easily come off, run or fade (this also applies to dangerous goods labels).

Other important features which may be required are:

- signal words (in red)
- dangerous goods label (diamond) showing dangerous goods class if applicable.
- UN number
- ingredient description
- directions for use
- risk phrases
- safety phrases
- hazard pictogram
- environmental phrase
- first aid instructions
- name, address and emergency contact number for manufacturer or supplier
- expiry date
- reference to MSDS.

Check your local requirements.

MSDS

MSDS generally come in two formats, US or European Union. However, an EU format MSDS would be acceptable in the USA if it contains all the information required by the US Hazard Communication standard 1910.1200. In Australia one company also specializes in a user-friendly, one page electronic format including pictograms. The US and EU versions are compared in the Appendix to this chapter.

Inventory and reporting requirements

Most of the inventory and reporting requirements for hazardous substances have been covered above. However, there are some other requirements if your company is a manufacturer or importer of such substances. These can include notification of generic names to the OHS authority, and notification of use of carcinogenic substances at a workplace. You need to read the carcinogenic substances section of your OHS regulations.

There is a requirement to record and make available the results of monitoring of hazardous substances, and to do the same with health surveillance records, respecting medical confidentiality. OHS regulations may also lay down periods for keeping records of chemical risk assessment, monitoring, health surveillance, induction and training, and the chemical register, and for handing them to the OHS authority if an organization ceases to operate. The OHS authority must then keep them for a stated period, e.g. 30 years.

Your OHS authority may need to give approval and impose conditions on use of carcinogenic substances. Record keeping and requirements re information to employees may also be more stringent for such substances.

Specialist facilities for the management of hazardous substances

Decanting and repackaging of hazardous substances

Decanting of hazardous substances generally means pouring them from larger containers into smaller containers for convenient use. For example, dippers or buckets could be filled; test-tubes could be filled; containers such as bottles for sale of a product such as kerosene could be filled on a packing line; or large closed containers of, say, cleaning solvent in a store could simply be emptied into smaller closed containers for dispatch to individual areas such as workshops.

The requirements will vary depending on the situation but include:

- Ensuring that where flammable solvents in say a 200 L drum are being decanted into smaller vessels, earthing clips are used to prevent fire from static electrical discharge.
- Bunding of the decanting point to contain spillages, with washdown facilities.
- Absorbent or neutralizing material on hand for spills.
- Suitable fire-fighting equipment to be available.
- Proper ventilation facilities to capture vapours given off.
- Appropriate personal protective equipment available including, if necessary, anti-static clothing.
- Unless the secondary container is to be used immediately, properly labelling it. There is debate on the meaning of the word 'immediately' but it may mean up to the end of a shift.
- Use of suitable pumps rather than buckets for transfer. These must be cleaned between uses unless 'dedicated'.
- Signage to warn of fire hazards.
- Container-tipping devices and holding racks to reduce manual handling injuries.
- Clean secondary containers which do not contain residues of substances which can react with the material decanted – for example, a chlorine bucket at a swimming pool then used to receive hydrochloric acid for pH control. Repackaging requires correct choice of containers and packaging, and the markings or labelling on them.

Spill collection procedures and equipment

Spill collection procedures can include:

- bunding and drainage to a suitable sump as a precaution
- use of sand to limit the spread of a liquid

- use of absorbent material, e.g. attapulgite clay to soak up material
- use of foam to blanket a flammable solvent
- shovelling up absorbent plus liquid into suitable containers for disposal at an approved site
- provision and use of appropriate PPE, including antistatic clothing if needed
- oil/water separators.

The MSDS (or Hazchem Code if applicable) should be used to decide on the approach to use – dilute and wash-down or contain; whether to use jets, fog, foam or dry agent to prevent fire or extinguish fire; whether to evacuate the area; and the type of PPE to use. More specific requirements for certain chemicals should be found in the MSDS.

Decontamination of empty containers

Firstly, some containers should just be disposed of at an approved site with no attempt to decontaminate. You will need to seek advice on this.

Where decontamination is an option, the decontamination requires treatment of the interior of the container with a suitable neutralizing solution (consult the manufacturer, importer or supplier), or solvent; collection of the neutralizing solution or solvent; and disposal of the washings suitably. They may go to a sump for treatment before entering a sewer, or be collected and discharged at an approved disposal site. For pesticide containers the triple rinse technique has been abandoned in favour of use of a recirculating pump.

Personnel undertaking decontamination need proper training; they should wear appropriate PPE and should ensure suitable fire precautions are taken. More specific information should be sought from MSDS and suppliers, importers or manufacturers. Accidents have occurred, for example, when people have cut containers, which have held isocyanate ingredients for polyurethane foam manufacture, fuel or bitumen (asphalt), with an oxyacetylene torch. Some containers such as gas cylinders are, of course, recycled through the suppliers.

Bulk transfer hazards

Hazards associated with bulk transfer will vary with the nature of the substance. For example, flammable liquids and LPG present the problem of possible fire, both from more obvious sources of ignition, but also in the case of flammable liquids from static electricity or static electrical build-up and discharge. Accidents have occurred when personnel on top of a bulk tank were engulfed in a cloud of vapour, as the ullage (the gas space above a liquid) was discharged rapidly through a top vent as the tank filled. So, appropriate venting as a vessel fills is important.

Static build-up due to the flow of dry flammable powders such as epoxy paints can also cause fires. Explosions can occur where excess powder paint is carried away in local exhaust ventilation.

Personnel who may be adequately protected for a small splash may not be adequately clad if there is a large spill during bulk transfer of, for example, anhydrous hydrofluoric acid, sulfuric acid, caustic soda solution, or sodium cyanide. Bulk transfer can involve transfer under some pressure, either gravitational or otherwise, so it is important that this pressure is monitored and in accordance with design, or hoses may burst. A major incident in the Italian port of Bari in 1990 involved a fire during discharge of propylene through a ship to shore line from a ship, the *Val Rosandra*. The response was interesting. It was safer to continue to allow the fire to burn rather than disconnect the ship. See Mathews in Further Reading at the end of this chapter.

It is important to refer to specific requirements in regulations for goods such as LPG, chlorine, flammable and combustible liquids, and cyanides, as they cannot all be detailed here. However, it may involve use of gas detectors; access to shut-off valves; ventilation; drainage; prohibited spaces under floors; and use of air pressure for certain forms of transfer.

Other case histories

Some instructive case histories are included here:

- A pipeline carrying natural gas underground from one part of Mexico City to another led to an explosion ripping up a street for six blocks because the natural gas leaked into the sewer running under the street.
- Leakage of natural gas from a pipeline running alongside an electric railway in Siberia led to a big explosion when an electric locomotive ignited the gas cloud.
- An LPG tanker leak led to an explosion in Spain near a camping area causing major loss of life and injury.
- The Bhopal incident in India in the mid-1980s resulted from inappropriate controls and venting during tank filling involving methyl isocyanate. Nearby shanty town dwellers experienced major lung damage, or death.
- A road tanker roll-over on the Nullarbor Plain in Australia fortunately discharged anhydrous hydrofluoric acid into limestone, a natural neutralizer.
- The Piper Alpha disaster in the North Sea in the mid-1980s, involving over 150 dead, was worsened when natural gas continued to flow from another gas platform to Piper Alpha (which was the junction or collecting platform) because no-one told the other platform to cut the flow; the emergency plans had apparently not included a multi-platform approach.
- Eroded O-rings on the Challenger spacecraft at Cape Kennedy apparently led to leakage of hydrogen during the transfer of liquid hydrogen to the rocket motors. The hydrogen then ignited in the wrong place.
- The explosion at the Longford (Victoria) natural gas plant which supplied the whole of Melbourne, cost two workers' lives and affected the state for weeks. Co-location of pipes led to damage to the alternative supply route.

Signage required for storage, handling and transport of hazardous substances, including dangerous goods

Interpreting signs for hazardous substances

Using the word 'signs' in a fairly broad way, these include:

- EIPs – emergency information panels (in some jurisdictions) or placards for vehicles carrying dangerous goods
- placarding for front entrances to dangerous goods premises, and entrances to bulk and package stores
- labelling of pipelines and their coded colouring
- safety signs, as noted earlier, of the type in prescribed standards, with differentiating colours, viz:
  - red circle with diagonal bar – prohibited activity
  - yellow triangle – caution
  - blue circle – PPE required
  - green rectangle – emergency safety equipment or recommended exits or movement to safe place
  - red, black and white danger signs
  - red and white fire safety signs
- the *marking* (dangerous goods) or *labelling* (hazardous substances) on containers of chemical products or formulations.

You may wish to refer to standards on safety signs for the occupational environment, and to standards on identification of the contents of piping, conduits and ducts, and safety signs suppliers' catalogues. Guidance notes from dangerous goods authorities can also provide additional information.

Determining required signage for a variety of hazardous substances situations

Situations may include:

- tanks and bulk stores
- hazardous substances contained in enclosed systems
- decanted substances.

Appropriate Hazchem or similar information, e.g. NFPA, is part of this.

Tanks and bulk stores

If the hazardous substance is a dangerous good, then the requirements for placarding of tanks and bulk stores are generally set out in your relevant local legislation.

Enclosed systems
Generally, OHS regulations on hazardous substances require that substances in enclosed systems, such as a pipe or piping system, or process or reactor vessel, have labels on the system, use the correct colour code, and indicate the direction of flow.

Decanted substances
For these, where they are not to be consumed immediately, the secondary container should carry the brand name, trade name, code name or code number specified by the supplier of the hazardous substance, and the risk and safety phrases which apply to the hazardous substance.

Hazchem
Where Hazchem is used, the required Hazchem information can be determined by referring to the list of dangerous goods in the local dangerous goods code. An explanation of the Hazchem Code is given earlier in this chapter. There is also a table in such codes to decide the required Hazchem for stores containing a number of dangerous goods.

Reference is also made to a US DOT website with similar information (see Further Reading at the end of this chapter).

Controls to minimize employee exposure

Management techniques for control

A variety of workplace situations are possible, including:

- cutting processes which produce dust
- products formed during chemical processes
- emissions from power sources during operation
- fumes from welding and heat cutting operations
- fugitive emissions (unintended leaks)
- chemicals used in manufacture.

Cutting processes
These processes can involve wood, stone or metal, or other formed products such as polyurethane foam. Wood cutting and machining usually involves the use of effective ventilation on the machines which captures nearly all the dust at source. Wood dust can produce constriction of the lung vessels, and eye and nasal irritation. Some wood dust has led to nasal cancer.

Stone, brick, and concrete cutting or chasing of concrete to install conduits can utilize water sprays on cutting wheels as a control technique. If this is not possible, effective fixed or portable ventilation systems can be connected to the tool, which include a suitable movable spring loaded hood.

Polyurethane foam or polystyrene foam may be cut with a hot wire, so once again effective capture at source by local exhaust ventilation is generally possible.

Grinding operations have, in the past, involved lung hazards (hard metal disease) from the abrasives in the wheel. Again, effective fixed or portable collection devices exist for grinders.

Products formed during chemical processes

Enclosed processes are to be preferred because collection and absorption of any emitted gases and vapours is more easily achieved. However, open or semi-open tanks are used for some processes such as metal plating or aluminium anodizing. Effective control of emissions can be achieved with ventilation, such as push–pull over the surface of the liquid or slot extraction round the lip of the tank.

Emissions from power sources

This covers a number of possibilities. For example, motor-powered machinery operating underground; fork-lifts carrying containers in a RORO (roll on roll off) vessel; and fork-lifts operating in cool or freezer storage. It could also include ozone emissions from electric motors, cars in a multilevel car park, and stack emissions from coal, gas, or oil fired electricity generation.

Controls on underground diesel machinery generally involve correct tuning of the motors with frequent checks and adequate mine airflows in the operational areas. There has been experimental use of liquid traps on diesel exhausts. The aim is to keep emissions of NOx, aldehydes and particulate containing polynuclear aromatic hydrocarbons down. Carbon monoxide (CO) is a relatively minor problem with diesel motors. However, for motor vehicles generally in an enclosed car park, CO is the major problem, and adequate mechanical ventilation must be provided. Emissions from petrol and LPG powered motor vehicles in the general environment are partly controlled by catalytic converters. They are also controlled by recycling crankcase gases, and by more careful control of fuel supply and fuel burning through better engine design and computerized fuel injection. Flexible exhaust collection hoods should be provided in vehicle testing workshops. Fork-lifts operating on LPG do produce a cleaner exhaust, but this still contains CO. So adequate ventilation in warehouses, cool stores, freezers and ships' holds is essential.

Ozone emissions from electric motors, such as those on underground trains, are controlled using a good flow of supplied fresh air, partly utilizing the trains to pull air in and move it about.

One source of fatalities has been poor exhaust piping and/or piping location on a *hookah* used to supply air to a diver, so that ingoing air was contaminated with CO.

Stack emissions from large power generating sources usually rely on height and placement to avoid problems, but as the acid rain issue in Europe shows, collection of the emissions of sulfur dioxide from sulfur-containing fuels needs more attention.

Welding and heat cutting

The composition of welding fumes can be partly controlled by the settings used for welding, providing a suitable weld still results. For hand-held welding equipment a movable hood, as part of either a portable or fixed installation, can be used. In the area of PPE, helmets/visors which rely on either supplied air or filtered air from a belt-mounted motor and filter can be used. For fixed welding installations, down-draught tables with a grating surface can be used. Flame cutting can be a dirty operation and cutting over water tables is used in some industries regularly cutting aluminium alloy or steel plating. Cleaning of coated or dirty surfaces before cutting will reduce emissions. Laser cutting of metal or plastic is usually in a machine with adequate built-in local exhaust ventilation.

Fugitive emissions

These emissions are those which may escape from enclosed or semi-enclosed systems. They generally occur, despite design, due to wear; lack of maintenance; thermal expansion and contraction; worn expansion collars or sound transmission control collars; and corrosion.

Some of the causes above suggest the controls to be used, but in addition there is a need to ensure adequate natural or mechanical ventilation, and if the emission is particularly hazardous, automatic detection systems which trigger an alarm. This is used in chlorine production plants, for example. pH control ensures fugitive emissions of hydrogen cyanide in carbon in pulp gold plants are minimized.

Chemicals used in manufacture

Such chemicals may be harmful by skin contact or skin absorption or may be inhaled. Some of the risk arises from the way the chemical is used or the way it changes during use. It may be purchased as a gas, liquid or solid. In solid form it may be a dust. In liquid form it may give off gas, vapour, mist or fog depending on how it is handled and the temperature changes. A solid may be ground to a dust (or a semi-solid may be repeatedly stretched to render it plastic) and get hot in the process, giving off vapour, e.g. latex or rubber being calendered so it becomes plastic and can be made thinner and cut. A resin may be powdered to make it easier to dissolve.

Controls can be arranged at the purchasing end by choosing the substance to be supplied and its quality, or by specifying how the substance is to be supplied, i.e. pre-ground, and the container size, reducing the need to decant – or supplied bulk and fed to the production area through piping. However, reticulation through piping can bring its own problems if checks aren't adequate; for example, supply of nitrous oxide must be kept separate from supply of oxygen to hospital bedsides. A fatality occurred because a nitrogen cylinder was connected to the breathing air supply in a factory. Controls can be introduced for the decanting, dispensing, addition, or mixing of the substances. They can occur during the reaction phase if there is one. There can be controls during purification, refining and packing off, e.g. the bagging of cement. And there is a need for controls when disposing of waste materials and cleaning emissions collection devices, such as bag houses, and disposal facilities.

No one particular method of control can be suggested here. The assessment should involve a full look at the process and application of the hierarchy of controls. For carcinogens such as MOCA used in polyurethane rubber making, complete changing rooms and showers with dirty and clean areas are used, and all clothing is changed when entering and exiting. The probability of skin absorption of MOCA due to contamination of surfaces is high if precautions are not suitable.

Engineering controls

Some of the engineering controls have been detailed above. However, others include:

- use of process control hardware and software to maintain correct temperature, pressure, flow, viscosity and liquid levels, and use of overflow collection
- placement of the operator in an air-supplied booth, e.g. while cleaning rust from rolls of steel sheet in acid baths prior to further processing. The alternative to this is to enclose the process which gives off dust or vapours, for example, and having a required procedure if entry to such an enclosure is required. Such a method is used for some abrasive cleaning of metals prior to coating or painting.

Personal protective equipment

Personal protective equipment (PPE) for chemical work will include suitable:

- headwear
- face protection
- body protection
- hand protection
- foot protection
- respiratory protection.

Overalls may be washed frequently, where appropriate (but not by being taken home) or disposable overalls may be used. Full impervious body suits are available for handling particular chemicals. For some situations aprons are appropriate. Gloves vary, with PVC, reinforced PVC, Viton and nitrile gloves offering different levels of protection for various solvents and other chemicals such as pesticides.

Once contaminated, some gloves and footwear can become the main source of exposure. For example, for farmers in air conditioned tractor cabs designed to filter out pesticide mist, their principal source of exposure in some cases came from the splash on their overalls which occurred while pouring from a drum and mixing. For some flammable vapours such as ether, care is needed to use clothing which minimizes static electricity. Clothing can also be chosen to minimize the retention of dust and to allow personnel to have a first shower while still wearing the clothing.

All PPE must be carefully selected prior to purchase, properly stored, issued for the correct purpose, maintained (e.g. air pressure in self-contained breathing apparatus (SCBA) air cylinders), disposed of or repaired, and disposable components replaced. Proper fitting and training are required. For example, the use of self-rescuers for mine fires. Supplied air must be checked to see that it complies with nationally accepted specifications for air quality.

For further information you are referred to standards on:

- eye protectors for industrial applications
- selection, use and maintenance of respiratory protective devices
- occupational protective footwear
- clothing for protection against hazardous chemicals
- industrial safety gloves and mittens.

Role of emergency personnel

Response needs

Workplaces dealing with hazardous substances need to consider the issue of emergency response from four points of view:

- minimization of injury to their own personnel
- minimization of injury to the public
- protection of the environment
- minimization of damage to the premises and equipment of the organization.

The hazards themselves can perhaps be best looked at from two points of view as far as emergencies are concerned:

- fire and explosion
- toxicity.

Note: Both can pose problems inside and outside the workplace. The risk assessment process should consider the profile of hazardous substances the organization purchases, uses, uses as reaction intermediates, produces and despatches, the nature of operations in which they are used, and the methods of waste disposal. Organizations despatching substances with particular hazards may maintain a responsibility to the public outside their own workplace, e.g. the extended safety programmes run by suppliers of liquefied chlorine or the 'stewardship' programmes run by some pesticide suppliers. Some larger companies offer 24-hour emergency response assistance for transport spills involving their own chemicals or those of companies contracted by them to provide emergency information and/or response.

Regulations such as the UK Control of Major Accident Hazard (COMAH) regulations mandate that for major hazard facilities the public must be informed of the risks in advance. US right to know legislation, for example, does a similar thing.

Internal emergency response capabilities

The organization must:

- Be able to detect early and respond appropriately to fire, remembering that stored hazardous substances can form a serious cocktail when subjected to fire, or mixed when containers shatter.
- Have the materials on hand to absorb or neutralize expected spills, e.g. stores of ferrous sulphate in areas where sodium cyanide is transported, and ozone injectors on a stream where cyanide contamination from transport spills in a watershed could occur.
- Have on site the first aid equipment and trained personnel to deal with the expected range of incidents involving hazardous substances, e.g. refrigerated calcium gluconate gel for hydrofluoric acid.
- Have accessible and unobstructed safety showers and eyewash facilities if necessary.
- Have available any necessary emergency respiratory protection such as SCBA, fully charged, ready to use by trained personnel.

- Have available, with any current tags if appropriate, fire-fighting equipment ready to use and ensure personnel are trained to use it.
- Have properly planned emergency response procedures, communications and responsibilities for personnel and ensure that these have been documented and tried in drills.
- Have spill collection, disposal and decontamination procedures planned and documented, with any necessary approvals obtained from public authorities.
- Have arrangements in place with medical services and hospitals for an emergency, indicating any special requirements for particular hazardous substances.
- Make arrangements to prevent spread of contamination of waterways from, for example, oil due to flooding or cyanide from tailings dams due to heavy rain.

It is important to identify for what emergencies and at what stage, extra support is sought from partner companies in a group response scheme, and from public authorities.

Emergency contact numbers

The list you compile will need to be kept up to date. It will need to include:

- police
- fire services
- dangerous goods authority
- OHS authority
- environmental protection authority
- local, borough, municipal or county government authority
- ambulance/paramedic services
- suppliers of hazardous substances purchased
- suppliers of extra fire and emergency equipment, and neutralizing or absorbing agents
- partners in any group response scheme
- media contacts
- weather forecasters
- suppliers of equipment to limit spread of contamination on waterways
- medical practitioners
- hospitals.

Emergency evacuation plans

Emergency evacuation plans may involve four stages:

- evacuation of non-essential personnel
- evacuation of essential personnel
- evacuation of emergency response personnel
- evacuation of the surrounding public.

Some may take place in parallel depending on circumstances.

The need for emergency evacuation may be internally generated or the hazardous substance emergency could arise from an external threat such as a bomb, sabotage or

breach of site security by unauthorized personnel. The evacuation plan must consider:
- the decision-making process which identifies when, where and at what stage evacuation is to occur
- the personnel in charge of evacuation
- the routes of evacuation and muster points
- the method of accounting for personnel
- the need for identified personnel to adjust plant controls to acceptable or safe mode, or to go through shut-down procedures, if possible, before evacuating
- the actions to be taken for further evacuation if plant cannot be left safe
- the point at which trained emergency personnel within the identified 'combat perimeter' are withdrawn to a new perimeter
- the structure for transition of command from company personnel to public authorities
- the decision-making process in relation to the risk to the surrounding public – this really applies to larger plants, those which may spread substances of great concern, or those which may give rise to a major explosion or cloud of irritating, choking or toxic smoke from fire – and which may have been built prior to the application of better planning requirements. Procedures must be agreed on in relation to notifying the public, securing their property and providing them with shelter, food and medical attention.

Some companies will need to consider the plans from the perspective of a variable public if they may be involved in a transport spill or fire along a distribution route. This is more difficult than the consultative arrangements which some companies have with a relatively fixed public near their premises.

Major hazard facilities

The reader is referred to appropriate books on preparing a safety case. See for example DISR (2000) in Further Reading at the end of this chapter.

Useful information on toxicity and confined spaces

Acute LD 50s of some chemical compounds:

Table 9.4 is useful to relate the toxicity (in LD_{50} units) of workplace substances to save well known substances.

Table 9.4

Compound	Acute LD 50 (mg/kg)
Ethyl alcohol	10 000
Sodium chloride	4000
Morphine sulphate	900
Strychnine sulphate	2
Nicotine	1
Tetrodotoxin (blowfish)	0.1
Dioxin (TCDD)	0.001
Botulinum toxin (botulism)	0.00001

Table 9.5 Oxygen levels in air and their effects

Oxygen level (%)	Effect
21	Normal
17	Normal level in exhaled air, OK for expired air resuscitation
12–16	Pulse increases, muscular coordination disturbed. Normal flames go out
10–14	Emotional disturbance, fatigue
6–10	Nausea, vomiting, loss of consciousness, collapse
Less than 6%	Gasping and collapse, heart stops

Effects of reduced oxygen levels

These can occur in closed spaces where rusting, etc., has consumed oxygen or where an inert gas (nitrogen, argon) has diluted the oxygen level. See Table 9.5.

Further reading

ACGIH (current biennial edition). *Industrial Ventilation*. Cincinnati: ACGIH, for illustrations on industrial ventilation solutions.

Bretherick, L. (1999). *Bretherick's Handbook of Reactive Chemical Hazards*. 6th edn. London: Butterworth-Heinemann.

Bryson, B. (2003). *A Short History of Nearly Everything*. London, Doubleday. pp. 134–5, (tetraethyllead manufacture).

Carson, P. and Mumford, C.J. (2002). *Hazardous Chemicals Handbook*. 2nd edn. Oxford, Butterworth-Heinemann.

CCH (looseleaf update). *Hands On Guide – Risk Management*. Sydney: CCH Australia.

(DISR) Department of Industry Science and Resources. (2000). *Guidelines for Preparation and Submission of Safety Cases*. Canberra: DISR.

European Commission. (1991). *Directive 91/326/EEC*. Brussels, EEC.

European Commission. (1967/1996). *Directive 67/548/EEC as updated by 96/54/EEC*. Brussels, EEC.

Garnett, P.J. (ed.). (1996). *Foundations of Chemistry*. 2nd edn. Melbourne: Longman Cheshire.

Hathaway, J. et al. (1996). *Chemical Hazards of the Workplace*. 4th edn. London: Van Nostrand Reinhold-Sax.

HMSO. (1999). *COMAH Regulations*. London: HMSO.

ILO. (1995). *Safety, Health and Welfare on Construction Sites – A Training Manual*. Geneva: ILO.

Keenan, R. (1999). Confined Space, Serial Killer. *Safeguard (New Zealand)*, March/April, pp. 18–25.

Lewis, R.J. Sr. (1995). *Sax's Dangerous Properties of Industrial Materials*. 9th edn. New York: Van Nostrand Reinhold.

Mathews, C. (1992). Inferno at the Pier. *Reader's Digest*, **140(839)**, 123–30.

O'Leary, D. (2001). Implementing COMAH in a Multisite Organisation. *The Safety and Health Practitioner*, **19(1)**, 31–3.

Tweeddale, M. (2003). *Managing Risk and Reliability of Process Plants*. Oxford: Butterworth-Heinemann.

UN. (2001). *Recommendations on the Transport of Dangerous Goods*. 12th Rev Edn. Geneva and New York: UN.

UN. (2001). *Recommendations on the Transport of Dangerous Goods*. 12th Rev Edn. *Manual of Tests and Criteria*. Geneva and New York: UN.

Winder, C. (1993). Are MSDS Good Enough? *Health and Safety at Work*, **11(2)**, 5–8.

Worksafe Australia. (1990). *Guidance Note for Emergency Service Manifests*. (NOHSC: 3010). Canberra: AGPS.

Worksafe Australia. (1996). *National Standard and National Code of Practice for the Control of Major Hazard Facilities*. (NOHSC: 1014, 2016) Canberra: AGPS.

Relevant acts, regulations, codes, standards, guidance notes or rules on dangerous goods and on control of hazardous substances used at work.

Safety signs suppliers' catalogues.

See also ecb.jrc.it/classification-labelling (no www) for classifying chemicals under the EU system, and hazmat.dot.gov/abhmis.htm (no www) for the US system on safe transportation of hazardous materials.

Activities

1. Describe the authorization procedures (if any) required for moving hazardous substances inside a suitable selected workplace, and outside the workplace.
2. What equipment is a road vehicle for dangerous goods transport expected to carry?
3. Draw a label in the format your jurisdiction requires for Alkaglo metal cleaner containing 20% sodium hydroxide solution, manufactured by Shine Nominees of 200 Traffic Road, Someplace. Phone no. XXX 234 YYY.
4. Construct a job safety analysis (JSA) for decanting mineral turps from a 100 litre drum into 750 ml polycarbonate bottles. Write a safe job procedure.
5. Explain what Hazchem '2 WE' means, or in the USA explain a National Fire Protection Association diamond showing clockwise from the top 3, 2, 0, 3*.
6. Select one hazardous substance used in a workplace and carry out an assessment of the risks associated with its use. What controls would you put in place?
7. In a suitable selected workplace find out and write down the arrangements to respond to hazardous substance emergencies.

Appendix 9.1 – European Union and US material safety data sheet requirements

EU SDS Directive 98/98/EC	US MSDS 29 CFR 1910.1200 para (g)
1. Identification of the substance/preparation and of the company/undertaking 1.1. Identification of the substance or preparation:	**(g)(2)(i)** The identity used on the label, and, except as provided for in paragraph (i) of this section on trade secrets: **(g)(2)(i)(A)** If the hazardous chemical is a single substance, its chemical and common name(s); – definite nil entry in any MSDS section if no information available – complex mixtures

1.2. Company/undertaking identification:
     – identification of manufacturer, importer or distributor
     – full address and telephone number of this person
   1.3. Emergency telephone number of the company and/or official advisory body
	(g)(2)(xii)
	The name, address and telephone number of the chemical manufacturer, importer, employer or other responsible party

2. Composition/information on ingredients
   – thresholds of constituents as per 67/548/EEC
   – those with recognized exposure limit values
   – the classification given in the form of the symbols and R (risk) phrases
   – confidential ingredients
	(g)(2)(i)(A)
	– details of mixtures and thresholds of constituents

3. Hazards identification – critical hazards to man and the environment	**(g)(2)(iv)** The health hazards of the hazardous chemical, including signs and symptoms

EU SDS Directive 98/98/EC	US MSDS 29 CFR 1910.1200 para (g)
— most important adverse human health effects and symptoms uses and possible misuses	of exposure, and any medical conditions which are generally recognized as being aggravated by exposure to the chemical **(g)(2)(v)** The primary route(s) of entry
4. First aid measures — whether delayed effects can be expected after exposure — any specific and immediate treatment required to be available at the workplace	**(g)(2)(x)** Emergency and first aid procedures
5. Fire-fighting measures	**(g)(2)(x)** Emergency and first aid procedures
6. Accidental release measures — personal precautions — environmental — methods for cleaning up	**(g)(2)(viii)** ... procedures for clean-up of spills and leaks
7. Handling and storage — including any special requirements such as the type of material used in the packaging/containers	**(g)(2)(viii)** Any generally applicable precautions for safe handling and use — including protective measures during repair and maintenance of contaminated equipment
8. Exposure controls/personal protection — any specific control parameters such as limit values or biological standards – recommended monitoring — type and quality of personal protective equipment	**(g)(2)(ix)** Any generally applicable control measures **(g)(2)(vi)** The OSHA permissible exposure limit, ACGIH Threshold Limit Value, and any other exposure limit
9. Physical and chemical properties	**(g)(2)(ii)** Physical and chemical characteristics of the hazardous chemical (such as vapour pressure, flash point)

EU SDS Directive 98/98/EC	US MSDS 29 CFR 1910.1200 para (g)
	(g)(2)(iii) The physical hazards of the hazardous chemical, including the potential for fire, explosion, and reactivity

10. Stability and reactivity
    – including specifically the need for and the presence of stabilizers, the possibility of a hazardous exothermic reaction, safety significance, if any, of a change in physical appearance of the substance or preparation, hazardous decomposition products, if any, formed upon contact with water
    – possibility of degradation to unstable products

11. Toxicological information
    – including known delayed and immediate effects and also chronic effects from short- and long-term exposure: for example sensitization, carcinogenicity, mutagenicity and reproductive toxicity including teratogenicity and narcosis

12. Ecological information

13. Disposal considerations

14. Transport information
    – including additional information provided for by the UN Recommendations and other international codes, e.g. IMDG code

15. Regulatory information
    – including labelling and limit values

EU SDS Directive 98/98/EC	US MSDS 29 CFR 1910.1200 para (g)
16. Other information which might be of importance for safety and health and for the protection of the environment, for example training advice — the date of issue of the data sheet	**(g)(2)(vii)** Whether the hazardous chemical is listed in the National Toxicology Program (NTP) Annual Report on Carcinogens or has been found to be a potential carcinogen in the International Agency for Research on Cancer (IARC) Monographs, or by OSHA **(g)(5)** — information accurately reflects the scientific evidence — new information to be added to the material safety data sheet within three months **(g)(2)(xi)** The date of preparation of the material safety data sheet or the last change to it
Also give the date of issue of the data sheet, if not stated elsewhere	

10

Work environment

WORKPLACE EXAMPLE

Two men in Australia were sent to repair a section of gas pipeline. To isolate the section of pipe, ice-plugging is used which requires the use of liquid nitrogen. This is fed into a collar on the pipe. The pipe was apparently in a trench at the bottom of a hill. It was early morning.

Fellow workers found the men unconscious in the trench. One worker died, while the other required intensive care.

Evidence pointed to lack of implementation of procedures, and disagreements between safety personnel and line management.

A job safety analysis had been carried out, and as a result a supervisor was apparently required to be on-site during the operation, but problems arose in implementing this. In addition the worker who recovered said he had had no induction training. Finally, the safety superintendent identified differences between the main contractor and a subcontractor over implementation of the safety management system.

Major characteristics of noise

Introduction

This chapter looks at sampling or monitoring what are called physical and chemical stressors which may be found in the work environment in order to evaluate the magnitude or size of the risk they present. Physical stressors include:

- temperature, humidity and airflow
- ultraviolet, infra-red, ionizing, radiofrequency, microwave and laser radiation
- noise
- vibration.

Chemical hazards in the air include dusts, fibres, gases, vapours and fumes.

Sound waves – difference from other waves

Vibrations in solid or semi-solid material, or movement in liquid, can produce a series of minute changes in the pressure of the air which is in contact with the material. When this series of small air pressure variations reaches the ear and triggers a message in the auditory nerve, the brain registers this as sound. Sound waves can also travel as vibrations through solid materials such as steel, and liquid materials such as water. We can hear the sound in steel if we put part of our skull in contact with the steel. This is known as bone conduction of the sound. Some of the pressure waves reaching us through the air can also travel to the inner ear by bone conduction rather than up the ear canal.

Sound waves are called 'longitudinal' waves, and are unlike the waves in water or light waves. The molecules of water in a water wave move up and down at right angles to the direction that the wave is travelling. The molecules of air in a sound wave move backwards and forwards in the line of travel of the sound wave. However, for convenience, we can consider a sound wave as a wave with side-to-side vibration (a 'transverse' wave).

Describing sound

The more pressure variation (amplitude), the higher the intensity of the sound. The more waves generated per second (Hertz) the higher the pitch or 'frequency'. The higher the pitch or frequency the smaller the length (wavelength) of each wave. In air at sea level the velocity of the sound wave is around 330 metres per second. The velocity of the sound equals frequency times the wavelength. The 'period' is the time taken for one individual complete sound wave to be generated. For example, if the frequency is 8 Hertz (8 cycles per second), the period is one eighth of a second. A baby can hear from around 10 Hz to about 20 000 Hz (20 kHz); an adult with good hearing can hear from 10 Hz to 8000 Hz.

Sound intensity, loudness and sound pressure level

The intensity of sound is a measure of the actual energy in the sound. Research shows that the amount of damage done to the ear by excessive noise depends on the amount of sound energy the ear receives. Loudness is a little different. It is our own idea of how loud one sound is compared to another, and we think differently about high and low pitched sounds when it comes to deciding on loudness. Noise is unwanted sound. Excessive doses of the sound energy in noise permanently damage the hair cells in the cochlea of the inner ear.

Sound pressure level

Pressure differences in the surrounding air make up sound. Sound pressure is measured in pascals. At the threshold of hearing, the pressure difference between the compressions and rarefactions (opposite of compressions) is very small – 0.000 02 Pascal. But the level of noise on the tarmac of a jet aircraft gives a pressure difference of 20 Pascal. This is a very large range of sound pressure.

Decibel scale

To obtain more manageable numbers, when we measure noise levels in a workplace we compare the pressure of the noise coming from the noise source with the pressure of the threshold at which normal people can hear sound. The result is obtained by using logarithms and multiplying by 20. Logarithms are used to provide more manageable numbers and the result is given in a scale created by the above calculation. The noise is measured in decibels (dB), i.e. one tenth of a Bel (named after Alexander Graham Bell, inventor of the telephone). Threshold noise is then 0 dB while the aircraft above measures 120 dB.

Using a logarithmic scale lessens the range required to express sound energy measurement; therefore, a small increase in decibels represents a large increase in sound energy. Each increase of 3 dB represents a doubling of sound energy; an increase of 10 dB represents a tenfold increase and 20 dB increase represents a 100 fold increase in sound energy – that is, 100 times more damaging to the ear. Therefore, 93 dB noise contains twice as much sound energy as 90 dB noise of the same frequency. This can do twice as much damage to your hearing over the same period of time. If you have two noise sources each producing 90 dB in a particular part of a workplace, together they will give 93 dB. Also 90 dB of high frequency noise contains more sound energy and can do more damage to the ear than 90 dB of low frequency noise. Four hours exposure to 88 dB noise delivers the same noise energy to the ear as eight hours at 85 dB (at the same frequency).

Adding sound pressure levels

It follows that adding two sound pressure levels is different from normal addition. For instance, as noted above, two machines in a workplace individually produce 90 dB at a certain spot; however, if they are switched on together they will give 93 dB. Adding two equal levels is easy – 92 and 92 give 95; 97 and 97 give 100. For levels like 91 and 94, calculation tables should be used.

Effect of distance from the source on sound

Like other waves in a free field (that is, a space with no reflections or reverberations) with the waves spreading in all directions, the intensity of the wave depends on how far it is from the source, the spot where the sound is produced. We say that the intensity decreases with the square of the distance. To explain this, the intensity twenty metres from a source will be a quarter (not a half) of that at 10 metres from a source. The reason: 20 m is two times 10 m, but the square of two is two times two, or four. So, as we said, the intensity is one quarter (1/4 or 1 over 4).

When sound pressure level is halved, the number of dB decreases by 3. This means that if the sound pressure level at 10 m is 96 dB, then at 20 m it will be one quarter of the sound pressure level, which is 90 dB; that is, half of 96 dB is 93 dB and a quarter (half of a half) of 96 dB is 90 dB.

Although sound pressure level drops with distance from the source (this can be one way of reducing noise exposure), the drop or sound pressure level reduction will not be as great in a shed or workshop with reflecting roof, floor and/or walls.

Nature of sound from different sources

Different noise sources produce different noise spectra (spectrums) or 'frequency distributions'. This means that if we measure the dB level at each of the frequencies of 63, 125, 500, 1000, 2000, 4000, 6000 and 8000 Hz, then the dB level at each frequency compared to the others will be different for different sounds. For instance, a jet turbine, with many blades cutting the air at high speeds, will have high dB levels in the higher frequencies. A calendering machine, with slow turning rollers driven by gears, will have highest dB levels in the low frequency range. The nature of noise also varies. A pneumatic hammer or chipper gives staccato 'impulse' noise, a ventilation fan may give a steady hum (steady state noise), while an aluminium extrusion docking cutter gives a series of intermittent, high-pitched, loud noises.

Control of excessive noise

Directivity, transmission and reflection of noise

In controlling noise it is worth understanding the concept of directivity. Noise is not usually generated with the same intensity in all directions from a machine. It will depend on the moving parts or fluid involved, the shape, and covering of the machine. For example, a directivity factor of two in a particular direction means that the sound pressure level at 1 m (for example) from the machine in that direction will be twice the level at 1 m that would be found if the machine gave out noise equally in all directions.

Directivity is also influenced by placement of a machine in a room or space with sound reflective walls and floor. Generally, the result is that a hard floor (usually unavoidable) doubles the sound energy upwards; therefore, against a floor/wall junction the sound energy is four times that from the machine alone, and eight times higher if the machine is in a corner. A machine measuring 90 dB in an anechoic chamber (a special room with no echoes) will give approximately 93, 96 or 99 dB for each of the three locations above, so machine placement can affect noise levels. A reverberant space will alter the effect described.

Transmission of noise can occur directly or by 'flanking'. For example, a wall between rooms may reduce noise but the noise can reach a ceiling cavity or the hollow under a floor and bypass the wall. A partition not reaching the ceiling can reduce noise but noise can bypass it by 'diffraction' over the top edge, and by reflection from the ceiling. Diffraction is more pronounced with low frequency noise. So a high brick wall shielding a house from traffic noise will screen out high-pitched noise more effectively.

If we consider the partition itself, or the ceiling or walls, we can reduce noise to some degree by having a rough, rather than flat, surface where outgoing reflections interfere with each other and the incoming waves. If the surface is covered in porous material, the noise energy is lost in the many tiny channels in the fabric.

Sound energy which reaches the partition material itself must be transmitted through to the other surface and re-radiated into the air. Transmission can be reduced by correct choice of materials. Density, which is the weight of a certain volume of material, is important. Heavier materials transmit less well. But composites and layers can achieve the same effect with less density. Re-radiation of the noise on the far side can be reduced by attaching porous materials.

Basic solutions for reduction of noise hazards

Elimination or substitution
Control measures can start with *elimination* of the noise source altogether. Perhaps the procedure can be eliminated altogether or, if this is not possible, another process can be substituted for the one in use. For instance, removal of surface scale on steel with a chipping hammer could be replaced by a chemical treatment, provided this does not result in another substantial risk.

Isolation, segregation and engineering controls
Generally, with chemical hazards, it is preferable to use the first two methods before the last. However, with noise, control at source is a key option and this involves engineering. For example, high frequency noise from a fan can be dropped to low frequency by cutting the number of blades or 'rpm' providing performance is maintained. Similarly, where gears mesh, a change in the number of gear teeth contacting per minute, can drop the frequency.

Pressure relief valve outlets can be re-engineered to reduce noise levels using mazes which allow gradual, rather than sudden, pressure reduction. High volume/low pressure nozzles are less noisy than the reverse. Flexible sleeves in pipework may limit the travel of vibrations giving rise to noise. Panels on a machine may be given more mass (weight) to reduce vibration and hence noise emission.

Isolation or segregation is somewhat different. If noise is coming from moving parts inside the machine, the cover panels, which are primarily designed to prevent transmission of noise through the panel, can be covered inside with porous, lightweight, sound-absorbent material. If a lightweight panel is vibrating and hence radiating noise, making it heavier may reduce noise emissions. Alternatively, the entire machine can be surrounded by an enclosure which minimizes the escape of sound to the outside. Another option is to isolate the operator in a sound control booth which may be easier than isolating the process. As explained earlier, distance from the source is a work practices control which may also be used to reduce exposure to noise.

Noise is often generated from floors, walls, etc., by vibrating machinery on the floor. The use of isolating mounts or damping mounts can reduce noise generation.

Ergonomic principles and control of noise, vibration and lighting hazards

Merits of different forms of acoustic control

The form of acoustic control will vary with the configuration of the work station, but it is important to use the design stage to ensure that design of equipment or tooling within a workplace, which makes up part of a work station, is always the primary method of attacking noise problems. This assumes that the process and the equipment associated with it can't be eliminated.

Acoustic control materials are used to reduce noise problems where the design of a machine or tool hasn't solved the problem, or to add to the reduction in noise resulting from good design. As noted earlier, acoustic control materials take two basic forms – porous material which reduces reflection of sound, known as reverberation, or material of a relatively high density which can reduce the vibration of sound-emitting surfaces or reduce the transmission of sound through a barrier, such as a wall or floor. Certain composites achieve the same effect with less density.

In the office situation, if it is open plan, use may be made of barriers of composite material, porous ceiling tiles and porous flooring to reduce noise transmission and reflection. Noise can come from conversation, older style printers, telephones and so on. Well designed partitioning in closed plan offices is also effective. Serious interference by outside noise may be coming via windows. Double glazing can reduce this.

In the workshop situation, the sound reflectivity of surfaces such as metal roofs and walls can be reduced by fitting porous material, such as heat insulation, which serves a double purpose. A workshop with an arched metal roof can focus noise along the centre of the floor.

It is important in both situations to consider flanking transmission mentioned earlier. Even a crack may be sufficient. The placement of noisy machines near reverberant surfaces such as walls, corners and floors can also increase overall noise levels.

Merits of different forms of acoustic control materials

A variety of equipment and tooling design configurations will need to be addressed. Many pieces of equipment have a primary source of noise, such as the motor, and secondary sources of noise, such as surfaces which vibrate as a result of transmission of motor vibration. If the surface has certain frequencies of sound at which it naturally vibrates (rather like a guitar string of a certain length, diameter and tension), and the sound transmitted from, say, the motor, is mainly of that frequency, the surface resonates, i.e. the intensity of the sound it gives off is maximized. The resonance of a flat piece of sheet metal can be reduced by various means. For example, cloth-covered lead or other relatively heavy material can be fastened to it, or a fold or groove can be stamped into it. An example is the use of such material to reduce the resonance of the floor of a car. Bitumenized cloth will also do this.

So acoustic control materials can include:

- some form of material of different elasticity between two metal parts, e.g. rubber mounts between a machine and an overhead metal platform
- extra density either built into or fastened onto parts of machinery which might resonate
- the use of fastenings which prevent rattling
- dense materials used to put an acoustic enclosure around a noisy machine. (The enclosure is even more effective if the inner surface has porous, sound-absorbent material fitted.)

In some situations, e.g. outdoors mineral processing activities well separated from areas such as offices likely to suffer annoyance from noise, the way to reduce reverberation from surfaces, and hence the noise workers on the plant are exposed to, is simply to leave it open. The sound reflectance of an open space in the surfaces surrounding a machine area is zero.

Again, it is useful to remember that machine design can reduce the noise level before acoustic control materials are employed. It should also be remembered that partial sound barriers are more effective against high frequency noise (a loud squeak) than low frequency noise (a rumble).

Conducting a noise survey

Calibrating and using a sound pressure level meter

Two basic instruments are used to assess the workplace for noise: the sound pressure level (SPL) meter, and the noise dose meter (NDM). The SPL meter in normal mode automatically adds the dB levels at various frequencies to give a total dB level. This is when it is in dBlin (dB linear) setting. Many SPL meters have either a built-in or add-on 'octave band analyser'. This allows you to measure the SPL at a range of frequencies, each at the centre of an octave band. Just as in music, going up one octave from a particular frequency means doubling the frequency.

The human ear hears, and is affected differently, by different frequencies so a correction scale called the ' "A" weighting' is applied. If a SPL meter is set on 'A' weighting it subtracts or adds dB at each frequency, then adds up the total. See Fig. 10.1 for the adjustments.

The action level for a workplace, that is the level over which noise control is required, is given in dBA, not dBlin, except for impulse noise such as that from a

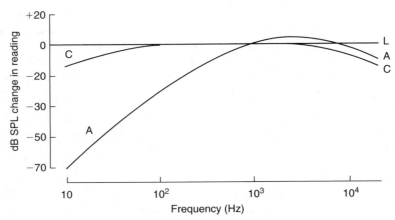

Figure 10.1 The effect of the linear (L), A and C weighting filters of a sound level meter on actual sound pressure level

piledriver. The action level for non-impulse noise is 85 dBA (8 hr average), and 140 dBlin (or peak) for impulse noise. In the USA the permissible level is 90 dBA, with a peak for impulse noise of 140 dB. In some countries the peak level for impulse noise is now 140 dBC ('C' is another weighting scale).

You must consult the manual with your SPL meter for correct setting up, field calibration with a portable calibrator, and battery maintenance. For measurements to check compliance with regulations or compulsory standards, the SPL meter must also be submitted for a calibration certificate to an approved station regularly (usually every two years).

An NDM is generally worn on the lapel or on the upper right pocket. It measures the total noise energy received over a period and can convert it to the equivalent steady dBA level for an 8 hour working day. This is called the L_{Aeq}, 8 h. For further instructions on use and calibration of a noise dose meter, consult the manual with your meter. You are not usually present continuously during measurements with an NDM. It is, therefore, possible for the wearer to falsify the reading by deliberately placing the NDM close to noise sources in a way which does not represent normal activity. Be aware of this possibility.

Measuring sound pressure level in a workplace

You will need to refer to the manufacturer's directions on correct operation of the controls for the particular SPL meter you are using. This includes tests such as battery condition and internal calibration. You can then fit a pistonphone to check calibration further. The choice of instrument settings may include slow or fast response (time over which the signal is averaged), dBlin, dBA, dBC, L_{Aeq}. You can measure at specific workplace stations or you can also work out a grid using convenient reference points (e.g. stanchions, overhead lighting) and measure at each grid point. Be careful to minimize reflections from the body by holding the SPL meter away from you. Keep careful readings as you go. Check calibration before, during and after measurements. If you are using a stand for the meter, vibrations from the floor can affect it.

Estimating a worker's exposure and calculating the daily noise dose (DND)

To work out the daily noise dose for a worker you need to either fit the worker with an NDM for a day, or use a questionnaire to find out where the person typically works, what is normally done there, and how long he or she spends at each task or location. A so-called 'nomogram' is then used to calculate the DND, based on the principle that the sound energy reaching the ear doubles for every 3 dBA increase. This allows you, for example, to work the L_{Aeq}, 8 h from information like the following:

Table 10.1 Noise exposure

Task	Time	Measured SPL dBA
Task 1	2 hours	89
Task 2	1 hour	95
Task 3	1 hour	93
Task 4	1 hour	60
Task 5	2 hours	92
Task 6	1 hour	91

You then need to use the nomogram below in Fig. 10.2 to work out the total noise dose, by adding up the fractions of dose each of the exposures above represent. US readers should note that the action level set by law in the USA is 90 dBA, and that US law specifies that the permissible exposure time is reduced by half for each 5 dBA increase in sound pressure level (29 CFR 1926.52).

Drawing a noise contour diagram

To draw a noise contour diagram, use graph paper and decide on a scale, say 1 cm = 2 m. Mark the grid points you used for the noise measurements on it, and write in the measured SPL. Mark on it machines and work stations. You should then try to draw lines linking points with the same SPL. Some lines will need to be estimated and will pass between measured points. The lines represent points with equal sound pressure levels. See Fig. 10.3.

You can then decide boundaries for hearing conservation areas, i.e. those over 85 dBA (US 90 dBA).

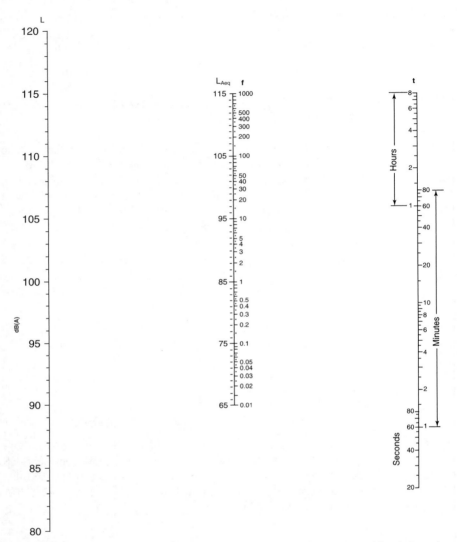

Figure 10.2 Nomogram for calculation of equivalent continuous sound level, based on action level of 85 dB(A). (Adapted from UK Health and Safety Executive based on 85dB(A) action level. Acknowledgement to Health and Safety Executive (1998). Reducing Noise at Work, L108. Crown copyright material is reproduced with the permission of the Controller of HMSO)

1. For each exposure use a ruler to connect sound level dB(A) (left-hand scale) with exposure duration t (right-hand scale) and read fractional exposure f on right-hand scale of centre line.
2. Add together values of f (fraction of noise dose) received during one day to obtain total fraction of daily noise dose.
3. Read equivalent continuous sound level L_{Aeq} on left-hand scale of centre line opposite total fraction of daily noise dose.

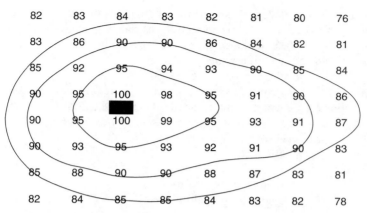

Figure 10.3 Noise contour diagram

Selection of personal hearing protection

Personal hearing protection is the least favoured method for hearing conservation, because it relies on constant wear in noisy areas, good fit and good maintenance. However, it may be the only option on a temporary basis or in some situations such as mines and construction sites. Recommendations for personal protection should be based on the method given in relevant standards or codes of practice. A full explanation of how to assess the suitability of a particular make and model of hearing protection using the manufacturer's information is given in appropriate codes or standards. See for example the UK HSE document above. You will need to have recorded the SPL for each octave band – that is, at the centre frequencies of 63, 125, 250, 500, 1000, 2000, 4000 and 8000 Hz – at a work station during your survey to be able to do the calculations needed to recommend suitable hearing protection.

Evaluating and reporting workplace noise

Some jurisdictions have specified methods for reporting on noise in a workplace. These should be followed. Where these do not exist, or where a full written report is required, include the following information in the report:

Introduction	– where?
	– when?
	– for whom?
	– why the noise survey?
Processes	– what is being done?
	– what is being used?
	– who does what?
	– how many?
	– diagram.
The survey	– how was it planned and carried out?
	– sampling points.

Table of results — include in table form the sampling points and sound levels found.
Contour map — useful in some circumstances showing lines of equal noise levels around machinery or in areas with variations.
Discussion — discuss the results
— explain any surprises.
Evaluation — in terms of the action level, are there problem jobs or areas?
Recommendations — suggested control measures. These should reflect the preferred order, but note what was said in *Basic Solutions* about engineering for control at source.
- elimination
- substitution
- segregation or isolation
- engineering modification
- work practices
- personal protection.

Preparing an action plan for a noise survey

Using the information given in the preceding material you can draw up an action plan to carry out and report on a noise survey in a workplace. Include in your action plan:

- review of previous survey results
- consultation
- walk through preliminary survey
- dates of survey
- work sites
- types of survey (whole workplace, work stations)
- collection of existing hearing protection information (e.g. PPE, marked hearing conservation areas)
- instrumentation
- calibration
- an analysis of noise levels in each octave band so as to decide on appropriate hearing protection, if necessary
- recording
- reporting
- who is to do the survey.

Measuring light levels

Introduction

Lighting is an important part of the work environment. It has a direct impact on safety; prevention of fatigue; prevention of eye problems; and product quality. Eye strain can be caused by subjecting the eye to bright light and glare. Eye strain may cause inflammation, headaches and general tiredness.

Lighting should ensure the safety of people while facilitating the performance of visual tasks and aiding the creation of an appropriate visual environment. The lighting system should, therefore, be so designed and installed as to effectively illuminate the task and provide a safe and comfortable visual environment. A 'luminaire' is a light source and its fittings.

A good lighting system should include the following considerations:

- illuminance
- avoidance of glare
- colour and contrast
- type of task and work area.

Units of measurement

The *intensity* of a light source is measured in 'candela'. This comes from the term 'standard candle'.

The flow or 'flux of light energy' from a light source is measured in 'lumens'. Lumen describes the quantity of light emitted by a source or received by a surface.

Linking each of these together, if a one candela source was surrounded by a totally clear glass sphere of one metre radius, then one lumen of light would pass through each square metre of the sphere's surface. Ignoring reflections, a light meter placed on the inside would read one 'lux'; that is, one lumen falling on each square metre of the surface.

So, the actual light falling on a surface is called the 'illuminance', measured in lux.

Measuring illuminance in a workplace

Recommended light levels for various situations and tasks are set out in the UK Illuminating Engineering Society standards and CIE and national standards.

Illuminance is measured with a light meter. This basically uses a material sensitive to light, such as cadmium sulfide, which generates a current; therefore the more light, the more current. Readings are taken to detail the light levels in working areas and on working surfaces. Specific points may be checked or a whole area may be surveyed using a grid. Care must be taken to avoid *shadowing* the instrument with your body. A so-called 'cosine-corrected' light meter has a plastic hemisphere over the light sensitive cell and this accounts for light coming from oblique directions, as well as that coming towards the surface at angles close to a right angle.

Reporting on light levels in a workplace

A report on light levels needs to include:

- reasons for the survey
- a description of the workplace, people, processes and existing lighting – natural and artificial
- an explanation of how the survey was done, and equipment used
- a drawing showing the task area and light sources

- table of results
- discussion
- recommendations.

Evaluating workplace light levels

The level of illuminance measured should be checked with statutory requirements and the appropriate standards for the location and task. Recommendations can deal with shadows, contrast, colour, glare (light coming into the eyes from directions near to that of the object the eye is looking at) and unwanted reflections. VDU work stations, in particular, need proper light levels, contrast and colour.

Recommendations for some tasks

Appropriate standards list the illuminance and the maximum glare index allowed for most tasks. Table 10.2 shows some examples of the illuminance required for some tasks. You are advised to refer to the full document for more information.

For 1, 5, 6 and 8 use localized lighting. For 7 avoid annoying reflections.

Optimizing the links between lighting and people in the workplace

There must be adequate illumination of the task. A person will find it difficult to see properly if the level of illumination is not sufficient for the task they are performing. Since people are attracted to light, the task area should be brighter than the surrounding area, so illuminance in the surrounding area should not exceed by much the illuminance on the task. High illuminance within the surrounding area is distracting. As the table shows, more illuminance is required for tasks involving fine detail – for example, gem cutting.

Where there is a large variation in illuminance from one area to another, the eye needs time to adjust. This means that the vision is impaired for a brief time. Sudden changes of illuminance should be avoided – for example, leaving a brightly lit room and entering a darkened passage. Impairment of vision due to rapid changes in light intensity makes working in an area unsafe until the vision returns to normal. Vehicle

Table 10.2 Recommended illuminance

Type of task or situation	Maintenance illuminance (lux)
1. Entrance halls	160
2. Corridors, stairs, lifts (elevators)	40
3. Toilets	80
4. Storerooms, packaging and dispatch, wrapping, labelling, filling	160
5. Gem cutting, polishing, setting	1200
6. Gas and arc welding	160
7. Tasks involving keyboarding, reading, filing, etc.	320
8. Study and sustained reading	240

tunnels have graduated lighting to overcome this problem. A task such as checking surface imperfections requires glancing light.

Glare

A safe and comfortable visual environment depends on no unwanted reflections and the absence of direct glare from lamps, luminaires and windows. Glare occurs when one part of the interior is brighter than the general light in the area and this bright light is close to the eye's line of vision. Glare can cause discomfort (discomfort glare) or impair vision (disability glare). Glare must be avoided in situations like stairs and keyboarding.

There are three types of glare:

1. Disability glare affects the ability to see clearly, e.g. undipped headlights of a car.
2. Discomfort glare causes discomfort by impairing the ability to see, e.g. windows, bright lamps, etc.
3. Reflected glare is from reflections in the task or its surroundings which interfere with visual efficiency and comfort by reducing contrast or causing distortion, and by reflecting off a shiny or polished surface.

Colour and contrast

The use of suitable colours allows for a more comfortable visual environment. Light sources with suitable colour characteristics can be used to complement the task or the surrounding area. Colour matching tasks need carefully chosen lighting.

Contrast is used in the task area to enhance visual ability. The contrast can be from a variation in the illuminance of the task or a variation in colour between the detail and its background. When the task is similar in form and texture to the background, it is difficult to distinguish between the detail in the task and the immediate surroundings.

Strobe effects

Lighting from a 50 or 60 Hz alternating current electrical source can result in 'strobe' effects on moving parts, making them appear to move slowly, move backwards or appear stationary. Machinists on lathes, for example, should be made aware of this.

Maintenance of light fittings

Dust, dirt and use will reduce the illumination from lamps over a period of time. General cleaning and replacement of light sources will ensure that illumination levels remain within standards.

Ergonomic design and selection and orientation of work station and workplace lighting and colour

Work station and workplace lighting can come from natural and installed sources. There is a need to be able to adjust natural sources through blinds and shutters so as to prevent glare and reduce problems where the eye takes time to adjust when moving from very bright to darker areas. Workers driving into the sun have an obvious safety problem. Sunglasses and sun vizors assist.

Installed sources of lighting need to be able to provide adequate light when natural light is not available, and this can be a requirement indoors during the day or at night, in dark areas such as cable pits, or outdoors at night, e.g. repair work on motorways or freeways carried out at times of low traffic, and 24-hour haul truck loading on mine-sites. For VDUs and other instrumentation there must be a balance between the screen brightness and other illumination. Glare from low level lighting able to be seen by the VDU operator's eye, including that of outside light from a window, must be avoided by suitable layout. Reflections of lights or windows on the screen must also be avoided. There are recommended levels for the lighting of the screen, the work desk and the copy the VDU operator is reading from.

Choice of colours is important too, for walls, partitions and desktops, and even for the letters and background on the screen. Certain colours such as blues and greens are restful, while others like red are less so. Neutral shades are favoured for desktops.

In the workshop, good lighting is needed in areas where pedestrians walk, including stairs and the head and foot of stairs. Lighting must be bright enough for fine work and provide the appropriate shadow and contrast so that the operator can clearly exercise correct judgement when using a tool on workpieces. For certain tasks, e.g. spray-paint or fabric colour matching, the correct selection of installed lighting is essential.

Air contaminant measurement

Occupational hygiene terms

Aerosols

Aerosols are suspensions of fine particles of liquids or solids in air. Particles in aerosols are generally below 0.01 micrometer. One way of measuring how long particles stay suspended is the falling speed (or terminal velocity). Respirable dust (including respirable fibres) is dust which stays suspended long enough to reach the deepest parts of the lungs, called the 'alveoli'. Dust is generally considered respirable if the particles have a falling speed of less than 3.06×10^{-1} cm per second (3 mm per second).

Another way of looking at this is to talk about 'equivalent aerodynamic diameter (e.a.d.)'. What we mean is that if a particle has an e.a.d. of 7 micrometres it falls at the same speed as a 7 micrometre sphere with a density of 1 g/ml. This approach means that we get away from size and shape and look at what the particle does; that is, how easily does it 'sediment' or fall out. Experiments have shown that in dust which reaches the alveoli, virtually no particle has an e.a.d. of more than 7 micrometres – 50% of particles with e.a.d. of 5 micrometres get that far, and nearly 100% of particles with e.a.d. of 0.1 micrometer or less do so.

If we are interested in what reaches the lower lung then sampling devices should collect dust on the same basis. Sampling devices have been designed which do this, such as the 'horizontal elutriator' as the reference device (heavy and bulky), and the 'mini-cyclone' (lightweight and portable).

You may have read that it is asbestos fibres of up to 3 micrometer real diameter which pose a threat to the lower lung. This is so because an asbestos fibre (a fibre is a particle at least three times as long as wide) of 3 micrometer diameter falls out at a speed the same as a 'spherical' particle of ordinary dust with an equivalent aerodynamic (not real) diameter of 7 micrometers (it is not a true sphere but is called spherical because it is very roughly the same diameter whichever way you measure it).

Some workplace dusts are neither fibrous nor 'spherical', but 'platy'; that is, roughly speaking, a thin disc. Talc is an example.

We call all dust which can enter the nose 'inspirable'; that is, dust with an e.a.d. from just above zero to approx. 30–50 micrometres e.a.d. Generally speaking, the health effects of 'spherical' particles of concern on the lung depend on the mass or weight of dust taken in, and this depends on the concentration in the air and the time over which it is breathed.

Particles of very small diameter (below 1 micrometre e.a.d.) don't make up much of the weight of respirable dust so their potential to cause disease is much less than particles at the upper end of the respirable range. Respirable crystalline silica (quartz) can cause silicosis or scarring (fibrosis) of the lung.

On the other hand, the health effects of fibrous dusts, such as white asbestos, are related more to the **number** of respirable size fibres which are breathed in. But with blue asbestos the relationship between the number of fibres and contraction of the disease mesothelioma is very variable.

Not all the dust and fibres which are breathed in stay in the lung. Finer particles are largely breathed out again, and the 'macrophages' (clean-up cells) and a moving hair and mucus system (the 'mucociliary escalator' in the lung vessels) carry out many of the particles deposited.

While we usually use the word 'fume' loosely, in occupational hygiene we mean very fine liquid or solid particles suspended in air as a result of a liquid or solid becoming a gas or vapour, then recondensing. The iron/iron oxide fume from welding is an example. This can reach the alveoli.

Gases and vapours

A 'gas' is the result of evaporating a liquid (or sometimes a solid) at temperatures above the boiling point. Solids which go directly (sublime) to gas include carbon dioxide (dry ice), iodine, and naphthalene (mothballs). 'Vapour' is again generally used loosely, but in occupational hygiene we mean a gas given off by a liquid (or solid) below the boiling point. Many industrial solvents evaporate quite quickly at temperatures below their boiling point leading to significant airborne levels, as does mercury. Water gives off water vapour at ordinary ambient temperatures.

Occupational hygiene survey terms

Generally, identification and a first up assessment of chemical hazards in a workplace are done using a 'walk-through survey'. The people who do the walk-through will vary with the type of workplace and the complexity of the work processes. They need to note:

- operations carried out
- areas where people work

- number and type of employees
- the chemical register
- chemicals or hazardous substances in use and, for example, dusts generated
- ventilation
- personal protection
- whether material safety data sheets (MSDS) are available to employees.

Questions should be asked about what happens during less obvious work, such as maintenance operations. The survey can be used to decide whether a more detailed assessment using sampling equipment is required.

Sampling

Sampling can be divided into two types:

- grab
- continuous.

'Grab sampling' involves sampling over a short time period, usually with some type of more or less instantaneous monitor, e.g. colorimetric (length of colour stain tube – as in a breath-alcohol test), electronic or tube-electronic. It can give useful information to back up a hunch formed during a walk-through survey about the level of exposure to an airborne vapour or gas. (There are even instantaneous dust and fibre devices, at high cost.) It does not give a good idea of average exposure over a working day.

Continuous sampling can be 'active' or 'passive'. Active involves pulling air through a sampling device with a battery-operated pump. Passive generally involves a lapel 'badge'; the gas or vapour molecules diffuse through a plastic membrane to be caught and held on 'activated carbon' or another absorbent.

Sampling used is mainly personal sampling; that is, sampling on the worker within the worker's breathing zone which is accepted as an imaginary hemisphere of 30 cm radius drawn from a point midway between the person's ears. However, background ('static' or 'positional') readings using a sampler at a fixed point can provide useful information in some cases.

Sampling which truly represents the work situation needs to be planned properly. There are tables which indicate how many people should be sampled and systems to decide whether sampling is to be in the morning or afternoon (or 12 hour day or night shift), and on which days of the week. (See also UK HSE HSG173 *Monitoring strategies for toxic substances*, US OSHA information and Tables 10.3 and 10.4 in this chapter). Such sampling only applies to more or less five-day-a-week continuous processes.

Exposure standards

There are three types of 'exposure standards' set up by the American Conference of Governmental Industrial Hygienists (ACGIH). See the section Information Sources in Chapter 9 and Further Reading at the end of this chapter.

These exposure standards called 'threshold limit values (TLVs)', are reviewed every two years by the ACGIH. Other standards are published by national OHS authorities. These standards include the US OSHA permissible exposure limits (PELs); the German maximum allowable concentrations (MAKs); the UK occupational exposure standards (OESs) and maximum exposure limits (MELs); the EC indicative occupational exposure limit values (IOELVs); and the Australian exposure standards (ESs). For additive exposure where two toxins affect the same organ, the formula in US 29 CFR 1910.1000(d)(2) may be useful. To convert mg/m^3 to parts per million use ppm = $(mg/m^3 \times 24.45)$/molecular weight of contaminant.

There has been some criticism of exposure standards as a guide to whether a workplace is acceptable. The aim should be to reduce levels well below the set standard if it is reasonably practicable to do so. Recommended exposure standards are not meant to be exact cut-offs between safe and unsafe but in some jurisdictions exposure standards have the force of law.

The 'Skin' notation for chemicals in the exposure standard lists indicates that absorption through the skin in significant quantities may occur.

Biological monitoring

Biological monitoring involves both invasive techniques for measuring the exposure of workers to hazardous chemicals and non-invasive ones. Invasive ones include blood and body-fat sampling. Non-invasive techniques include nail clippings, hair, urine, and breast milk samples. The actual toxin (poison) may be measured, e.g. lead, or a metabolite (that is the toxin after the body has modified it; for example the body produces hippuric acid after toluene, a common solvent, has been breathed). These are specific tests. Less specific ones include measuring a protein in urine for cadmium metal poisoning, or measuring enzymes in the liver. Their level rises when the liver goes to the body's defence and tries to 'detox' a toxin. With some pesticides, a drop in a blood enzyme indicates overexposure.

To decide what the results mean, there are tables of Biological Exposure Indices (BEIs) published by the ACGIH (see Further Reading at the end of this chapter). The German MAK Commission also publishes BAT values.

Sampling devices

Colorimetric tubes

The detector or colorimetric tube was mentioned earlier. It contains chemicals which develop a long or short colour stain depending on the concentration (strength) of vapour or gas detected, or which rely on colour matching. Although the discussion further back was about grab samples, there are long-term detector tubes available for sampling over eight hours.

Colorimetric or colour indicator (detector) tubes should be used as in the manufacturer's directions and thrown out on the use-by date. Some tubes give a colour with another chemical (called 'cross sensitivity'), so the manufacturer's notes should be checked. A new type as noted earlier still uses detector chemicals but measures changes electronically, not by colour.

Some of the chemicals in the tubes are harmful so proper disposal is essential. They must be read immediately as the stain can fade, discolour or spread, and they should only be used with the pump the manufacturer designed for them.

Direct reading instruments

Direct reading instruments are available. The results from them may not always line up with established standard methods, so cross-checks will be needed.

Some direct reading instruments are:

- mercury vapour meters (use ultraviolet light)
- electronic gas monitors e.g. carbon monoxide, hydrogen sulphide, formaldehyde, chlorine (based on chemical cells)
- FAM (fibrous aerosol monitor)
- portable gas chromatographs (chemical vapour and gases)
- photo-ionization gas and vapour detectors
- photo-acoustic gas and vapour detectors
- portable electronic dust monitors
- colour-change continuous paper-tape devices with electronic readout
- computer chips, which linked to detector tubes, can identify gases and measure their concentration.

Explosimeters are also direct reading but are aimed at higher levels of gas and vapour than general occupational exposure levels.

Charcoal tubes

While not really a detector tube, a tube containing grains of activated charcoal can be used to collect vapour or gas from air pumped through it. It requires further analysis to give a result. Gas liquid chromatography (GLC) or high performance liquid chromatography (HPLC) is used for this.

In using charcoal tubes filled with activated charcoal granules, some figure work is needed if time is not to be wasted going back for a second try. An educated guess at the concentration is needed (perhaps from your detector tube results). You should consult a tube manufacturer's sheet to ensure the chemical you are collecting can:

- be collected on charcoal and not partially pass through it
- be 'eluted' off the charcoal for analysis in a laboratory. Most tubes have two charcoal layers and if the vapour collected 'breaks-through' layer one, the analyst will find it in layer two. This suggests a re-think. It may be due to 'overloading' (that is, you underestimated the amount of vapour present or sampled too long), or it may be that charcoal is unsuitable. Other adsorbents such as silica or ion exchange resin are available.

The other person you need to consult before sampling is the laboratory analyst. She or he can tell you how much vapour must be collected to give a meaningful result.

Example

The exposure standard for toluene is $380\,mg/m^3$. Let's say there is $400\,mg/m^3$ present. Let's also assume that the analyst needs 10 ml in the 10 ml of solvent she or he uses to elute (strip) your charcoal tube. The usual instrument used for analysis is a gas chromatograph. The analyst says this will require an injection of 5 microlitres of solvent containing 5 micrograms of the contaminant (which she or he can achieve if there is

10 mg/10 ml of solvent.) Some gas chromatographs using capillary columns now require only nanogram quantities.

Sampling rates for solvents on charcoal are generally about 100 ml of air per minute, so to run 1 litre through the tube requires 10 minutes sampling. As there are 1000 L in 1 cubic metre, at this point you have collected 0.4 mg of toluene. If you want 10 mg you will need to sample 25 times as long; that is, 250 minutes or just over half an eight hour shift.

Be careful to keep the 5 ml or so of the workplace solvent being analysed, which the analyst will require, well separated from your samples, even though they are capped.

Passive badges

Passive monitoring with a badge requires similar calculations beforehand. There are also some liquid-based passive monitors which allow vapour molecules to reach the liquid through very fine glass tubes (capillary tubes).

Liquid impingers

Another method of collecting some airborne chemicals is the 'liquid impinger' in which air is sucked through a liquid absorbent. These are most efficient at collection if a chemical reaction takes place in the impinger, rather than simply dissolving the vapour or gas contaminant. They are also used for some mists – for example, acid mist. Modern impingers are available which are break-resistant and spill-proof (the wearer could do a somersault). An alternative to the liquid impinger is the use of a chemically-treated filter paper or fritted glass disc through which air is drawn.

Methods for airborne dust and fibre contaminants

Guides, or manuals, or standards such as those published by the US NIOSH, UK HSE or Australian Standards respectively, should be consulted for methods of measuring inspirable dust, respirable dust, asbestos, synthetic mineral fibres, and organic vapours. Inspirable dust measurements are useful for dusts like lead-containing dust where dust in the mouth and nose may be swallowed.

Once again, it is important to estimate likely workplace concentrations so that filter loadings are not too low or too high. Low loadings may be unavoidable with 'para-occupational' fibre sampling, done for example after clearance of asbestos from a building.

If you are interested in the reasons for choice of filters (e.g. glass fibre, PVC) consult manufacturers' information and occupational hygiene texts, e.g. Waldron and Harrington (see Further Reading at the end of this chapter). Some filters used in gravimetric (weight) methods build up static charge which affects weighing accuracy. A 'destaticking' device may be needed. Others absorb moisture and require placement in a desiccator before weighing empty, and again before weighing loaded with sample. Special filters such as silver membrane are used for collecting coke oven emissions, as they don't dissolve when solvent is used to strip the contaminant from them.

Dust and fibre samples must be carefully transported to the laboratory with the dust or fibre on top of the filter. You must include field blanks in the filters which go to the laboratory.

Sampling strategy for representative data

Sampling which truly represents the work situation needs to be planned properly.

Sampling strategies – their strengths and weaknesses

The primary method of choice for sampling of chemical stressors is personal sampling. If it is performed over a working day, or a representative part of a working day, it allows assessment of what the workers are actually exposed to as they move around and go from task to task.

Not all stressors allow for this approach to sampling. The sampling equipment must be robust, light, portable and not interfere unduly with the workers' activities.

Exposure standards for stressors are generally based on personal exposure of workers. For personal sampling, the length of sampling, selection of workers and number of samples per worker are important to get statistically valid results. That is, results which reflect reality as well as possible.

The same general statement about length, selection and number also applies to area samples. These are samples generally taken at a fixed point in an area. For example, in monitoring temperature, humidity and airflow in a workroom, unless there are local sources of difference in temperature, e.g, a furnace, in humidity, e.g. an industrial ironing press, or airflow, e.g. nearness to an air outlet, an area sample may be adequate and is suited to the current lack of availability of personal heat stress monitors, although one is being developed.

Area (or static) sampling can also be useful for checking the hazard from an operation, process or activity (both initially and down the track); routine monitoring (on the theory that if area results haven't changed, personal results won't have – which is not completely correct); and checking changes in operations, processes or activities. An electronic chlorine monitor in a chlorine production area will show any change in the level of chlorine escaping (fugitive emissions).

The timing of sampling and the associated number of samples is important. Assessment against standards for time-weighted average requires a different approach to assessment against standards for peak exposure or short-term exposure, such as might occur when a process vessel is opened at times during the working day. Such exposure is called 'cyclic'. The types of standard reflect whether a stressor has an irritant, acute or chronic effect. For noise sampling, the peak value is sampled when the noise is impulse noise, e.g. from a pile-driver. Once again instrument limitations may dictate area sampling only; for example, a fibrous aerosol monitor (FAM) is not something which can be worn but gives instantaneous and immediate values for the level of airborne fibres.

The information in Fig. 10.4 expands a little more on choice of sampling patterns, how many samples to take and when to sample (Coffee might replace tea!).

Cyclic exposure requires sampling over one full cycle and partial consecutive samples taken during periods of expected vapour emission.

Statistically, consecutive samples covering the full work period are better than one full period sample and yield more information about sources of highs and lows, providing the sampler is actively observing during the sampling period. For enforcement

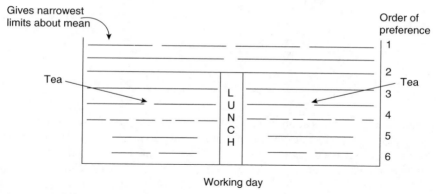

Figure 10.4 Sampling patterns

or compliance purposes specified sampling protocols applying in that jurisdiction should be consulted.

If only one sample result is known, it is possible to calculate to a given level of confidence, whether that result indicates that the standard has been exceeded (if the coefficient of variation of the method is known).

With noise sampling, the general approach is to identify particular activities and the time spent on each, and then measure each with a sound level meter. A noise dosemeter can be used, but results as noted before can be affected by worker misuse out of sight of the person sampling. Ionizing radiation is usually monitored continually while at work using personal devices. Heat and cold stress from natural sources can be highly variable throughout the working day, and day by day, so this must be taken into account in a sampling strategy to obtain representative results.

How many samples

A system devised in the USA suggests the following:

Table 10.3 Sample sizes

No. in Group	No. of Samples
<7	7
8	7
9	8
10	9
11–12	10
13–14	11
15–17	12
18–20	13
21–24	14
25–29	15
30–37	16
38–49	17
50	18

Table 10.4 Random sampling

	M	T	W	T	F
AM	1	3	5	7	9
PM	2	4	6	8	10

The aim of this is to sample the top 10% with a confidence greater than 90%. For from 8–50 people, select at random the number of people shown. This ensures with 90% confidence that sufficient in the top 10% are sampled.

(*With acknowledgement to Leidel and Busch in Cralley, L.J. and Cralley, L.V. (eds) Patty's Industrial Hygiene, 2nd edn Vol. 3A, copyright © 1985, John Wiley and Sons Inc. Reprinted by permission of John Wiley and Sons, Inc.*)

When to sample
Table 10.4 can help you to pick sampling periods using random numbers to pick one period per week. This can be extended to seven days for continuous operations and two 12 hour shifts can be split into six periods of four hours (with appropriate corrections to the time-weighted exposure standard).

Detailed survey
A detailed survey can be used where the extent and pattern of exposure cannot be confidently assessed by a basic survey, exposure is highly variable between employees doing similar tasks, carcinogens or lung sensitizers are in use, the initial appraisal and basic survey suggest that the TWA results may be close to the exposure standard (or a lower in-house standard) and the cost of extra controls cannot be justified without evidence of variability of exposure, or when undertaking major maintenance or one-off jobs, e.g. decommissioning.

(*With acknowledgement to UK Health and Safety Executive. (1997). Monitoring Strategies for Toxic Substances HSG173 – Crown copyright material is reproduced with the permission of the Controller of HMSO.*)

Sample size
The sample size needed for air contaminants affects the size of samples. Here is a guide.

$$\frac{10 \times \text{sensitivity of the analytical method } (\mu g)}{\text{exposure standard } (mg/m^3)} = \text{no. of litres}$$

You will need to get the sensitivity figure from the analyst, if one is involved.

Other considerations
Some other issues need to be considered in deciding on sampling. They include:

- cost of sampling equipment
- analytical method and consequently analytical facilities required
- personnel to collect the samples (e.g. degree of skill and time needed).

Too many samples can be collected for the analytical time available. The cost of the exercise has to be related to the magnitude of the problem and the costs of its alleviation. Even if sampling strategies are good, the results obtained depend on the method and equipment; for example, if a solution designed to turn green in contact with ethanol vapour also turns green in contact with methanol vapour, and if both could be present, your result doesn't accurately reflect just the ethanol.

Portable sampling pump flow can fluctuate. If it varies more than 10%, discard the sample.

There are three other issues in the use of sampling methods:

- accuracy
- precision
- reliability.

A method can be accurate, i.e. the results obtained are close to the true value. It can be precise, that is, re-measuring the same sample gives results which agree well. It can be reliable, that is, it will give results for a sample on a different day (where there has been no change in the sample) which are the same as those on the first occasion. On the other hand, the method may fail on one or more of those counts.

Finally a flowchart is included to provide a guide to monitoring. While it deals with air contaminants, it can be applied more generally. See Fig. 10.5.

Setting up, calibrating and adjusting a personal sampling pump

It is essential that, with all pump sampling, the flow rate is accurately calibrated with a detachable inline flow meter. The in-pump flow meter, where present, can be checked at the same time and then used for mid-sample check and re-adjustment. Flow rate must be checked just before completing sampling, and if the flow rate has dropped more than 10%, the sample should be ignored. Some pumps automatically keep flow steady. Remember *intrinsically safe* pumps (look for a recognized compliance plate on the pump) are needed in flammable areas such as oil refineries and coal mines.

To protect the pump an inline moisture trap is recommended. **Under no circumstances connect the inlet end of the impinger to the pump**. If you do, immediately pull distilled water through the pump, and then acetone if available (take care – flammable!). The pump must then be dismantled and repaired.

Active sampling, with samples ready for a laboratory

You will find the following useful when sampling:

- record book – essential
- scissors

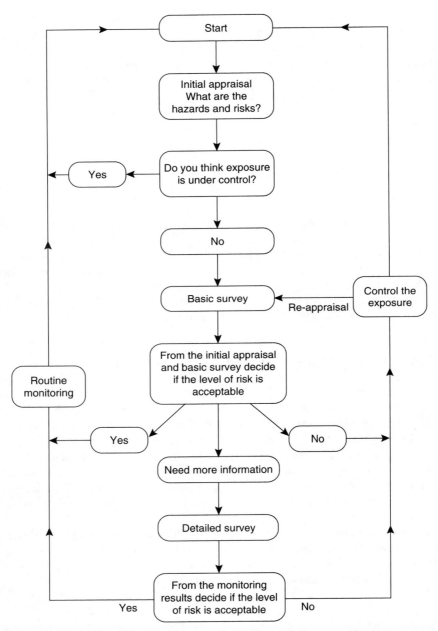

Figure 10.5 Monitoring flowchart: A structural approach to assessing exposure to substances hazardous to health by inhalation (from UK Health and Safety Executive (1997): *Monitoring Strategies for Toxic Substances HSG173* – Crown copyright material is reproduced with the permission of the Controller of HMSO)

- spare plastic tubing
- paper packaging tape
- safety pins
- permanent marker
- a robust belt or harness to hold the portable pump
- spare pump batteries.

It is essential to gain the acceptance of management, unions and workers before sampling commences. This may include guarantees about making results available.

The portable pumps generally rely on rechargeable nickel-cadmium batteries. Although these may have the capacity to run eight hours (depending on filter resistance), if they are only run for half shifts repeatedly before recharge they may acquire a memory effect; that is, only run for four hours even though fully charged. To avoid this, after use the batteries should be discharged in a circuit with an appropriate resistance, or as a second preference, the pump left running until it stops, and then recharged.

Types and applications of analytical methods

The method of analysis in the laboratory following sampling for airborne chemical contaminants can involve the use of a number of instruments.

You may encounter the following types of laboratory instrumentation:

- atomic absorption spectrophotometer
- gas liquid chromatograph
- mass spectrometer
- high performance liquid chromatograph
- infra-red spectrophotometer.

A basic understanding of how these work is useful. It is set out below.

Operation of AAS, GLC, MS, HPLC and IR

Atomic absorption spectrophotometry (AAS)
This technique was originally developed by Australia's CSIRO to accurately measure essential elements such as cobalt, manganese, copper and selenium in soils, because certain Australian soils were deficient. It allows accurate and specific measurement of metals and semi-metallic elements (such as arsenic, selenium and antimony) at part per million levels (less if various pre-concentration steps are employed). The metals of interest in occupational health include cadmium, mercury, vanadium, cobalt, zinc, iron, manganese, copper, nickel, chromium, tin, thallium, lead and possibly tellurium and osmium. The three semi-metals above are also of interest.

The basis of AAS is that a solution of the metal being analysed is mixed with the gases going into a flame in a spray chamber. Once in the flame, a fraction of the chemical compound the metal is part of is broken up into its different elements by heat.

A hollow cathode lamp (containing atoms of the metal concerned) is used to pass a narrow beam of light through the flame.

The heated atoms in the lamp of the metal concerned give out light at wavelengths specific to that metal (although one metal can interfere with another). The neutral atoms of the metal in the flame (not ions of the metal) absorb some of this light in proportion to the concentration of metal in the solution. Standards are also put through, and modern AASs carry out all the calculations. Both they, HPLCs and GLCs can also come fitted with automatic sample injection for a number of samples.

Various techniques such as a pulsed signal are used to distinguish the light of the lamp from 'background' effects of light from the flame. To overcome lamp flicker, a double beam is created using a splitting mirror and the instrument compares the sample beam with the reference beam.

The so-called Zeemann effect is also used to give better results.

Depending on the element to be analysed, rich and lean flames may be used. The normal flame uses air and acetylene, but acetylene and nitrous oxide can be used (with due care – it is important that the flame is never too lean so as to avoid an explosion, and this is achieved by appropriate procedures during switchover from air/acetylene and back to air/acetylene).

Pre-concentration of metals by reacting them with a metal coordinating or complexing compound (such as those used for visible spectrophotometry) and extraction of the complex into a solvent, can reduce interference from other metals in the sample and achieve lower levels of detection. It can also be used to separate metals directly from a sample such as urine.

Some elements ionize in the flame very easily and such metal ions are not wanted. An ionization suppressor such as lanthanum salt which is preferentially ionized can be added.

Three other variations of AAS are commonly used in OHS work. One is called 'graphite furnace', the other 'hydride generation'. The first allows direct analysis of, for example, lead in blood by drying, ashing and then vaporizing the blood in a heated graphite furnace through which the hollow cathode light beam passes.

The second, used for arsenic and antimony uses an extremely powerful reducing agent, sodium borohydride, which converts arsenic and antimony in solution into arsine and stibine gas respectively. This is then swept into a heated silica tube where it decomposes to neutral atoms of arsenic (or antimony) in the light beam.

For mercury, the borohydride technique is used to generate elemental mercury vapour from a mercury solution. This is then passed into the mercury light beam without any need for a flame to decompose mercury compounds (so it is called cold vapour AAS).

AAS can be used to analyse filters which have collected various metals in welding fume, or to analyse collected samples of dust containing a toxic metal, as well as the biological samples mentioned above.

As with other forms of analysis, it is important to find out the practical detection limits so that sufficient air is sampled. The detection limit is twice the standard deviation (in this case twice the variation of the 'background' or 'noise' mentioned above).

Gas liquid chromatograph (GLC)

Chromatography was described in the last century by Nicholas Tswett. He poured an extract of plant colours through a column of chalk. Some colours were more 'attracted' to the chalk (had more affinity for the chalk) than others, and so the colours most attracted stayed near the top of the column while those less attracted moved to the bottom. A series of colour bands could be seen. Continuing to pour the right liquid through the chalk, the bands would move down and then drip out. Each could be collected separately. Because these were colours, the idea was called 'chromato' (Greek for colour) 'graphy' (Greek for writing). You can do something similar by making a heavy mark with a black ballpoint on blotting paper and then dropping a spot of methylated spirit on the mark – you will see the ink dyes separate.

Gas liquid chromatography works similarly. There is a long column filled with small granules coated with a chemical compound which can take temperatures of up to 350°C. An alternative type of column (capillary column) is a column with a capillary through it. The sides of the capillary are coated with the chemical compound. The compound can be, for example, one of various types of silicones. The column sits in a heated oven. The compound (and the granules if present) are called the 'stationary phase'.

When a sample is injected onto the column, it vaporizes. Instead of a liquid, as in Tswett's experiment, a 'carrier' gas, usually nitrogen, flows through the column. Some substances in the sample are more attracted to the chemical compound in the column, others less. The result again is bands of each substance moving through the column at different speeds. As each one reaches the end of the column (*elutes*), it passes through a detector which measures how much there is of it by comparing it with a standard amount of the same substance put through the column earlier. Standard substances are also used to find out how long each substance stays in (is retained by) the column – the 'retention time'. This time varies with the type of column, length of column, carrier gas flow and column temperature.

In itself a GLC can't identify each substance, but by educated guesses and the injection of standard substances and standard amounts of them, each can be tentatively identified and the amount measured.

Repeating the exercise using another carefully chosen column is regarded as reasonable proof of the identity of a substance if its retention time again matches the retention time of a known standard. Note that a substance may have a retention time greater than another substance on one column, but the reverse may apply on another column. Organic compounds are described as having greater or lesser degrees of polarity. In polar compounds there is some separation of the positive and negative charges in the molecule. Organic compounds of lower polarity move through more quickly on high polarity columns and organic compounds of higher polarity move more quickly on low polarity columns. Substances containing phosphorus and nitrogen, and substances containing chlorine can be separately identified by using different types of detectors.

Where two substances in a mixture are difficult to separate by attempting to obtain different retention times, temperature programming can be used, i.e. injection may be at, say, 150°C, and the GLC increases the temperature at, say, 5°C/min to 225°C. Easier identification (at a price) is achievable with a mass spectrometer linked to a gas liquid chromatograph.

Gas liquid chromatography is very good at separating and identifying many organic compounds, i.e. those containing carbon, hydrogen, and often oxygen and/or nitrogen and/or sulphur and/or chlorine and/or phosphorus.

Mass spectrometer (MS)

Some gas liquid chromatographs (GLCs) are now coupled to a mass spectrometer. As each compound leaves the GLC it enters the spectrometer. This breaks the molecule into fragments and measures the molecular mass of each fragment compared to one sixteenth the mass of an oxygen atom. It also measures the relative amounts of each fragment. It compares this with a computer database to identify the chemical compound.

High pressure liquid chromatograph (HPLC)

This is based on Tswett's liquid chromatograph. However, today it uses a variety of 'adsorbents' or stationary phases other than chalk; the liquid is pumped through under high pressure, and the liquid composition can be changed during the analysis. This allows for better separation of chemicals in a mixture and a faster passage through the column. Once again, there is a detector at the outlet from the column. The retention time is different for different compounds and the detector measures the amount of substance coming off the column.

Infra-red spectrophotometer

This operates by passing infra-red (heat) radiation through the sample. The heat radiation is made of electromagnetic waves with a range of wavelengths. Just as a certain string in a piano will vibrate if a sound of the right wavelength strikes it, so the covalent bonds between atoms will do the same with infra-red radiation. If we select the right wavelength of radiation we can stimulate the bond we want, e.g. the bond between silicon and oxygen in quartz. The bond's vibration absorbs some infra-red energy. So if we have two beams of infra-red light, one passing through the sample, one not, we can measure the energy lost from the sample beam. This is a measure of the amount of the substance we are analysing which is in the sample beam. We select the correct wavelength to analyse a bond as specific as possible to the substance we want to measure. Fourier transform IR (FTIR) extends the usefulness of IR analysis.

Other techniques

These include:

Wet chemistry

For other air contaminants and wipe samples, so-called 'wet methods' of chemical analysis may be employed. Full details can be found in publications such as the US NIOSH *Manual of Sampling Methods*, publications from the UK Health and Safety Executive, *Annals of Occupational Hygiene*, *Applied Industrial Hygiene*, *AIHA Journal*, and analytical chemistry journals.

Biological samples

Standard biomedical science methods of plating-out, incubating and plate counting are used for dip-slides.

X-ray diffraction (XRD) and fluorescence (XRF)
These methods can be used for the identification of the hazardous components of mineral dust and for calculating the percentage present. IR can also be used for this in some cases.

ICRP – inductively coupled radiofrequency plasma spectrophotometry
This allows rapid determination of many metallic elements simultaneously.

Evaluation, exposure controls and reporting on chemical contaminant monitoring

Reporting on a survey and proposing control measures

The report
The report on a workplace occupational hygiene (hazardous substances) survey should include:

- reason for the report
- date
- place and description – diagram
- description of operations
- workers involved
- test methods used
- table of results
- discussion of results (use existing standards and regulations to assist judgements)
- conclusion
- recommendations – these can be placed at the front of the report. They will propose varying options for elimination, substitution or other control of the hazard.

Chemical contaminants may also be absorbed into the body through skin absorption or ingestion of chemicals from the hands when there is poor personal hygiene. Your report should include control measures which take into account all three routes of absorption – inhalation by nose and mouth, skin absorption and ingestion.

Controls
If your survey has shown unsatisfactory exposure levels for people in the workplace, consider whether the process can be *eliminated*. If not, can something else be *substituted*? For example, solvent hazards from paints have been overcome in some cases with water-based paints.

'Segregation' or 'isolation' is the next choice. Can the worker be isolated from the process in an air-fed booth, or can the process be totally enclosed with the worker outside? For example, sealed systems are used for some chemical processes, although the sealing needs constant checking and maintenance.

'Engineering' is a further method of control and a common engineering approach is ventilation.

Ventilation

In many cases ventilation is the method of choice for airborne contaminant control. This offers a number of choices:

- improved natural ventilation – this can be subject to wind conditions, and weather (doors, windows open or closed). In winter, closed doors improve comfort but increase contaminant levels, although the cooler temperature may reduce vapour levels given off
- combination of natural (e.g. roof or wall register, that is, a screened opening) and mechanical
- mechanical.

Main types of mechanical ventilation

Mechanical ventilation can consist of:

- general dilution
- directed dilution
- local exhaust.

General dilution

General dilution is achieved usually by relatively large diameter axial fans, some with a housing, either pushing air in, pulling it out, or both. Often the push fans are in the opposite wall to the pull fans. This is satisfactory for relatively low emission rates of relatively low-toxicity materials.

Directed dilution

Better control can be achieved with directed dilution if there is a point source of contaminant emission. See Fig. 10.6.

In some cases, even a pedestal fan can provide 'spot' directed dilution (and thermal comfort if required).

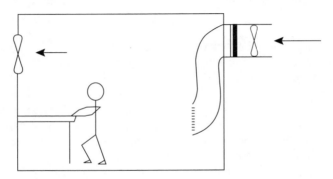

Figure 10.6 Directed dilution ventilation

Work environment 431

Local exhaust

Control at source is the best choice where emissions are relatively high, or the materials are relatively toxic or irritant (low exposure standard), or have a bad smell. Control at source is achieved by local exhaust ventilation.

Local exhaust uses 'captor hoods' or 'receptor hoods'. With captor hoods the contaminant source is outside the hood, e.g. a handpiece welding fume extractor, and with receptor hoods it is inside, e.g. spraypainting booth, laboratory fume hood. These take many forms and a full review of design is not given here; however, the principal points to note are given here.

The hood itself can be one of a number of types:

- a thin slot
- a flared round end
- a rectangular aperture
- a hood or fume hood with access at one side; this side may have a transparent gate which can move up and down
- a hood overhead above a table or tank; this has limitations due to draughts.

For small jobs, such as tissue work in medical laboratories, or shot blasting of small parts, a fully enclosed ventilated 'glove box' can be used. Figure 10.7 shows a receptor hood in use at the top of an adhesive mixer. Note also the use of a simple manual handling aid.

When using captor hoods, the suction ability or vacuum which creates a *capture velocity* decreases rapidly as the distance away from the intake 'face' increases. A hood with a flanged edge (the flange can be achieved by flush mounting in a wall) requires less air throughput and saves noise, power and costs.

Figure 10.7 Receptor hood

A captor hood must generate sufficient capture velocity at the point of contaminant generation. A receptor hood must have sufficient face velocity (velocities vary from 0.5 m/s, e.g. laboratory fume hood, to 10 m/s, e.g. disc grinder).

Some slot hoods are mounted to be movable and may also be attached to a portable suction unit (fan) and appropriate filter.

Key features of a local exhaust system

The hoods themselves have been described above.

'Ducting' links the hood to the fan. If dust, mists or vapours which require trapping are involved there will be a collector device between hood and fan. This depends on what is to be trapped but can be bag filters, a wet scrubber, a cyclone or an electrostatic precipitator.

Ducting may have several branches. These may have different resistances and need *balancing* with 'blast gates' (butterfly valves) or be designed for balance (this approach makes additions harder). The ducting must increase in size as more branches join and the joints must be well designed (no right angles). For dusts, the 'transport velocity' in the duct must be sufficient to keep the dust in suspension (10 m/s–22 m/s depending on the substance involved).

Fan choice is important. Centrifugal fans operate well against the relatively high resistance the ductwork presents. Backward curved centrifugal fans are generally most desirable as they are less noisy. Variable ratio connections to the motor allow design adjustments.

Finally, the exhaust stack needs to offer adequate clearance above the roofline (and adjacent rooflines) and be of proper design. A conical top is not the best design – there are others to prevent rain or snow intake.

Assessing and periodic checking of a local exhaust system

The easiest check on system performance is to check the static pressure at the hood throat (or throats), the point where the duct leaves the hood. This is simply the difference between pressure in the duct and atmosphere, and can be measured with a manometer – at its simplest a U-tube of liquid, although electronic versions are available. The ductwork can have removable caps at pressure check points. Records should be kept of ventilation system checks.

Work practices

A further level of control is to improve work practices, and the procedures which describe them and form a basis for training. Some examples are:

- Pump a liquid from a supplier's drum into a vat, rather than using a bucket.
- Place lids on vats and mixing vessels when access is not required.
- Have covered containers for disposal of contaminated waste.

- Keep all surfaces clean (wet or dry sweeping, vacuuming).
- Correct adjustment of butterfly valves on ventilation.

Selecting personal protective equipment for a chemical hazard

Even though other preferred controls may have been chosen for a chemical hazard, it may still be necessary to use personal protective equipment (PPE). This is especially so in case of emergencies (e.g. splash; or in underground mines, fire).

The PPE can include:

- Resistant clothing suitable for the chemical involved. This can be a combination of headgear, eyewear, footwear, gloves and apron. In some cases the respiratory protection, headgear and eyewear form one unit. For further information on selection of these refer to manufacturers' information and to the relevant national standards.
- Respiratory protection. This can consist of:
    - dust, vapour or mist half-face (*orinasal*) mask, disposable or with replaceable filter
    - powered air purifying respirator (PAPR) (portable pump on belt), with face shield and neck semi-seal or full face hood
    - positive pressure air-supplied respirator (PPASR)
    - self-contained breathing apparatus (SCBA)
    - self-rescue devices with built-in air regeneration.

In all cases, fitting, cleaning, maintenance, training and proper storage are necessary. The simpler face-pieces are not for shared use. In certain circumstances, a bearded worker can use a mouthpiece respirator with a nose-clip. For information on selection and use of respiratory equipment refer to manufacturers' information and to standards or codes of practice (in the USA, MSHA and OHSA).

Once controls are introduced, it will be necessary to monitor that they are effective and that their effectiveness continues. Is ventilation equipment checked? Are chemical purchasing guidelines followed? Are new workers trained? Is PPE maintained or replaced?

Thermal comfort and heat stress

Introduction

Workplace climate problems can range from excessive cold through what is generally called 'thermal comfort' at moderate temperatures to natural high temperature and/or 'hot work processes'. Seven main factors are involved:

- humidity of the air
- airflow

- temperature
- radiant heat energy
- clothing
- work rate
- fitness/age/health status.

Forms of heat transfer

Heat transfer between the body and the environment takes place through:

- conduction – body contact
- convection – airflow over body
- radiation – the body radiates heat to cooler surroundings or absorbs radiant heat energy (e.g. from the sun).

Evaporation and latent heat

Losses and gains of heat from conduction are relatively small. Convection is more important and cools the body unless the air temperature exceeds skin temperature, when convection will heat the body.

In hot conditions, evaporation of sweat is a major factor. This relies on what is called 'latent heat'. This is the heat energy needed to break down the attraction of water molecules in liquid water to form a vapour with almost no attraction between molecules. This heat energy does not raise the temperature of the water. In fact, as airflow over the sweat strips off water molecules from the surface, this heat energy (latent heat of evaporation) can only be obtained by drawing heat from the skin. This cools the body.

If we think about what may affect body temperature we can look at the following:

M	±	R	±	C	=	E
metabolic heat		radiant heat		convection heat transfer		evaporation

That is, in hot conditions, cooling from sweat evaporation must equal metabolic heat, radiant heat and convection heat together if body temperature is not to rise. Metabolic heat increases with heavy workload. If the body is in cool air and cool surroundings then cooling from evaporation need only equal metabolic heat minus radiant heat and convection.

There is a limit to the ability of the body to sweat, and evaporation depends on airflow, humidity and clothing. If **M + R + C** are more than **E**, then body temperature rises. A rise of more than 1°C in core body temperature is considered undesirable, and 2°C can produce heat stroke and other heat-induced conditions.

Relative humidity

Humidity is an important factor in thermal comfort and heat stress. At a certain temperature the air can only hold a certain amount of water vapour. As the air temperature goes down, the air can hold less water. This is why water droplets (dew) form overnight when air temperature drops. The percentage of water vapour actually in the air at a certain temperature compared to the maximum amount which could be held at that temperature is called the 'relative humidity'. This can be measured using wet and dry bulb thermometers (see 'Thermal environment measuring devices').

Heat stress and thermal comfort indices

In order to decide whether thermal conditions are acceptable for a particular type of work, a number of indices (indexes) have been developed. These were developed in a number of ways, for example:

- work on submarine crews (Bedford)
- work in US army training camps (Yaglou)
- work on students in rooms with precise temperature/ humidity control (Kansas State University).

The most important indices are:

- WBGT (wet bulb-globe temperature) index
- the Belding and Hatch Heat Stress Index (HSI), as modified by McKarns and Brief
- CET (Corrected Effective Temperature) Index developed by ASHRAE/KSU (The American Society of Heating, Refrigeration and Airconditioning Engineers and Kansas State University).

There are two other indices called the Fangers PMV (Predicted Mean Comfort Vote) and the P_4SR (predicted 4 hour sweat rate).

The WBGT, see Table 10.5, is the most favoured in Australia, although with outdoor workers, some unions work simply on a temperature of 37.5°C as a basis for walking off the job. The Belding and Hatch index has the value that predictions can be made of what will happen to the index when humidity, temperature or airflow are altered.

The WBGT and HSI are suitable for hot conditions and heavier work; the CET is more useful for lighter work in indoor environments, although there is a modification to take workload into account.

Table 10.5 Recommended maximum WBGT indices

Work-rest regime (each hour)	Light work	Moderate work	Heavy work
Continuous work	30.0	26.7	25.0
75% work – 25% rest	30.6	28.0	25.9
50% work – 50% rest	31.4	29.4	27.9
25% work – 75% rest	32.2	31.1	30.0

Thermal environment measuring devices

The basic instruments needed to assess thermal comfort or heat stress are:

- dry bulb thermometer
- wet bulb thermometer – some indices require natural ventilation, some require aspirated (blown air) ventilation
- black-globe thermometer – to measure the effect of radiant heat energy
- anemometer to measure linear airflow or Kata thermometer to measure random airflow.

An alternative is the Wibget (see below).

Methods of measuring factors in the heat balance equation

While there are quite accurate ways to measure metabolic heat generation by measuring oxygen usage, generally, the metabolic rate for an activity can be obtained from standard tables.

To calculate E, R and C the environmental factors measured are as noted dry bulb, black-globe and wet bulb temperatures and air velocity.

Dry and wet bulb temperatures can be measured with an aspirated or whirling hygrometer, and using a chart, the vapour pressure of the water in the atmosphere can be read off using the dry and wet bulb values. A wet bulb thermometer has the mercury in the bulb surrounded by wet cloth – the drier the air, the faster the evaporation of water and the lower the wet bulb temperature.

A globe thermometer is a thermometer inserted into a matt black copper globe. This provides a measure of radiant (infra-red) heat energy. A simple equation gives the mean radiant temperature from the globe temperature. Air velocity, if it is in one clear direction, can be measured with a vane anemometer; otherwise a Kata thermometer or hot-wire anemometer can be used. The Kata thermometer is warmed and then the time for temperature drop between two points observed. A chart gives the airflow.

For WBGT there is a combined instrument available, an electronic WBGT meter (mini-Wibget) using thermistors for temperature measurement. Time must be allowed for readings to stabilize as far as possible, although conditions fluctuate. A personal heat strain monitor is under development.

The formulae for calculating WBGT are set out here:

WBGT for indoor environments:

$$WBGT = 0.7 \times \text{wet bulb temperature} + 0.3 \times \text{globe temperature}.$$

WBGT for outdoor environments:

$$WBGT = 0.7 \times \text{wet bulb temperature} + 0.2 \times \text{globe temperature} + 0.1 \times \text{dry bulb temperature}.$$

Some of the original work on clothing was done for the military with khaki. The reason for the difference in the indoor and outdoor WBGT equations relates to the fact that short wave IR radiation, a significant component of solar radiation, is reflected by clothing, whereas long wave IR, the major component of indoor IR, is absorbed.

The Belding and Hatch HSI (as modified by McKarns and Brief) can be calculated or obtained from charts and takes into account clothing and work rate, and from it can be obtained recommended durations of work and rest periods, as with WBGT.

The calculation of the HSI is as follows:

For a lightly clothed man, with mean radiant temperature t_r, black-globe temperature t_g, and dry bulb temperature t_a (all in °C), air velocity v in metres/min, and vapour pressure of water in air, p_w in mbar:

- $R = 7.93 (t_r - 35)$
- $C = 8.1 V^{0.6} (t_a - 35)$
- Maximum heat loss due to sweating, E_{max}, is $12.5 V^{0.6} (56 - p_w)$ watt
- Mean radiant temperature $t_r = t_g + 1.82 V^{0.5}(t_g - t_a)$.

The vapour pressure of water in air is found from a psychrometric chart using t_a and the wet bulb temperature.

Some further revisions, also for lightly clothed men, to what are known as the Fort Knox coefficients (the 7.93, 8.1 and 12.5) above, have been made. An extra 17% can be added to each of those three figures.

The metabolic rate, M will vary with the amount of work the body is doing. Light work involves production of 115–230 W, moderate work 230–400 W, and heavy work 400–575 W.

The evaporation required is given by $E_{req} = M + R + C$ watt

$$\text{Maximum time in the hot environment before resting} = \frac{4400}{E_{req} - E_{max}} \text{ minutes}$$

$$\text{Time needed to cool off} = \frac{4400}{E_{max} - E_{req}} \text{ minutes}$$

(Note: the thermal conditions in the cooling off area must be measured to do the second calculation, using the equations above.)

Reporting thermal environment results

Reports on thermal comfort surveys follow similar lines to noise and chemical contaminant surveys. However, in addition, descriptions of clothing and estimates of metabolic rate based on work being done, are important. The report also needs to refer to fitness, acclimatization, solar protection, rest periods, and fluid intake.

As shown, both the WBGT and HSI allow estimates to be made of the recovery time needed between bouts of active work in hot environments.

Finally, there are work situations where the Siple and Passel (US Antarctic explorers) 'wind chill index' may be valuable for assessing work in cold conditions.

Vibration

Whole-body and localized (segmental) vibration

Vibration can be a source of discomfort, fatigue or permanent medical conditions. Both whole-body and segmental (e.g. hand-arm vibration) are important. Many of us are aware of the effect of side-to-side roll of about 0.25 Hz (hertz or cycle per second) on a boat.

Lower frequencies cause most problems in whole-body vibration because, for example, internal organs tend to resonate to those frequencies. Higher frequencies, such as those from poorly designed chain saws can, for example, cause the loss of blood supply and nerve activity in fingers (called 'Raynaud's' syndrome or 'vibration white finger'). Cold conditions worsen this. Vibration from travelling over rough ground in a vehicle (e.g. in a skid-steer loader) for long periods can cause spinal damage.

A large number of jobs give rise to problems; e.g. working with hand tools, or working near machinery with 'eccentric' drives, such as ore crushers.

Parameters used to describe vibration

Vibration can occur in a number of 'planes' or about a number of 'axes' (plural of axis). For example, side to side, up and down, forward and back. Each of these can be in a straight line or circular (as in the case of the roll of a boat). Think of pitch, roll and yaw, which give a combined effect.

Apart from the direction/s of the vibration, a second 'parameter' or way of describing vibration is the 'intensity', that is how far the body or part of it, is moved each side of a fixed point by the vibration.

The 'frequency', that is the number of vibrations per second, is also important.

The major parameter used to measure vibration is 'acceleration'; that is, the acceleration of a vibrating part as it moves in a particular direction. This takes account of intensity and frequency.

Ergonomic design problems and localized vibration in power tools

Power tools and power equipment of various kinds can be subject to vibration. The tool can be hand held, such as a building sand pad compactor, a chain saw or jackhammer, or it can be ridden in or on, such as a dozer or a blast-hole drill in an open-cut mine.

The vibration of hand-held tools can, as noted earlier, cause vibration white finger or Raynaud's syndrome. Good design can reduce transmission of vibration to the hands.

Jackhammer design, for example, has been improved by including counterbalancing parts which move 180 degrees out of phase with moving parts.

Ergonomic design problems and whole-body vibration

Whole-body vibration can occur when driving vehicles or using machinery. Whole-body vibration from equipment an operator is working in or on can produce headaches, affect vision and digestion, and cause progressive deterioration of the spine. The spine can also be more acutely affected by 'impulse' vibration, such as jolting from riding vehicles or machinery over rough ground. Pitching, rolling or yawing can produce feelings of nausea.

Various solutions have been devised to reduce vibration in hand-held tools and in machinery and vehicles which are worked in or ridden in. For example, the 'coupling' which occurs between moving parts and those parts in contact with the operator, which is needed to transmit vibration to the operator, has been reduced by sandwiching material of less stiffness than steel, e.g. rubber.

Ergonomic reduction of vehicle seat vibration

Careful design of a vehicle can reduce discomfort, fatigue and damage from vibration to parts of the body such as the spine. A number of aspects need to be considered:

- operating conditions, e.g. surfaces traversed
- type of vehicle, e.g. utility, earthmover, skid-steer front end loader
- suspension and damping – mounting, springs, shock absorption, including tyres
- seat mounting, design and damping.

In vehicles and mobile machinery in which an operator is seated, the springing and damping of the seating should be adjustable so that it can be set up to minimize the vibration transmitted from the terrain or from the work, e.g. rock drilling.

Although it is not directly a part of seating, transmission of vibration to controls, e.g. pedals, levers, steering wheel must also be considered.

Non-ionizing radiation

Introduction

The radiation in our environment, apart from alpha and beta radiation (which is usually thought of as small particles), consists of waves which are partly electric, partly magnetic.

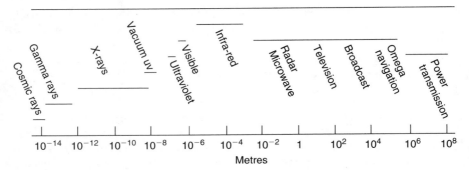

Figure 10.8 Wavelengths of various components of the electromagnetic spectrum (with acknowledgement to Grantham, D. (2002). *Occupational Health and Hygiene*, 2nd edn. Melbourne, AIOH)

The different types of *wave* radiation are only different because the wavelength and frequency change. The speed of the waves is always the speed of light, so the larger the length of each wave, the lower the frequency, because:

Frequency × wavelength = speed of light, 3×10^8 m/s.

Non-ionizing radiation in the electromagnetic spectrum

We can draw a spectrum of the different types of wave radiation showing where one type merges into the next. See Fig. 10.8. Notice where radio frequency, microwave, infra-red and visible radiation are.

These are all described as non-ionizing radiation. That is, they cannot knock electrons off the atoms in the body to upset the chemical bonds in the compounds there. However, microwave radiation of the right frequency, 2.45 GHz (giga = one thousand million), can heat internal tissues (cook them) just as a microwave oven does, and it may cause mutations by coagulating DNA. (Note: very high frequency, therefore very small [micro] waves.)

UV, laser, IR, microwave and RF radiation in workplaces

Ultraviolet radiation

Ultraviolet radiation is broken into three parts: UV-A, UV-B and UV-C. UV-A (the UV closest to visible violet light) is considered relatively safe and is used in solarium tanning lamps but it can cause sunburn and photochemical cataracts on the cornea of the eye. UV-B and UV-C are higher frequency and contain more energy; UV-C particularly having enough energy to disrupt chemical bonds in the DNA of skin cells and cause cancers such as melanoma. UV-B and UV-C can also cause corneal injury (photo-keratitis). Low intensity UV-B has been used in banks to reveal invisible signatures in passbooks. Intense UV is produced in arc welding, especially shield-gas welding of aluminium, but even gas welding with an oxy-acetylene flame produces hazardous UV.

Lasers

Visible radiation (visible light) can cause physical injury including eye injury when it is coherent and intense, as in lasers. Another factor is that the laser beam does not spread as a normal light beam does. Note, however, that there are also ultraviolet lasers which are extremely dangerous – they can cut concrete. There are also infra-red lasers.

Workplace uses for lasers vary. They include surveying, setting levels in tunnelling, high accuracy computer-controlled cutting of metal and plastic, and surgery, including eye surgery.

Infra-red radiation

Infra-red radiation can, of course, burn skin but near visible IR (IR-A) can also cause cataracts on the cornea of the eyes and burns on the retina. Intense sources of IR are found in furnaces, molten glass 'lehrs' and during molten metal pouring and casting. Long wave IR, IR-C, tends to burn only the cornea.

Microwave radiation

Microwave radiation of certain frequencies especially, as noted, around 2.45 GHz, can cause localized heating in the body due to the interaction of the radiation with water molecules. This can be particularly damaging to the aqueous humour in the eye.

Radiofrequency radiation

Military and communications personnel may be exposed to RF radiation heating in the body. Some sealing devices for plastics also use RF radiation. RF can produce local heating in the body, so the eyes face a particular risk. There is debate at present about the effects of mobile phones on tissues in the head, such as the potential for brain tumours.

Laser use, regulation and control methods

There are four classes of lasers, around which regulations are built. Check your local regulations. In most jurisdictions these will be regulations under radiation safety legislation. These may be administered by your local occupational health and safety authority or by a department such as health. The classes are:

1. Inherently safe. Totally enclosed or of very low power or interlocked (if open, off).
2. Eye protection given by aversion response. Accidental viewing not hazardous.
3. (a) 5 mW. May be hazardous especially if optical aids used.
   (b) Direct viewing hazardous – diffuse reflection not hazardous.
   Note: a direct reflection would be hazardous.
4. Direct viewing and diffuse reflection dangerous. A fire hazard.

When the eye focuses the laser beam, it concentrates the intensity by 100 000 at a point on the retina. Threshold Limit Values (TLV) are set for lasers by the ACGIH in the TLVs for Physical Agents, but the emphasis is on prevention; that is for example, room interlocks, and specially made goggles for the more critical lasers, and other controls.

Checking a microwave oven for compliance

One very common use of microwave radiation, both domestically and in the workplace, is of course the microwave oven. These are constructed with an interlock on the door which shuts off radiation if the door is open, and also have shielding in the design. However, as the molecular bonds in water are particularly easily vibrated, causing heating of fluids such as the aqueous humour between the cornea and retina of the eye, microwave ovens should be checked as frequently as regulations require. A relatively simple test device which responds with an induced current to microwave leakage (no battery needed) can be used to test the oven. The test consists of measurement at a distance of 30 cm and testing at 5 cm from the edge of the door. (A 5 cm spacer is provided to assist.) Results are in mW/cm^2. These should be compared with the current standards set by radiation safety regulations in your jurisdiction.

VLF electromagnetic fields

The effects of very low frequency electromagnetic fields such as near the electrolytic cells in aluminium smelters, and near high voltage power lines and equipment, are still being debated.

Ionizing radiation

Key types

Ionizing radiation disrupts atoms in the body's DNA and can cause mutations and cancer. In higher doses it directly upsets the body's biochemistry. The key types of ionizing radiations are:

- alpha particles (these are actually the nucleus of helium atoms)
- beta particles (these are actually electrons)
- gamma radiation
- X-rays
- cosmic rays
- neutrons

} none of these can be focused by electronic means

Cosmic rays, as their name suggests, arise in space and we are largely shielded by the atmosphere. Somewhat higher occupational exposure occurs in airline staff, astronauts and at high altitude.

Workplace sources of ionizing radiation

All types of radiation except cosmic may be present in nuclear power stations or non-power nuclear reactors, such as those operated for scientific or medical research or medical isotope production. Each type of radiation (except cosmic) can be given off from radioactive ore mining and production. This includes uranium ore, but also thorium containing minerals such as monazite, which is present in mineral sands mined to produce white pigment.

Different radioactive isotopes are fissile (i.e. they break-down):

- in different ways
- at different speeds
- give different 'daughter' isotopes, some of which are radioactive themselves.

Radon is a radioactive gaseous element which is found in some underground mines. It breaks down to give solid radioactive daughters which can remain in the lung.

Radioactive sources are used for medical treatment of cancer (cobalt 60), for industrial X-rays of metal components, for thickness gauges, for chemical process control, and to 'anti-static' plastic sheet. Radioactive isotopes are used in chemical 'tracer analysis', while X-rays are also used medically for diagnosis.

Internal and external risk

External exposure to radiation is easy to understand, but internal exposure occurs as a result of inhaling radioactive dust or gas, swallowing material containing radioactive particles, taking material into cuts or punctures, or absorption (in some cases) through the skin, e.g. water containing the tritium isotope of hydrogen.

The severity of the risk is shown in Table 10.6.

Note that alpha particles are most easily stopped by shielding, while gamma and X-radiation are least easily stopped. Neutrons of similar energy to other forms of radiation are more difficult to shield, and the internal hazard varies with their energy level.

Dose, quality factor and dose equivalent

Just as with other contaminants, dose is also important when considering the ill effects of radiation. Absorbed dose is today measured in gray (1 Gy = 1 Joule/kg) and submultiples

Table 10.6 Severity of radiation hazards

Severity	External hazard	Internal hazard
Least severe	Alpha	Gamma, X
Moderately severe	Beta	Beta
Most severe	Gamma	Alpha

Table 10.7 Radiation weighting factors

Radiation	Radiation weighting factor (W_R)
Photons of all energies (includes gamma)	1
Electrons and muons of all energies (includes X)	1
Neutrons less than 10 keV	5
Neutrons 10 keV to 100 KeV	10
Neutrons 100 keV to 2 MeV	20
Neutrons greater then 2 MeV to 20 MeV	10
Neutrons greater than 20 MeV	5
Protons other than recoil greater than 20 MeV	5
Alpha, fission fragments, heavy nuclei	20

Table 10.8 Tissue weighting factors

Tissue	Tissue weighting factor
Bladder	0.05
Bone marrow red	0.12
Bone surface	0.01
Breasts	0.05
Colon	0.12
Gonads	0.2
Liver	0.05
Lung	0.12
Oesophagus	0.05
Skin	0.01
Stomach	0.12
Thyroid	0.05
Remainder	0.05

such as milligray (mGy). But as you saw above, the severity (or quality) of the hazard is different for different types of radiation, so we give radiation different radiation weighting factors (W_R). See Table 10.7.

When we multiply absorbed dose (Gy) by radiation weighting factor, we get the dose equivalent (measured in sieverts – Sv). There is another factor 'N' which takes account of, for example, non-uniform spread of radiation but we will ignore it here – N usually equals 1.

There is also a tissue weighting factor to allow conversion of the dose equivalent absorbed by one tissue from a beam of radiation to be converted to a whole-body equivalent. See Table 10.8.

Typical low doses of radiation may be in millisievert quantities. Even living in a granite area can give 1 or 2 millisieverts annually. Permitted radiation dose equivalent standards are given in mSv.

Instruments to measure for ionizing radiation

A variety of instruments are available to detect and measure ionizing radiation. In this book their operation and use will not be described; they will simply be listed.

They include:

- Geiger-Müller tube
- ionization chamber
- proportional counter
- scintillation counter.

Film and TLD badges

Two other devices, because of their simplicity, may not be thought of as instruments. However, they are capable of giving quite accurate indications of dose. They are:

- film dose meters (film badges)
- thermoluminescent dose meters (which work like a badge).

A film badge can be designed with different outer shields to measure different types of radiation. When the film is developed (usually monthly) it gives a good indication of exposure to certain types of ionizing radiation. Thermoluminescent (TLD) dose meters absorb radiation and if they are later heated under controlled conditions the absorbed energy is given out as light which can be measured, again to work out the dose received.

Legislation on ionizing radiation safety

Legislation spelling out the precautions for dealing with ionizing radiation is largely uniform because it is based on international agreements, the key body being the International Commission for Radiological Protection (ICRP). However, you should check on the act, regulations, standards and codes applying in your jurisdiction and on which government authority administers them.

You need to check on licences and permits for operators; for use of substances; for use of instruments; and for selling, obtaining, storing and disposing of radioactive substances or devices. In particular, find out what aspects, if any, apply to your workplace.

Criterion and three principles of radiation protection

A criterion and three principles for control are used to ensure that radiation exposure is kept to a minimum. The overriding one is the ALARA criterion. That is, that radiation dose should be **as low as reasonably achievable**. In other words, no unnecessary exposure should occur.

The three principles used are:

- time
- distance
- shielding.

Dose is a product of time and concentration, so limiting the time of exposure reduces the dose.

Another aspect of time is the *decay* of radioactive isotopes. The 'half-life' is the time it takes for the isotope to drop to half the radioactive intensity. This can be microseconds to hundreds of thousands of years. Some isotopes used in medicine have half-lives of four hours. Plutonium (a synthetic radioactive element) has a half-life of 24 110 years (^{239}Pu) or 370 000 years (^{242}Pu). It is regularly shipped from East Asia to Europe and back again.

Iodine 131, which the thyroid takes up, is a result of uncontrolled emissions from nuclear power stations and some atomic bombs. It will be concentrated by the thyroid making its possible effect greater in a critical part of the body. It has a half-life of 8.04 days.

Strontium 90, also a result of bombs and uncontrolled power station emissions, behaves chemically like calcium, so it finds its way into cows' milk, and then, particularly, into children. The rapid cell division in growing children means that there is more likelihood that radiation-caused interference with cell DNA will lead to cell defects (mutations, cancers). It has a half-life of 28 years.

Distance is important. Where radiation spreads in all directions from a source, its intensity drops rapidly with distance from the source. At two metres, the intensity is 1/4 that at one metre. At three metres, it is 1/9th that at one metre. (This is the inverse square law.) So remote handling of materials, even just using long tongs, can greatly reduce dose. Note that this will not apply to a collimated (narrow and straight) or focused beam of radiation. Also, in some cases, so-called scattering and secondary radiation can affect the principle.

The third principle is shielding. Various materials absorb different types of radiation. Thin tissue paper stops alpha particles, while lead of varying thickness (depending on radiation intensity) is needed for gamma and X-ray radiation. Concrete and brick can shield but require greater thickness. Special barium concrete is more effective.

Neutrons are absorbed by water, or paraffin wax, so water is used for this purpose in nuclear power stations. The breakdown or fission of uranium which generates the energy results in many neutrons (high neutron fluxes or flows).

In medicine, another criterion also applies: the medical benefit must outweigh the risk from radiation exposure.

Further reading

American Conference of Governmental Industrial Hygienists (ACGIH) (latest year). *Threshold Limit Values for Chemical Substances and Physical Agents and Biological Exposure Indices*. Cincinnati:ACGIH.

American Conference of Governmental Industrial Hygienists (ACGIH) (latest biennial edition). *Industrial Ventilation*. Cincinnati:ACGIH.

American Industrial Hygiene Association Journal.

Annals of Occupational Hygiene (UK).

Applied Industrial Hygiene (USA).

Ashburner, I. (1990). Vapours, Gases and Fumes. *Health and Safety at Work*, **12(7)**, 18–20.

Baxter, K. and Rogers, G. (1997). Pumping Ions. *The Safety and Health Practitioner*, **15(1)**, 42–46.
Bernard, T.E., Dukes-Dobos, F.N. and Ramsey, J.D. (1994). Evaluation and Control of Hot Working Environments. *International Journal of Industrial Ergonomics*, **14**, 129–38.
Bruel and Kjaer. *Measuring Sound*. Naerum, Denmark: Bruel and Kjaer.
Cember, M. (1996). *Introduction to Health Physics*. 3rd edn. New York: McGraw.
Editorial. (2001). Adjusting Exposure Limits to Non-traditional Work Schedules. *Accident Prevention (Canada)*. September/October, p. 7.
Golding, D. (1990). *The Differential Susceptibility of Workers to Occupational Hazards*. London: Garland Publishing Inc.
Grantham, D. (2002). *Occupational Health and Hygiene*. 2nd edn. Melbourne, Australia: Institute of Occupational Hygienists.
Harrington, J.M. and Gill, F. (1999). *Occupational Health*. 4th edn. Oxford: Blackwell Science.
Harris, R.L. (2000). *Patty's Industrial Hygiene*. 5th edn. New York: Wiley Interscience.
Health and Safety Executive (UK). (1995). *Sound Solutions – Techniques to Reduce Noise at Work*. London: HMSO.
Health and Safety Executive (UK) (various years). *Methods for Determining Hazardous Substances Series*. London: HMSO.
International Standards Organization. (1982). International Standard 7243. *Hot environments – Estimation of the heat stress on working man, based on the WBGT index* (Wet bulb globe temperature). Geneva, ISO. Instrumentation and measurement see ISO7726.
Kroemer, K. and Grandjean, E. (1997). *Fitting the Task to the Human*. 5th edn. Chapters 18, 19. London: Taylor and Francis.
National Institute of Occupational Health and Safety (US) (various years). *Manual of Sampling Methods*. Washington: US Government Printer. Available online.
Plog, B.M., Niland, J. and Quinlan, P.J. (eds). (1996). *Fundamentals of Industrial Hygiene*. 4th edn. Chicago: National Safety Council.
Roach, S.A. and Rappaport, S.M. (1990). But They are not Thresholds: A Critical Analysis of the Documentation of Threshold Limit Values. *American Journal of Industrial Medicine*, **17**, 727–53.
SKC (latest edition). *Comprehensive Catalog and Air Sampling Guide*. International Edition. Eighty Four, PA, USA: SKC Inc.
Standards Australia. (1993–4). Australian Standard 1680.2.1–2.5. *Interior lighting*. Sydney: Standards Australia.
Tranter, M. (1999). *Occupational Hygiene and Risk Management*. Alstonville, NSW, Australia: OH&S Press. (Box 737, Alstonville, Australia 2477).
Waldron, H.A. (ed.) (1997). *Occupational Health Practice*, 4th edn. London: Butterworths. (Monitoring including gas liquid chromatography, pp. 311–26, atomic absorption spectrophotometry, pp. 330–3.)
Waldron, H.A. and Harrington, J.M. (1980). *Occupational Hygiene*. London: Blackwell Scientific.
Worksafe Australia. (1994). *Joint National Standard for Safe Working in a Confined Space*. (NOHSC:1009). Canberra: AGPS.

Ziem, G.E. and Castleman, B.I. (1989). Threshold Limit Values: Historical Perspectives and Current Practice. *Journal of Occupational Medicine*, **31(11)**, 910–18.

National, province, state or territory acts, regulations, codes of practice, rules, standards, and guidance notes on occupational exposure standards (chemical and physical), noise, lighting, hazardous substances, heat and cold stress, and ionizing and non-ionizing radiation.

Activities

1. What methods would you use to carry out a survey in a selected workplace for physical, biological and chemical stressors?
2. Draw up a checklist and carry out the survey in Activity 1 in a suitable selected workplace. Classify the types of stressors.
3. List five sampling methods which can be used for such stressors and the strengths and weaknesses of the methods.
4. Take part in the measurement of a workplace stressor (chemical or physical, e.g. noise).
5. Describe and illustrate in detail control measures which can be used for three different physical, chemical and biological stressors.
6. Visit a laboratory which measures workplace contaminants and comment on the use and limitations of either an AA, GLC, HPLC or other complex instrumentation.

11

Ergonomics

WORKPLACE EXAMPLE

Some men were at work shifting sea containers in Malaysia. After fitting the twist lock to a container before it was lifted one of the men disappeared from view. The crane on the quay moved, the man was dragged over six metres and died at the scene. A subsequent investigation reported that there was a high probability that he had been resting against the wheel of the crane.

The investigation also found that there were no rest areas and no guidance for employees. The recommendations included stricter supervisory control, a warning buzzer on the quay crane, warning signs, and an area where a worker could rest in comfort and safety. This is of particular importance for workers involved in manual labour carried out in a hot and/or humid environment.

(With acknowledgement to the Malaysian Department of Occupational Safety and Health.)

Origins and history

What is ergonomics?

Ergonomics comes from the Greek words 'ergon' (work) and 'nemein' (to arrange or manage). Today we understand it to mean a study of the *fit* between a person and the elements of the task they are required to perform. This is not confined to the workplace – it could, for example, apply to any car driver. In the United States this study has, until recently, been described as 'human factors engineering'. Ergonomics aims to design the task for the person, not the other way round. The ancient Greek story of Procrustes illustrates the point. Procrustes used to cut a visitor to fit the bed he provided, not adjust the bed. In ergonomics the layout, management system, methods, equipment and environment are all considered in relation to the inherent capabilities and limitations of human beings.

Research and development work in the United Kingdom, America and Australia

Some early work on ergonomics was done by a Polish researcher, Jastrzebowski, in the nineteenth century. But even prior to this some of the principles were recognized by Bernardino Ramazzini when he wrote his book *On the Diseases of Occupations* in 1700. The term 'ergonomics' was coined by Jastrzebowski in 1857. It was reintroduced in the UK in 1950, but the study of the person within the system in the United States, known as human factors, which has a variation of emphasis, goes back well beyond that.

The heat stress to which submarine and tank crews were subjected was studied during the Second World War. Work and physiological demands were also considered. Efficient armaments production in World War 1, when large numbers of extra women became part of the workforce, demanded attention to methods of designing production. Working hours and overtime in relation to production, and relationship of temperature to accidents, were studied. The interpretation of radar and sonar signals by the operators involved new ergonomic considerations, because the senses were *extended*. Extending the sight sense with infra-red night goggles deserves care, as the 1996 Blackhawk helicopter disaster in the Australian Army showed.

Work-study, including time and motion studies, looked at efficient ways to design production. Tayloristic production-line methods paid too little attention to the whole person and their needs and, although they are now recognized as a mistake, they may have been the best approach in, for example, the World War II production of Liberty ships.

Studies of human performance have been many and varied. They include human response to controls (dials, levers, switches, displays), human performance under high and low gravity (for fighter aircraft and orbiting space stations), human behaviour in relation to complex systems (nuclear power station operation), and effective sorting (mail exchanges). More recently the interaction of humans and computer terminals has received a lot of attention.

Matching the physical dimensions of a workstation with the dynamic and static physical dimensions of people is known as anthropometry. Anthropometry has been used to design many things for effective and compatible use by the majority of people – for example, bench and chair heights – but it is only relatively recently that children have had purpose-designed toilet pedestals, for example. Biomechanics has been used to study the muscles, tendons, ligaments, joints and bones used in lifting, pushing, pulling and twisting. Anthropometry and biomechanics have been utilized together to examine the design of, for example, hand tools.

Heat stress has also been studied in army personnel undergoing rigorous training, initially by Yaglou and Minard in the USA (and continuing to this day at Fort Detrick); in UK submarine crews by Bedford; and in bushfire fighters in Australia by Brotherhood and Budd. Performance in Antarctic cold has been studied by the latter researchers and United States researchers such as Siple and Passel.

Shift work, its psychological effects and the fatigue associated with it, have also been the subject of intensive study by researchers such as Moore-Ede in the USA, and Wallace in Australia. Repetitive tasks such as writer's cramp have been reported on since a late nineteenth century study of people working all day with quill pens.

Clearly, the needs of physically or mentally challenged people form an area of special consideration in ergonomics – for example, rehabilitees, and those in 'sheltered workshops'. A visit to a centre displaying items for the challenged reveals the thought which has gone into overcoming the problems associated with reduced sight, reduced movement, reduced hearing, reduced motor coordination and so on.

Scope and levels of ergonomic activity

Three different levels of work which involve ergonomics can be found in the workplace. They are:

- work station design
- workplace design
- job design.

Work station design involves the disciplines mentioned above – anthropometry and biomechanics (which involves functional anatomy and physics) – as well as engineering and psychology. Psychology is involved because it looks at how people receive information or stimuli; how they process this, or these, in the brain; how they make decisions; and then how they act upon them. Experience and training can improve perception and decision making. Perception of some stimuli, such as visual ones, is affected by old age.

Workplace design involves fitting work stations into the overall physical design of a workplace. This will include issues such as noise, temperature, lighting, colour and rational movement of people and materials.

Job design takes into account what has been said above, but also looks at the way tasks are broken up, the decision-making processes and the operations of work groups. Important considerations are level of mental stimulation (under or over); conflict between different people and different work areas; meaning in the work; and degree of control (consultation). Motivation and hence safety culture are important elements in this. The psychology of errors and types of errors are important aspects of safety. Physiology plays a part in examining the way people perform, particularly in extremes of temperature and humidity.

Development of ergonomics for the work station, workplace and job design levels

The earlier description of some areas of ergonomics in which research and development has taken place can be analysed further using the levels just identified; that is – work station, workplace and job design. Clearly many areas of what we consider mainstream 'workplace safety' fit these ergonomic levels. Think of, for example, machine guarding using mesh to prevent hand entry; pedestrian zone markings in workshops and safe job procedures; and the maintenance of vigilance and attention to detail, which can be affected by, for example, heat, noise and vibration. So some of the trends developed over time in the safety arena can now be seen just as surely as ergonomic advances.

The person–machine model

Components of the ergonomics person–machine model

While there are five major components within the ergonomic system – the person, the machine or equipment or tools, the job procedures, the materials, and the environment, which together with the management system, accomplish the task – in this section we will concentrate on the person–machine interface or model. This involves considering some of the attributes of, and issues relating to us as persons shown in the table below.

There is information on human attributes in Chapter 1, especially the Human Factor parameters under 'Technical Specifications'.

Table 11.1 compares the things which people do best and those which machines do best.

Table 11.1 Comparison of human and machine attributes

Person vs machine	
People excel in:	*Machines excel in*:
Detecting certain stimuli of low energy levels.	Monitoring (both people and machines).
Sensing an extremely wide variety of stimuli.	Performing routine, repetitive, or very precise operations.
Perceiving patterns and making generalizations about them.	Responding very quickly to control signals.
Detecting signals in high noise levels.	Exerting great force, smoothly and with precision.
Storing large amounts of information for long periods – and recalling relevant facts at appropriate moments.	Storing and recalling large amounts of information in short time-periods.
Exercising judgement when events cannot be completely defined.	Performing complex and rapid computation with high accuracy.
Selecting own inputs.	Sensitivity to stimuli beyond the range of human sensitivity (such as infra-red, radio waves).
Improvizing and adopting flexible procedures.	Doing many different things at one time.
Reacting to unexpected low-probability events.	Reasoning deductively – going from general to specifics.
Applying originality in solving problems, i.e. coming up with alternate solutions.	Being insensitive to extraneous factors.
Profiting from experience and altering course of action.	Operating very rapidly, continuously, and precisely the same way over a long period.
Performing fine manipulation, especially where misalignment appears unexpectedly.	Operating in environments which are hostile to people or beyond human tolerance.
Continuing to perform even when overloaded.	
Reasoning inductively – specifics to general.	

(*With acknowledgement to Woodson, W.E. and Conover, D. Human Engineering Guide for Equipment Designers, copyright* © *1964 Wesley Woodson, reprinted by permission of University of California Press.*)

Integration of human, machine and environmental factors in work stations and workplaces

There are a large number of examples of work station and workplace scenarios which illustrate the integration of human, machine and environmental factors. Three examples of those will be given here:

1. Pliers, paint scrapers and the pistol-grips on electric drills all require attention to the tendons, muscles and joints of the hand and arm used to grip and control the device. A close examination is needed to ensure that the position of the parts of the hand doesn't result in early fatigue and cramp, and inappropriate direction of application of force through the forearms. Grip will be affected by sweating due to environment and physiological (work effort) factors. Chain saws and building site soil compactors add vibration to this list of environmental factors.
2. Control of an aircraft involves integration of a complex range of information about the machine and its spatial position in the environment. The vertical angle in relation to the line of flight governs stall; lift is affected by speed and flaps; and information about height and banking angle may be partly visual and partly through pressure receptors in the skin, but also relies on instruments. The psychology of perception becomes important. The Air New Zealand aircraft disaster on Antarctica's Mt Erebus in 1986 resulted from an inability to distinguish sky from landscape in certain weather conditions (plus an error in programming the flight control computer). Wind shear on landing requires split-second ability to respond.
3. Assembly and soldering of coloured wiring or connections to printed circuit boards with further assembly, requires the ability to distinguish colours clearly – remembering that a significant percentage of men, in particular, are colour-blind – and to maintain care and vigilance during repetitive work. Noise and heat or humidity will affect concentration, but heat and humidity are often controlled not so much for the operator's comfort as to protect the electronics.

Differentiation between manual, mechanical and automatic systems

Within the framework of the human–machine model, we can distinguish manual, mechanical and automatic systems of work for accomplishing a particular task.

The electrical/electronic assembly job just mentioned is mainly a manual system, although mechanical jigs may be used to hold parts.

Mechanical scenarios are many. They include metal presses, buzzers for dressing or planing timber, lathes and drill presses. Automatic or semi-automatic systems include contact or spot-welding machines on vehicle production lines, either controlled by an operator or fully robotic, automatic control of chemical process facilities such as oil refining (with a manual override), car cruise control and aircraft autopilots. Mechanical systems place large amounts of energy at the operator's disposal so a high-powered car and an inexperienced driver can be a fatal combination.

Anthropometry

Static and dynamic anthropometry

Static anthropometry is concerned with a variety of measurements of the human at rest, such as width, girth, height, weight, heel to back of knee (popliteal) height, etc.

Dynamic anthropometry deals with the range of movement of people and also links with biomechanics when considering speed of movement in reaction to an event. 'Reach' is an important aspect of dynamic anthropometry. Strength is an important aspect of biomechanics and varies more widely than anthropometric data, e.g. compare the strength of 18 and 79 year olds.

Static and dynamic anthropometric data collection

Static and dynamic anthropometric data is collected by taking measurements from general or selected populations which result in generally available tables to assist design. These are updated from time to time.

General user populations and specific user populations

Static and dynamic anthropometric considerations are used to design for general users or specific users. Suitability for 95% of the general population (i.e. within two standard deviations from the mean) is often the aim in design. For a chair, the popliteal height for US men in this range is from 39–49 cm and for women 36–45 cm, for ages 18–79 years old.

In specific user groups such as schoolchildren, chairs designed for Year 1 must be different from those for Year 7. As an extreme example of the problem, consider the Efe and Dinka of central Africa mentioned by Pheasant (see Further Reading at the end of this chapter). Both live in the same country, but one group are pygmies from the jungle and the other are very tall savannah dwellers. Imagine the dilemma a central African ergonomist is faced with. As another example, ethnic differences in facial characteristics have an effect on the design of respirators for an effective facial seal.

Key stages of the ergonomics design cycle

The key stages of the ergonomics design cycle are:

- establish and define the user population
- locate or create the data relating to that population
- work out what the equipment or work station is required to do
- try prototypes out on the user population

- make sure that you are satisfied with the safety and effectiveness of the design in the hands of the user group
- work out what percentage of people the design actually suits
- consider whether the above information indicates the desired or a satisfactory result
- test the design in use
- monitor results of the test and adjust as necessary before full production
- ensure a design commensurate with the level of training which can be realistically provided
- ensure a design suited to the operational environment, e.g. extreme heat or dust
- ensure that the full design parameters are set, e.g. noise levels
- ensure that equipment gives the operator adequate *feedback* on its operation and continuing safety, examples being tool *feel*, car road-holding, various warning alarms.

Use of static and dynamic anthropometric data in the ergonomics design cycle

Different stages of the design cycle above will require different anthropometric data. The controls on a machine may be designed to suit the dimensions of the hand, but the operation of them may require consideration of reach distances. The shape and size of a tool may be good, but the weight may not be suited to data on strengths, particularly when it is held in an extended position. For example, a tool such as an underground miner's airleg drill is very heavy but manageable when held correctly. (Note: strength is a biomechanical rather than anthropometric issue.)

Relative merits of methods of collecting and applying anthropometric data

Review of anthropometric data collection methods

Anthropometric data collection to define human requirements for work station design and redesign requirements, involves assembling data based on measurements of *populations* of people. The measurements can include:

- Structural anthropometric data – measuring bodily dimensions of people in fixed positions, either from one part of the body to another or to a fixed external point, e.g. the floor.
- Functional data – the movement of a part of the body in relation to a fixed point, e.g. the reach distance of a seated person.
- Newtonian data – this looks at the body as a system of forces, levers and load bearing structures and allows best handling of objects, and use of displays and controls by

selected appropriate *regions*, e.g. the position of the body and the leg when force is to be applied to a pedal.

Data of these types is analysed by normal statistical techniques.

We take much of this for granted, but it applies everywhere, not only in workplaces. We don't necessarily need a maximum height doorway but we do need a minimum height. Car driving seats must suit most of the population within the range of adjustment. Clothing and shoes must be made in size ranges rather than one piece being designed or adjustable for most people. Stools must bear weight adequately. (Sumo wrestlers, please don't sit!) Racial differences pose difficulties in a global market. The Efe and Dinka (one very short, the other tall) of central Africa mentioned earlier illustrate the difficulties. Some occupations – e.g. fighter pilot – of necessity restrict the range of people acceptable on economic, payload/fuel load, aerodynamic, trim and design grounds. The author Roald Dahl, a six footer (183 cm), described how he had to be prised out of a World War II fighter after flying it from Egypt to Greece. Those selected for professional basketball may well experience ergonomic problems in normal houses, offices and aircraft seats. (Those seats may also pose the threat of deep vein thrombosis.)

Limitations of applying only static anthropometric data

In defining human requirements in work station designs, it is fairly clear that static data alone is not enough. As one part of the body moves, so do others, and the forces on body parts change. In reach, for example, we need to know the arc within which the hand moves. Certain positions put maximum stress on muscles (and use a lot of energy); other positions are less stressful and use less energy.

Reliability and validity problems in collecting static and dynamic anthropometric data

Reliability when carrying out research to measure something refers to the degree to which the measure produces consistent results if the research is repeated. For example, on a purely physical measure, such as popliteal height (floor to back of knee) reliability might be good. However, let us assume that we have set up a work station with certain recorded adjustments and the operator, when asked, says the overall feel is good. Then we pretend to re-adjust the work station, but actually use the same settings without the operator knowing. We again ask about the overall feel. The operator says it doesn't feel quite right. In other words, on subjective questions in particular, reliability can be a big issue. Another way of looking at reliability is to try to measure something such as upper body strength with two different measures. If the statistical results match well, we can say the test was reliable.

Validity looks at the matter from another angle. To measure anything we have to decide what measure we will use. Once again, we could measure popliteal heights to decide what height a chair squab should be. But, if we use a standard barefoot measure, this may not provide comfort if people in reality wear shoes with a range of heel heights.

In other words, we would need to think again about the exact type of measure we use. So it is important to have exact standards covering how a measurement is to be done.

The time when the data was collected can affect reliability. Human populations change – for example, as people in Asia achieve greater wealth, diet changes and so do body dimensions. However, this time change is normally described as instability. Many surveys are carried out on a particular group of people for a particular purpose, and so the data is not reliable when used for other purposes. Data on children is important but has often been collected on a particular group and may not apply to other groups who live on a different diet because of different socio-economic position, or who are predominantly from a different ethnic group.

Reliability issues in specific design scenarios

The reliability of anthropometric data will affect the success of the outcome if that data is used for specific design scenarios. Data which is not valid, such as that collated according to one set of standards for measurement, while another is used when it is applied, can lead to costly mistakes; for example, if a region of reach distance is measured allowing for forward movement of the torso, and this is then used in a situation where forward movement of the torso is very limited because the torso is strapped in place, as it might be in a pilot's seat. Handreach improves if feet can move; seated reach is better if there is no backrest. So collecting data in one way and using it in another is undesirable (and a person using the data may not be aware of all the conditions applied when it was collected). The data may be unreliable simply because not enough attention was paid to how one body movement affects other body positions, and so when the tests are repeated, the results are not statistically the same.

In other words, if validity is poor, reliability can be poor as a result because when the research is repeated, the same results may not be obtained.

Using anthropometric data

Applying anthropometric data to a variety of workplace scenarios for the purpose of defining optimum use requirements presents several challenges.

The range of possibilities for applying this data is quite large. However, you may wish to take something quite simple like a chair and decide, if it is not adjustable in any way, what the dimensions should be to suit a certain proportion of the normal population. Or you could take a certain desk and work out what proportion of people will be comfortable working with it.

You could then consider, if you were only allowed in terms of price to allow one adjustment, what that adjustment would be. There is more than just height to consider with a desk, particularly once you put the user on a chair. Another possibility is to consider the design of kitchens or laundries, domestic and commercial.

One accident occurred because Mexican women workers were using mechanical presses designed in Germany for men, and the travel limit stops were tampered with simply so that the women could operate the machine.

Common forms of occupational overuse syndrome (OOS) and preventative ergonomic strategies

Occupational overuse syndrome – OOS

OOS, or occupational overuse syndrome, is a condition of generalized or particular mental or physical fatigue, which may be accompanied by physical pain, and results from overuse of particular parts of the body to perform work tasks.

Particular types of OOS include carpal tunnel syndrome due to swelling of tissues in the wrist from activities such as excessive keyboarding, and *tenosynovitis* where the synovial membranes in which tendons are covered, such as the tendons in the hands, become swollen and inflamed.

Common forms of OOS

An early form of OOS in people employed as scribes in the nineteenth century was 'writer's cramp'. A number of musculoskeletal disorders of the upper limbs (arms) have well-known sources related to tool design, wrist posture, or task design, and the solutions are known.

However, there are a number of syndromes (conditions) the underlying pathology of which is unknown; that is, there is no well-identified reason for the condition.

RSI (repetitive strain injury) refers to the cause being repetitive work, but the causes of the syndrome are broader than that and can include chronic muscular overloading with perhaps a psychological aspect as well.

Use of the terms varies, but often RSI or OOS refers to the arm. However, it can extend to the shoulder and neck. Particularly with the advent of VDU (VDT or computer) work stations with very small key-touch movements and high stroke counts, the problem became a big issue. Sometimes inflammation could be observed along with pain; sometimes there was just pain, with aching, weakness or fatigue also noted. Where there is a psychological aspect to the problem, simply adjusting the anthropometry and physical ergonomics won't be enough; there will be a need to look at organizational structure, and management and task design.

Some further work suggests that the nervous system may be the key. For example, movement in one part of the body can affect nerves at a distant point in the body. The condition is also connected with long periods of static muscle loading; that is, there is little movement, the muscles compress blood vessels, restrict oxygen supply to nerves and muscles, and restrict the removal of waste products.

Types of OOS which can be directly observed by a doctor are:

- carpal tunnel syndrome, noted earlier, where a ring of soft tissue around the tunnel through which the nerves and tendons of the wrist pass becomes inflamed
- tenosynovitis, also noted earlier, where the smooth lubricated synovial membrane round the tendons in the hand becomes inflamed

- tendonitis – where tendons become thicker and inflamed
- peritendonitis – similar to tendonitis but occurring where tendon joins muscle
- tennis elbow – muscles/tendon injury at the elbow, often due to twisting movements of the forearm.

One other type of OOS is:

- tension neck – the shoulder joint and neck muscles are affected by long periods in a relatively fixed position, or indirectly due to static loading in the arm-hand structure.

Back strain also may be involved.

Risk factors in the development of OOS

Risk factors which can lead to OOS include:

- monotonous, fast-paced work with little control over pace of work, pauses or rest breaks
- parts of the body held in positions which require continuous static loading on muscles
- use of parts of the body to carry out repeated, fine, precise movements
- use of parts of the body in repeated strained positions, e.g. poor tool design or, in microelectronics task design, can impose unnatural configurations on the hands.

In some cases the effort and type of continuous mental activity may contribute to the problem, e.g. monotony mentioned above, or the amount of demand for precise detail, e.g. prolonged 'spell-checking'. It is important to use known risk factors to identify possible problem areas, assess them and institute control measures, e.g. maximum key-stroke counts for any particular work day, rest breaks, etc.

Work station, workplace and job design factors and OOS

A number of factors contribute to the types of conditions which are covered by the term OOS. These can be:

- Poor design of the task or equipment. The relatively rapid change from lever action typewriter to button press keyboards was done without sufficient thought about the functional anatomy involved and the likely effects of certain demands on that anatomy. This was combined with the unexplained effects of putting operators in front of print and figures of variable sharpness, contrast and light levels for long periods.
- Awkward positions can place unusual strains on the system *skeleton – muscle – tendon – joint*, leading to inflammation or pain. An example is a situation where a clerical worker has a well-designed VDU work station, but has a poorly designed typewriter work station (for labels and envelopes) where the legs cannot be placed under the desk because that is not the *official* work station.

Enhancing occupational safety and health

The first factor mainly looked at the work station aspects, but also touched on job design. Job design issues can involve how narrowly focused a job is. If a VDU operator does nothing else but operate the VDU, then there is a possibility that tension will cause build-up of waste products and constriction of blood vessels if the operator perceives no possibility of a change of pace or task, or becomes bored. Management has a role to play here. Many keyboard staff are expected to overcome the lack of good planning and scheduling skills by superiors by working to tight deadlines, so the issue of control over the task one does, how one does it, and when one does it, may be involved. Tension may also arise from tight demands on people doing their own keyboarding with poor keyboarding skills.

Other issues could best be described as *workplace* factors although in a sense all three factors fit that description. These could include poor airflow, crowding, noise, unsuitable humidity and/or temperature, and poor or badly placed lighting.

Although the examples given relate to VDU work stations, OOS can result from many types of jobs. These include electronic component assembly or soldering, hand sewing wire mattress mesh to a steel frame, and food preparation, to name a few. Shortsightedness is also thought to be growing problem from long hours at a VDU.

Key elements of an OOS prevention programme

These key elements of an OOS prevention programme to some degree flow directly from the factors which lead to OOS. Clearly, though, it is not always easy to reduce the risk while maintaining staffing levels or output. In overall productivity terms though, avoiding OOS problems (with their potentially long periods off work or the need for other duties by affected staff) should pay off.

The solutions include:

- equipment redesign
- genuine commitment by the employer (and managers) to resolving the problems
- regular breaks from a task
- use of 'pause gymnastics'
- adjusting work stations – some people think it is not worth spending time to do this
- early reporting when a problem is seen to be developing. This can only happen if the employee works in an atmosphere of trust where such reporting is treated seriously and not as *whingeing* – in other words, good consultation mechanisms exist
- job redesign to provide variety
- semi-autonomous work groups where the group have flexibility in how they get a task done
- use of a quality management approach, i.e. quality circles, etc.
- attention to issues like lighting, noise, temperature, etc.
- adequate information and training for those likely to be at risk.

Good posture and physical fitness will help, and posture can be assisted by correct choice of the type of equipment, such as fixed and mobile machines, chairs and desks.

Ergonomic preventative strategies to control the risk of OOS

The ergonomic preventative strategies for keyboarding include consideration of the following points:

Work systems: organization and design
- approaches to job design
- organizational and technological change
- aspects of computer systems design
- supervision
- work practices
- ergonomic factors in work design.

Workplace: organization and design
- work posture
- work station arrangement
- equipment design and positioning
- environment conditions.

Training and education
- target groups
- types of programmes.

(*With acknowledgement to WorkSafe Australia (1996) – see Further Reading at the end of this chapter.*)

Note the importance of work being correctly prepared by authors – for example, managers – prior to keyboarding by, for example, a keyboarding operator in a keyboarding pool. This minimizes unnecessary rework. Other occupations can also give rise to OOS.

Action plan for assessing manual handling and sources of OOS

The preparation of an action plan involves:

- consultation
- examination of existing injury data
- walk-through survey
- identification of problem areas
- prioritizing work stations and jobs to be assessed
- preparation of checklists
- assessment of sites, e.g. dates, times, etc.
- evaluation

- reporting with recommendations on control options including training
- consultation on the report.

Methods for identification, assessment and control of manual handling hazards

Biomechanical principles used for the identification and assessment of manual handling hazards

Biomechanics is a particular area of study which plays an important part in ergonomics. It views the body from the point of view of physics. It looks at how and where forces are applied by and to the muscles and tendons to overcome gravity and move loads. The skeletal system is viewed as a system of levers to which the forces are applied.

Manual handling

'Manual handling' means any activity requiring the use of force by a person to lift, lower, push, pull, carry or otherwise move, hold or restrain any animate object (e.g. a bull or a hospital patient) or an inanimate object (e.g. a wool bale, carton, shovel-full of sand or a concrete slab). Although the use of large wrenches has risks, this is not considered manual handling.

Manual handling of loads and objects is a very important source of workplace injury. The way load 'stresses' are applied to the body can result in a variety of *strain* injuries. We know that the further from the fulcrum a force is applied to a lever, the greater the 'moment' set up. In the case of a load being picked up manually, to get the load to move, the muscles must apply a slightly greater moment than the moment the force of gravity on the load creates. If the fulcrum is the hips, then the centre of gravity of a bulky object will be further away horizontally from the hips than that of a less bulky object. So even if both objects weigh the same, one puts a greater moment or stress on the muscles, tendons and skeleton.

A second aspect is to consider the biomechanics when different choices are made in picking up a load. For example, if we flex the knees, keep the load close to the body and then raise the load, we make use of the strong muscles of the upper leg. If we bend over, keeping our legs straight, and pick up the load, we are relying on weaker muscles in the lower back. These muscles and their attachments are more easily overloaded.

A third aspect is twisting movements of the skeleton, especially the spine. This relies on rotation of vertebrae, each of which is separated by a disc. Twisting movements of the spine with a held load can lead to rupture of a disc and leakage of fluid, putting pressure on a nerve, and so causing pain (see Fig. 8.7).

Risks in manual handling practices

Back pain is the greatest single cause of time loss attributed to work in industry in many countries. The onset of symptoms often bears no relation to a particular incident.

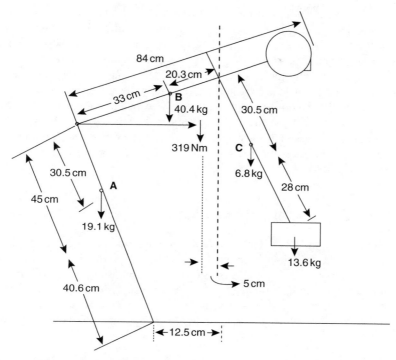

Figure 11.1 Biomechanics of lifting. This shows the forces acting on the legs, torso and arms when lifting using the lower back, not thigh, muscles. The torque on the hip joint is 319 Nm, which is very severe (adapted from US DHEW-NIOSH (1973): *The Industrial Environment – Its Evaluation and Control*)

It is frequently recurrent and the consequence of repeated and cumulative accidental and postural stresses encountered not only at work, but also in the home, in recreation and in travel.

We can study the musculoskeletal system of the body, which we use for holding the body in a particular position, (e.g. standing, sitting, squatting) and for doing various activities with the body (e.g. running, walking, lifting, shovelling, dancing, playing tennis, etc.). The system is made up of bones, cartilage, tendons, muscles, and ligaments and is controlled by nerves. It can be broken down into a series of levers. For example, Fig. 11.1 shows the lever system the body uses for lifting, and also shows why a thigh lift is better than a back lift.

Work effort framework of stress, strain, fatigue and injury

The work effort framework can be thought of as four parts – stress, strain, fatigue and injury. If we are thinking about the musculoskeletal system, the 'stress' as noted earlier is the load we apply to a part of the system. The 'strain', a certain level of which is acceptable, is the response of the system to the stress applied. Even if some type of load is not being moved, a load may still be imposed by part of the body, such as an arm. Holding it against the force of gravity will still require the muscles to use up body energy by converting glucose to carbon dioxide. So physical 'fatigue' can set in, and this may be felt as an ache or pain as lactic acid builds up in the muscles.

'Injury' results either more or less instantaneously when the strain of muscles, tendons or ligaments cannot match the stress or load applied, and these tissues are stretched or torn. Injury can also result from repeated lower level overloading, i.e. cumulative trauma disorder – CTD. Note that the word 'strain' or 'sprain' is also applied to the injury.

Static and dynamic muscle loading

As indicated above, muscle tendon and ligament loadings can be either 'static', e.g. holding a part of the body in a fixed position, or 'dynamic' where the body is moving a part of itself or moving an object. Dynamic damage can also occur if there is sudden deceleration of, for example, the leg, such as when a sportsman's leg encounters the ground too fast and at an angle where muscle support is poor so that there is damage to knee ligaments or even a broken leg bone – or where a construction site worker falls.

Strength and endurance

'Strength' generally describes the maximum force which a particular part of a person's musculoskeletal system can exert momentarily. Think, for example, of a weightlifter who quickly lifts a bar with maximum weight. This may be done without injury or it may produce injury. Strength varies depending on the way the body applies force – the foot exerts pressure best when the leg is outstretched rather than bent.

'Endurance' describes the ability of various muscles to continue to provide the strain necessary to support a load or carry out an activity. It depends on the ability of the muscle to accept and process enough oxygen from the blood to continue providing energy. This ability is assisted by fitness training. Of course, endurance is assisted if a person can tolerate the pain in the muscles from lactic acid build-up. Lack of blood supply to the muscles – such as when more blood flow is switched to the digestive system after a meal – can produce, for example, cramps while swimming.

Principles of biomechanics in lifting

The fundamental principles of lifting biomechanics involve identifying:

- the muscle, tendon and ligament groups associated with particular activities
- the bone attachments to these groups

- the fulcrum or pivot point around which the 'bony levers' move, e.g. lifting with the back involves the hip as a fulcrum; lifting with the legs involves the knee as a fulcrum.

Once this information is known, it is possible to measure the capacity and performance of these lever systems to perform, identify problem areas, and either develop better methods for using the body or provide an alternative method, e.g. using mechanical aids for lifting.

Causes of manual handling injuries

Manual handling injuries can result from personal factors such as:

- lack of physical strength or endurance for the job in hand, e.g. endurance is a problem if a person can't keep up a static muscle strain and falls as a result
- lack of fitness
- failure to warm up or limber up before a manual handling task.

Back injuries are sustained most frequently when lifting or setting down loads, but they also occur:

- in carrying, stacking, pushing, pulling, rolling, sliding and wheeling of loads
- in the operation of levers and other mechanical devices or where a large wrench suddenly unfreezes a nut
- in the maintenance of unbalanced postures while performing these tasks
- where normal movements of the spine aggravate existing damage to the discs between vertebrae.

The injuries may occur, for instance, in order pickers in retailers' warehouses, youths employed in supermarket trolley collecting, and in office workers moving files or books.

Many jobs require performance of tasks in the seated position. Constant sitting in poorly designed chairs, in stressful postures, to carry out visual or manipulative tasks is a common source of recurrent or persistent back pain. Appropriate workplace and chair design promote maintenance of correct posture and the prevention of postural strain.

Aggravating factors are:

- gross overloading of the spine
- lifting with a bent back or with the object held well out from the body or to one side
- loss of balance while lifting
- using a jerking or twisting lifting motion
- unexpected weight bearing
- restricted room to manoeuvre
- inadequate grip on heavy, awkward, hot, corrosive, sharp or slippery objects
- unstable footwear; slippery, uneven or littered floor surfaces
- uncoordinated team lifting
- work by a person unfit to lift.

Stress factors

The occurrence and severity of low back and other musculoskeletal injuries increases directly with increase in the distance of the centre of gravity of the load from the handler's body, whether this increase is occasioned by bulk of load or workplace layout. Refer again to Fig. 11.1.

Gender, age, body weight, height, state of health and fitness, training, spinal structure and other personal attributes all influence ease and safety of handling. Protective clothing, nature of floor surface, access to the load, space constraints and general workplace environment also affect ease and safety of handling. The lifting capacity of a given worker decreases progressively with the increase in the initial height of lift above the floor.

Requirements for manual handling in legislation, codes and standards

For this refer to the regulations on manual handling in the piece of OHS legislation under which you work. The key requirements are policy, consultation, design and risk identification, assessment and control.

Assessment of manual handling tasks

Issues

Assessment of manual handling risks includes the following issues:

- actions and movements
- workplace and work station layout
- working posture and position
- duration and frequency of manual handling
- location of loads and distances moved
- weights and forces
- characteristics of loads and equipment
- work organization
- skills and experience
- age – young and old both require extra consideration
- clothing – does clothing or PPE interfere in manual handling?
- special needs – created by, for example, disability, handicap or impairment, returning to work after illness, or pregnancy.

(*With acknowledgement to Worksafe Australia (1990) – see Further Reading at the end of this chapter.*)

We might also include gender and fitness level. Snook and Ciriello in the USA have also developed a checklist for manual handling (see Further Reading at the end of this chapter).

A checklist for the investigation of manual handling tasks is very useful. It should refer to matters shown in the bullet points above. It is best used at the stage of designing the job to fit the worker.

It is impossible to specify with any accuracy, maximum weights that may be handled in industry because of the variety of factors in the load, the handler and the handling environment, which affect ease and safety of handling. However, the Worksafe Australia *Code of Practice for Manual Handling*, for example, recommends that no person, whatever their fitness level or strength, should be required to handle a load above 55 kg without some form of assistance.

Manual handling task assessment and general and specific user populations

Care needs to be taken when assessing manual handling tasks. As mentioned before, the youth of a worker; the level of physical strength and build of workers; the health status of workers; pregnancy; and physical impairment must all be considered. People less than 18 years old are still developing physically and single lift loads should not be more than 16 kg. Older workers may have a loss of physical strength but often overcome that by using skill and experience in how they perform a task. People whose jobs do not usually require them to lift heavy objects are more prone to injury if they do. From middle age on, low back problems are more likely because of age effects on the lumbar spine.

Stress on the trunk from lifting heavy loads should be avoided by women, given that it can increase menstrual flow and menstrual pain, and that there is also the possibility of osteoporosis in women at menopause, reducing the strength of bone. This is of particular importance for women workers, such as nurses, in the health care sector. Research suggests training won't work – manual lifting of patients is inherently unsafe. Task redesign is the answer, and often involves mechanical aids.

Safe manual handling task requirements

Such requirements vary – for example, for seated and standing work. The spinal column itself is actually quite weak. The reason it can take considerable loads is that it is part of a whole system – a surround of muscle layers together with what is in the chest and abdomen. When a person leans forward, the muscles in the back become more active, then when they lean back the muscles to the front respond. However, when a person leans to lift a load the muscles towards the front also squeeze to increase the pressure in the abdomen and so provide extra support. It should be noted that use of back belts can adversely increase pressure inside the abdomen, and continual use may weaken supportive muscles.

The trunk can actually fail in several ways when a load is lifted. When the ring-shaped fibres in a disc tear and allow soft material in the disc to move towards the rear, it is called a 'slipped disc', but it is not actually the disk which has slipped. Such an injury can prevent any more strenuous work or leisure activities – a serious problem. So in the USA, for example, the National Institute of Occupational Safety and Health produces a work practices guide for manual handling. This is based on an equation which lets the safe load be calculated. A seated person faces different risks to a standing person. When a person is seated, the spine is actually already under some stress. The leg muscles are not in a position to help provide a reaction against the ground, so the muscles in the trunk have to do more.

In lifting, the greatest stress occurs at the beginning when the movement of the load must be initiated and accelerated. Preparatory tightening of the abdominal muscles increases pressure within the abdomen, which takes some of the leverage off the spinal muscles and reduces the compressive force on the discs. Controlled movement, in which the shoulders and hips are raised at the same time during the initial stages of the task, imposes the least strain on the back. Jerking and snatching at the load are potentially hazardous. Strain injury in manual handling of loads is quite unlikely to occur when the handler has prepared properly.

Bending down while seated to get the heavy contents of a bottom drawer or a heavy file placed on the floor and to the side of the chair, for example, creates a real risk of back damage because the file, down and at the side, requires twisting as well. Moving objects from one side to the other while seated is a greater risk than standing and doing the same task.

Prevention of manual handling injuries

Key elements in a manual handling programme

The three-stage approach of risk identification, risk assessment and risk control should be used. The control measures agreed on are then the requirements for particular manual handling tasks.

These include:

- job redesign
- mechanical handling equipment or mechanical assistance
- provision of training
- auditing manual handling
- investigation of accidents and injury involving manual handling.

Job redesign includes:

- modifying the article to be moved
- modifying the layout of a work station or workplace
- rearranging the flow of material
- changing the workers' actions, movements or muscular forces used
- modifying the task by using mechanical assistance
- modifying the task by using other people to help with lifting.

Ergonomics

Figure 11.2 Airleg mines (photo courtesy Elaine Cullen, US NIOSH, Spokane)

A task requiring good control of equipment balance to avoid muscular strain is shown in Fig. 11.2. All-up weight of an airleg drill like that shown can reach 50 kg.

Prevention strategy

Because back injuries and manual handling accidents have the potential to cause severe and prolonged disability, a prevention strategy must be adopted.

- Ensure that jobs involving physical effort are correctly planned and carried out.
- Ensure that employees receive adequate instruction in the skills of manual handling, including follow-up instructions and briefing for any particular task.
- Identify handling hazards. Initiate corrective measures and opportunities for employees to participate in suggesting these corrective measures.
- Insist that, when protective equipment is used, it is compatible with the task and does not create any impairment to performance.
- Consider marking the weight on packs which are to be moved manually.

Before commencement of any manual handling procedure (lifting, lowering, pushing, pulling, supporting or carrying), first assess the situation in accordance with the relevant

Table 11.2 Manual handling adjustment

Change	To
Lifting	Lowering
Lowering	Carrying
Carrying	Pulling
Pulling	Pushing

(*Source unknown*)

principles of correctly handling materials. In a familiar handling situation, where all the characteristics of the load and surroundings are known, this step is carried out automatically; in unfamiliar circumstances, however, or where the loads are variable, a deliberate appraisal should be made.

Whether seated, as in, say, postal sorting, or standing, it is important that the initial and final height of an object to be lifted is carefully controlled. When standing, the preferred height of an object to be lifted should be between knee and elbow, and no lift should be greater than shoulder height. Slippery floors increase risk, especially in pulling and pushing jobs, and a cramped space causes problems because choices of how to move or place oneself to move, for example, an oxygen cylinder in a welding set, are restricted.

A mechanic under a vehicle is much better off if he/she is lying in a properly designed trolley which provides support; and rollers and trolleys, for example, assist with sideways movement of loads and pushing and pulling jobs. Pulling has a generally greater risk of causing a fall and even a crush injury from the load if it moves suddenly, compared to pushing. Things like handles, hooks and loops allow a better grip and reduce the risk of the load moving the wrong way and causing workers to be injured as they move to save what is being lifted from damage.

A useful thought is shown in Table 11.2.

Team lifting and mechanical lifting in patient care situations are very important because the object being lifted, the patient, has limbs which can fall under gravity or be moved and change the weight distribution. There can be sudden changes of muscular assistance from the patient due, for instance, to faintness from movement after a period lying in bed, particularly if they are not generally mobile and used to supporting themselves, or parts of themselves. They may move a limb to maintain balance, or they may move themselves the wrong way, perhaps in order to relieve pain generated by movement.

Some specific control strategies to prevent back injury are given below.

Prevention of back injury

The prevention of back injury requires, whenever practicable, the elimination of manual handling through the provision of mechanical aids. Mechanical aids reduce the need to select a worker for the job, to rely on a worker's willingness to adopt recommended methods of lifting, and to find a suitable replacement for an absent worker.

When manual handling of loads is unavoidable, ergonomic job design and careful selection of workers are prime requirements for the prevention of back injury.

To reduce the stress of lifting there are two prime requirements:

1. Ensure that the body is as close as possible to the centre of gravity of the load, to reduce the additional load imposed by displacement of the centre of gravity of the head and chest due to the upper part of the body bending over.
2. Keep the back straight, though not vertical.

To meet these requirements, the adoption of the bent-knee position is commonly recommended. However, the ability to lift with the legs is reduced if the knees are bent beyond a right angle.

To get close to the load, the feet should straddle it before bending the knees for lifting. When the load is so large that it cannot be straddled, another pair of hands, or mechanical assistance, should be sought.

Positioning the feet to facilitate transport of the load in the desired direction, and pivoting on the feet to change direction, will avoid twisting the spine during transport.

Whenever possible, all loads should be transferred mechanically on a pallet or in a container, or by other bulk mode. When manual handling is unavoidable, the task should be made subject to ergonomic job analysis and designed so as to fit the capacities of the workers.

Manual handling procedures

The following handling techniques thus form the key factors in the pattern for recommended posture and movement in manually handling materials and loads.

Placement of feet

Before commencing handling procedures, the feet should be placed comfortably apart, one foot forward of the load. This position assists in maintaining balance.

Bent knees – straight back

The knees should be bent before the hands are lowered to lift or set down loads.

When loads must be raised above waist level, the knees should first be bent, then straightened to give momentum to the upward movement.

When a load is taken from a shelf or received from another person, conveyor or chute, the knees should be bent so that the force of the load can be absorbed.

The back should be kept straight at all times. However, a 'straight back' does not necessarily mean a vertical back. The spinal column may be inclined from the vertical providing the head is held up and the chin pulled in to ensure the lumbar spinal curve is flattened, thus keeping the back straight.

Arms close to the body – correct grip

The hands should, if possible, hang between the thighs when lifting or lowering a load. The load should be carried close to the body with the elbows by the sides, and not held out from the body when being raised, lowered or supported.

On lifting, the hand on the same side of the body as the forward foot should grasp the side of the load furthest from the body. The other hand should grasp the opposite side of the load at the most convenient point adjacent to the body. The load should be grasped with both hands, using the whole length of the fingers and part of the palm of the hand, not just the fingertips.

Posture – head erect

Raising the head at the commencement of lifting automatically assists in keeping a 'straight back' during the lift. When a change of direction of movement is required during handling, or it is necessary to look around, the body should be turned by pivoting on the feet. Never twist the spinal column. Avoid jerky actions when raising or moving loads by applying force slowly to the load through the shoulders, arms, hands and legs.

Limitation of loads

Two or more persons should share the load if it is:

- excessively heavy or long
- too bulky or awkward to be readily grasped by one person
- likely to cause loss of balance or vision to one person
- greater than 55 kg when no mechanical assistance is available.

Where team lifts are used, coordination is essential and an appointed team leader should call signals. Gross disparity in the height of members of a team should be avoided.

Aids to manual handling

Appropriate mechanical aids should be used wherever practicable to enable employees to handle materials more efficiently and with less physical effort. Simple and inexpensive devices can often help with awkward handling tasks, and avoid possible personal injury. The worker is often the best source of information on how mechanical aids can be applied.

In some countries, taking the load on the head is commonly used, especially by women. Working conditions in agriculture, especially non-commercial agriculture, may be largely unregulated. See Fig.11.3.

Assessing the energy cost of work

There is a fairly straightforward measure of energy expenditure for typical tasks. The energy expenditure is given in either kcal/hr or watts. Table 11.3 shows this.

Energy is provided in the body by a combination of processes which are independent of oxygen and those which are oxygen-dependent. Dynamic muscular activity is oxygen-dependent, while static muscular activity isn't. Moderate-intensity, continuous work

Figure 11.3 African workers (photo courtesy of Pasi Toivonen, Finland)

Table 11.3 Energy cost of work

Activity	kcal/hr	W
Sedentary (seated)	100	116
Light work	100–200	116–232
Moderate work	200–350	232–407
Heavy work	350–500	407–580

depends on processes using oxygen. Each litre of oxygen produces about 4.8 kcal of energy.

The aerobic capacity or VO_2 *max* describes a person's ability to use oxygen. If a person tries to work at a rate above his/her VO_2 max, in a short time performance drops greatly. VO_2 max can only be improved a little by training, but the removal of lactate (a muscle waste product of oxygen-independent energy production) can be greatly improved by training. If a task requires a VO_2 of 2 L per minute, a worker with VO_2 max of 5 L per minute is obviously better off than one with VO_2 of 2 L per minute. One kilogram of muscle can produce about 0.22 kW, so an average man could theoretically produce 7.5 kW in a burst. Champions can only produce about 1.5 kW in reality. One litre of oxygen produces about 0.075 kW.

Enhancing occupational safety and health

The classic method of measuring energy usage by measuring oxygen usage involved a bag to collect exhaled air. There is now more convenient electronic instrumentation which measures exhaled airflow and oxygen content as the air passes through.

Another way of assessing energy expenditure is to measure heart rate. Because VO_2 varies between people, firstly the link between VO_2 and heart rate has to be established by laboratory tests. Once this is done, heart rate alone can be used in the field to measure energy usage.

There is also a subjective scale called the Borg RPE scale where workers are asked to rate their level of exertion from 6–20, based on heart rates of 60–200 beats/min. A surprisingly good match between the rating and the heart rate has been found.

Although not an energy cost measure, stressful situations where there is low physical activity can be assessed using certain types of measurement of CO_2 in expired air. Where breathing due to stress exceeds the O_2 demand of the body, the level of CO_2 in the blood drops. Blood vessels constrict and this can cause dizziness and heart palpitations.

Finally, it is useful to remember that in hot conditions sweating uses energy and this is part of the body's total energy requirement.

(*With acknowledgment for some ideas in the preceding pages to Bridger, R.S. Introduction to Ergonomics, copyright © 1995, published by McGraw Hill and used with their permission.*)

Using energy cost of work to define safe task selection and rotation

The use of energy cost data is particularly important for workers who do prolonged manual labour. Simple tests such as a step test (step on and off 24 times/minute for 9 minutes) are used and the final heart rate measurement determines the type of work which can be allocated. See Table 11.4 below.

Task rotation can obviously be used to vary the level of energy expenditure required. Allowance of a rest time should also be used. A formula has been proposed for calculating the rest allowance of a worker after a period of sustained heavy work. Those in a non-manual job who comment critically on outside manual workers leaning on their shovels should be reminded that this is a necessary part of doing such a job in a healthy way. Interestingly, high daily energy expenditure in an individual is statistically likely to make him or her healthier.

Table 11.4 Work allocation from step test

Heart rate	Type of work
Less than 120	Strenuous, especially if hot as well.
121–140	Less arduous, and not hot.

(*Source unknown*)

Ergonomic principles associated with integration of controls and displays

Orientation of controls and displays in the person–machine system

Controls and displays play a critical part in the two-way communication which is essential if there is to be effective use by persons of VDUs, machinery, tools and equipment. The interaction involves perception; cognition (processing the information); decision making and action or response, usually through the psychomotor system, i.e. 'brain–nerves–muscles'.

It is a common experience that the way information is presented can affect the way we perceive it and so the interpretation we put on it. Some ways of presenting information require less attention to interpretation than others. For example, an analogue display (white needle on black circular dial) of vehicle speed is easily interpreted at a glance, while a display in digit form, with digits anywhere between 57 and 62 during the observation, takes more effort.

As far as controls are concerned, some are easier to adjust than others. Rotating controls on volume are easier to set than linear controls or push-button digital controls.

In critical situations, people have been known to become convinced that a display is 'lying' (and if it malfunctions it may), and so not react appropriately.

So within the person–machine system, the following are important if error is to be avoided:

- the presentation of information
- the layout of displays presenting a variety of information which may need to be put together (or integrated)
- the design of controls
- the layout of those controls which are used for a response.

Consider the types of controls and displays needed, for example, on:

- a kitchen device
- a fixed machine in a workshop
- mobile equipment
- a vehicle.

Design criteria to optimize integration of displays and controls

This is a complex area and can only be touched on in this text. In relation to displays, some key issues are:

- 'Display' is perhaps a misnomer because information can also be passed on by non-visual stimuli, such as different types of audible signals.

- Avoiding over-complexity, e.g. with a computer display. If necessary, the operator can start with the main screen and, if further information on a particular detail is required, bring that up separately.
- Avoiding more graduations of the scale than are practically necessary, and ensuring readability at the designed reading distance.
- Avoiding too many symbols or icons which require separate study to recognize effectively. Computer programs have 'balloon help' to overcome this.
- Colour can assist but use it logically, e.g. red for hot, blue for cold. Also, remember that 5–8% of people have colour vision problems.
- Grouping complex displays into sub-systems – on a board define the boundaries of the sub-systems with lines or coloured backgrounds.
- Using analogue displays if they are more useable and less likely to lead to perceptual errors, even if digital is 'flavour of the month'.
- If parts of a display appear to belong together, make sure they do belong together.

Broad design criteria include considering:

- figure and ground; that is, what people see as the subject and what they see as the background
- that the brain will fill in missing detail, rightly or wrongly, based on previous experience, to form a complete picture
- proximity; that is, as above, ensuring that what goes together logically, is together
- symmetry, which means a viewer will attempt to balance what is seen
- continuity – even if two different parts of a display overlap, the parts of one may be mentally put together to form a complete picture to the temporary exclusion of the parts of the other.

As far as controls are concerned, key issues include:

- clear labelling, including separation of function by colour (remembering the colour vision problem) and/or shape, e.g. controls on a dashboard are best recognized by *feel* in the dark, if the eyes are not to be taken off the road
- design and layout, so that accidental striking of the wrong control is minimized
- standardization of commonly used controls, unless the new layout is clearly superior; for example, in a vehicle gearshift, don't put reverse in the position in the H pattern which most other manufacturers use for first gear
- two-handed operation – for example, where there is a risk of hand injury in a machine
- logical connection of control position and movement to function, e.g. four controls for four burners (with the same spatial layout) on a stove; turning a knob clockwise to increase something
- for some controls, e.g. steering wheels, a degree of resistance to movement in the direction more likely to lead to loss of control, which also provides tactile (feeling) feedback on the position of the control, the wheel
- consideration of which type of control – button, toggle, lever, wheel, rotatable knob – best suits the function. Consider, for example, controlling a car's acceleration, braking and steering with one joystick – would this be suitable?

(*With acknowledgement to Bridger*, Introduction to Ergonomics, *vide supra*.)

Merits of using digital versus analogue displays

As mentioned above, there are advantages in analogue display for a control such as a speedometer, whereas a digital display is suited to an odometer. Digital displays are not as good as analogue for showing rates of change, such as acceleration for example. Modern microprocessor technology in many cases now allows both forms of display to be shown simultaneously, or for the operator to select the one preferred. Digital displays, such as the volume indicator bars on a TV set, showing the effect of manipulating a control may not be as easy to interpret as the position of circular rotation of a marked spot on a knob. Digital displays, however, can provide more accurate information where that is required.

Job design structure

Parameters of job design structure

Ergonomics can have an impact on this structure to optimize the relationship between people and their work. Job design is one of the factors which determine how humans function within the overall system by which an organization works. It is the fourth step in the following analysis of the human functions component:

- task analysis
- skills and knowledge requirement
- workload predictions
- job design
- selection, induction and training.

Before these steps are considered, when a workplace and worksystem are at the design stage, decisions are made on which functions humans will perform and which functions *machines* – e.g. plant, information systems, equipment, mobile equipment, robots – will carry out. In an existing organization, gradual adjustments can be made, especially where equipment is to be replaced.

However, it will be necessary to make further decisions beyond just *people versus machines*. For example, energy sources, level of automation, environmental issues, and the types of people – e.g. those in sheltered workshops, newly industrialized workers, those selected through affirmative action – may be some of the issues to be considered.

Once decisions are made about which functions within the work system are to be dealt with by machines and which by people (this can be altered, if necessary, if development of both go on together), then jobs can be designed on:

- the job-centred approach, or
- the person-centred approach.

The first makes the best of the job content to create a satisfying job, the second is based on motivating people to work well because the work meets their needs.

Of course, the 'machine' aspect can be designed to complement the job and, in the process, there may be modifications to the job.

Using ergonomics in job design

Ergonomics has been said to be the science of fitting the task to the person. Even if this is done successfully, so as to take account of human abilities and limitations in a way which makes the task relatively challenging and satisfying, that does not necessarily ensure a person's motivation and hence a good level of productivity. If ergonomics fails to take into account the psychological and personal needs of people, it will fail to produce the desired results.

For this reason the question of motivation, which is extensively studied in management, becomes important. Key studies in motivation were those of Maslow and Herzberg (see Further Reading at the end of this chapter). Hackman and others developed the job diagnostic survey based on aspects of motivation (see Further Reading at the end of this chapter). Motivation in turn depends on factors such as work organization. For example, one can have a group of workers with specific skills controlled by a hierarchy of supervision – the classic 'production line' (F.W. Taylor) system.

On the other hand, work may require teams which have a mixture of skills and decide much of the day-to-day way they go about their work, e.g. the design team for a new building.

Jobs and job descriptions

A job as a whole consists of the task or set of tasks which are required to be carried out by a person holding a particular position in an organization. This will not necessarily just include the specified tasks; in reality, it may also include tasks which for one reason or another, 'fall to the lot' of that person. Job descriptions or position descriptions attempt to write down the functions or tasks required by a person in that position, and usually include the estimated percentage of time different tasks will take. One problem with job descriptions is that they may interfere with the organization's desire for an adaptable and multi-skilled workforce. They may also be difficult to keep up to date as demands change.

On the positive side, they provide a basis on which to assess required skills, and provide training. This helps to avoid people being called upon to do things for which they are not skilled and where they may make errors. The job descriptions can also resolve role conflicts between workers and between workers and supervisors. They also act as a basis for fair periodic assessment of work performance.

Job design practices

Such practices can include:

– job rotation
– job enlargement
– job enrichment.

Many industrial jobs are simple and repetitive and so fail to make use of a person's full potential, and can lead to fatigue and boredom, although this varies with the personality of the person.

The three approaches above have tried to overcome this.

'Job rotation' involves changing the jobs carried out between different people in a workgroup, so that there is more variety. It can be useful, but some people do not like constant change. It does fit in with multi-skilling, which allows an operation to continue with less staff, because one staff member has been given the skills to *cover* for another staff member and this can be used in an emergency, such as absence due to illness.

'Job enlargement' has usually involved giving a person the skills to carry out any one of a number of roles or jobs which are required to complete a recognized piece of work. It is felt that the sense of seeing something completed rather than just tackling one part is more satisfying and therefore achieves better motivation. The negative view is that the person now has several tasks, not just one task, all lacking any real meaning.

'Job enrichment' involves including in the job a higher level of responsibility, by including some of what would traditionally be supervisor tasks. This might include consultation on, or workgroup responsibility for, job scheduling or planning, or it could include an aspect such as quality control. One industrial relations problem is that it may lead to demands for higher pay, although that may be accommodated by a flatter structure which achieves higher productivity.

Quite often the informal organization in the workplace may have made these adjustments based on the particular mix of people working there.

Elements of a good job design structure

Karasek (see Further Reading at the end of this chapter) outlined five factors which he considered could lead to undesirable stress at work. These were:

- conflict, such as role conflict, personality clashes or poor interpersonal relations
- overwork or underwork
- lack of meaning
- lack of predictability
- lack of control.

The former Australian trade union official, John Halfpenny, cited the following sources of stress:

- boredom
- job dissatisfaction
- too much or too little work
- insecurity
- responsibility for others
- lack of participation
- alienation from production
- lack of social support

- piecework
- bonus production
- migrant workers
- sexual harassment
- for women – two jobs, one an unpaid domestic one.

So, a good job design structure can firstly be based on addressing these problems.

Job or position descriptions if used properly can reduce 'role conflict'. If they are matched against the changing demands of the work they can avoid having people who lack enough to do or have too much to do – in reality many jobs are a mixture of these factors. 'Overwork' or 'underwork' are issues for management to adjust as required. Despite criticisms, 'lack of meaning' can be dealt with through job enlargement and through the degree to which the organization successfully manages to motivate employees – through a mix of factors such as management commitment, management competence, organizational design and human resources policies. 'Predictability' is a more difficult issue and it depends on the type of environment in which the organization operates, but again a commitment to consultation can reduce the stresses this puts on employees. 'Lack of control' can be dealt with to a degree through job enrichment, so that a worker is not a slave to rhythms which he or she has no chance of altering.

In summary, a good job design structure might include:

- optimum interaction between the person and the machine
- a recognition of the needs of the individual
- inbuilt factors which improve motivation
- clear definition of the tasks which the job requires and hence the skills to perform them
- adequate recognition of achievement or performance
- effective integration within the system of work organization in the enterprise.

Ergonomic principles and design and redesign of work stations

Musculoskeletal principles of seated posture

Working posture in general can be considered in terms of task requirements, workspace design and a range of personal factors. Enforced sitting for most of a work day should be considered an occupational risk factor.

The key musculoskeletal issues involve consideration of:

- the position of the spine
- the effect on the discs between the vertebrae
- the effect of, and on, posture
- fatigue associated with static muscle loadings
- the effect on muscles of reduced blood supply
- the effects of restricted movement (enforced by, for example, the need to view a VDU) on the head and neck

- the different muscle groups which are used for lifting while seated because the full set of muscles in the body cannot be employed.

Progressive or even acute spinal damage can occur from jarring while seated at the work station in off-terrain vehicles. Persistent pressure on discs due to spinal position while seated can be a source of fatigue and discomfort. Measurements have been made and show least pressure on the disc between the third and fourth lumbar (lower back) vertebrae when a person is leaning backwards. The pressure increases in the writing position, even more when typing and more again when lifting a weight from the desk or table. When sitting, the upper edge of the pelvis tilts backwards and changes the bow (from bow and arrow) shape of the spine, called the lordosis, into a backward curve (kyphosis). People find a slight bend forwards comfortable because it requires less muscle strain. There is a conflict between the muscles, which prefer the curvature, and the discs, which require an upright posture.

Discs require nourishment, and this is achieved in part by frequent changes of pressure – such as can occur when posture is adjusted. Further work has shown that the best angle between thigh and back is 120 degrees. To make work at a desk feasible while retaining this angle, the knee chair has been introduced.

Fixed posture increases the static loading (the work muscles must do to hold the body in one place) on the back and shoulder muscles. It can also reduce blood flow to the legs and so lead to swelling and discomfort.

(This para and the last bullet point below are adapted from Sanders M.S. and McCormick, E.J. *Human Factors in Engineering and Design*, 7th ed. Copyright © 1992, published by McGraw-Hill and used with their permission.)

The following recommendations are made in relation to seating:

- Promote lumbar lordosis (the bow shape in the spine) – a lumbar support between seat back and spine helps.
- Minimize pressure on discs. Reclined backrests, lumbar support and arm rests all help; arm rests reduce the work shoulder muscles need to do.
- Minimize the static loading on the back muscles. An angle between squab and backrest of around 110 degrees helps.
- Reduce fixed posture. Take breaks away from, for example, VDUs and VDU work if possible.
- Use chairs which can be readily adjusted, and show employees how to use them: seat height and slope, depth and width, contouring and cushioning and seat-back design all need to be considered. Different people prefer different seat designs.

Specific ergonomic design features

Design features which should be included in pieces of equipment are considered here. The equipment discussed includes:

- non-adjustable and adjustable chairs which are to be used for seated work at benches and tables
- document holders and foot stools

- storage facilities for office and industrial-based environments
- non-adjustable and adjustable tables
- non-adjustable and adjustable benches.

Chairs

A non-adjustable chair should be such that its design allows comfort for sustained work for as wide a range of people of different anthropometry as possible. Broad rather than narrow design and avoidance of a contoured squab will assist. Contoured squabs can reduce comfort in women using sanitary pads. The backrest should be set at a height above the squab which suits a majority of people. Generally, spinal length doesn't vary as much as long-bone length. If necessary, for short people, additional comfort may be possible with a suitable foot rest. Seat depth should also be considered as some people have shorter thighs. However, the height of the work table, desk or bench must also be considered, and if necessary either height added under the legs, or the chair raised on a platform. Care must be taken to prevent trips or falls.

Clearly, adjustable chairs are preferable. A good chair allows for adjustment of squab height; the angle of the squab; and the height; movement backwards or forwards; and degree of tilt of the backrest. However, with a shaped squab, the back-forward movement of the backrest is not a total answer if people have short thighs. Adjustment from a sitting position should be easily achieved by the user, especially if the user of the chair changes. If a chair is more or less for one person, then less 'instant' controls may be acceptable.

These comments refer to conventional adjustable chairs. However, a knee-chair offers certain advantages such as reduced spinal compression, and can also be adjustable. The required adjustments are somewhat different from those for conventional chairs. For certain types of work, such as drafting and some types of assembly work, an adjustable stool or even a sit-stand support, may be a better choice.

If the bench, desk or table cannot be adjusted for height, a foot support may be required. Some stools have height adjustment. Figure 11.4 gives useful information on the ergonomics of a working table or desk.

Document holders

Document holders preferably need to be on a counterbalanced arm and be adjustable for tilt, or if fixed to a monitor, on a swing arm, so that in either case the distance of the document from the eye can be varied to suit the operator. Some swing arm types (fixed to a monitor) don't allow tilt adjustment. Both types allow a document to be at screen height.

Footstools

Footstools, if required, should be adjustable for height and tilt. However, moving a footstool of fixed tilt or the feet on it back or forwards, if it has a reasonable front-to-back width, can achieve a lot.

Storage facilities

Storage facilities for office and industrial-based environments vary widely. Where manual access is required, the height of materials accessed or placed from a standing position should vary from knee to shoulder height (and no more) without some form of assistance,

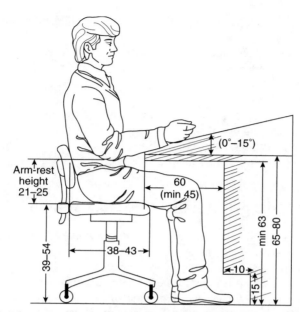

Figure 11.4 Working height for sedentary jobs. These measurements show the variation advisable for adapting the work station to different individuals. For exacting vision requirements, the target of vision may need to be lifted still higher (with acknowledgement to Arbetsmiljofonden, Sweden)

such as a librarian's platform or a well-designed storage area self-supporting ladder. Putting heavy items in bottom drawers or in over shoulder height locations, even if a platform is available, should be avoided. Multilevel stacking and storage requires the use of a lift truck. The size will vary and includes 'operator off', 'operator standing' and 'operator sitting' types with various capacities. A platform or self-supporting ladder used with a height adjustable trolley fitted with wheel brakes is a possible alternative in some cases.

Benches, desks, tables, etc.

Non-adjustable and adjustable tables, stoves, sinks, baths, machine tools, desks and benches can be considered from an ergonomic viewpoint. The issues involved have been partly dealt with in relation to chairs. For work performed standing, or with stools, or with sit-stand supports, this will further affect placement and adjustability of other parts of the work station. The height, reach distance, and hence shape, and angles of tilt can be important, depending on the type of work. Thighs must be able to fit comfortably under the table, or desk if seated. Reach distance is affected by late pregnancy.

Other considerations

Consider the issues in:

- male, female and child use, and use by ethnic groups of varying anthropometry, of a domestic kitchen (sink, preparation area, stove, and eye- or shin-level oven), and the same tasks in a commercial kitchen

- male child use of wall (pear-shape) urinals and use by children of both sexes of toilet pedestals
- pigeon-hole sorting operations
- tool use on work benches
- fabric cutting
- sheet-metal marking
- male and female use of an ironing board, washing machine, and rotary clothing drier, and the hanging out of washing
- cleaning a bath or shower recess
- bathing patients in health care facilities.

Analysis of tasks performed at the work station

The variety of tasks and work stations within the workplace is considerable. It is not possible here to do more than remind the reader to consider the biomechanical, anthropometric, physiological, and psychological features which apply to the task and the work station at which it is performed. The reader then needs to refer to further books and papers on ergonomics which provide guidance in relation to the type of situation being reviewed. This might cover the design of displays or controls, specific solutions which have been developed, e.g. better store checkouts, improved handtools, better chairs, better keyboards, or counter-balance devices which can assist an operator to turn a work-piece. See Further Reading.

Human error

Psychological aspects of a task

Psychology, as noted earlier, plays a large part in ergonomics. Because generally we are successful in completing a task, error is not as common as is sometimes thought. Yet no consideration of ergonomic interactions, or of the causation of accidents or the investigation of accidents can afford to ignore the role of human error. The psychology of error has been studied but not as much as might be expected. Komaki, Norman, Rasmussen and Reason are among those who have studied it (see Further Reading at the end of this chapter). Design, as Norman showed, can play a large part in whether or not error occurs.

Rasmussen developed what is known as the skills-rules-knowledge (SRK) framework. Errors are divided into skill-based, rule-based and knowledge-based errors. So we can make an error in the application of a skill, e.g. failing to 'dunk' the ball in basketball. Or we can make an error in the application of a rule, e.g. misinterpreting the offside rule in soccer. We can use either the wrong rule or misapply a rule. The third possibility is that the wrong knowledge is applied or knowledge is applied wrongly.

Tasks using skills and rules probably exceed those requiring a call on knowledge by about 1000:1. However, when we do call on knowledge, the possibility of error is usually greater.

We tend to use similarity matching (which previous challenge does our current challenge most closely resemble?) or frequency gambling (which solution do we call up most often?). However these approaches can fail where the new challenge is sufficiently different from the ones we have previously responded to.

Reason divided error into slips, or lapses, and mistakes. Slips or lapses involve the correct choice of a method to achieve a task, but a failure to carry it out correctly. Mistakes on the other hand involve a flaw in the plan which means that even if it is carried out correctly, the objective isn't achieved. Slips and lapses are often skill- and rule-based errors, whereas mistakes are often knowledge-based errors.

Habits can intrude on newly learnt procedures to cause error, e.g. switching from a trafficator on the left of a steering column to one on the right. The visual sense is extremely strong and can sometimes cause errors in making rational choices. (We all know of different optical illusions such as the apparent oasis in the desert, and of the apparent movement in our train when it's really the one alongside which is moving.)

Procedures need to be devised with the possibility of error in mind, e.g. any procedure which has a loop in it where the operator runs a check before moving on raises the possibility that the operator will re-enter the procedure at the next step regardless of the test result.

Reason believed that after initially training people in a procedure, as error is almost certain to occur sooner or later, the next step is to train them to recover from anticipated errors. An understanding of human error and its various forms must form part of any consideration of a safety management system, a procedure and an accident investigation.

Further reading

Arbetsmiljofonden (The Swedish Work Environment Fund). *Make the Job Easier – Human Proportions.* 20 pp. Stockholm: arbetsmiljofonden.

Armstrong, S. (1997). These Boots Aren't Made For Working (multicultural workforce fit). *Safeguard (New Zealand)*, November/December, pp. 28–9.

Barron, S. et al. (eds). (2001). *Simple Solutions: Ergonomics for Farm Workers.* Cincinnati, US DHHS NIOSH.

Bernard, T.E., Dukes-Dobos, F.N. and Ramsey, J.D. (1994). Evaluation and Control of Hot Working Environments. *International Journal of Industrial Ergonomics*, **14**, 129–38.

Bridger, R.S. (1995). *Introduction to Ergonomics.* Singapore: McGraw-Hill. esp. Chs 12, 13, 16.

Carrasco, C. et al. (1995). *Research Update: Packing Products for Customers: An Ergonomic Evaluation of Three Supermarket Checkouts.* Issue 95014. Sydney: Worksafe Australia.

Casey, S. (1998). *Set Phasers on Stun*, 2nd edn. Santa Barbara: Aegean Publishing Co. (20 true tales of person-design-error).

Department of Transport WA. (1998). *Fatigue Management for Commercial Vehicle Drivers Code of Practice*. Perth: Department of Transport.

Dul, J. and Weerdmeester, B.A. (1993). *Ergonomics for Beginners*. London: Taylor and Francis.

Editorial. (2002). Program Reduces Work-related MSDs (strengthening and stretching routine for construction workers). *Professional Safety*, **47(9)**, 13.

Editorial. (2002). Accommodating Older Workers. *Accident Prevention (Canada)*, May/June, p. 21.

Grant, K.A. et al. (1993). Ergonomic Evaluation of Checkstand Designs in the Retail Food Industry. *Applied Occupational and Environmental Hygiene*, **8(11)**, 929–36.

Hackman, J.R. et al. (1975). A New Strategy for Job Enrichment. *California Management Review*, Summer, **17(4)**, 57–71 together with the Job Diagnostic Survey.

Herzberg, F. (1966). *Work and the Nature of Man*. New York: World Publishing Co.

Karasek, R.A. (Jr.) (1979). Job Characteristics and Mental Strain. Implications for Job Re-Design. *Admin. Science Quarterly*, **24**, 285–308.

Kilbom, A. (1994). Repetitive Work of the Upper Extremity. *International Journal of Industrial Ergonomics*, **14**, 51–7 & 59–86.

Komaki, J. et al. (1978). Effect of Training and Feedback: Component Analysis of a Behavioural Safety Program. *Journal of Applied Psychology*, **65(3)**, 434–45.

Kroemer, K. and Grandjean, E. (1997). *Fitting the Task to the Human*. 5th edn. London: Taylor and Francis.

Lodge, D. (2000). Safe for Southpaws (left-handed ergonomics). *Safeguard (New Zealand)*, July/August, pp. 52–4.

MARCSTA and Circadian Technologies 2001. *Managing a Shiftwork Lifestyle – A Personal Approach*. Perth: MARCSTA.

Maslow, A.H. (1943). A Theory of Human Motivation. *Psychological Review*, **50**, 376–96.

McConnell, J.V. and Philipchalk, R.P. (1994). *Understanding Human Behaviour*. 8th edn. Fort Worth: Harcourt Brace Jovanovich.

Norman, D. (1988). *The Psychology of Everyday Things*. New York: Basic Books.

Osborne, D.J. (1991). *Ergonomics at Work*. 2nd edn. New York: John Wiley & Sons Ltd.

Patterson, S. (2002). Tips for Working at Home. *Accident Prevention (Canada)*, November/December, p. 10.

Pheasant, S. (1991). *Ergonomics, Work and Health*. London: Macmillan Academic and Professional.

Randolph, R.F. and Peters, R.H. (1989). *Management and Behavioural Factors Associated with Safety and Productivity in the US Mining Industry*. US Department of the Interior, Bureau of Mines, Pennsylvania, USA.

Rasmussen, J. (1983). Skill, Rules and Knowledge: Signals, Signs, and Symbols, and Other Distinctions in Human Performance Models. *IEEE Transactions on Systems, Man and Cybernetics* SMC-13(3).

Reason, J. (1991). *Human Error*. Cambridge: CUP.

Robertson, M.M. et al. (2003). Telecommuting: Managing The Safety Of The Worker In The Home Office Environment. *Professional Safety*, **48(4)**, 30–6.

Salvendy, G. (1987). *Handbook of Human Factors*. New York: John Wiley & Sons Ltd.
Sanders, M.S. (1993). *Human Factors in Engineering and Design*. 7th edn. New York: McGraw-Hill.
Sirois, W. (2000). Achieving Shiftwork Excellence: Maximizing Health, Safety and Operating Efficiency in Round-the-Clock Operations. *Proceedings of Minesafe International Conference*, 4–8 September. Perth, Chamber of Minerals and Energy of Western Australia, 441–54.
Snook, S.H. and Ciriello, V.M. (1991). The Design of Manual Handling Tasks. *Ergonomics*, **34**, 1197–213.
Soljak, M. (1999). Call Centre Comfort (ergonomics). *Safeguard (New Zealand)*, January/February, pp. 28–30.
Stevens, M.A. (1990). Avoiding Manual Handling Injuries at Work. *The Safety and Health Practitioner*, **8(1)**, 8–10.
Torunski, E. (2001). Wake up to Napping. *Accident Prevention (Canada)*, March/April, p. 27.
US DHEW-NIOSH. (1973). *The Industrial Environment – Its Evaluation and Control* Washington, US Govt. Printer.
Waters, T.P. et al. (1993). Revised NIOSH Equation for The Design And Evaluation of Manual Lifting Tasks. *Ergonomics*, **36**, 749–76.
Winn, G.L. et al. (2002). Texpert – A Tool for Safety Professionals and Design Engineers. (software for safer products). *Professional Safety*, **47(10)**, 32–7.
Worksafe Australia. (1990). *National Standard for Manual Handling and National Code of Practice for Manual Handling*. NOHSC:NS001–1990 CP004–1990 Canberra: AGPS.
Worksafe Australia. (1994). *National Code of Practice for the Prevention of Occupational Overuse Syndrome*. NOHSC:2013. Canberra: AGPS.
Worksafe Australia. (1996). *Guidance Note for the Prevention of Occupational Overuse Syndrome in Keyboard Employment*. NOHSC:3005. Canberra: AGPS.

Activities

1. Select an instrument, tool, piece of equipment or vehicle and explain how human error in its use may interact with design features to increase the likelihood of an accident.
2. Select a particular task and work station (for preference not a VDU), and make the anthropometric measurements necessary to see if the work station is appropriately designed for the operator.
3. Select a suitable workplace and identify an area where OOS might be a possible risk. Write a short report on preventative measures.
4. In a suitable workplace of your choice, identify equipment and tools used which have been designed with the use of acoustic control material, or with engineering controls on noise.
5. Draw a layout of your preferred office and consider the most appropriate location in which to operate a computer. Discuss how you would reduce screen glare, and

ensure adequate lighting for all office requirements. What colour would you choose for walls and carpeting – why?
6. In a workplace of your choice with a range of tasks, list five tasks which involve different levels of energy expenditure between 100 and 500 watts.
7. Identify three jobs in a suitable selected workplace which would benefit from job redesign. Explain how you would go about this and the expected benefits.

12

Workers' compensation and rehabilitation

WORKPLACE EXAMPLE

A government-owned collieries company in India has 67 mines with around 100 000 miners. The roof of a mine collapsed killing ten miners. The mine was 15 years old and the accident occurred between 1.30 and 2 am. When workers in the third shift carried out a blast to break the coal, the roof collapsed. Two miners who remained alive were admitted to hospital.

An inquiry was ordered, and the government also announced that each miner's family would receive 600 000 rupees (approx. USD 15 000) compensation, with half paid by the colliery company. In addition their children would receive free education and the families would receive free collieries company housing.

The tragedy followed an earlier one in the company's mines the same year when 17 miners drowned. The wall of the 7 Life Extension project had collapsed and the mine was flooded. There had apparently been worker complaints of lack of safety management by the company.

A judicial inquiry by a High Court judge was just about to hold a hearing in the state when the new tragedy struck.

(Source: IndoAsian News Service).

Development of employer's liability

Introduction

Risks and hazards within our environment are not always controlled. Systems to compensate employees who have a work-related injury or illness have, therefore, been introduced and are covered under statute law.

Difference between common law and statute law applied to workers' compensation

Two distinct arrangements in law have developed for those who have been injured, killed or who developed a disease due to work. In tort law, which allows a person to sue for damage caused to them or their property, there was the gradual development of a body of case law (law based on precedents set by court cases) which allowed an employee to sue an employer, and succeed in being awarded damages if the employer was found to be negligent. This is the common law approach.

Beginning in Germany in 1884, with Britain following in 1897, statute law was passed which was based on strict liability. This statute law made the employer liable to pay for loss of income and the impairment of capacity to earn an income regardless of whether the employer was negligent. (That is the meaning of strict liability.) In those jurisdictions where common law claims are still allowed, damages are paid for pain, suffering and loss of enjoyment of life.

The aim of the workers' compensation system is for the employer to provide workers who are injured (through the course of their work) with compensation for such items as:

- loss of wages
- medical expenses
- vocational rehabilitation
- permanent disability
- travelling expenses
- payments to dependants if a worker dies.

Whilst being able to compensate for the above listed items, the intent of the legislation is also to promote the worker's return to gainful employment as soon as possible. This is primarily achieved with the use of medical practitioners and vocational rehabilitation provided by appropriately qualified people. The medical practitioners provide the information on the injured worker's fitness for work and provide the ongoing medical treatment. Vocational rehabilitation will attempt to match the injured worker's abilities with the work environment. This may include such activities as 'graduated return to work programmes' (work hardening), and counselling. Workers' compensation is covered under statute law and is payable to the employee, or in the case of the death of the employee, it will be paid to his or her dependants. In most jurisdictions it is compulsory for all employers to have workers' compensation insurance (some are permitted to be self-insured). In many jurisdictions an additional right to sue a negligent employer in common law also exists.

Provisions exist to ensure there is no 'double dipping', i.e. payments are not made twice for the same thing, e.g. medical examinations. Strict liability applies in many jurisdictions to a common law suit where the employer has breached a duty written down in a statute.

Each jurisdiction, as noted in an earlier chapter, has its own act and regulations covering workers' compensation and rehabilitation. Although there are many similarities in the schemes, differences for the employer and employee may be found in areas of compensation values awarded, insurance provisions and rehabilitation requirements. It is, therefore, very important that you read your own legislation and interpret it correctly, especially if you have responsibilities under it.

Note that in the USA the OSH Act 1970 set up a National Commission on State Workmen's Compensation Laws, with a limited lifetime. In Canada there is the Association of Workers' Compensation Boards of Canada.

Principle of duty of care in the development of employers' liability for compensation under common law

The principle of duty of care in the development of the employers' liability for compensation under common law was that the moral rule – 'love your neighbour as yourself' – became the legal rule, 'you must not injure your neighbour.'

A no-fault system

The responsibility of the employer to pay statutory compensation under the terms of such legislation is **not** dependent upon the employee showing that the employer has done anything wrong (with exceptions in some countries). The employer's liability under the legislation is called 'strict liability'. This means, as noted earlier, that proof of fault or negligence is not required in order to receive workers' compensation. And generally, the employer is liable irrespective of any wrongdoing on the part of the employee.

Insurance providers

Included in the legislation is a requirement for employers to provide insurance for employees. Insurance may be sourced through:

- government-based insurance schemes
- private insurers
- self-insured schemes.

Depending upon which insurance provider is prescribed under legislation, costs and the procedures to commence a workers' compensation claim will vary.

The amount of insurance required to be paid may be calculated on:

- an organization's claims experience
- a percentage of payroll for the given employees
- the type of industry classification in which the employees are placed, or
- some other basis.

Once a premium or insured sum is calculated for the organization, it is in the employer's best interest to minimize costs related to workers' compensation and manage claims effectively so that employees return to work as soon as practicable. A good claims record with low accrued costs may result in the insurance provider reducing the insurance premium in the next financial year. This system is similar in some aspects to car insurance. If, for example, you have had no claims in the last year or two, your insurer may reduce your insurance costs the following year.

Employer's liability for compensation under workers' compensation legislation

Types of employees for whom the employer can be liable

It will be useful to obtain further information from guides or brochures on the workers' compensation scheme in your jurisdiction.

'Workers' compensation' basically applies to anyone who has entered into or works under a contract of service, and whether the contract is expressed or implied, oral or in writing. Certain other people may be included; for example, independent contractors may be included in certain circumstances. You will need to refer to your own legislation to see who is 'deemed' to be an employee. There have been many court cases on who is an employee, the degree of responsibility that an employer should expect of skilled and highly skilled employees, intermittent duties, salesmen on commission, agents, e.g. insurance agents, and company directors.

Criteria for determining an employer's liability under workers' compensation legislation

The relevant legislation will also contain a definition of 'employer', including the legal personal representative of a deceased employer. Various tests are applied to determine employee eligibility in relation to a particular employer.

An employer's responsibility for injuries sustained on journeys to and from work is conditional and should also be checked in your local legislation.

An employer may also not be liable if there has been serious and wilful misconduct by an employee, injury has been self-inflicted, there has been misrepresentation about disease, or there has been fraud. Drunkenness (and presumably other substance abuse) is an issue here. There may, however, still be liability if there was death or serious and permanent disablement. Diseases or deafness, in particular, bring up the question of which employer is liable.

Another issue is to determine if an injury or disease is work-related, i.e. did it arise in the course of employment or as part of employment, e.g. camping out, working at home, in a hotel, at a conference, attendance at a vocational college, etc.

Total, partial and permanent incapacity

'Incapacity' generally means 'physical incapacity for actually doing work in the labour market in which the employee works or may reasonably be expected to work'. Both the carrying out of work or seeking and obtaining work may be prevented by incapacity.

'Total incapacity' refers to the situation where no suitable work is available due to incapacity, while partial incapacity is where some such work is available but the person is able to earn less than if there was no incapacity.

'Permanent incapacity' refers to a part of the body being permanently or wholly useless. So, in deciding if an eye or foot or other member is deemed 'lost', it is if it has been made permanently or wholly useless.

Calculation of weekly benefits for totally and partially incapacitated workers

Weekly payments, while they are based on an employee's pre-injury rate of pay, vary between jurisdictions in relation to the time period for which a certain rate is paid, and in allowing for people who were earning more than the base-rate or had other entitlements.

A payment for total incapacity granted soon after injury may be reduced when the injured person is able to return to some *light*, but not pre-injury, work.

Benefits for permanent incapacity

Lump sums are also provided, in addition to other payments, for specific injuries; these are for non-economic loss, e.g. loss of a finger. Other losses such as loss of ability to have sex, may also be covered by a lump sum.

Situations under which benefits are reduced or withdrawn

Benefits are withdrawn where a victim recovers and is declared fit. A lump sum settlement can result in an end to weekly benefits; however, the worker is not usually obliged to accept this. In addition, benefits can generally be withdrawn if the injured person fails to undergo appropriate medical and other consultations. Usually a tribunal of some form has to approve such a suspension.

A lump sum redemption may also end the right to sue under common law. If a person successfully sues at common law and gets a lump sum settlement, weekly benefits end.

Someone who leaves a jurisdiction without going through the formalities can lose benefits, or she/he may be able to take a lump sum on leaving. Generally, benefits cease at retirement age. Weekly payments cease when a prescribed total is reached; however, in some jurisdictions this can be appealed against.

Process for resolving disputes in workers' compensation

Circumstances under which a workers' compensation matter would be disputed

It is clear from what you have just read that there are many steps in the workers' compensation and rehabilitation process, as well as in common law suits for injury sustained at work, where disputes can occur.

Some possible issues are:

- the actual effects of the injury and the degree of pain (pain can't be seen)
- whether a person was deemed an employee
- whether a person is fit for light duties
- whether an injury or illness was work-related
- delays in the start of weekly payments
- whether an injury was a new injury, a pre-existing injury or an aggravation of pre-existing injury
- under which employer the injury was sustained
- whether there was serious or wilful misconduct

A worker may be paid out of an uninsured liability fund if the employer has failed to insure and the fund then proceeds to recover the money from the employer.

In the case of a common law suit, contributory negligence may be raised by the employer.

Roles of employers, insurance companies and workers' compensation authorities in disputed workers' compensation matters

Each jurisdiction has a particular method for resolving different types of disputes. In the Canadian Province of Ontario, for example, there is the Workplace Safety and Insurance Appeals Tribunal. You will need to read your relevant workers' compensation authority guides. In those juridictions where private companies provide workers' compensation insurance, the workers' compensation authority generally provides forums to resolve disputes. To reduce costs, strict legal procedures and the use of lawyers for the earlier steps in dispute resolution may not be required. Of course, common law suits may be resolved by a court or by an out-of-court settlement.

Procedure for disputing workers' compensation matters

As noted, this varies from one piece of legislation to another and you will need to refer to a workers' compensation authority guide which sets out the rights of employees. The system in Western Australia, for example, has three tiers, with the first two designed to exclude lawyers. However, it is difficult to see how many a worker or their family is sufficiently well informed and articulate enough to argue their own case against insurers who have ongoing access to legal assistance.

Role of relevant government agencies or compensation tribunals in disputes

Once again, this varies from jurisdiction to jurisdiction and you will need to refer to the information material you obtain from your relevant workers' compensation authority.

For example, medical panels are usually set up by workers' compensation authorities to resolve medical disputes.

Importance of early intervention and return-to-work strategies as part of the rehabilitation process

Early intervention and return-to-work strategies include:

- Encouraging re-integration of employees as soon as possible through continuing social interaction, thus preventing the onset of 'illness behaviour', e.g. dependence, lack of motivation and inability to complete routine tasks.
- Using work as a therapy.
- Reducing the cost to compensation schemes.
- Reducing the chance of a person becoming an economic cost, rather than an economic contributor, to society.
- Reducing the chances of loss of valuable skills to an organization.
- Being socially sound in ensuring that people continue to find purpose and challenge.
- Recognizing that for many people, men in particular, work is an essential part of maintaining emotional and psychological health.

Principles of rehabilitation applying to injured workers

Rehabilitation

Workers' compensation legislation in general promotes the concept of:

- employers keeping an injured worker's position open for a given time
- provision for payment of rehabilitation providers, e.g. physiotherapists, occupational therapists and ergonomists
- encouragement of the return to work of employees as soon as practicable following a workplace injury or illness
- reduced workers' compensation payments to the incapacitated employee over time.

Whatever the legislative requirements within your jurisdiction, it is clear that rehabilitation is a management tool to assist employees who are incapacitated to return to work as early as practicable.

Physical, psychological and social impact

The effects of work-related illness or injury vary widely, and are in part related to the severity of the condition. Without adequate attention and proper advice and assistance, physical and mental impairment will actually be worsened, e.g. lack of appropriate

Enhancing occupational safety and health

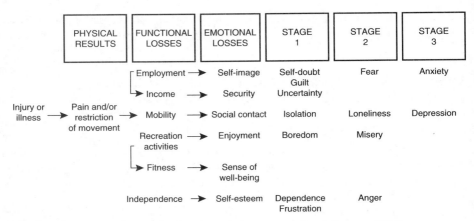

Figure 12.1 Effects of injury (with acknowledgement to Daryl Cooper, 1986)

support for exercise and physiotherapy, or access to it. Some physical impairment involves big adjustments in lifestyle. Psychologically, it has been found that, in general, the longer people are away from the workplace, the less their chances of successful reintegration. However, psychological issues from, for example, loss of a body part or function, or disfigurement, unless properly dealt with, can lead to loss of self-esteem and confidence. This can damage social and family relationships. Depression can lead to despair and in some cases, suicide. Some of the effects are set out in Fig. 12.1.

The process of adjustment to, and acceptance of, the fact that things may never be the same, and to proceed confidently and acceptingly to achievable outcomes will, in some cases, require a high degree of best practice in rehabilitation processes and strong support from the employer, workmates, family and friends. Reactions of victims vary a lot. The person affected will almost inevitably have to go through the process of discovering who are fair-weather friends and who are not, but may also strike up new social relationships in the process of rehabilitation. There may be anger at the employer for letting the accident happen; the employee may not want to return to a workplace they see as dangerous; and they may not want to be seen as leaning on others by doing reduced duties. They also worry about the family's financial future. Acquiring a work-related disease which carries a poor chance of averting early death will require the very best of palliative support for the person concerned.

Medical and rehabilitation models for managing recovery of workers

The old *medical* model was a partly passive approach which saw an injured worker recovering from the injury (or not recovering from the injury) accompanied by visits to a treating physician, with accompanying restorative surgery where appropriate. Eventually the worker was either pronounced fit for work (with varying degrees of

Workers' compensation and rehabilitation

knowledge by the physician concerned about the workplace involved) or unfit for a return to such work. Vocational retraining might then follow.

The rehabilitation model involves best practice in pursuing an active approach to returning the worker to work where at all possible.

It involves:

- early injury reporting
- early referral to rehabilitation
- coordination of injury management by a team with a variety of suitable expertise (occupational therapist, psychologist, physician, etc.)
- a planned, graduated return to work with suitable duties if possible or, if necessary, where reasonably practicable, a modified approach to work procedures and a modified work station
- if necessary, assessment and evaluation of functional capacity to identify suitable alternative vocations and appropriate retraining.

The aim is to ensure the employee remains, to the best possible level, a functioning member of society. The chances of successful reintegration into an employee's work area are substantially reduced if the employee is off work for three or more months.

How is rehabilitation managed?

Communication and liaison between the following parties is crucial:

- incapacitated employee
- employer and rehabilitation coordinator
- employee's supervisor
- treating doctor
- insurance company
- union, if applicable
- employee advocates
- other health professionals.

These participants are often referred to as the rehabilitation team within a workplace.

Sometimes it is not possible to provide required resources for rehabilitation from in-house sources. Many external professionals may be used to assist in rehabilitation management and may include:

- occupational health physicians
- occupational therapists
- physiotherapists
- ergonomists
- risk assessors
- specialist doctors
- psychologists.

Role of rehabilitation in the workers' compensation system

Requirements for rehabilitation in workers' compensation legislation

Workers' compensation legislation as noted now generally places a strong emphasis on rehabilitation. Some of this legislation now provides that up to a certain percentage of the maximum amount, or *prescribed* amount of compensation, is to be used for rehabilitation. Without rehabilitation, workers lose confidence and self-esteem and this can threaten relationships as observed earlier. They also become an unnecessary cost to general medical insurance and social security services.

Process for providing rehabilitation in accordance with legislative requirements

You will need to refer to local guides on how rehabilitation is handled in your jurisdiction. However, larger employers will have a rehabilitation programme and usually a rehabilitation coordinator or will contract a coordinator (see next section). Generally, such people must be accredited to do this work. Every employer should include rehabilitation as part of their health and safety policy. Early return to work where possible it is emphasized, is a principal focus of rehabilitation programmes.

Role of vocational retraining in rehabilitating people at work with a work-related illness or injury

Vocational retraining may be required either:
- where it is reasonably practicable for the employer to modify the workplace and work practices to allow a person being rehabilitated to continue doing the work, or
- where a functional assessment of the worker indicates that they will need to train for different work. An occupational therapist (in consultation with a physician) will recommend types of jobs which are within the ability of the person, based on the functional assessment, and recommend appropriate courses of training for the new duties. Some work may be in 'sheltered workshops', e.g. if there has been brain damage.

Some legislation provides that entitlements aren't lost where a return to work is not successful and some provide subsidies for trial placements of workers being rehabilitated.

Resolving disputes over rehabilitation

Some of the dispute procedures for workers' compensation as a whole will apply also to the rehabilitation component. Government facilities and arrangements vary between jurisdictions, so you will need to check on these local factors.

Workers' compensation benefit system and provision of rehabilitation

Rehabilitation is not only designed to benefit the worker for the reasons given earlier, but also to manage the costs of workers' compensation. For this reason, as also explained earlier, a percentage of the prescribed maximum payments may now be set aside for rehabilitation.

Factors in an effective injury management system

The rehabilitation process and its role

Workers' compensation legislation has recognized the value of maintaining the human asset in employment. This is demonstrated by the requirement for occupational rehabilitation in the legislation. The number of workers who have experienced a lost-time injury and require rehabilitation is only a small percentage, but they make up a significant proportion of the compensation costs. As already noted, occupational rehabilitation aims to optimize an injured employee's return to work. This is achieved by the communication and cooperation of a team of people who are all required to interact with the worker. This interaction aims to provide recovery not only for the physical injuries, but also mental, social and vocational aspects.

Some of the team members who may be involved in occupational rehabilitation were given in the last section 'Principles of rehabilitation applying to injured workers'.

Effective coordination of such a team by the rehabilitation provider is essential.

Elements of rehabilitation programmes

The aim of rehabilitation is to return an injured employee to the same job and duties as were being carried out before the injury. This is achieved in a rehabilitation programme by assessing the physical and mental limitations of the employee and then comparing these to the job demands. But if the job demands are beyond the employee's limitations, then a programme of restricted work duties and reduced work hours may be required. The elements of such a return-to-work programme would be decided in consultation with the employee, his/her manager, the treating medical practitioner and the rehabilitation coordinator. The return to work programme must be formalized and well documented with copies of the documentation going to all parties involved.

If the pre-injury work duties are unsuitable for the injured worker, then alternate duties as noted earlier will be required. The physical and mental abilities of the injured worker are defined by the functional capacity evaluation. This may be performed by an occupational therapist with, if needed, a psychologist. If alternate duties cannot be found in the same position, then alternate duties within the company may be required. Vocational training may be required to help with the placement of the injured worker into an alternate job outside the company or organization.

So a rehabilitation programme established for an incapacitated employee may provide:

- selected duties – specific tasks which can be upgraded to match the employee's capabilities
- modified duties – modifications to usual tasks – for example, providing an adjustable height bench to meet the need of the employee
- a graduated return to work – working hours are gradually increased to meet the employee's capacity until usual working hours are resumed
- job retraining with another organization – if the employer is not able to provide suitable work options, job retraining with another organization may be another option
- management commitment and supportive policies.

When a rehabilitation programme is being managed at the workplace, it needs to be dynamic and worthwhile for the organization. This means that a programme must have a time-frame in which the employee's job tasks are upgraded to meet their capabilities. Work provided also needs to be productive and meaningful. This will maximize the employee's participation and productivity in the programme.

Roles of rehabilitation professionals in the rehabilitation process

All of the team listed earlier, including the rehabilitation coordinator from the approved provider, are involved in the rehabilitation process. Rehabilitation coordinators have primary responsibility in the rehabilitation process and are involved in all of the elements of rehabilitation described in this section, 'Factors in an effective injury management system'. It is their role to ensure that the appropriately qualified people are available to assist the injured worker according to the needs of that person and the stage in the rehabilitation process that he or she has reached.

Issues in determining selected duties in a rehabilitation programme (or plan)

Based on the medical opinion of the treating doctor who should consider any functional evaluation, an indication will be given on the medical certificates if restricted hours or alternate duties are required for the injured worker. Some of the workers' compensation medical certificates will recommend vocational rehabilitation. Elements of a good rehabilitation programme (or plan) will include the following:

- a contact letter to the injured employee explaining rehabilitation goals
- an outline of the rehabilitation programme
- duties that can and cannot be performed
- hours to be worked
- days to be worked
- a review date for the injured worker's rehabilitation plan
- a copy of the rehabilitation plan sent to the supervisor
- the details of the rehabilitation plan as discussed with the treating doctor and the supervisor
- documented progress notes filed as a reference for the case.

An example rehabilitation programme is given in Fig. 12.2 for an enrolled nurse with a sprain/strain back injury.

REHABILITATION PROGRAM

Progressive Outline

Name: JONES, Mary
Injury: Back sprain/strain
Occupation: Enrolled Nurse
Department: Ward 28
Injury Date: 5 Jan 04

Week	Duties to perform	Duties to avoid	Hours to work
10 Jan 04	• Patient observations • Making max. of 3 beds • Short dressings • Clerical duties I.V. therapy	• Lifting or transferring patients • Lifting or moving heavy equipment • Showering or walking patients • Remaining in flexed postures	5 hrs/day 3 days/ week Mon, Wed, Fri.
	Progress notes: Mary is coping well with restricted duties and reduced hours. Not experiencing any back pain. Will review progress in one week. To see her G.P., Dr. Jones, in two weeks.		

Figure 12.2 Example rehabilitation programme

Assistance in rehabilitation of employees
Communication with injured workers on their needs

Determining duties
When the injured employee is fit to return to work, careful review of their limitations and injuries will need to be examined in relation to the tasks which can be performed. The tasks should be as productive for the employer as possible whilst providing job satisfaction for the employee. Considerations for the range of tasks will include:

- the ergonomic requirements of the employee
- pain limitations
- whether work station modifications need to be made
- intensity of the work
- rest breaks
- support from the supervisor
- support from other employees.

As the employee recovers from the injury, regular review will match the employee's increasing ability with an increase in task variety and, ideally, a return to pre-injury duties. Regular liaison with the treating doctor will help confirm the suitability of the alternate duties.

Work-hardening programmes may act as a conditioning tool for employees who are not ready for the workplace yet, but who require intense physical conditioning before a return to work.

Reviewing your rehabilitation programme
To help maintain the effectiveness of a rehabilitation programme, a number of elements should be regularly reviewed. This will include, but is not limited to, the following:
Documentation:

- Are written reports kept on each rehabilitation case?
- Are letters sent to both the employee and supervisor about the rehabilitation programme?
- Is any literature given to the employee about rehabilitation in the organization reviewed, to ensure it is current?
- Are the rehabilitation programmes read and signed by the employee and the supervisor?
- Has all of the statutory notification been completed as per the legislative requirements?

Outcomes:

- Rehabilitation providers will have different ways of measuring their performance and some performance indicators are determined by statutory bodies.

Return-to-work rates can provide useful information for employers. These are measured by calculating the number of finalized rehabilitation cases which have returned to work in a year or other nominated time-frame. Performance can be benchmarked and measured over consecutive years.

Negotiation of premium level with the insurer

Calculation of the workers' compensation premium

Determining a specific rate of premium to be paid in relation to workers' compensation insurance is negotiable between an organization and an insurer. In many ways, it is similar to a home owner who is looking to insure his/her home. The home owner may request quotes from many insurers to try and get the best deal. Often workers' compensation legislation has provision for recommended rates of insurance. These rates are based on a number of factors and are reviewed regularly by the relevant committee in a jurisdiction. Different rates are set for various industry classifications and, when these rates are viewed, it can be seen which industries have a higher insurance risk. Information provided by insurers to the committees which determine the industry rates includes the amount of money paid on claims, the number of outstanding claims, and totals of wages paid by employers. Industry premium rates may be permitted to be exceeded by a set percentage. The rates can also be reduced. Many schemes offer the reduction of insurance premiums if there is a demonstrated safe track record. This lowering of the insurance premium for an organization reduces their operating expenses and thus increases profitability. This is a frequently used argument when safety professionals are trying to justify expenditure on accident prevention to management.

Premium calculation review with the insurer

When trying to negotiate an insurance premium for your organization certain information must be obtained, for example:

- total wages paid for the year
- cost of claims paid
- estimates for costs on outstanding claims
- industry premium rate
- evidence of significant improvements in safety and health in problem areas (areas with highest numbers of claims)
- other performance indicators such as frequency and duration rates
- trends for the above information from past years.

With the above information a basis for premium review and negotiation can begin. In jurisdictions which don't have a mandated single insurer, insurance brokers can also be engaged to act on behalf of an organization with respect to premium negotiation. It is important to ensure that employees have been placed in the correct occupational classifications.

Workers' compensation premium disputes with an insurer

There are a number of approaches which can be taken to resolving these types of disputes. Firstly, consult your relevant workers' compensation legislation to see if any mechanisms are laid down for resolving such disputes. Some jurisdictions have a premium rates committee and it may be possible to take up the issue of the existing rates for industry or occupational classifications, or the classifications themselves if there has been a substantial change in the way your workforce operates compared with past practices, e.g. the technology used may be demonstrably safer, or the task content of occupational categories may have changed. You may be able to raise the fact that you have introduced a significantly revised safety plan. You can refer to the fact that your organization holds a range of other insurances with the insurance company concerned. You can check that the assigned occupational classifications are correct. As mentioned above, a good broker can do much of this work. If you are still not satisfied, then in those jurisdictions without a single mandated insurer, generally you have the choice of switching to another approved workers' compensation insurance company.

Effective claims management

Legal requirements for recording and reporting work-related injury and illness

Statutory bodies within each jurisdiction have different workers' compensation reporting requirements. This allows for the monitoring of industry by government and will influence policy and resource allocation. At a local level, organizations can plan and monitor their own risk management strategies by effective information monitoring systems. Standards are an attempt to help industry standardize injury recording. They outline the elements of injury recording which will allow for an analysis of injury trends. Example accident report forms and performance indicators are included. You are advised to inquire about the workplace injury and disease reporting requirements in your jurisdiction under the workers' compensation and OHS acts.

Effective information systems and their role

Workers' compensation and accident data can be collected manually or on a computerized information system. The size of an organization will usually determine what accident data is collected. Certainly, workers' compensation claims information will be maintained by organizations, as the claim will have to be processed by the employer. Companies can obtain statistical workers' compensation data from their insurer in the form of printed reports.

Use of workers' compensation data to identify and assess risks

Insurers' reports can present information on occupational injury trends, which can be used as part of the organization's risk management strategy. Many off-the-shelf information systems are available to help organizations manage their occupational safety and health accident data. These systems often involve a package of safety management software, including such items as chemicals management systems and workplace auditing tools. In-house database systems can be created by using spreadsheet software packages. Creating your own database in this way allows you to tailor it to your specific needs.

Workers' compensation data – risk control strategies

Whatever the source of workers' compensation data is, it provides valuable information on specific injury trends. These trends include:

- distribution of injuries by body location, e.g. back, hand, eye
- distribution of injuries by their nature, e.g. sprains, burns, lacerations
- distribution of injuries by their agency, e.g. chemical, biological, machinery
- distribution of injuries by their mechanism, e.g. falls from a height, repetitive movement, being hit by a moving object
- distribution of injuries by occupation
- workers' compensation cost allocation by departments within an organization
- performance indicators trends for an organization by years.

The latter are usually expressed as rates and include:

- frequency rate
- incidence rate
- duration (average days lost per lost-time injury)
- severity (percentage of injuries which were fatal or where time lost was greater than sixty days).

As these indicators are standardized they can be used to compare different parts of the organization or used in comparisons with other organizations.

Whilst workers' compensation data is useful to identify occupational injury risks, it must be remembered that it is a retrospective analysis, i.e. an event has occurred to stimulate investigation. A proactive occupational health programme will recognize this and include other risk management tools such as workplace inspections and job safety analysis. Reports which are aimed at identifying occupational injury trends as outlined above, must not only be meaningful to management, but must be simple and visual. Examples of some graphical presentations are given in Figs 12.3 and 12.4.

Line graphs are useful for displaying trends over time. As can be seen, the Maintenance section of this company had a peak in compensation costs during 2002. This type of information can allow management to understand the situation at a glance.

Enhancing occupational safety and health

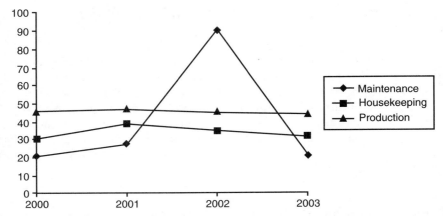

Figure 12.3 Metals Manufacturing Pty Ltd – departmental compensation costs (the vertical axis shows annual claims costs in $000s)

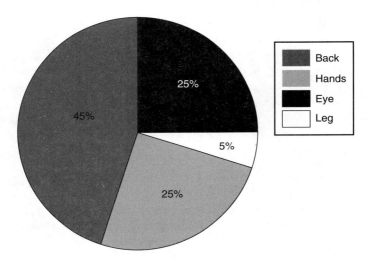

Figure 12.4 Metals Manufacturing Pty Ltd – distribution of injuries by body location, 2003

Pie charts are useful for displaying the distribution of data within one specific group. As can be seen here, during 2003, back injuries accounted for 45% of all injuries in relation to location on the body. This would suggest that emphasis needs to be placed on manual handling risk reduction strategies.

Risk control strategies and premium reduction

Insurance premiums are based on factors such as industry rates, claims experience, risks which insurance companies perceive they need to cover, and possible future losses.

Employers who can demonstrate effective risk control will aim to have their premiums discounted. Effective risk control is demonstrated by factors such as:

- a decrease in the number of claims
- a decrease in the cost of claims
- implementation of policy and procedural controls
- elimination or reduction of high-risk situations
- formulation of a specific business plan to target injury reduction
- specific training, e.g. a back care programme.

Other strategies which have been used for risk control, and thus premium reduction, include:

- outsourcing of tasks which account for the highest claims cost
- a threshold below which the employer directly pays the cost of the claims, e.g. 5000 currency units.

Ultimately a combination of both retrospective analysis and proactivity will reduce injury rates and, therefore, workers' compensation premiums.

Further reading

Allworth, E. (1998). Case Study in Occupational Rehabilitation. *Journal of Occupational Health and Safety – Aust. NZ*, **4(3)**, 219–23.

Barth, P.S. (1995). Compensating Workers for Occupational Disease: An International Perspective. *International Journal of Occupational and Environmental Health*, **1(2)**, 145–58.

Brown-Haysom, J. (2001). Back On The Job (back injury recovery). *Safeguard (New Zealand)*, November/December, 26–31.

Dewees, D., Duff, D. and Trebilcock, M. (1996). *Exploring the Domain of Accident Law*. New York and Oxford: OUP.

Fraser, R.D. (1996). Compensation and Recovery from Injury. *Medical Journal of Australia*, **165(7)**, 71–2.

Hopkins, A. (1994). The Impact of Workers' Compensation Premium Incentives in Health and Safety. *Journal of Occupational Health and Safety – Aust. NZ*, **10(2)**, 129–36.

ILO. (serial). *Bulletin of Labour Statistics*, ILO, Geneva.

Kenny, D. (1994). The Relationship Between Workers' Compensation and Occupational Rehabilitation. *Journal of Occupational Health and Safety – Aust. NZ*, **10(2)**, 157–64.

Larson, T.J. (1991). We Need Applied Prevention not Statistics. *Journal of Occupational Health and Safety – Aust. NZ*, **7(4)**, 287–94.

NIOSH. (2000). *Worker Health Chartbook*. Cincinnati, US DHHS.

Taylor, G.A. (1997). Workers' Compensation in East Asia. *ICOH Quarterly Newsletter*, **16(2)**, 11–21.

Williams, C.A. (1991). *An International Comparison of Workers' Compensation*. Boston: Kluwer.

Wood, G., Morrison, D. and MacDonald, S. (1995). Rehabilitation Programs and Return to Work Outcomes. *Journal of Occupational Health and Safety – Aust. NZ*, **11(2)**, 125–37.

plus national, state, provincial, or territory workers' compensation authority information packages.

Activities

1. List and explain four things which are taken into account when a court is deciding whether an employer has been negligent in relation to an employee who has suffered a workplace injury. (Answer this even if common law claims have been ended in your jurisdiction.) Explain three problems which might arise with producing evidence about these things.
2. Which persons *employed* in a workplace does an employer have liability for under your legislation? Who is deemed to be the employer in a contractor/sub-contractor situation? Is workers' compensation insurance compulsory?
3. When are benefits under your workers' compensation legislation stopped?
4. How are weekly benefits calculated under your legislation? What specific benefits apply for permanent incapacity?
5. Using a workplace you are familiar with, follow up a case of serious injury. (Obtain necessary permission.) Comment, based on some of the issues covered in this chapter.
6. Select, with appropriate permission, an actual case of injury involving rehabilitation and write up how it was handled. Explain the various steps and the reasoning for them.
7. What problems involving various professional providers did or might arise in Question 5? Explain your solutions to these problems.
8. Draw a flowchart for the workers' compensation dispute resolution procedures under your relevant legislation.
9. Obtain a claims report for a selected organization (with their permission) from their insurer. Identify and assess the risks and explain your risk control strategy.

13

Health and safety training

> **WORKPLACE EXAMPLE**
>
> A barge was undergoing repair and maintenance in Singapore. A worker entering a ballast tank lost consciousness. Two other workers who went in to rescue him were similarly affected.
>
> All three were rescued and recovered. The tank had been closed before the accident and was rusted inside. Rusting consumes oxygen. The oxygen level in an adjacent unopened tank was tested and the result was 4.5%. The normal level of oxygen in air is 21%. The barge owner was fined as there were no safe work procedures and permit-to-work system as the law required. A second fine was applied because the workers had not been sent to a safety training course which had the Chief Inspector of Factories' approval.
>
> (From Ministry of Manpower, Singapore.)

Health and safety training programmes

Education and training needs

Education
Employee education aims to develop knowledge and understanding, rather than knowledge and skill, for a defined activity through various methods which provide an understanding of traditions, ideas and concepts. It involves verbal as well as other communication channels which are fundamental to learning. Education on matters such as safety, hazard management, and emergency procedures is vital to safety management.

Training
Training, on the other hand, is the planned and systematic sequence of instruction, under competent supervision, designed to develop or improve the predetermined skills, knowledge and abilities required by an individual to perform a task to a particular

standard. Training can involve various techniques including on the job coaching or mentoring, demonstrations, group or individual exercises, role playing, case studies and displays to name a few. Some can be on-the-job. An example of on-the-job training would be in accident investigation.

Advantages of a systematic approach

A systematic approach to training is a well organized, ordered manner of instruction with the aim of allowing participants to gain first-hand knowledge and experience dealing with situations they face on a daily basis. Good safety training cannot be achieved by a hit or miss approach based simply on experience and imagination. There has to be a carefully thought out and rigorously executed range of safety training approaches. The effectiveness of the training programme in achieving outcomes needs to be assessed through proper evaluation.

There are many advantages to a systematic approach to training, some of which include:

1. Meeting goals and objectives more efficiently and effectively
   If problem areas can be targeted then emphasis on these will allow management to deal with pertinent issues (i.e. use the Pareto 80/20 rule approach) and not waste valuable time and resources.
2. Greater employee satisfaction
   This arises from a strengthened competency base and, therefore, boosts self-esteem and confidence of employees, with additional advantages of greater opportunity for advancement, job enlargement and to learn more.
3. A more versatile workforce
   Training allows the workforce to be flexible and respond to changing needs as workers acquire new knowledge and skills which they can put into practice. Workers are also better equipped to adapt to inevitable dilemmas by using skills gained from case studies and exercises performed during training sessions.
4. Cost savings
   Training results in minimal interference to production as workers are better equipped to deal with the unexpected; there are reduced hidden costs; reduced errors, accidents and absenteeism; reduced labour turnover; reduced recruitment costs; and reduced set-up and change-over times.
5. Enhanced company image
   A company's policy statement on commitment to safety, health and welfare of employees is well demonstrated by conducting systematic training programmes; the quality and quantity of job applicants might also improve as a result of the company being seen in a positive light.

Role of health and safety training in safety management

Effective health and safety training supports organizational objectives and plays an important role in safety management. The management system creates a safety culture that reinforces safe and healthy work practices while training helps provide the knowledge, skills and practice necessary to sustain this culture. The objectives of health and safety management cannot be achieved without systematic training to identify, assess and control hazards and to put safety in the forefront of every person, activity and situation by using defined methods. Training creates an environment within which positive change can occur and provides a forum for discussion to improve performance.

Current approaches to training

There is an important emphasis now in a number of countries on competency-based training, with an emphasis on getting away from the training room and assessing people's actual performance on the job. There may be special units in OHS written for nationally recognized training packages for different industries and for different levels of supervision and management. The Australian National Occupational Health and Safety Commission, for example, has also developed generic competencies in OHS for employees, supervisors and managers, which are a useful basis for integrating OHS into other training, and are now incorporated in some national industry training packages.

Inductions

A training-needs analysis allows optimal use of money and resources. A lot of training plans follow from job descriptions (including future roles for employees) and from procedures.

Induction training for all new employees is especially important. Training needs to consider cultural differences in perception of risk, and the needs of workers who don't speak the local commonly used language. In industries with a high turnover such as mining or construction, a common OHS induction module has been developed in some countries, e.g. Australia, leading to a portable qualification recognized across the industry. The UK is considering such a 'passport'. In Australia the module is based on a core OHS module which has been developed to sit within 'training packages' for different industries.

Writing lesson plans

Lesson plans outline what is to be taught and the methods to be used in a standardized format. They can be used in the traditional training approach, but also in on-the-job methods. They are written to help the instructor to:

- present material in the proper order
- avoid omission of essential material
- conduct the sessions according to a timetable
- place proper emphasis on items to be covered
- provide for trainee participation
- gain confidence
- assemble all equipment needed.

Lesson plans should be written so that information can be viewed easily; that is, handwrite or type legibly allowing enough space to insert breaks, messages, overheads, etc., during the course of training. It is important to remember that lesson plans are not a script; they should contain only the bare minimum of what is required; pertinent information to be revealed; and what is to be accomplished at the end of each segment. A variety of lesson plan formats exists. It is best to draw up a lesson plan that you are comfortable with by integrating the best features of different lesson plans that you have at your disposal.

The bottom line when designing lesson plans is to:

- keep information flowing in an ordered fashion
- have a clear objective of the message you want to convey to the trainees by the end of the session
- keep the participants interested.

Preparing a training register

It is essential that a register recording employee participation in training programmes is kept so that new or untrained workers are identified; those needing retraining are listed; and proof exists that the company has fulfilled its obligation under training and occupational health and safety legislation to provide adequate information, instruction and training to workers. As a minimum, the record should contain the trainee's name, the job for which he/she is being trained, the date and period of time when training took place, the training given, and comments by the trainer or instructor. Advantages of this approach are its usefulness when considering the future development of the individual and possible improvements to the training system. Figure 13.1 is an example of a Training Programme Record.

Young workers

Training for young workers has to address the 'immortality factor', i.e. 'we are immune from harm' and its effect on attitudes and behaviour.

Date	Time	Course details Objectives met	Participant's name	Signature	Trainer's signature	Comments/ assessment

Figure 13.1 Training programme record

Training needs analysis, and design, conduct and evaluation of an OHS training programme

Undertaking a training-needs analysis

Before going further into this matter, we need to focus on the words 'for a given workplace'. The reason is that we will assume that training-needs analysis at the organizational level has been carried out. This first requires that areas of priority for the organization have been analysed to see if they are necessary, and then job descriptions have been prepared. Another way of looking at it is to consider that a number of tasks are required to be undertaken by the organization to achieve its objectives. Once we have decided what the tasks are and how we divide them up into manageable jobs for people, we can write a job description. This is an important step, because multi-skilling requires us to look at which tasks will be assigned to a job rather than asking, for example for a carpenter's job, what the tasks are. Sometimes, preparing the job description allows us to decide if a particular occupation is necessary for the organization, and this may feed back into re-evaluating the way tasks are divided up.

If the process above has taken place, and the job descriptions have been written, then the knowledge and skills needed to carry out particular tasks can be identified in the job specifications. Skills may be cognitive, affective or psychomotor skills – that is, reasoning, sensitivity or feeling (as for a musician) or effective movements, respectively. You will need to identify aptitudes and particular personal traits which suit the person to the job.

If you have a wide range of people to choose from, the job specification need not contain too many fundamental items of knowledge and skill. This is the role of personnel selection, e.g. if you don't want to have to train people in a certain level of reading skills, you could select accordingly. If you want to select people with fundamentally good attitudes to safety, good targeted selection questions can help you to do that. On the other hand, aiming to provide job opportunities for disadvantaged groups may require attention to fundamentals, such as literacy.

The next step is to compare the knowledge and skills the job specification has with the existing knowledge and skills of the person who will do the job. This is called gap analysis. The training needs are then identified and training objectives set. The idea is set out in Fig. 13.2.

You can prepare questionnaires or forms to analyse trainees' characteristics and to analyse a trainee's current knowledge and skills against the different job requirements. This will tell you the training required. You can then write training objectives. Job procedures are a good basis for designing training.

Designing an OHS training programme

The programme needs to meet the organization's individual needs, and apply the legal requirements for health and safety training to the programme's design.

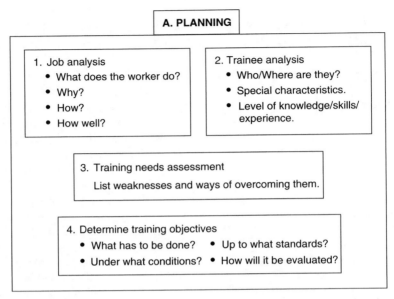

Figure 13.2 Planning training (From Moss, *The Trainers Handbook*. Reprinted courtesy Geoffrey Moss)

Objectives

The training objectives should describe the type of behavioural change which will occur; that is, the skills which will be observed on the job. The objectives show:

- what has to be done
- under what conditions, and with what
- to what standards

and must be accompanied by an effective way of evaluating the results of the training. This approach is known as competency-based training (CBT), mentioned earlier.

So far we have looked at training in general rather than OHS training but, remember, many health and safety skills should be an integral part of job skills. For a better idea of the types of OHS skills required by different job levels and categories you are referred to Appendix 13.1 at the end of this chapter.

In writing objectives you will find that a set of action words based on what is called 'Bloom's Taxonomy' assist. These relate to increasingly complex levels of activity, as shown here:

Knowledge
Comprehension
Application
Analysis
Synthesis
Evaluation.

Table 13.1 Bloom's Taxonomy

Knowledge	Comprehension	Application	Analysis	Synthesis	Evaluation
define	translate	interpret	distinguish	compose	judge
repeat	restate	apply	analyse	plan	appraise
record	discuss	employ	differentiate	propose	evaluate
list	describe	use	appraise	design	rate
recall	recognize	demonstrate	calculate	formulate	compare
name	explain	dramatize	experiment	arrange	value
relate	express	practice	test	assemble	revise
underline	identify	illustrate	compare	collect	score
	locate	operate	contrast	construct	select
	report	schedule	criticize	create	choose
	review	show	diagram	set up	assess
	tell	sketch	inspect	organize	estimate
			debate	manage	measure
			inventory	prepare	criticize
			question	summarize	
			relate		
			solve		
			examine		
			categorize		

Competency-based training focuses on the last four levels. A copy of Bloom's Taxonomy is shown in Table 13.1.

We will assume here that the organization is satisfied with the way tasks are currently performed and divided up, and with the job descriptions and hence the job specifications, or that it has made the necessary changes and carried out the training-needs analysis. This will affect the content of the programme. Two further aspects of organizational needs will be delivery and costs of training, which are interrelated.

Induction training

Delivery and costs and their interrelationship were mentioned above. For induction training, for example, people may be coming on site daily. It is no use running induction training once a fortnight, so the organization may be looking for a self-paced package with built-in assessment so that people can do it on any day at any time. In the interactive multimedia mode of delivery this can be fairly expensive, but may still be a cost-effective option for some organizations. Simpler print-based computer packages and print and paper-based materials may be sufficient. Some industries now have generic inductions with portability (that is, mutual recognition), and then only an add-on site and workplace specific induction may be required.

The same content can be delivered on- and off-the-job. On-the-job may involve one-on-one training under a supervisor or other experienced employee, but these should have had 'train-the-trainer' training, i.e. training in how to train, and assess competency.

Legal aspects

The legal aspects are going to vary depending on the type and level of training and the type of trainee. Training for an open-cut mine manager, for example, would require extensive knowledge of the mining safety legislation.

Enhancing occupational safety and health

For anyone in a workplace the minimum requirements in the training should include:

- appropriate knowledge of legislation
- knowledge of employee entitlements under the legislation – training, including safe procedures; safe systems; safe workplace (so far as is reasonably practicable); election of safety and health representatives where applicable; consultation; right to refuse unsafe work; and information, e.g. MSDS
- duties of employees
- reporting accidents
- hazard identification and risk assessment, e.g. job safety analysis
- specific hazards relevant to the workplace concerned.

For those working largely with chemicals, the hazardous substances requirements of the relevant act and regulations must be addressed. In Manitoba, for example, this would involve the Workplace Hazardous Materials Information System (WHMIS) Regulations 52/88. Many employees will need to be familiar with the requirements in the relevant act and regulations appropriate to plant, equipment and tools.

You should now be able to write the objectives for a short training course based on your previous activities in this learning outcome.

The training course itself

It is necessary to emphasize here once again that a *training course* need not consist entirely of trainees sitting and doing activities in a training room. This might be all of, part of, or none of the course. There might be a face-to-face off-the-job segment, an on-the-job segment and some individual work at a time selected by the trainee, all of which is part of the course.

Once you have decided on the objectives, the four key steps in preparing the training are shown in Fig. 13.3.

Conducting a training session

Within a training course conducted in a *training room*, you need to vary the training techniques to suit the objectives, which may include knowledge transfer; problem solving; skills development (of which problem solving is a part); and change in attitudes. You may decide to move out of the training room to do a hazards identification exercise in the workplace, remembering the safety of your trainees. Some training techniques are set out below.

Which training ways should you use?

To transfer knowledge, use:

- group discussions (questions and answers)
- group or individual exercises

Health and safety training

```
┌─────────────────────────────────────────────────────┐
│                  B. PREPARATION                     │
├─────────────────────────────────────────────────────┤
│  1. Select and organize content                     │
│       • Study sources of information                │
│       • Decide on content                           │
│       • Organize content in logical sequence        │
│                                                     │
│  2. Select training techniques, methods, aids       │
│       • Decide on appropriate techniques            │
│       • Select suitable methods                     │
│       • Decide on training aids required            │
│                                                     │
│  3. Prepare lesson plans                            │
│       • Decide on how each lesson is to be presented│
│       • Set out each lesson step by step            │
│       • Allocate times for each activity            │
│                                                     │
│  4. Plan evaluation                                 │
│       • Decide on information required              │
│       • Decide when this should be collected        │
│       • Study methods of gathering information      │
│       • Select method to be used                    │
│       • Prepare questions which have to be answered │
└─────────────────────────────────────────────────────┘
```

Figure 13.3 Preparation of training (From Moss, *The Trainers Handbook*. Reprinted courtesy of Geoffrey Moss)

- lectures (with handouts)
- forums
- panel discussions
- films, videos, etc.

To practise problem solving, use:

- case studies
- brainstorming
- discussion groups
- exercises, etc.

To develop skills, use:

- demonstrations for manual skills
- role playing for interpersonal skills
- peer teaching
- programmed instructions, etc.

To change attitudes, use:
- debates
- displays
- role playing (for clarifying how others feel)
- group discussion (for group attitudes)
- individual exercises
- demonstrations
- campaigns, etc.

(*From Moss, G. (1989).* The Trainers Handbook, *Sydney, CCH, reprinted courtesy of Geoffrey Moss.*)

Behaviour, including safety behaviour, is influenced by attitudes, but often it is easier to change behaviour than attitudes.

For advice on making training interesting and effective you are referred to Further Reading at the end of this chapter.

You should now be able to put together a lesson plan, overheads or computer slide show and handouts for the training room segment of your training programme and present the segment.

Frank Bird Jr. suggests you keep five Ps in mind in a presentation: prepare, personalize, picturize, pinpoint, prescribe. He also suggested FIDO – frequency – how often; intensity – how vivid; duration – how long; and over-and-over – spaced repetition of the issue.

In countries such as Australia you will also find assistance in materials which have been produced under the national training agenda. These include OHS competencies for specific occupational areas and levels of responsibility. In the UK the New Vocational Qualifications relating to safety are of assistance.

Training and development models

A shift away from the traditional prescriptive type of safety laws towards self-regulation has required a complementary shift in the training and education models used in the learning process.

The bulk of workplace OHS learning is aimed at adults in industry, which requires the use of adult learning models.

The adult learning model will centre on the learner, and what they want to know, rather than traditional learning where a teacher instructs on what they want you to know. This approach to learning places a high demand on the facilitator but is highly motivational and rewarding for the learner.

At the core of adult learning remains the attraction of self-discovery, the capacity to be in a self paced learning environment and a suite of competency-based assessments that relate to the work activity.

Learning styles

Competence involves a combination of knowledge and skill.

Learning styles which affect the development of competence can be broadly characterized as:

- auditory – this learner learns best by listening
- visual – this learner learns best by seeing
- kinaesthetic – this learner learns best by doing (touch and movement), or
- a combination of those styles suited to the task to be learnt.

For example, learning to play a musical instrument can involve listening, reading music (seeing), and touch and movement (selecting the correct stops on a guitar and strumming or picking).

Keep in mind that a person with all senses intact learns around 83% by sight, 11% by hearing and 6% through other senses. Clearly this will vary for certain tasks such as, for example, wine appreciation, or perfumery.

The order of value of learning experiences is:

- real experiences
- artificial experiences
- demonstrations
- exhibits
- films/TV
- pictures
- written words
- spoken words.

Clearly, artificial experiences have a place in safety and health training where failure to perform a task correctly could lead to serious injury or health problems.

Further work has been done on learning styles and the Myers-Briggs Type Indicator uses over 100 items to assess this. Learners can be characterized as extroverted versus introverted; sensing versus intuitive; thinking versus feeling; and judging versus perceptive. This has implications for the method of learning adopted by the trainer.

Barriers to learning

Depending on the situation, the following items can either assist or be barriers to learning:

- communication skills: language, literacy and numeracy
- physical impairment
- previous experience and learning
- learning style
- cultural background
- motivation
- personality traits or attitudes.

The first item above, communication skills, is of great importance and will involve varying degrees and types of reading, writing, speaking, listening and numeracy.

Aids to adult learning

Adult learners generally do best if:

- The material is meaningful and can be related to the learner's knowledge and experience – Ausubel's 'advance organizer' method of training started with a situation learners could use to provide a frame on which to hang the new material to come.
- There is active participation.
- Learning is holistic, starting with an overall view and then bringing in the components and dealing with them.
- Using as many senses as possible.
- Opportunities to practise what has been learnt are provided and their learning is reinforced through encouragement – some adult students have had negative educational experiences years before and need to discover that it needn't be like that.

Feedback on how a learner is doing and reward for undergoing the training also assist the learner.

Adult learning characteristics

Remember to keep these characteristics in mind in planning and running training. Generally, adult learners:

- Use life situations, events and objects to organize units of learning, rather than subject matter or course content.
- Are real-life centred in terms of orientation to content and skills that can be immediately applied to areas of personal interest or need.
- Use analysis of their own and others' practical experience as a core problem-solving and learning methodology, rather than imposed external methods and algorithms associated with standard textbooks and courses.
- Exhibit great differences in style, place and pace of learning. These differences increase with age and experience as peoples' individual differences become greater.
- Tend to prefer project-oriented problem solving that makes extensive use of tools and resources.

(*With acknowledgement to Henry Cole, IMMR/BRASH, University of Kentucky for these characteristics.*)

Assessment methods

Choice of methods

Judgements about a learner's competence can be made by an assessor in many ways. The methods chosen need to be those which are most suited to the competencies involved, and most relevant to the learner's situation. A combination of methods, not just one, is needed. Sometimes simulation will be the method of choice where access to the workplace is difficult or assessment on the job is difficult.

The instruments used for assessment need to be able to collect representative, authentic and sufficient evidence. It must be realized that the assessment instrument only takes a sample of the learning of the candidate and infers from that sample the learner's competencies in a more general context.

Assessment methods include, first and foremost, evidence of prior learning. It is an important principle of competency-based learning that a person does not have to be endlessly reassessed on the same competencies.

Assuming the learner does not have evidence of prior learning, other methods of assessment include:

- direct observation on the job by the assessor
- indirect observation by someone qualified who will provide written evidence to the formal assessor
- skills demonstration
- simulation
- questioning.

In a competency-based training (CBT) course the assessment methods to evaluate whether the required skills transfer or skills acquisition has taken place are usually specified. They may involve having the trainee demonstrate to an experienced person the skill passed on – for example, the series of steps from start to finish in using a danger tag. This is best done under operating conditions because certain visual and other cues, which will assist the trainee, may not otherwise be present, i.e. *describing* is not *doing*. (Think of a driver doing a driving test as against describing how to drive.) For CBT modules already written, these conditions are usually specified, e.g. given access to a conveyor belt, demonstrate the series of steps from start to finish in using a danger tag while cleaning a drive roller.

Note that in a CBT course, the Element of Competency (objective) is already broken into criteria for assessing transfer of specific skills or knowledge (performance criteria). Some recognized CBT courses base the Elements of Competency in them on standards of performance (required behaviour) put together by a group with industry input and review.

Aids to assessment

Some of the aids for each particular method of assessment are shown here:

- direct observation – use of a checklist, log book, peer assessment, research task, work experience
- indirect observation – evidence from supervisors, workmates, clients, or from a portfolio
- skills demonstration – work sample, practical project, structured task
- simulation – case studies, simulated client, simulated workplace, simulated task
- questioning – oral or written questions, case studies, interviews, group assessment, short answer, multiple choice, essay, true–false, matching.

Key competencies

It is important that the assessment is fair and transparent, so that the person being assessed knows how they are to be assessed, when and why. In addition to specific competencies, there are some key competencies which will be required to a greater or lesser extent, and at varying levels, for the performance of many tasks. These are:

- collecting, analysing and organizing ideas and information
- communicating ideas and information
- planning and organizing activities
- working with others and in teams
- using mathematical ideas and techniques
- solving problems
- using technology.

An area of competency also has a number of dimensions which need to form part of the training and assessment:

- task skills – being able to carry out the task at an acceptable level
- task management skills – managing several tasks together
- contingency management skills – reacting properly when things go wrong
- role environment skills – meeting the expectations and responsibilities of the workplace
- transfer skills – being able to transfer skills and knowledge to new situations.

Contingency management skills, that is being able to recover from or minimize the effects of a developing unwanted incident, are particularly important when safety and health is part of the competency, as it often is.

Clearly, in deciding on an assessment method, consideration needs to be given to the preferred learning style of the learner and also to the types of skill being assessed – psychomotor, cognitive, or kinaesthetic.

Some learners can quite easily show an assessor how something is to be done if they are on the job, and they are aided (legitimately), as noted earlier, by the visual, tactile, olfactory (smell) and auditory cues on that job. If on the other hand they are asked away from the job and the cues to explain how they *would* do something, it may be much harder. It can be even harder again if the explanation is to be written and they do not have strong writing skills. The conditions in a Range Statement for a competency in a training package assist because they may say for example 'given access to a TR 32 lathe, demonstrate how to safely machine a 5 cm mild steel spindle'. It may be possible to further generalize if a TR 32 lathe has no unusual features compared to lathes generally, and say 'given access to a lathe, demonstrate how to safely machine a 5 cm mild steel spindle'. If the safety aspect can be generalized in this way, it eases the learning load, widens the area of application of the competency, and reduces the assessment load.

Assessments should be valid (i.e. test what it is intended they should test), and reliable (i.e. should produce the same result in the hands of a second assessor). They should also be holistic – that is, allow the testing of a combination of skills, knowledge and understanding. Safety and health skills need to be assessed as part of the doing of the primary task. They should also take into account the level of skill expected of a learner appropriate to their level of responsibility in the workplace.

Finally, assessment needs to be flexible, suitable for the situation and meet cost constraints.

Options for training delivery

Trends

Currently, the most significant group turning to formal OHS education is found in the work environment, that is people already employed in some capacity requiring OHS skills and seeking formal qualifications.

The trend is away from short courses to a longer term learning commitment. There are several learning opportunities available to meet the needs of those seeking formalized education programmes. There remains the traditional method of institutional attendance. However, people are finding the commitment financially demanding. The requirement to give up their job to study full-time is, in most cases, not an option. Part-time education goes some way to satisfying the demand but the courses generally offer extended time lines and persistence could give way to alternative attractions.

The more progressive view is to see the workplace as the adult classroom and deliver the course directly into the learner's work environment. Information technology has made this option viable and when supported by relevant learning materials and, competency-based curriculum, courses can be delivered in real time. This learning and delivery style provides an immediate benefit to the learner, the industry and the community, and the style may well engender a whole-of-life learning approach to a person's work.

(*With acknowledgement for some of the ideas in this section above to Moss, G. (1989). The Trainers Handbook,* Sydney, CCH; *State Training Board of Western Australia. (1997). Framework for Competency-based Assessment,* Perth, author; *Brightman, H.J. GSU Master Teacher Program: On Learning Styles,* Georgia, author.)

Evaluating a training session or programme

Areas of interest

After conducting a training session, you will want to find out how people responded to the session in a number of areas, e.g.:

- the quality of the presentation itself – interest, variety, relevance
- the effectiveness of the session in transferring knowledge and skills.

If the session was part of a training programme with other segments, e.g. on-the-job, you will need to evaluate the whole programme as well as parts of the programme such as the session. Evaluation of the programme itself can be by way of a properly constructed questionnaire. The questionnaire can include questions on:

- whether programme objectives were explained
- whether the objectives were met
- the amount of material covered
- the programme length
- how valuable the trainee found the programme
- whether the trainee would recommend it to others
- impression of the training provider
- information given prior to starting the programme
- information about what the trainee needed to provide
- information given about support facilities
- suitability of equipment provided
- suitability of training venues or areas
- suitability of session timing
- fulfilment of expectations
- preparation of sessions by trainer
- quality of sessions, participation, feedback, counselling
- quality and content of notes provided
- use of training aids
- trainer's knowledge and skills
- ability to relate to the trainer
- trainer's encouragement of questions and the trainer's responses
- overall rating of trainer/s
- use of real life and background examples.
- suitability and fairness of assessment methods

Behaviours after training

The final step is to evaluate the behaviour/s or use of the skills coming from the training programme. This outcome depends on factors other than just the training, so remember that observing the use of safe procedures on the job depends on more than just the training and so the training cannot be held totally responsible for poor behavioural (skills enhancement) results. For example, poor management or supervisor attitude or example will influence continuing performance of the safety skills acquired in the training.

The importance of procedures

Are the designer and the user on the same wavelength?

It is clear from what has been said earlier in this chapter that good competency-based training depends on good procedures.

Health and safety training

'Following the correct procedure' is a common enough phrase. Procedures play an enormous part in our lives, both in terms of getting things done, but also in getting things done safely. They include instructions for correct turn indicator use on a traffic roundabout; procedures for filling in a tax return; instructions issued for setting up and using a new videocassette player; and work instructions or standard operating procedures on a worksite.

Writing good procedures as we have noted is particularly relevant in relation to competency-based training (CBT). This is because CBT focuses on acquiring or possessing the skills to perform a task properly. This in turn involves the learning of appropriate procedures. The degree to which actual work practices mirror procedures is a constant source of concern. Too literal adherence can unnecessarily slow up work – too slack and errors occur.

Procedures describe what are known to psychologists as action sequences. Norman (see Further Reading at the end of this chapter) commented that surprisingly little is known about the nature of action sequences.

Procedures, together with other documentation, labels and the equipment itself, form what is called the 'system image'. This transmits the designer's conceptual model of the operation of the equipment (or a system) to the user, who forms a mental model. An unclear system image results in a mental model in the user which is ill-matched to the designer's concept. Norman comments that surprisingly little is also known about the properties of mental models. We might take as an example a comparison of a manual and an automatic vehicle. The manual model requires a mental model which visualizes an engine operating within a certain rpm range. It also requires visualizing a mechanism (the clutch) which disengages the engine from the final drive. Synchromesh has already simplified the mental model we require, but an automatic transmission virtually allows the driver not to need a mental model for matching engine performance to vehicle movement. However, a mental model for the operation and limitations of c.v. joints is needed for the many drivers who now drive a front wheel drive vehicle. Lack of awareness is likely to cause early damage from tight turns under acceleration.

In a further example, an electrician in a roof space must match lines of wiring to a mental model of the house electrical circuit.

What's in an action sequence?

An action sequence is regarded as having seven steps. These steps are:

- forming the goal
- forming the intention
- specifying an action
- executing the action
- perceiving the state of the world
- interpreting the state of the world
- evaluating the outcome.

Have we successfully achieved a desired outcome? We perceive where we are at now after the action sequence, interpret this in the light of our expectations, and then compare or evaluate it bringing in both our intentions and our goals.

However, the stages are not necessarily separate. Numerous sequences may be involved, and the complete activity may go on for days, months or even years. Feedback from one activity informs and directs associated activities. For example, consider a repair job on a vehicle or piece of machinery. It may be discovered after dismantling that a part will need to be replaced or remetalled and the remetalling tested. That sets off a second action sequence and so on.

Let's consider an example of an action sequence. Assume that we want a weed-free garden bed. That is our goal. We form an intention to hoe it. Our expectation is that the hoe will remove all the weeds. We plan a sequence of actions with the hoe – start at the left of the bed, hoe the area within reach, step backwards, repeat, step forward and sideways, repeat, and so on. When we reach the end of the bed we rest on the hoe. We perceive the bed. We then interpret what we see against our expectations – the bed appears to be hoed. Next we evaluate the hoed bed to see if our intention has been followed by a successful strategy – has the bed been fully hoed? Further, has even full hoeing resulted in the achievement of our goal, which is complete removal of all weeds?

Not all activities involve the degree of conscious planning and reflection given in this garden bed example. A daily routine may take us near a shop and so while we are there we think 'Why not step in and buy bread since I'm here?' These opportunistic departures have positive and negative impacts. The positive side is that perhaps we meet someone we need to meet and sort out something. The negative side, during a worksite procedure, may be that valuable time is lost or the original intention is forgotten or delayed, holding up work or the acquisition of a needed tool.

What level is an action on?

Procedures can specify action at a number of levels. We will take as an example spreading gypsum on a farm, using a machine spreader.

The levels are illustrated by these actions:

- spreading gypsum
- setting up the machine
- filling hopper with gypsum
- lifting a bag of gypsum
- squatting and placing left hand on one side of the bag and right hand under it
- tensing the muscles of the thigh to lift the bag.

An upside down tree-like picture of these levels can be drawn in which tensing the thigh muscles is just one twig.

So specification of an action can be high level through to low level. A procedure may need to consider the level at which a step is specified. People often attempt to correct a problem but at the wrong level. This may arise through restricted ability to apply interpretation and evaluation skills to the work done. Error correction seems to start generally at a low level and work upwards. Norman gives an example of a person in a car park who tries a number of manoeuvres with a key because it won't open their car, only to finally realize that the problem is high level – it isn't their car after all.

It follows that training people to remember to shift levels when solving a problem is important – sometimes it can be the way to overcome a mindset.

Knowledge in the head and knowledge in the world

These two aspects of knowledge are of particular importance to the assessment of competence. Behaviour follows from information in a person's memory, together with information in the world impinging on the person's senses. This is of particular importance in the application of kinaesthetic skills – skills involving dexterous use of the body in relation to equipment, machines, tools and material. ('Material' may include a sports ball in this context). Another example might be the application by an artist of paint to a canvas in a particular way.

Knowledge in the head has certain limitations inherent in the nature of memory and forgetting. The world or the environment supplies many of the cues which, if they are absent, place too many demands on knowledge in the head. In assembling an item, the user is guided by observations such as the size of a bolt in relation to a hole; left or right hand thread; support legs which usually face down not up; indentations on the surface; and the like.

With these 'world' cues, the precision of the knowledge needed in the head need not be great, and yet competent behaviour will still result.

Society also has some conventions which further determine actions. Given a choice of placing a white or blue light on the top of a police car, the assembler would choose blue (or blue and red), because of these conventions.

So here we see demonstrated the critical value of on-the-job training and assessment. The trainer is usually assisted by the environment in delivering the training, although constraints such as safety may demand off-the-job training first. The assessment mirrors the situation where adequate performance is demanded in the real world. So a training room situation devoid of cues, where the trainee is asked to describe how they would do something, and where sometimes the assessor is even warned not to provide any prompts, is seen as being undesirable in competency-based testing, where the on-the-job situation quite reasonably and normally does provide prompts. This squares with our understanding of how knowledge in the head and in the world combine to allow a person to display competence.

A clear example is in the use of proprioception, the sense of where the body and its parts are in space and where the pull of gravity is, particularly for sportspersons, gymnasts and ballet dancers. This can be developed through exercises like balancing on a board on a ball. It is the key to learning to ride a bicycle, of which it is often said 'once learnt, never forgotten'. Compare the difference between assessment where a person is asked to describe riding a bike and assessment where they demonstrate the outcome – riding a bicycle. A second example lies in the difference between describing how to write Chinese characters, and the almost automatic hand movements that write them once they are learnt. They cannot in general be reproduced simply by studying what they look like. Writing each character forms a step in the procedure. A loop at each step describes the order in which strokes are written. Likewise a sentence in any language is constructed based on a procedure involving grammar and syntax.

Designing for competence

The quality of outcomes will suffer if there is an attempt to use procedures to overcome poor equipment or system design. Thus, because equipment design, procedures, and competent use are related, what sort of design features should we look for?

The first, Norman notes, is visibility – a user should be able to tell by looking what the state of the device is (e.g. on-off, in gear, out of gear) and what alternatives exist.

A simple example is a door. Some doors have been designed not to look like doors, so that it is by no means clear what or where one presses (or does one press?) to open it.

Norman notes that the second design feature is that there should be a good conceptual model. Someone trying to adjust temperatures in a refrigerator's main compartment and freezer after being led to form a mental model based on separate thermostats will become frustrated if the actual design only incorporates one thermostat and the user doesn't know where it is. From a safety perspective, if the user's mental model doesn't match the designer's mental model, the user may find themselves repeatedly in a situation of deviation from the desired system state (i.e. the temperature will never be right). This is very important for food quality or in a hazardous chemical process for example.

A third feature of good design is what are called mappings. Actions and results, controls and effects, the system state and what is visible should link in a logical fashion. As an example, if there are three parallel airport runways and three switches in a row to activate the lights, the middle switch should control the middle runway lights. It would invite problems to do it otherwise, even if the switches were labelled 06 L, 06 R, 06 M. And remember you may confuse those three choices more easily than say A, B and C.

Feedback is the fourth key to design. The result of an action should swiftly be made clear so that the action, if wrong, can be adjusted or corrected. Modern PCs try and assist us, if we have the sound on, by using sounds, not just the screen, to tell us if what we are doing is correct, or if there is a problem. If you can easily and without effort use something unfamiliar correctly, it has been well designed.

Sound from a tool has the ability to convey a variety of feedback – scraping, vibration, and screeching all carry a meaning to the experienced user.

The cultural cues which modify behaviour, mentioned earlier, sometimes sit at odds with design. One car built with safety as a keynote nevertheless does not have the normal placement (for right-hand drive countries) of the indicator stalk on the right side of the steering column (because it is built in a left-hand drive country). In a critical situation, windscreen wipers are no substitute for the indicator! 'Forcing functions' designed into equipment can also prevent undesired actions being taken, or actions being taken unconsciously, e.g. a deliberate left-hand thread on a gas cylinder should prevent fitting of the wrong regulator.

Getting it wrong

Procedures are of course intended to get something done safely and without error.

The deviation of a work practice from the procedure may indicate error on the part of the person carrying out the procedure, but it can also indicate an error in drawing up the procedure. For example, if a contractor repeatedly breaches a procedure on a site, the normal reaction may be to criticize the contractor. However, there may be more value in finding out why the breach repeatedly occurs.

As we have just done, people often talk of breaches or violations of procedures. But it needs to be borne in mind, as noted in Chapter 11 that there are three types of error according to Rasmussen (see Further Reading at the end of this chapter) – skill-based, rule-based and knowledge-based (s, r and k) and so the reasons for the breaches of procedure can be quite different. As an example, we tend to use skill-based and rule-based decisions a thousand times more than knowledge-based decisions in driving a car.

James Reason (see Further Reading at the end of this chapter) explains types of error in more detail. Understanding these can help to reduce their occurrence. Errors in skills result from variability in time, space or force coordination. Errors with rules arise from misclassification of the stored rules, thus using the wrong rule, or from incorrectly recalling a rule. Knowledge is primarily used in new situations where actions have to be consciously planned. Errors can arise in the use of knowledge because of limited knowledge resources, or incomplete or incorrect knowledge. Each type of error may be subject to the effects of memory. So any training on procedures should assume error may occur, try to minimize error from the above sources, and include methods to recognize error and recover from it.

Skill- and rule-based decisions are the basis of routine activities. As a result, there are more absolute numbers of errors in such decisions, but the percentage of errors is much lower than in knowledge-based decisions. The types of errors in activities based on skill- and rule-based decisions are more likely to be what are called 'slips' and 'lapses'. Slips come from attentional failure, lapses from memory failure – they are two types of unintended actions. The types of errors in knowledge-based decisions are more likely to be what are distinguished as mistakes – errors in planning a path to an outcome. Mistakes are a type of intended action, and may be rule-based or knowledge-based. They occur when the selection of an objective and the means to achieve it are faulty, whether or not the actions go as planned.

The funnel

It is a paradox of procedures that the more tightly we prescribe them (visualized by Hudson and Verschuur – see Further Reading at the end of this chapter – as a narrowing funnel), the more likelihood there is that there will be deviations. Thinking persons adapt procedures to circumstances, and what one person sees as a sensible adaptation or initiative may be seen by another as an unauthorized departure from what is laid down. There are dangers here. For example, in a laboratory with trained chemists, suitable changes were made to analytical methods to improve them. However one technician, not a trained chemist, obtained poor results. It was found that he had changed the analytical procedure. He was asked how he made the decision to change it. He said he had no theoretical basis for doing so, lacking the necessary training, it just seemed a good idea! The moral is that initiatives should be welcome, within the scope of skills and knowledge, but that scope should be defined or the person should get approval for the change from someone with appropriate training first. The procedures should reflect that.

Tasks may be wide and deep, that is with lots of steps and lots of sidesteps or loops. They may be shallow (a couple of steps with lots of possibilities such as choosing an ice cream flavour). They can be narrow (a linear series of steps with few possibilities of, or a need to move sideways). An example of a narrow procedure is getting on a particular bus and paying your fare. Procedures involving many steps and loops inherently involve more opportunities for error.

Procedures with loops invite the possibility that the user will re-enter the mainstream either upstream of the entry point, or downstream of the entry point. They will either repeat a step unnecessarily after making an inappropriate check (the loop) or miss out a mainstream step. People who become expert in an area tend to do so by 'chunking', that is learning to group individual items into certain critical patterns. This is why people who enter a familiar room and say it has changed, often can't immediately give details.

Remember that much of the thought which we routinely use is subconscious and relies on pattern matching – matching previously observed patterns with a current situation, and the match is not always all that accurate. We tend to group similar patterns in our memory even though there may be significant differences, and only place the unusual separately. That may cause us to give them both about the same weighting. That is a two-edged sword because it may make people forget that something works well 99% of the time. On the other hand, remembering the unusual incident may be important in safety terms, given that things in general operate safely much more than they operate unsafely.

Developing procedures

What these ideas suggest for those developing procedures is that firstly a complex procedure should be broken down into a series of manageable bites. Some of the bites, or at least some of the loops may, once learnt, operate largely in subconscious (automatic) mode. Chunking will also occur, so that a series of steps and checks proceed smoothly and quickly. What is needed is to remember how many bites, the entry/exit points to the loops, and a system for remembering or recording which loops have been completed. There is also a need to build in a method whereby straying from a loop into a subloop is a conscious and recorded decision, if the user is not to totally lose track of the mainstream. A piano piece can be learnt by breaking it down into a series of steps, learning the steps and then putting them together. Perhaps the best example of straying into subloops in piano music was the comedy pianist Victor Borge. The rhythm and key of one composer would flow effortlessly into a piece by a second composer and so on, although in this case the slip was probably feigned.

Procedures are often set out in manuals. However, manuals may be written after equipment is designed, rather than before. Good design would see the manual come first. Good design might also decide against incorporating some of the 'features' often added to equipment which few people can remember how to use. As a result people automatically default most of the time to the common usage mode.

Once again we should also try to anticipate where s, r and k errors are most likely to occur in procedures and either alter the procedure or improve the training on the procedure.

Effective procedures

Effective outcomes in safety and efficacy for a task depend on effective training and this depends on well constructed procedures. As much as we might not like to, we must recognize that on many occasions the procedure and the training must overcome poor equipment design, and possibly poor educational, literary or verbal standards.

A good procedure, which assists the development of a good mental model by the user, forms the basis for the development of effective training. A good procedure for a complex task should incorporate and facilitate the use of chunking. It should ensure that there is minimal potential to stray from the steps and loops. A good procedure will also include:

- the indicators needed to accurately interpret and evaluate progress in a task
- recognition of error, and how to recover from it
- a means to enable the user to recognize when to suspend the procedure and seek outside assistance.

Further reading

Bird, F.E. (Jr.) (c.1980). *Mine Safety and Loss Control*. Loganville, Ga: Institute Press.

Bone, D. (1988). *Four Key Elements of Good Listening. The Business of Listening*, pp. 12–21, 60–2. California: Crisp Publications Inc.

Boyce, M. (1996). NVQs – Is This What We Have Been Waiting For? *The Safety and Health Practitioner*, **14(2)**, 12–13 (vocational training in OHS for practitioners).

Brooman, G. in Williams, S. (ed.) (1994). *Successful Training Strategies*. Ch. 1, Sect. 6, pp. 1–9. Ontario: Southam Information and Technology Group.

Business Services Australia. (2003). *National OHS Practitioner Competencies*. Melbourne: Business Services Australia.

CCH Australia (2000). *Planning Occupational Health and Safety*. 5th edn. Sydney: CCH Australia.

Cohen, A. and Colligan, M.J. (1998). *Assessing Occupational Safety and Health Training*. Cincinnati: US DHHS NIOSH.

Collegan, M. (ed.) (1994). Occupational Safety and Health Training. *Occupational Medicine, State of the Art Reviews*, **9**.

Eitington, J.E. (1996). *The Winning Trainer*. 3rd edn. Houston: Gulf Publishing Co.

Else, D. (1992). Enhanced Cohesion and Coordination of Occupational Health and Safety Training in Australia. *Report to the Minister for Industrial Relations*. Ballarat: VIOSH.

Fettig, A. (1990). Ten Ways to Make Your Meetings Sizzle! *The World's Greatest Safety Meeting Idea Book*, pp. 10–16, 20–7. USA: Growth Unlimited Inc.

Gebrewold, F. and Sigwart, P.F. (1997). Performance Objectives. Key to Better Safety Instruction. *Professional Safety*, **42(8)**, 25–7.

Godbey, J.F. et al. (2002). Training Managers for a Safer Workplace. *Professional Safety*, **47(7)**, 28–31.

Hollands, R. et al. (2000). Equipment Operation/Safety Training Using Virtual Reality and SAFE-VR. *Proceedings Minesafe International Conference*, 4–8 September. Perth, Chamber of Minerals and Energy of Western Australia, 165–77.

Hudson, P.T.W. and Verschuur, W.L.G. (1994). A Review of the Necessity for Procedures and Problems Associated with Them. *Department of Experimental and Theoretical Psychology*. Leiden University.

International Social Security Association. (1996). Role, Tasks, Skills and Training of Health and Safety Experts. *International Committee of ISSA – Education and Training for Prevention*, Paris.

Keith, N. (2001). Training and Due Diligence. *Accident Prevention (Canada)*, September/October, p. 10.

La Montagne, A.D. et al. (1992). Participatory Workplace Health and Safety Training Program for Ethylene Oxide. *American Journal of Industrial Medicine*, **22**, 651–64.

Landow, D. (1998). Arresting the Fall Guys (fall protection training for construction workers). *Safeguard (New Zealand)*, September/October, pp. 18–20.

Moss, G. (1989). *The Trainers Handbook*, Sydney: CCH Australia.

Norman, D. (1988). *The Psychology of Everyday Things*. New York: Basic Books.

Parker Brown, M. and Nguyen-Scott, N. (1992). Evaluating a Training-for-Action Job Health and Safety Program. *American Journal of Industrial Medicine*, **22**, 739–49.

Peel, M. (1988). *Meeting Hardware: How to Make Meetings Work*, pp. 35–46, 101–6. London: Kogan Page Ltd.

Rasmussen, J. (1983). Skill, Rules and Knowledge: Signals, Signs, and Symbols, and Other Distinctions in Human Performance Models. *IEEE Transactions on Systems, Man and Cybernetics* SMC-**13(3)**.

Reason, J. (1991). *Human Error*. Cambridge, CUP.

Rylatt, A. and Lohan, K. (1995). *Creating Training Miracles*. Sydney: Prentice Hall.

Taylor, G.A. (1995). Degree Level Education in Occupational Health and Safety in Australia. *Journal of Occupational Health and Safety – ANZ*, **11(4)**, 359–71 and **11(5)**, 479.

Taylor, G.A. (1996). Toolbox Meetings with a Difference. *Safety in Australia*, **19(2)**, 14–15.

US Government. *Training Curriculum Guidelines on Hazardous Substances*. 29 CFR 1910.120. Appendix E.

WorkCover NSW. (1998). *Guidelines for Writing Work Method Statements in Plain English*. Sydney: WorkCover.

Worksafe Australia. (1998). *National Guidelines for Integrating Occupational Health and Safety Competencies into National Industry Competency Standards*. 2nd edn (NOHSC: 7025). Canberra: AGPS.

Worksafe Australia. (1992). *National Guidelines for OHS Competency Standards for the Operation of Loadshifting Equipment and Other Types of Specified Equipment.* (NOHSC: 7019). Canberra: AGPS.

Plus www.ntis.gov.au (for Australian training packages including the OHS modules in those packages).

Activities

1. List the skills needed to carry out a procedure in a workplace safely and healthily. Will the trainee be required to do this or supervise others doing it?
2. Compare these skills with the existing training for this procedure and identify any gaps.
3. Carry out a training-needs analysis to identify any of the extra skills you have identified in which a group of workers need further training. (It may be that the group have already acquired the skills despite the gaps you identified.)
4. Identify the conditions under which these skills must be performed, e.g. night work, repetitive, at height, etc.
5. What attitudes and knowledge are required along with the skills you have identified in questions 1–3?
6. Write a training programme to address the extra skills (including safety and health aspects) which are still required after considering your data from questions 1–3. Which learning style will best suit the skills – kinaesthetic (physically doing), auditory or visual?
7. How will you ensure that the programme provides training to deal with contingencies (i.e. things out of the ordinary)? Will the programme include how to transfer those skills to a related application?
8. How will you ensure that assessment after the training validly checks competencies in those skills?
9. If possible, trial the training you have planned.

Appendix 13.1 – OHS competencies and performance criteria for key workplace parties

	Competency	Performance criteria
National core competencies	**Employees and Trainees** Identify hazards which may be present in a workplace.	Hazard is reported.
	List preferred order of controls for hazards present in a workplace.	Safe-place controls are suggested as first option and behavioural controls as the last.
	Apply knowledge of rights and responsibilities under common law.	Responsibilities are met and infringements of rights recognized.
Occupation specific	Recognize and predict hazards associated with the occupation.	Perform tasks safely.
	Recognize unforeseen hazards and irregularities.	Employ appropriate action and report as appropriate when contingency arises.
	Apply preferred order of controls to hazards associated with specific occupation.	Others are questioned if elimination of hazard is not suggested as the preferred control strategy.
Workplace specific	Raise issues via appropriate people and mechanisms.	OHS issues are raised with the OHS representative and supervisor using locally agreed procedures.
	Use appropriate systems of work, permits and procedures wherever necessary.	Systems of work and emergency procedures are followed and where necessary permits to work are used.
Jurisdiction specific	Apply knowledge of rights and responsibilities and mechanisms for consultation under regulations.	Responsibilities are met and issues raised through consultative mechanisms in the workplace.

National core competencies	**Health and Safety Representatives**	
	Identify and predict by observation, inspection and sound questioning, hazards which may be present in workplaces.	Hazards identified, reported and followed up.
	Identify and predict hazards which may be associated with specific occupations and tasks. Undertake rudimentary risk assessment.	Hazards identified, reported and followed up.
	Investigate accidents.	Questioning of the decisions of others is based on an assessment of risk. Notification of all accidents is received. Investigations identify causes and cost-effective remedial action sought and/or recommended.
	Obtain and apply comprehensive information, legislation, codes, etc. relevant to hazards and concerns.	Information is interpreted and applied in the resolution of problems.
	Question the application of the preferred order of controls for any hazards in the workplace. Communicate effectively with workers, supervisors, managers and Government inspectors.	All discussions about control incorporate the appropriate hierarchy of controls. Consultative processes are effectively employed and issues successfully resolved.
	Question at decision making, selection, design and planning stages of new plant and equipment. Employ effective problem-solving skills.	OHS implications are discussed at planning and design stage. Cost-effective solutions to problems are discussed using a hierarchical approach.
	Question ways in which control strategies may predictably fail.	Predictable ways in which controls may fail are discussed prior to implementation.
	Question regarding integration of OHS into the management systems of the enterprise.	Opportunities to integrate OHS into management systems identified.

(*Continued*)

	Competency	Performance Criteria
	Represent workforce, question and negotiate. Question the effectiveness of controls.	Effective representation of workforce. Controls are monitored and improved as necessary. Successful solutions are documented and shared.
Occupation specific	—	—
Workplace specific	List hazards which are associated with specific occupations and tasks.	Hazards are predicted and identified.
	OHS issues are raised and resolved via the appropriate consultative mechanisms.	OHS issues are successfully resolved with the assistance of, or with the knowledge of, the Health and Safety Representative.
	Implementation of systems, permit procedures, etc., as necessary, is monitored.	Suitability and use of systems, permit to work procedures, etc. are questioned.
Jurisdiction specific	Apply knowledge of rights and responsibilities and mechanisms for consultation under regulations.	Responsibilities are met and issues raised through consultative mechanisms in the workplace.
	Apply knowledge of local legislation, codes, standards, etc.	Any activities fall within the requirements of local legislation, codes, standards, etc.
	OHS Committee Members	
National core competencies	In addition to the competencies of the health and safety representative, committee members require the following:	
	Communicate ideas and opinions to other committee members.	The safety committee effectively and efficiently resolves issues and recommends solutions to problems.

	Apply committee skills.	Action required following committee decisions clearly identified and those allocated responsibility monitored by the committee.
	Assess, set and monitor priorities.	Action plan to address issues of concern developed for consideration by management.
Occupation specific	—	—
Workplace specific	Describe the role of the safety committee and its members.	The safety committee functions effectively and efficiently within the overall committee structure and the organization as a whole.
Jurisdiction specific	Apply knowledge of rights and responsibilities and mechanisms for consultation under regulations.	Responsibilities are met and issues raised through consultative mechanisms in the workplace.
	Apply knowledge of local legislation, codes, standards, etc.	Any activities fall within the requirements of local legislation, codes, standards, etc.
	Supervisors	
National core competencies	Identify and predict hazards which may be present in workplaces.	Hazards identified, reported and controlled.
	Identify and predict hazards which may be associated with specific occupations and tasks.	Hazards identified, reported and controlled.
	Undertake rudimentary risk assessment.	Action is decided on the basis of risk.
	Investigate accidents.	Notification of all accidents is received. Investigations identify causes and cost-effective remedial measures implemented.

(*Continued*)

	Competency	Performance Criteria
National core competencies	Apply the preferred order of controls for any hazard in the workplace.	All decision-making processes apply the appropriate hierarchy of controls.
	Communicate effectively with workers, health and safety representatives, managers and Government inspectors.	Consultative processes are effectively employed and issues successfully resolved.
	Consult workers and OHS representatives at selection, design and planning stages of new plant and equipment. Employ effective problem-solving skills.	OHS implications are discussed at planning and design stage. Cost-effective solutions to problems are implemented using a hierarchical approach. Unresolved issues are referred to the appropriate manager.
	Identify ways of integrating OHS into systems and procedures.	OHS integration into systems and procedures.
	Predict ways in which control strategies may fail.	Control strategies implemented incorporate mechanisms to prevent failure.
	Evaluate the effectiveness of controls.	Controls are monitored and improved as necessary. Successful solutions are shared.
	Formulate cases for allocation to control hazards.	Cases accepted for resource allocation by managers.
Occupation specific	Recognize and predict hazards which are associated with specific occupation and tasks for which responsible.	Hazards identified are raised for discussion and controlled.
	Inspect plant, process and procedures for which responsible.	Hazards identified are raised for discussion and controlled.
	Identify training needs and ensure relevant training and monitoring of its effectiveness.	Employees operate and adhere to systems of work and receive refresher/update training as necessary.

Workplace specific	List hazards which are associated with specific occupations and tasks.	Hazards are predicted and identified and controlled.
	Resolve OHS issues via the appropriate consultative mechanisms.	OHS issues are successfully resolved with the assistance of, or with the knowledge of, the OHS representative and workers.
	Implement and monitor systems, permit procedures, etc.	Systems, permit to work procedures, etc. are successful and fully utilized.
Jurisdiction specific	Apply knowledge of rights and responsibilities and mechanisms for consultation under regulations.	Responsibilities are met and issues raised through consultative mechanisms in the workplace.
	Apply knowledge of local legislation, codes, standards, etc.	Any activities fall within the requirements of local legislation, codes, standards, etc.
National core competencies	**Employers and Managers**	
	Identify hazards which may be present in a workplace.	Hazard is reported to the appropriate supervisor and followed up.
	List preferred order of controls for hazards present in a workplace.	Safe-place controls are suggested as first option and behavioural controls as the last.
	Apply knowledge of rights and responsibilities under common law.	Responsibilities are met.
		New plant, equipment, processes, etc. employ safe-place controls to eliminate or minimize risk.
		Responsible employees provide regular reports demonstrating favourable safety performance.
		All managers, supervisors, OHS committees and representatives and employees understand responsibilities and level of accountability.

(*Continued*)

	Competency	Performance Criteria
Occupation specific	—	OHS on senior management agenda. All systems modified/developed to integrate OHS needs. Regular reviews of robustness of systems in place.
Workplace specific	Establish and monitor OHS arrangements. Establish and monitor consultative mechanisms.	The OHS performance of the organization is good. Issues are effectively and efficiently resolved via the consultative mechanisms.
	Monitor supervisory performance.	Supervisors successfully manage OHS in their areas of responsibility.
Jurisdiction specific	Apply knowledge of rights and responsibilities and mechanisms for consultation under regulations.	Responsibilities are met and issues raised through consultative mechanisms in the workplace.
	Apply knowledge of local legislation, codes, standards, etc.	Any activities fall within the requirements of local legislation, codes, standards, etc.
National core competencies	**Design and Technical Decision Makers** Eliminate or minimize risk when designing processes, plant and equipment. Obtain and apply comprehensive information, legislation, codes, etc. relevant to hazards and concerns.	Hazards are predicted and risk controlled by the application of a hierarchical approach. Information is interpreted and applied in the design.
	List preferred order of controls for hazards associated with processes, plant and equipment being designed.	Safe-place controls are implemented as the first option and behavioural controls as the last.

	Consult with users and OHS representatives at design stage of new plant, equipment and processes.	Users are satisfied with the operation of new plant, equipment and processes.
National core competencies	Obtain information concerning previous OHS experience with the plant, equipment or process being designed.	Designs utilize the experience of others.
	Predict ways in which designs and control strategies may fail. Identify the training need of users, supervisors, maintenance staff, etc. for safe use of new plant, equipment and processes. Review decisions.	Designs and control strategies implemented incorporate mechanisms to prevent failure. Users, supervisors, maintenance staff, etc. receive the training necessary to achieve the relevant competency standard. Decisions are subject to question before implemented. Decisions can be defended.
	Question the decisions of others.	Concerns over decisions of others are effectively communicated and defended.
Occupation specific	Identify the constraints placed upon a design by the particular occupational group.	Designs suit occupational group.
Workplace specific	Identify the constraints placed upon a design by the particular workplace.	Designs suit specific workplace.
Jurisdiction specific	Apply knowledge of rights and responsibilities and mechanisms for consultation under regulations.	Responsibilities are met and issues raised through consultative mechanisms in the workplace.
	Apply knowledge of local legislation, codes, standards, etc.	Designs meet with the requirements of local legislation, codes, standards, etc.

(With acknowledgement to Else, D. (1992). University of Ballarat, Victoria. Reprinted by permission.)

14

Health and safety management systems

WORKPLACE EXAMPLE

A colliery in South Africa addressed the human factor in safety systems. It was an underground bord and pillar mine with a workforce of 510 people. There were eleven home languages spoken and education levels were low.

The challenge the company faced was to unite management, the unions, employees and government in tackling safety.

The objectives for the next year were zero fatalities, reportable injuries less than two and lost-time injuries less than five.

Among other things the company initiated Simunye (Zulu for 'we are one') employee participation, and a core driver was the daily safety bulletin issue by senior management at 5.30 am each morning. Another initiative was flashing traffic lights with a safety reminder at strategic points around the mine site.

The company also instituted shock awareness meetings where senior management 'added energy' to the first ten minutes of each of the three shifts by spontaneously addressing their department on a current topic requiring urgent attention. There was also a phone hotline for anonymous reporting on a safety and security matter.

Good results were celebrated with a Simunye event for workers and their families, with a braai (BBQ), top band and dancing, fireworks, team-building aerial photograph, and industrial theatre on HIV/AIDS.

(With acknowledgement to John Standish-White, Greenside Colliery.)

Options for management of OHS in an organization

Systems

Most of us are familiar with a variety of what we call 'systems'. A system is a collection of individual component parts which interact with each other to achieve a particular objective or move in a particular direction. We are all familiar with natural systems such as eco-systems, or the various systems within the human body such as the nervous system and circulatory system.

Let's take as an example a car. The larger system is the car on the road in the external environment. A sensor (the driver) observes the external environment and makes adjustments to speed and direction. So the driver also acts as a controller. This is what is called a 'feedback loop'. The engine is a part of the system, and within the engine there are sub-systems, e.g. ignition, fuel and electrical.

Basic principles of an OHS quality management system

It is useful to consider this from two angles. First of all, the occupational health and safety (OHS) management system and then, secondly, to consider what is involved when the word 'quality' is added. Depending on the company or organization involved, the way in which an OHS management system is viewed will vary. Some companies see it as an integral part of the management system and, while certain components may specifically address OHS, there is no separate OHS system. Small enterprises, in particular, will probably not have a separate OHS management system. However, in larger concerns there is generally an identifiable OHS management system. Apart from the need to systematically manage health and safety, there are good legal reasons for doing this so as to be able to demonstrate compliance with the employer's duty of care under OHS law.

While there are 'invisible' components to any management system, the visible components of an OHS management system may include:

- a written OHS policy (which should flow from the vision, mission statement or business plan)
- guidelines which expand on particular areas of the policy (e.g. manual handling, noise)
- a safety management plan
- procedures which describe how the key parts of the system are to operate
- a hazard register
- at the operating level, safe work procedures, operational procedures or work instructions
- training packages
- checklists and auditing documents.

In larger concerns the overall way these fit together will be described in a manual and will be supported by computer-based data collection, collation, evaluation, and reporting. This may hold data on, for example, hazard evaluations and accident investigations.

So far, so good. Now let's consider what that word 'quality' adds to an OHS management system. In some cases it will add nothing because the system already contains all the ingredients of a quality system. But let's assume here that it doesn't. What does quality add? The overall quality approach is generally considered to be made up of three parts which fit together as shown here.

| Quality + Continuous = Quality |
| Assurance (QA)   Improvement (CI)   Management (QM) |

CI is also called quality improvement (QI). Quality looks at a system as a series of processes, each process having suppliers and customers. It looks at each process and identifies who the suppliers and customers are. These include the company as a supplier of a product or service to an external person or organization (that is, the company's major process), but it may also include many internal suppliers and customers, for example, the information technology part of a company provides a variety of internal customers with the information needed to manage properly. Such customers include you if you are the health and safety coordinator, manager or officer.

A second major feature of the quality approach is that it recognizes that most problems are problems with the system itself, not with the individuals within it. It also emphasizes a team approach.

The third major feature is that it incorporates the CI or QI feedback loop. The results of new initiatives or change are monitored, and the results are used to formulate new or altered action. The cycle is usually called the PDCA cycle – plan, do, check, act.

QA looks at the situation somewhat differently. It identifies four of the areas described above as parts of an OHS management system. These are the manual, procedures, work instructions and records. These are used to audit the system and so generate improvements. There has been some criticism of this because some companies have concentrated on acquiring the documentation without ensuring that it is applied in the workplace. The records are the evidence on which the auditor decides how well the job has been done, and a good auditor needs to check the reality *on the ground* against the documentation in various ways.

The quality approach started with an American writer, W. Edwards Deming, whose approach was taken up enthusiastically by Japanese manufacturers. It is reflected in the International Standards Organization ISO 9000 series, and for environmental management in the ISO 14000 series. More recently the approach has been taken up for OHS in OHSAS 18001:1999 (adopted in the UK and Canada), and Australian Standard 4801:2000, together with the implementation standards BS 8800 and AS/NZS 4804:2001. See also the Section 'Planning integration … '.

Integrated management systems (IMS)

Over the past fifteen years there has been an increase in systems management which targets the areas of quality, environment and worksafety (QEW) or quality, safety and environment (QSE). (The word 'worksafety' is used in place of the traditional use of occupational health and safety.)

All three disciplines have their own standards, procedures and audits. There is strong appeal in the integration of QEW notwithstanding that each discipline requires its own specialist operational knowledge. The reason for this is that integration of these systems should lead to more efficient management, administration and cost-effectiveness as well as better outcomes.

Health and safety management systems

Currently, as noted, quality can be referenced through ISO 9001, environment through ISO 14001 and worksafety standards are generally developed around an acceptable international protocol, e.g. OHSAS 18001, or a national one, e.g. AS 4801.

Common areas of interest

A review of these standards identifies several sections where the management responsibilities and actions are similar, or at least based around the same concepts.

The following are examples of where the QEW procedures share similar administrative requirements.

- policy development that builds sustainable commitment
- formal documentation for recording planning strategies and reviews
- document control
- work systems control
- monitoring standards
- reporting and correcting deficiencies
- collecting and using data
- auditing of systems
- development of skills and competencies.

Competent organizational people capable of providing an effective response in relation to an integrated management system are going to be in short supply. (That comment assumes that the response is more than a documentation activity.)

Many people conducting procedural audits in the QEW functions lack qualifications, technical skills and the practical experience to validate corporate documentation.

While the management of the environment and the attendant issues have, primarily, been the domain of the trained professional, safety and quality management have often been 'in addition' activities. It is important that this deficiency is addressed by organizations.

Prevailing management culture in an organization

A few words about different types of management culture may be useful here. The Macquarie Dictionary defines 'culture' as: *the sum total of ways of living built up by a group of human beings.* In the case of a workplace the 'group' would be those in the workplace, and the 'ways of living' would be the ways the organization in that workplace actually goes about running its affairs. In larger workplaces, there may be different cultures operating alongside each other. Senior management may have one culture, middle management another and the shopfloor another.

Most firms in our society share a common management culture, although some individual firms may be different.

The common management culture usually has a number of features:

- selection and marketing of products and services which people want, or for which a need can be created

- reliance on, and commitment to, continuing economic growth
- utilization of technology to enhance organizational results
- in profit-centred organizations at least, maximizing profits to shareholders.

In industrial societies, these are the key elements of such a culture. There are constraints on this culture. One is environmental – the growing concern about the greenhouse effect, global warming, as well as air, soil, water quality and the health of inhabitants.

Of special importance from an OHS perspective though, are the needs and aspirations of employees. The cynical view is that management generally has been forced to pay attention to these needs and aspirations only so as to continue to operate, and because people are needed to look after the organization's interests in areas such as production, legal, financial and taxation requirements. Management first and foremost usually pays people well who are good at financial management because it serves the organization's interests to do so. A management culture which aims at employee's interests first would be considerably different. There would be a recognition for example, that the public economic world interacts with the private world of the family.

In the current international liberal economic regime, unemployment, threat of job loss, or a work-only focus with seventy hours at work a week, is viewed by some as having nothing to do with marriage stability, says Peter McDonald of the Australian National University in a quote in September 2001. Nor, some would assert, does the cost of, and time taken to earn qualifications have any effect on marriage and having children.

Conflict between management and workers need not necessarily arise. If employees are committed to the same values as management, they may accept that management has the right to set the priorities. However, in many cases it is likely that employees will dispute management's choice of priorities, for example, production levels, cost-cutting and downsizing before safety.

In respect of health and safety, the desired culture of the organization is set out in the OHS policy. The actual culture will depend on the extent of management commitment and the effectiveness with which the OHS policy is implemented through an OHS system built into the organization's management system. One offshore oil and gas company is promoting a culture of love (not romantic).

The OHS laws are designed to ensure management pays attention to health and safety values. Consultation, as required by these laws, is designed to ensure that employee aspirations in regard to health and safety are taken into account in managing the organization.

Leadership styles are also important in shaping management culture and those interested further in this are invited to look at relevant texts and journals dealing with theories of leadership. Crucial to leadership is the way in which leaders motivate others to help the leader achieve their goals. Some basic styles of leadership are:

- highly directive or autocratic
- democratic or consultative
- laissez-faire – leaving people to do their own thing.

Styles of leadership and management culture need to be varied to suit the environment in which an organization operates and the competency levels of employees. For example, consultation on mine health and safety is good in an overall sense, but a mine rescue may need a highly directive command structure. On the other hand, it might seem obvious to say that an aircraft must be firmly in the hands of the pilot. Yet investigations of some recent aircraft accidents suggest that this culture taken too far has prevented co-pilots pointing out something the pilot has overlooked.

Tools for assessing organizational structure and needs in OHS

There are certain basic features which are common to nearly all organizations. They include:

- the people within them (and to a degree those outside who contract to them)
- the technology or tools employed
- the task or purposes of the organization
- the financial basis, the knowledge base, and
- the organizational structure which is used to link these parts.

Viewing or 'mapping' an organization to identify the organizational structure is particularly important when the organization is a complex one.

OHS management essentially is change management, just as every other aspect of management in an evolving responsive organization should be. People generally try to change organizations through changes in structure, in technology or in people's behaviour.

In larger organizations it is not generally difficult to discover the organizational structure. This is usually set out in organization charts, and individual job descriptions show which position a particular position answers to, and which people answer to the position concerned. Formal power, authority and responsibility is partly defined by such charts. That is not too difficult for a straightforward hierarchy. However, where organizations have opted for a 'matrix-style' with many horizontal lines of communication, not just vertical, it is more difficult to work out. Such an approach is often seen as more flexible and responsive. But even if the hierarchy or matrix can be more or less drawn up, it is important to be able to identify the informal organizational structure; that is, the informal centres of power and influence. To proceed without identifying, recognizing and using them (or trying to remove them if they are considered a negative influence) is to risk failure. Put simply, if you are joining an organization as the OHS officer or coordinator, you must 'win friends and influence people' to be successful. Many jurisdictions don't confer a statutory role on the OHS officer. The Malaysian OSH act does though in S.29.

If we look at this topic in a narrower way, it may also require you to look at 'organizational structure' in health and safety management. Some existing structures in companies are:

- safety, and emergency response staff member, and health and first aid staff member, separately answering to a human resources manager, or a line manager
- safety, health, emergency and first aid staff member, answering as above

- a health and safety team as an identifiable unit with a health and safety manager or coordinator (who in some cases is a medical practitioner)
- an employee or manager who takes on safety (and maybe also environment) as part of his/her duties
- a safety and emergency team with a coordinator, with the various members working in operational departments under the departmental manager, together with a health and first aid team working under a medical practitioner
- in large industrial groups, an OHS (or maybe health, safety and environment) coordinator/manager, with separate OHS teams in operating divisions or units. In this situation variable divisional cultures may reflect the type of work the division does (one may be offshore oil and gas, another information technology). Geographic separation, where it applies, may raise problems of communication.

Some organizations have tried the approach of fully integrating health and safety into general management with no separate OHS people or without even a person who has prime charge of OHS issues. In smaller companies this is the norm. (Remember that OHS law requires the manager at all times to have prime carriage of OHS issues, not the OHS officer/coordinator/manager.) However, the general experience in larger organizations is that a *functional* position with a person in it, who is a source of OHS knowledge, expertise and who is up-to-date on OHS issues, is needed. This is a person who twists arms, markets his or her value and services and prods management consciences as needed. In global operations, 'syncretism' – the effect of local culture on the central organization's desired culture, may be strong. The attitude can be 'Corporate HQ is a long way away and what they don't know won't hurt them. I'm working in the back of beyond mainly to look after me.'

The next part to consider is identifying needs. This has really been partly addressed in other chapters, for example, Chapter 4. Clearly, a careful analysis of accident data and investigations (from a similar company if your own is new), and of the hazard register is a good way of assessing OHS management needs. In a world where constant change is the norm, assessment of training needs (gaps in training) is always an important part of this. Avoid the tendency for the OHS unit or team to become self-serving; it is there to reduce accidents and the severity of injury, and to promote and protect the health of workers.

A second popular way of assessing needs now is to measure the organization's safety culture. Some companies and the UK Health and Safety Executive offer do-it-yourself safety culture survey material. This asks participants to respond to statements like: 'I am kept well-informed on safety and health risks'. When responses are collated appropriate steps can be taken to address problem areas.

Perception surveys can uncover some useful data about where to focus the safety effort. One study found that procedural and engineered factors had little effect on safety, although hazard control technology did have a positive effect, as did financial resources. The study also found that a good measure was to be found in the responses received to questions about the quality of the management system and impact of this on safety behaviour. The study found that OHS programmes which dealt with worker and supervisor attitudes and behaviour were successful (see Bailey and Petersen in Further Reading at the end of this chapter).

OHS needs, and priorities of an organization and its sub-groups

To address this question of needs, it is important to collect good data from accidents, audits, inspections and perception surveys, and from similar industries (benchmarking) and similar types of industrial operations (e.g. welding), and analyse it carefully. The effectiveness of audits will depend on how they are constructed and how well they succeed in evaluating the management of issues, not just the issues themselves. Another important approach is to use workers' compensation claims data associated with the injury-producing accidents, and this can be broken up by part of the organization or type of work. This is a very important way to address priorities. However, accidents are not necessarily a good basis on their own for decision making. Statistically, in any one company the database may be too small.

Any one organization's experience of serious injury is usually not common, thankfully. Use of a larger population from across an industry sector can help, but this suffers from the fact that operations are generally not exactly alike and may not draw on the same pool of people and expertise. So, a second proactive approach is to use good hazard identification and structured risk analysis, using the principles in Kinney and Wiruth (see Appendix 4.1) to set priorities for control. It is important when doing this to use the technique in a *holistic* way – that is, don't just look at an isolated physical hazard but review it as a system, with the components 'people, tools, materials, environment and work methods' in mind.

A third aspect is to address people's perceptions of relative risk. Even if these are not rational, they must be addressed in a programme or there will be disrespect or distrust for any programme. This is not to say that scarce resources should be misdirected, but that training, for example, must consider realigning inaccurate perceptions. For example, additional driver training for anyone using the roads as their workplace is very important. Yet these workers may be focusing their attention on much lower-perceived risks with a long-term potential, such as radiation from using a mobile phone, rather than the risk of driving while using a hand-held mobile phone.

Assessment tools, such as those developed by OHS companies and consultancies and OHS authorities, can help.

Structuring an OHS system to meet organizational needs

Obviously there is no one simple answer to this aspect. However, there are two essential objectives which any system must meet, however large or small the organization, or however profitable. The first is compliance with the OHS and workers' compensation laws, and the second is achievement of the lowest accident and injury rates possible. The system chosen will need to reflect the simplicity, or complexity, of the organization and the climate of intrinsic risk in which it operates.

Consider the difference in systems needed for:

- a shopping centre butcher compared to an abattoir
- a corner deli compared to a food supermarket

- a service station compared to an offshore oil and gas platform
- an IT and paper-based public service office compared to an underground mine.

If a system is to both address simplicity or complexity, and intrinsic risk, and straddle provincial, state or national boundaries it will need to reflect:

- Variations in legislation and the approach of regulatory authorities (bearing in mind that most organizations try to adhere to the requirements of the country with the highest standard or to international best practice when operating in lower standard areas).
- Variations in health status, education standards and attitude to risk of local inhabitants, especially in underdeveloped countries. In such a situation, the OHS system may have to pay more attention to off-the-job safety, health and security (note in French: safety = sécurité) than it might do in a more developed country.
- Effective communication in spite of geographic spread.

There are certain universal elements to an OHS system you can look for:

- a policy – even the smallest organization can have a written health and safety policy
- evidence that a hazard and risk analysis has been done
- evidence, even if it is only verbal (in the case of a small organization), that there are certain procedures (i.e. agreed ways of doing things) and that there is training in using these procedures, either on or off the job
- evidence of regular checks or inspections.

In larger organizations and those with particular problems, you would look to see that the organization either employs or contracts-in the right kind of expertise – for example, a psychologist for psychological counselling of employees who are having problems adapting to changes in the organizational environment, or a process control engineer where process control is critical to safety. You would also expect complete documentation of the system, the audits and continuous improvement actions, set out in a manual or in the computer-based equivalent. 'Hands-on' safety will generally be handled at a lower level in the larger organization.

Evaluating OHS management systems against organizational needs

Off-the-shelf OHS management systems are widely available. But it would be a mistake to think that these *are* the OHS management system. The OHS management system is the total collection of all aspects of the organization's activities which influences the number of accidents and injuries which occur while the organization is doing its work. What the off-the-shelf packages can do is make collecting, collating and handling data relating to OHS and training easier. They can also help keep the organization up to date and assist with effective management of injured workers and compensation claims. They are useful in the same way that a computer-based financial package for,

Health and safety management systems

say, a small business is useful. They can prompt certain action at the correct time, and on a regular basis, and analyse trends.

You will find OHS software suppliers are willing to assist you to obtain more information. Some also have demonstration discs or internet files available. See 'Reviewing an OHS management system' later in the chapter.

Strategies to integrate OHS into organizational quality management systems

Development, implementation and evaluation of an OHS plan

We will assume that you are going to introduce an occupational health and safety plan in a medium-to-large organization. If the organization is smaller, then you will need to make appropriate adjustments to the scope and detail involved in the plan.

The steps suggested are:

1. Forming a working party
2. Developing a hazard list or register if it is not available
3. Assigning risk priorities
4. Deciding on the purposes of the OHS plan
5. Developing plan objectives
6. Planning for review
7. Target or goal setting
8. Deciding on specific programmes
9. Allocating resources
10. Assigning responsibilities
11. Implementing the plan
12. Plan monitoring and review.

The work programme usually refers to an operational activity in a specific area of OHS which implements an OHS plan. Certainly we will use 'programme' in this sense (see later in this chapter). The same steps shown above can be used for these specific programmes.

Forming a working party

The working party could be the health and safety committee, if one is already in place. Or it may be set up by the health and safety committee, or management, or the OHS officer or manager, with suitable members. Membership may be broad if the focus of the programme has yet to be agreed upon. Or if the organization already has a good idea of the problem areas, the membership can, as a result, be more focused.

Developing a hazard register

Such a list or register may already exist, or the working party may decide either to start from scratch or to refresh it. Even a working party with a narrower focus on an individual programme will still need a hazard register within its areas of focus.

Assigning risk priorities
Priorities can be assigned to the identified hazards by the working party, using accepted methods of assessing the level of risk associated with the hazard.

Deciding on the purpose of the OHS plan
This step is clearly more relevant where the working party has started with a broad-brush approach looking at many hazards. The purpose of a plan should be to reduce the level of risk associated with a range of hazards.

Developing plan objectives
This will follow from the last step. For instance, it may be that one objective is to set up a programme to improve compliance with hot-work procedures, as a result of some burns or near misses.

Planning for review
Review after the plan has been going for some time can only be really effective if the working party has some baseline data to go on. In the hot-work permit example, the working party needs to know about existing levels of compliance with procedures, so as to assess later whether that programme has had the desired impact.

Target or goal setting
The working party may decide that the best way to achieve the objective above is to provide an increased level of training or to retrain people. This is based on a review of the reasons people identified for breaching the hot-work procedures. The target can be: retrain all relevant staff within three months.

Deciding on specific programmes
Using our example again, the target or goal is retraining, so in this step decisions need to be made on whether hot-work procedures first of all need review, and then on which type and level of training is going to be used, for whom, where and when.

Allocating resources
Here the working party needs to consider people (including outside assistance if needed), equipment and the budget needed to reach the targets, and then get management approval for it.

Assigning responsibilities
In this step the working party needs to decide who will do what; for example, who is to review the procedures; who is to develop and present the training.

Implementing the plan
At this point the plan swings into action, and specific individual programmes start up to implement it.

Plan monitoring and review

The working party should hold regular meetings, or require reports, to ensure that implementation is smooth and effective, and to deal with any unforeseen problems, or to change tack.

If the plan has a definite planned life, there should be a thorough final review. This can look at the successful or unsuccessful aspects, and record the reasons why the plan did or did not live up to expectations. This will assist in designing the extension of the plan or of new plans; perhaps the programmes were unsuited to the goals desired.

For a project which has a definite lifetime, such as the construction of a process plant where complex engineering is involved, there are a number of other steps which can be taken. The process described can either be planned to end with the commissioning after a construction project is complete, or to end when an engineering project has reached the end of its life. It is important to plan safe decommissioning in the last case. Other steps could be chosen as end points, with a new programme thereafter. The end points chosen might be end-of-construction, or pre-commissioning, for example.

There is one reactive rather than proactive part in all this, and that is where a hazard, in spite of the plan, or during the development of the plan, leads to an accident. A thorough investigation must be done and steps taken to prevent it happening again.

The programme examples given above revolve around tackling specific risks. However, the objectives may be broader, for example:

- contractor compliance with the principal's contract requirements
- aiming for a general upgrade to achieve an award under a government incentive or assessment system
- establishment of safety in purchasing
- improved permit to work systems
- building existing standards in regulations or codes into safe work procedures.

In this case the working party may need to set up sub-groups and coordinate and integrate their activities.

An activity where a permit to work system is essential is shown in Fig. 14.1.

The OHS policy

Writing a full OHS policy is important, and is a legal requirement in, for example, the UK, Malaysia and South Australia. Opinions on what such a policy should include vary, with some organizations seeing it, as noted in Chapter 2, more or less as a very general statement of commitment (in effect, a mission or vision statement for OHS which supports the organization's main mission or vision statement).

As a second step, some organizations go on to define some key areas which they consider the policy should specifically address; these include, for example – depending on the type of workplace – noise, eye care, manual handling, etc.

The policy should be the result of consultation with employees and must have the commitment of top management, so it is appropriate that it is signed by the management. The policy is the basis for setting objectives and targets for OHS in an organization.

Figure 14.1 Confined space entry requires clear procedures, training, permit-to-work and emergency response (photo courtesy of Joe Maglizza and MARCSTA, Australia)

OHS manual

Policies may form part of an OHS management system manual. The manual can include:

- definitions
- guidelines on how the policy can be put into practice
- procedures which describe actions to deal with issues in the way the policy intends.

The guidelines can show who is responsible for implementing the various parts of the policy, who can alter the policy, operative dates for the policy; and when it should be revisited. Regulations or codes of practice which support the policy can be shown.

Procedures not only show how something is to be done, but form a basis for training in doing it. A procedure can include forms (e.g. a job safety analysis form), and also refer to other documents such as technical instructions or operating manuals for a piece of equipment.

Elements of an OHS system

Introduction

There has been in the past some confusion between an OHS system, plan and programme. With the adoption of standards noted earlier for OHS management systems based on the ISO 9000/14000 structure for quality and environment systems, this is a lot clearer.

Under the standard the 'system' consists of five elements, which together should ensure continuous (or continual) improvement. They are: policy; planning; implementation; measurement and evaluation; and management review.

Health and safety management systems

There are of course other existing OHS management systems, such as the NOSA Five Star, and ISRS/DNV, but the new OHS standards make integration of OHS into the mainstream management, which the quality system addresses, a lot easier, just as ISO 14000 does the same for environmental management.

AS/NZS 4804, for example, describes the following elements as the keys to implementation of an OHS management system:

- capability which includes the resources available to the system
- support action which includes effective communication and information management
- hazard identification, risk assessment and control of risks, including design and purchasing
- preparedness for contingencies and a response to them.

AS/NZS 4804 goes on to provide considerable further information and explanation on each of these elements. For instance, capability includes aspects mentioned earlier, such as integrating any OHS management system into the main management system, consultation, motivation, competence and training, and supply of goods and services. OHSAS 18002:2002 also provides such guidelines. Figure 14.2 shows the relationships between the safety system, the safety plan and the safety programmes.

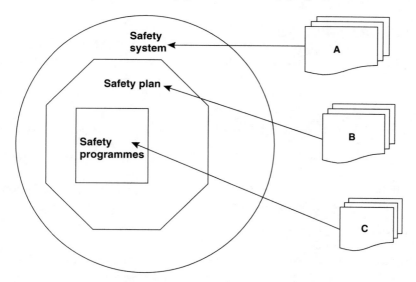

A: Policy, planning, implementation, measurement and evaluation, management review ⟶ continuous improvement

B: Consultation, supervision, procedures, hazard register, job safety analysis, accident investigation and reporting, selection and training, health surveillance, emergency response, etc.

C: Wellness, manual handling, hearing conservation, stress management, hazardous substances, permits and tagout/lockout, etc.

Figure 14.2 Safety system, plan and programmes

OHS plan

The OHS plan is the way in which an organization goes about ensuring that injuries, ill-health, disability and disease are minimized. It sets objectives and targets, and describes who is responsible for achieving them. It also describes how they can be achieved and sets deadlines.

The plan needs to consider issues such as:

- commitment
- consultation
- communication and recording
- hazard and risk management
- accountability
- procedures (strategic, day-to-day and emergency). Emergency procedures proved to be extremely important in reducing the overall loss of life in the September 2001 plane attacks on New York's World Trade Center.
- auditing and continuous improvement
- training
- investigation of accidents
- selection of personnel.

OHS programme

The overall OHS plan usually results in a series of programmes, each often running over, say, one year, which aim to attack specific OHS issues. (e.g. confined spaces, fatigue or lock-out).

Links between the OHS plan and organizational structure

Here we use the word 'plan' rather than 'programme' in line with our previous explanation.

It is important to link the OHS plan to the structure of the organization concerned. There are three key reasons for doing this:

- the size of the organization concerned
- the complexity of the organization concerned and the type of risks it faces
- the internal policy in relation to whether OHS is run completely through line staff, or whether an OHS team is used which supports line staff, or provides services not provided by line staff.

The organizational structure is based on the main management system. AS/NZS 4804 also describes the elements of the main management system for the organization into which the OHS system, as well as quality and environmental systems, can be integrated. These include:

- policies
- use of resources

Health and safety management systems

- control of operations and administration
- training and development
- accountability for actions and events within the organization
- appraisal and rewards
- the area of information and communication which includes measuring and monitoring (linked to appraisal), collection, data handling and data supply.

(*Readers are referred to the two relevant standards for further information and explanation.*)

Two examples are:

- professional development for future managers would include OHS and environmental components, rather than treating them as stand-alone training and development issues
- use of positive performance indicators for work areas which include OHS and environmental management (where relevant) as a mainstream expectation, not an add-on.

Implementing an OHS plan

A good health and safety plan provides a clear set of guidelines for activities which, if rigorously followed, will reduce accidents and cases of occupational disease. The key to success is the manner in which the plan is implemented and maintained.

Senior management must visibly support the programme by:

- providing resources such as time, money and personnel
- ensuring that employees receive training as required
- making all applicable health and safety information available to all employees
- including health and safety performance as part of employee evaluations
- attending health and safety meetings.

The programme must be communicated to all employees. Special emphasis should be given to new workers, newly appointed supervisors and new members of health and safety committees. Revisions of policies and procedures should be clearly communicated to all employees.

Companies listed on some stock exchanges now have to include OHS and environmental management in the corporate governance section of their annual reports.

Dow Jones, the international stock market index, now include the 'triple bottom line' concept – financial, OHS and environmental results. Baskets of 'ethical stocks' are based on this.

Evaluation procedures for OHS initiatives

The bottom-line method of evaluation for the success of OHS initiatives within the organizational structure, has traditionally been to use the standard methods of measuring injury rates given in standards such as Australian Standard 1885 or regulations such as the US 29 CFR 1904. However, this suffers from a number of problems:

- it uses a negative event, i.e. an injury, to measure the outcome of initiatives
- even within a well-controlled environment there remains an element of chance, so that an apparent undesirable blip in injury figures may be just chance variation
- safety initiatives must be prioritized, so the severity of injury becomes one of the factors to weigh up. However, it can be up to a year (using the AS 1885 criteria) before the full severity of injury in terms of time off is known, and in reality it may take more time than that to see the full picture.

Before moving on, however, it must be stressed that in some organizations and areas, injury figures, even if not statistically validated, are sufficiently above the average for the industry or type of work to warrant immediate action without waiting until 'all the votes are in'. A word of caution is needed, though, in using internal comparisons. While the figures for a construction firm on a site may be high in comparison to the figures for their office staff, they may not be high in comparison to those for on-site work in best practice construction companies. So any new initiatives need thorough planning and innovation to be successful – a 'gut' response may not work.

The current approach to evaluation has a number of facets:

- the development and use of appropriate positive performance indicators (PPIs) which focus on measuring how well various injury-prevention steps are implemented
- linking the PPI results to the appraisal, award and accountability structure in the organization, e.g. the PPIs for a supervisor should be part of their job description
- in some organizations where injury rates are so low that injuries are quite rare events, behaviour-based systems are used to monitor positive and negative behaviours.

These PPIs should not be confused with the old safety incentive systems where individuals or work groups were rewarded for lack of lost-time injuries. Incentives for achieving the PPIs do not present the same temptation to hide the truth. It must be remembered that PPIs at the end of the day generally measure inputs to accident prevention, as do behaviour-based approaches. Periodically, a thorough review must be undertaken to ensure that there is an adequate linkage between PPI results and incidents and injuries. Safety culture surveys can also provide an appropriate range of PPIs because they also measure factors which are now recognized as inputs to achieving better safety on the job.

Proposing and defending a strategy for management of change

Identifying the need for change

This section deals with managing change as it relates to the OHS needs of an organization. It is useful at this point to note that change in the way OHS is handled in an organization may not occur because of OHS concerns at all. It may be part of a wider move by an organization to use new management strategies. As an example, where a company decides to improve shareholder value by focusing on *core* business, reducing its wages workforce and using contractors to provide a variety of services it used to provide itself, then there will be a definite need for change in relation to the way OHS is tackled. So that will be one reason for change regardless of the success or otherwise of the accident prevention strategy to date. The US 29 CFR 1925 recognizes the setting of health and safety standards for contractors to government.

Unlike other areas of study, many management theories are built on the experience of a consultant going into a single company, trying an idea at a particular time in a particular situation and then drawing conclusions which he or she claims can be generally applied, often before enough time has elapsed to assess mid-term results. Few companies look to the longer term. J. M. Keynes, the economist, said, 'we are all dead in the long run'. The short-term result the consultant has achieved may be genuine, but it may not be sustained, because of the 'Hawthorne effect' – that is, people often perform better while there is external interest in what they are doing and how they do it.

There is considerable pressure now for short-term results by company boards of directors and shareholders. The mid-term consequences of downsizing, such as reduction in in-house training and coaching, and loss of knowledge, expertise and corporate memory, are often ignored. The effects of fatigue, from longer hours worked by remaining staff, on client relations, quality of judgement, and concerns of staff about family responsibilities, are often forgotten. In a study of 4000 Australian companies published in 2001, Dawkins and Littler of the Melbourne Institute of Applied Economic and Social Research pointed out the downside to downsizing, including loss of employee empowerment and hence employee commitment and loyalty (see Further Reading at the end of this chapter).

Some contemporary management strategies which have a short-term focus only on shareholder value may create an organizational culture with a detrimental impact on OHS because of the dilution of commitment and motivation. This can come about through:

- staff turnover rates
- increased casual, part-time or contract employment
- growth in demands on staff without staff increase
- loss of critical people.

Another point to remember is that a lot of people work in not-for-profit organizations, so care must be exercised in using the effect of injury on profits, as such, when making a case for managing safety and risk. Not-for-profit organizations still need to contain costs to break even or to accumulate reserves.

A health and safety practitioner who has been newly employed to act in a consulting role to line management to improve safety, needs to remember the old idea of 'teach a person to fish, don't give them a fish', if the practitioner wants them to be self-sufficient once he or she is gone. There is a need to avoid the understandable desire of some practitioners to create a constituency, or climate, of continuing dependency on them.

What has been written above is simply meant as a caution against change for change's sake. However, if senior management have committed themselves to change, you, as an OHS practitioner or line manager, must support it. Your focus then becomes how best to position the OHS programme in the changing environment. On the other hand, you may be taking the initiative in changing either the OHS programme as such or the management of OHS within the organization.

This could be:

- because you have been newly employed to do so
- because external events, such as prosecution for a serious OHS offence, have caused a rethink
- because after a period in the organization you are convinced OHS initiatives or the OHS programme have run out of steam and are losing their impact
- because you have enough foresight to see that the changing external environment (legislative or business) will require it.

Where current injury rates are unacceptable, or only good luck has made up for the lack of a decent OHS plan, then the OHS practitioner, or the line managers with responsibility for OHS, have to adopt roles as change agents. This is not easy. There is excitement at the challenge, but also the stress of finding a path around the opposition. Having accepted that change must occur, you need to adopt a strategy that is likely to succeed. There is no one strategy that is going to fit all circumstances and organizations, but you may want to follow these general principles:

- use your network of professional contacts to find out what others are doing and how they have tackled it
- collect a full set of facts on the existing OHS situation in the company and analyse them, i.e. a needs-analysis
- if you are faced with responding to changes in legislation, seek as much advice as you can get from the regulators about how the *words* will be interpreted on the ground (this is not always an easy task as the regulators are often finding their way in overseeing the application of new requirements)
- discuss the current situation with a variety of stakeholders in the company.

At this stage there are a number of possibilities. Two are:

- make up your own mind about where you want change and begin to sell the changes
- form a working group of stakeholders to consider the current situation and come up with the changes they see as necessary.

Be aware that self-interest is an inevitable reaction from many people; for example, 'Where do I fit into this? Does it reduce my power in the organization? Does it mean more work for me? Is my neck going to be on the line?' You need to recognize this and be upfront about it.

Selling the changes certainly requires that you win the hearts and minds of senior management. But remember that word 'consultation'. A top-down approach is unlikely to achieve as much as following up the senior management involvement with consultation down the line. Ensure that commitment to change becomes jointly owned. A sense of ownership is a powerful indicator of eventual success. Failure to consult is

Health and safety management systems

likely to lead to hidden resistance. By nature many people are conservative and don't take easily to change, not even that which is intended to help them. Where there has been one management scheme after another, there is often a climate of considerable scepticism about yet another one.

It is going to be your job to win over and keep the sceptics on your side. When it comes to expenditure on safety recommended by joint working groups, your credibility is going to depend on how well you succeed in helping to convince management to implement safety initiatives coming from down the line or from the shop floor. You may also like to refer back to the section 'Strategies to integrate...'" because some of the information there on the strategy for OHS plan creation is relevant to a strategy for change of OHS plans.

Potential barriers to change

A Buddhist writer who has worked in the West for many years, Sogyal Rinpoche, has said: 'The only thing we really have is nowness, is now.' In fact change is constant and continuous in everything; nothing, including us, is exactly the same as it was a moment ago. Acceptance of that undeniable fact is an essential requirement for anyone to adequately cope with all aspects of life.

Having said that, we cope psychologically with change – individually and in groups – by a variety of mechanisms. Just as a person might build a house on an island in a flowing river and provide it with protection of the foundations by means of retaining walls, people try to build themselves islands of stability:

- organizational – an effective department or work unit with a role in which they feel comfortable
- emotional – a family or 'significant other'
- spiritual – Christ, Allah, Yahweh, Krishna, Buddha or some other belief system as a constant
- financial – home ownership.

Having done this, many people tend to perceive change as a threat. (We will deal with the Chinese 'threat-opportunity' idea later.) Yet if they look more closely at people's 'island of stability', it in fact does change or respond to change, for example:

- the emotional island – the children grow and leave
- the financial island – the government changes the incentives for superannuation saving
- the home – it may now interface more often with the outside world through the Internet.

Let us now look more closely at the organizational setting and particularly at OHS in that setting.

Management of organizations continues to go through a succession of changes in approach. It is essential that an organization does this if it is to survive, because the external environment in which it operates is changing. However, in both government and private industry some change is done for fashion or political gain, and often with a rapidity or repetition and lack of infrastructure change which is dysfunctional. However, beware of 'short termism'.

Competition may change. For example:

- a new business sets up nearby providing similar goods or services
- the government opens a sector of business up to competition, where before entrance to that business area was by licence
- a new treaty allows foreign firms more freedom to operate
- even organizations which may not see themselves in competition, e.g. a government prison service, may find that government concerns, real or imagined, lead to the government considering competitive private options
- Internet allows customers to access overseas suppliers
- the equipment available for the organization to do business changes, e.g. the extensive introduction of information technology and computerized control devices. Contrary to the ergonomic approach of altering the task to match human limitations, many organizations simply eliminate the human in favour of the electronic chip, e.g. banks replace a person handling a banknote with an ATM which changes 'a blip in a chip'.

As a consequence of this the organizational structure changes. Much of middle management has been eliminated by the introduction of IT equipment, which can make many decisions middle management used to make. Typing pools disappear with everyone on their own terminal, although UK surveys show that male executives on computers are fairly inefficient compared with trained male and female computer task specialists. In some situations, computers mean a lack of mental stimulus and challenge for their operators.

A second consideration is that in parts of Western industry in particular, manufacture, called the 'rust-belt' industry in the USA, has been replaced by so-called 'post-industrial' activity (services).

Globalization has led to manufacturing competition from developing parts of the world, and rust-belt labour conditions which include OHS (even though a good idea) become uncompetitive. One shoe company sold US$70 shoes which are produced offshore for US$3, and was accused of poor labour, including OHS, conditions in its offshore plants.

Bureaucratic structures with fixed lines of authority and responsibility, and long communication lines through the hierarchy to the outside world and back, have often been found to be unable to respond to change. Loose arrangements with fewer imposed controls have been introduced. Peter Senge in *The Fifth Discipline* has written on the knowledge-based organization, emphasizing that skills alone have their limitations. Knowledge allows people to alter skills and so grow to meet constant change effectively. (Some years ago in Venezuela a government minister, Alberto Machada, ran a national programme on 'learning how to learn'.)

Looking at the above, it is possible now to sketch in some of the primary barriers to change.

Health and safety management systems

These can include:

- the negative benchmarking approach, e.g. 'The Handex company tried this and it didn't work'
- plain simple human inertia – 'Things are OK, why upset the applecart?'
- a lack of understanding by individuals of the importance of change for the organization
- people who see that their job in the organization (or at least their power) is threatened
- people who will need to move from a work group or form a new work group and don't want to
- people who feel inadequate to cope with changed demands and responsibilities, e.g. multi-skilling
- negative attitudes based on various ideas, e.g. 'It won't work.'
- the need for a team approach to achieve change, which may require extensive consultation and devolution of authority which some managers find uncomfortable or incompatible with their leadership style
- lack of effective communication of the need for, and planned mechanisms of, change
- lack of financial and human resources to service the change adequately
- sharp divides between the social cultures at the top and bottom of an organization
- a cynical attitude by employees expressed as 'OK, so I accept that there will be change but I will wait and see how it goes for the time being, because I've seen all this before.'

These lead to secondary barriers to change, for example:

- individual and organizational resistance to change and in extreme cases, sabotage (an obvious OHS risk)
- financial organizations who refuse to provide finance for the changes
- in new growth areas it is hard to access or pay for scarce outside expertise (this is happening in the offshore oil and gas industry and may affect safety)
- structural barriers to changes in working patterns and remuneration, e.g. industrial awards and union restrictions (some union requirements are necessary; they are intended to ensure the welfare of their members)
- in a world where contracting-out has become commonplace, legal structures such as workers' compensation (these are based on the contracting situation in particular industries such as construction and are not well suited to the newer areas of contracted service provision)
- the psychological barriers to changing attitudes.

In relation to OHS specifically, it is important for the person with responsibility for OHS to become familiar with some of the management literature (e.g. Peter Drucker's *Managing In A Time Of Great Change*) and then see how OHS can be best delivered in a changed structure and approach.

The very barriers to change which others erect could well be barriers to delivering OHS – something which is in their own interests – more effectively. In an organization, particularly if it has a poor OHS record, if you are the responsible OHS person, you may see that the barriers to change others erect are getting in the way of better delivery of OHS. On the other hand, certain changes, e.g. to shifts, may be potentially detrimental to the health of individuals and so not assist the OHS effort. So as an OHS practitioner, you will already be used to working with the consultation and empowerment of workers if it has been introduced by Robens-style legislation in your jurisdiction, but you should recognize that others may be uncomfortable with such change. Many an OHS person is already a change agent, so it may be wise for them to embrace management change and work to build into the changes the elements needed to achieve their OHS objectives.

Benefits resulting from change

This section is almost a mirror image of the previous one. The previous one considered the issue of the threats involved in change. But the Chinese character for 'crisis' can mean 'threat' or 'opportunity'. Change is also an opportunity to achieve positive things in, and for, an organization.

A major advantage of change is that it ensures survival of the organization in the marketplace, and hence jobs for those workers who are not offered redundancy packages. This applies both to profit-making and non profit-making organizations, such as the Red Cross.

It has already been found in a number of situations that achieving improved approaches, attitudes, and behaviours in OHS has a positive effect in other areas of operation. While some people instinctively see the improved adherence to detail, procedures and the thinking ahead which safety requires as a drag on productivity, in fact others have found that these are factors in improving quality and avoiding costly production errors. The Australian National Occupational Health and Safety Commission and the Journal of OHS-ANZ jointly published some years ago a number of case studies showing that change to achieve best practice benefits productivity, profitability and OHS. (See Blewett in Further Reading.)

It is not altogether surprising that changes to accident prevention strategies can achieve overall improvement, because the factors in an OHS plan given earlier in 'Strategies to integrate ...' such as:

- commitment
- consultation
- communication
- hazard management (for this, read *identifying, evaluating and controlling business risks*)
- auditing and continuous improvement
- procedures
- accountability
- training
- investigation (in this case, of 'quality' or 'service')
- employee selection

apply just as well to other aspects of effective operation of the organization.

Many people respond positively to change. It can offer them opportunities for further training, new challenges and different work locations. It may offer the opportunity to work with different people or gain a promotion. Change tends to keep positive people actively thinking about what the organization's objectives are, where it is going, and how they can contribute. It can improve people's ability to work together to solve problems. Some people have been waiting for change for years, and for many it is the organization which attempts to stay the same even when it should change, which creates poor morale. Antagonistic attitudes will develop and change will be impeded where there is a lack of honest and frequent communication about change.

Impact of change

The discussion above dealt with the benefits of change, but it would be inaccurate to say that the impact of change is all positive. We can look at change from two angles – the change which is happening on a daily basis outside the organization and outside your own state, province or country; and the changes in response to those changes inside the organization. The capitalist economy in which most of us live is based on ensuring the delivery of goods and services to the standard the customer wants at the best price to the customer. Competition is an essential part of ensuring that this occurs. But there is a quality angle to it because of the requirement 'to the standard the customer wants'. The actual product or service must meet those standards. In addition, some customers will also want to buy only those products which are produced within certain ethical criteria: the 'Rugmark', for example, certifies no child labour; tuna are sold on the basis that they were caught in a way which protects dolphins from driftnets.

Competition means that someone is always trying to come up with a new or better product and, if they do, other competing organizations are forced to respond. In the OHS arena, if an organization in the construction industry sets the benchmark for safety performance, others must change to try to follow suit and achieve similar performance or they risk losing contracts with those organizations which value safety performance as a criterion for awarding contracts. They also risk increased scrutiny by government regulators and higher compensation premiums which adversely affect their competitiveness.

Some proposed changes on the world scene at the moment appear to have an undesirable aspect. The 'level playing field' is a fine concept but, if it means that organizations in countries with low safety and environment standards can export to countries with higher standards, and the costs of good OHS and environment performance by firms in the latter countries render them uncompetitive, then such an aspect is undesirable. Unfortunately some governments seem unwilling to link trade agreements to ethical concerns.

One of the changes in some parts of the world, in response to increased competition in mineral products prices worldwide, has been a big swing to the use of contractors. A recent Australian inquiry into mining safety focused on the effects of this change. If the pressures on the contractor to achieve production *no matter what* are too high, then it is fairly certain that safety will suffer. If a contractor loses tenders because the contractor is serious about safety and environment, but the principal isn't, that is undesirable.

Change is indeed a two-edged sword. For the employees who are retained by an organization, the upgraded information technology equipment may well make their jobs more enjoyable, safer and easier. Certainly new information technology equipment may improve the effectiveness of the accident prevention effort. But this is often achieved at the cost of other employees' jobs. A proportion of the employees who lose their job, however, actually find that the forced change leads to new opportunities, some of which are better.

Developing a strategy to address the need for change

You should now be convinced that you can't escape or ignore change; it will happen anyway, and you can either respond to it in a planned way or let the way you and the organization evolve to meet it happen by chance. I suggest the first approach is more effective. It should be noted that evolving change won't stop just because the organization is involved in planned change. The moment someone takes a job with another company, there has been an evolutionary change within the organization they previously worked in.

Different types of change can occur in an organization. Some are planned and some of these may result from a response to rare and unexpected events.

The desire by motivated individuals to see the organization do things better is another driver of change. This can spring from the person's attitude alone, or it can be the result of a good fit between the individual's interests and those of the organization. The organization can create the right climate for it, but also go further, actually encouraging it.

At times change comes out of a looming crisis. A badly performing product line can be dropped, or a badly performing division can be sold off to a suitable buyer. The manager of a badly performing division may be moved on or dropped altogether.

The crisis may be an acute rather than a looming, slowly emerging event. It might be sudden loss of a major power source due to equipment failure on an isolated mine site, or it could be excessive rain, drought or snowfall. The changes come about as a result of the way the organization responds.

The change needed to manage a looming crisis may be an appointment of a new CEO, a new top management team, restructuring and change or departure of departmental managers.

Change can consume large amounts of time and energy, particularly if there is little time to bed it down before further change occurs or is forced on the organization.

There are a number of approaches organizations adopt to achieve change. The first is to change the structure of the organization. This can certainly give a fairly rapid appearance of change, but it may or may not achieve the real changes desired, especially at the grass roots.

Secondly, technology is also often favoured as a means of change. Email and the Internet have made rapid inroads into the way we communicate and bank, for example.

Health and safety management systems

Voice recognition and voice reproduction software is widely used. Automated switchboards to handle telephone calls have become commonplace.

Not seeking out more efficient methods to use is seen as a human weakness. On many criteria used to compare people with machines, machines come out better. But market research may reveal customers still want to deal with a person not a machine.

The third approach to change is a people-based one. Training, coaching, workshops, and psychological techniques are used to try and change people's perceptions, attitudes and behaviour.

A fourth approach is to change the tasks of the organization. For example, some small mining companies were reinvented as e-commerce companies during the 1999–2000 'net-boom'. One even switched to 'adult products'.

What type of typical strategy is used to ensure change that responds to deviations, opportunities, or anticipated opportunities? (Note that quite often change is introduced to respond to what has already happened, not what, it has been decided, should happen.)

Earlier in this chapter, in the section 'Strategies to integrate…' we discussed the steps used to introduce an OHS plan. An essentially similar planned approach can be adopted to introduce any form of change, whether it is OHS change or not.

Safety culture surveys were mentioned earlier and can be used to gauge perceptions, attitudes and values. Introducing change may need methods of facilitation where people's existing perceptions, attitudes and values can be opened up and examined in a supportive environment. The new or altered paradigm (set of values, attitudes and beliefs) can then be introduced. The third step is to gain adoption and ownership of the new paradigm and lock it in.

A new paradigm may meet with resistance, nonchalance or acceptance. The first two of these problems need to be anticipated and methods developed in advance to overcome them.

Change, however, is rarely as orderly as it is has been made to sound. It does involve interactions between people and between parts of organizations. Change causes people angst (anxiety). So change won't just happen. There will be a need to take any chance which presents itself to inch forward. Official and unofficial sources of power must be identified and either brought on side or outflanked. There will be a need to become strongly identified with the change, ('She's always on about team lifting') but also know how to win support from others. Don't wait to be pushed from outside if you can move steadily ahead under your own steam, but know when to call on external help (e.g. an OHS authority) if change is stalled.

Pilot projects can also be used to show the sceptics that the change is worth supporting. An understanding of how and why people acquire new knowledge and skills will also help in moving forward. It is important to develop a change management procedure and identify what level of changes should trigger the use of the procedure.

Change can involve items as diverse as organizational structure, new materials, operating conditions, equipment, tools, control software, procedures, work instructions, products, processes, workplace layout, responsibility or repairs. The essential step from the OHS angle is to ensure that the changes are properly described and reviewed. The review needs to identify new hazards, assess the altered risk, and ensure changes are made to the control measures necessary. There is a need to try and avoid an increase in risk as a result of the change. Formal approval of the changes from a responsible person is important, and once carried out, there should be a further follow-up review. Design and purchasing processes need to reflect this approach, and any new approvals required by law obtained.

Reviewing an occupational health and safety management system

Principles of an OHS management system

The essential requirements of the new international and national standard-based OHS management systems are as noted earlier the same as those for the ISO 9001 quality and ISO 14001 environmental management systems. A comparison of those and AS 4801 is provided in Appendix C to AS 4801. The essential requirements, with the corresponding OHSAS 18001 requirements in brackets, are:

Requirement 1 – Establish and maintain an OHS management system
Requirement 2 – Policy (OHS policy)
Requirement 3 – Planning
Requirement 4 – Implementation (implementation and operation)
Requirement 5 – Measurement and evaluation (checking and corrective action)
Requirement 6 – Management review.

Together these form a cycle of continuous (or continual) improvement. The sub-elements of OHSAS 18001 and AS 4801 are essentially the same. Neither OHSAS 18001 nor AS 4801 state specific OHS performance criteria, nor do they give detailed specifications for the design of a management system.

Three other standards, however, do so. OHSAS 18002 and the earlier BS 8800:1996 show how to tackle the various elements in an OHS management system, and how to integrate them into the everyday management of an organization. Likewise AS/NZS 4804, the standard linked to AS 4801, provides guidance on what to do to implement an OHS management system.

Identifying existing systems, structures and programmes

While the level of detail will vary a lot depending on the size of an organization, the type of work it does, and where it does it (e.g. offshore waters, metropolitan shopping centres, or rafts on whitewater), the essential parts of the system described above should exist if the duty of care is to be adequately met and the organization is to adopt a philosophy of quality in its operations.

Perhaps before you tackle the OHS system as such, you should compare the existing overall management system with the ISO cycle. This requires discretion as it may not be welcomed. It remains true, however, that it is going to be difficult to introduce

Health and safety management systems

a continuous improvement-based OHS system to a mainstream management system which doesn't work that way. It will not be possible to fully *integrate* such an OHS system into such an organization. There are a number of systems (other than safety, environment and quality as such) which exist in an organization, and you will need to re-examine them in the light of the standard.

Examples might be:

- the purchasing system
- the system of contracted services management
- the plant maintenance system
- the records and documentation system
- the human resources system
- the training system.

You may need to look at the existing structure of the organization and consider how well the OHS management system elements in the standard will operate (or how well they won't operate) within that structure. For example, you may have electrical and mechanical maintenance sections. So you are examining the 'purchase of services' element in conjunction with the 'fabrication, installation and commissioning' elements, for the OHS management system. Let's assume you are also aware of the design, supply and commissioning requirements for plant or equipment in your OHS regulations.

You need to ask how adequately the existing structures for electrical and mechanical maintenance address the requirements of the OHS management standard (and if you are taking a holistic approach, ISO 9000 and 14000 as well).

Many organizations have a training programme, and as part of the structure, a training section. This is linked to meeting identified human resource requirements such as a constant supply of qualified competent staff, and the need for performance-based appraisal. You need to ask how well the existing training structures and programmes fit in with the relevant element in the standard.

Scope and level of the review

The review of an OHS management system, like a number of other issues, may be done for a variety of reasons and at a variety of stages. For example, the organization may have made a decision to embrace a quality management philosophy, and is involved in a significant process of change (refer to the last section 'Proposing and defending a strategy for management of change'). The scope of your review of OHS (or the OHS review you are involved in) may then have considerable breadth and depth. Or it may be a two-stage process where depth is left until later while the main shape of the new systems (including OHS) is thrashed out.

A second scenario is that you have reached the review and improvement stage of the cycle of implementation of a system based on continuous improvement. This may involve more of a look at specific areas in which problems have arisen, perhaps going into them in considerable depth. You have probably identified *potential* problems which arose in the course of an audit, because *actual* problems are usually fixed as they arise.

For example, your audit may have identified quite old equipment which despite maintenance is coming to the end of its useful life, and the audit highlights the need for a fairly extensive refit. This time around, the purchasing needs to meet the demands of elements 'purchase of services' and 'fabrication, installation and commissioning' mentioned above in 'Identifying Existing Systems …'. The question then is – how will you achieve it? In this case the review will need to address that.

On the other hand, the end-of-cycle audit may simply produce a review which seeks to fine tune across a broad range of areas. An example might be better documentation of completion of on-the-job training. Another good example might be that, in some organizations, serious problems have occurred because of an over-reliance on technical systems to ensure safety (a fault some engineers have). This leads to a lack of communication or appreciation of, or a lack of 'mesh' with, the main management system. An example of this type is perhaps the RAAF's Mirage A3-40 crash in 1986. There was engineering debate about when a fatigue crack in metal posed a significant risk. On the management side, there was a need to adequately define the acceptable safe-life philosophy. There was a similar debate over how much loss of insulating tile was tolerable after the space shuttle *Columbia* tragedy. In a recent triple fatality in a mine, avoiding excess build-up of water pressure against a barrier depended on accurately monitoring the percentage of fines in the slurry being used to backfill a stope, and understanding its significance.

Evaluating existing practices, standards and conditions for a part of the organization

The key to evaluation is the word 'standards'. To *evaluate* we need to compare what we find against some type of standard. For instance, you may need to collect data using a checklist which allows you to evaluate how well the OHS management system in an organization lines up against the new standard for OHS management. Another way an in-depth review may arise is if there has been a serious incident, injury or fatality. This may be confined to one particular area, at least initially, and this may be gone through with a fine toothcomb. However, it may lead to a broader scope, if it appears that a significant deficiency at senior management level has been a contributing factor to the incident, injury or fatality, because this deficiency may have the potential to affect safety in any area of the organization.

The word 'practices' in this section is not intended to refer to lower level issues such as actual work practices, compared to work procedures. Here, the focus is on the practices associated with the OHS management system. In other words, assuming the system has been developed and documented, responsibility and accountability assigned, and finance provided, what are the actual *system practices*? How well is it actually being implemented and working? We *evaluate* these system practices against the relevant sections of the OHS management standard. For example, the standard may call for 'monitoring and measurement'. We could select the steam generation part of an organization. We then thoroughly check, by inspection of records, and interviews, the inspection, testing and monitoring of pressure vessels. Where a problem is identified (non-conformance) we would follow through to see what corrective action was proposed, when it was done, and whether it was effective.

In relation to an OHS management system, evaluation of the 'conditions' against standards doesn't refer to the physical conditions in which work is performed in a workplace. The word 'conditions' here really refers to the management environment in which the system operates in that part of the organization. To evaluate this you would need to consider the identified staffing needs of the area concerned; the identified experience and training personnel require, including training in the implementation and operation of the quality, OHS and environment systems; and the financial requirements of that part of the organization. This includes quality, OHS and environment systems costs. You can then evaluate, by observation and interview, the actual situation and compare it with the identified requirements for effective management (an effective management environment).

You will find evaluating an off-the-shelf OHS management system against the organizational needs, which arise from the considerations in the above paragraph and from assessing compensation data below, is a useful exercise which you may wish to try. This can assist you in deciding which system to buy. Providers of such systems are generally quite willing to assist in such trials.

Cost–benefit analysis for new acquisitions, refurbishments or maintenance

Identifying critical specifications for equipment, facilities or processes

Specifications for equipment, facilities and processes play a key role in ensuring effective management of the risks associated with occupational health and safety.

Identification of the key aspects of equipment, facilities and processes relating to health and safety needs to be carried out not only on each aspect individually, but also by considering the interrelationships between them and how equipment, facilities and processes in a workplace fit into the overall system of work in the workplace.

Equipment and facilities review involves consideration at the planning, design, maintenance and repair stages.

The Facility Description section of the safety case approach is designed to assist in identifying areas where specifications are critical. See DISR (2000) in Further Reading at the end of this chapter.

As an example, a critical specification for equipment may be the noise level it generates. Another may be the lock-out and de-energization requirements. Facilities may need to meet certain minimum dimensions, certain minimum structural requirements such as fire resistance level (FRL), and certain ventilation requirements. In turn, equipment may only operate within accepted standards of risk if specified process parameters such as maximum temperature or pressure are observed.

Assistance in identifying the specifications and checking that they are adhered to may be found in legal requirements, standards, industry norms, supplier's manuals, etc.

IEC 61508 may be of particular assistance. Critical specifications include not just the functional ones but the required levels of probability of failure.

Identifying perceived and actual costs

The phrase 'it's running over budget' is often heard, and there can be a number of reasons for this when looking at a work area, work process or equipment. Safety and health are essential aspects which must be taken into account when assessing costs in a workplace. If there is too great a difference between the costs perceived to apply to a piece of equipment or a process, either before it is installed or started up, or in annual estimates as compared to the actual costs of operation, there can be heavy pressures to achieve production at the expense of health and safety.

The reasons for cost blow-outs are many and varied, but may include:

- inadequate design of equipment or system
- failure to purchase or design for the operating environment (dust, salt, moisture, heat, cold, vibration)
- failure to provide adequate training in maintenance and repair
- effect of foreign exchange currency fluctuations on cost of imported equipment, parts or raw materials
- damage from injury or non-injury producing incidents.

On the other hand, good management and proper planning may actually result in actual costs running below those perceived at an earlier time. Effective health and safety management reduces the risk not only to people but to an organization's financial results. The systematic planning of health and safety has important spin-offs for an organization overall, if it forms an integral part of the overall system of management.

Performing a 'cradle to grave' cost analysis for a process or equipment

It is also important to perform a 'cradle to grave' cost analysis for a process or piece of equipment.

While it is obviously of value to an organization to carry out such a cost analysis for all aspects of a process or equipment, let us confine ourselves here to health and safety-related aspects.

IEC 61508, while it is focused on electrical/electronic/programmable controls, nevertheless identifies fourteen phases of a safety lifecycle which can be applied to any process or piece of equipment.

These phases include:

- concept (e.g. equipment and operating environment)
- scope of the hazard and risk issues, e.g. just OHS or including environmental aspects

Health and safety management systems

- hazard and risk analysis
- specification of safety requirements
- allocating the safety functions to different parts of the equipment together with the required integrity level (probability of failure)
- operation and maintenance planning
- planning validation of the specified safety requirements
- planning of installation and commissioning
- creating the safety system in which the equipment will operate
- installation and commissioning
- safety validation of the equipment in the system
- operation, maintenance and repair
- modification and retrofit
- decommissioning or disposal.

Note that the actual installation and commissioning and then operation are not reached until the tenth and twelfth steps.

We could also include, earlier in the process, planning for the last three steps. Poor location of equipment can add a lot to maintenance or modification costs.

Decommissioning also deserves consideration at the start. It can involve removal and disposal of hazardous materials to a safe waste site. As an extreme example one nuclear facility, Hanford in the USA, has simply been mothballed under guard as it is too expensive to remove.

So far this discussion has looked at the steps in 'cradle to grave'. A cost analysis covering all those steps is a valuable exercise.

Some costs will occur early in the whole process, some years later. So the cost analysis may be quite detailed, and involve estimating future cash flows and the future value of money.

Decommissioning in particular involves spending money on an asset which is no longer producing money, and in some cases, such as a mine, involves environmental restoration required by law.

Cost–benefit analysis

Apart from the moral and statutory requirements which may provide the emphasis for good occupational health and safety practice by employers, the benefit to be gained for the cost is constantly being demanded. To enable a full understanding of this, the life of a piece of equipment, component or process must be examined. This will generally involve the following issues:

- design
- entry/purchase into the organization
- commissioning of the piece of equipment, component, process
- useful life of the piece of equipment, component, process
- ongoing maintenance/repair of the piece of equipment, component, process
- decommissioning of the piece of equipment, component, process
- exit/disposal from the organization.

In many cases the design is not dictated by the purchaser, but influences the decision to buy the product based on the requirements of the task. Once the full life of a component or process has been understood, as identified in point form above, then a cost–benefit calculation can be performed. This involves two sides of an equation. One side entails the costs relating to the reduction of the risk or problem and the other side entails the estimated monetary value of the gains from controlling the risk or problem – the benefit. The following is an example of costs and benefits associated with a safety plan:

Cost of Risk/Problem
Safety staff salaries
Paid time for safety committee
Paid time for safety inspections
Paid time for safety training
Workplace improvements costs

Value of benefit
Direct
Reduced LTIs
Reduced insurance costs
Reduced administrative costs
Increased productivity

Indirect
Increased morale
Reduced staff turnover
Reduction in statutory fines
Reduction in first aid costs

Adding up the total for each side determines if the money value of the benefit is greater than the cost of reducing the risk.

Comparing and recommending purchase or refurbishment based on risk management options

Often there is a choice between keeping a system or equipment going or replacing the equipment and even changing the system of operation. One of the factors involved here is the temptation of automation. This can be a mixed blessing. From the point of view of safety, there must be a good fit between the risk reduction aspects of the electrical, electronic or programmable electronic equipment and other equipment external to that which also has a safety function.

The 'bathtub curve' is a longstanding concept and describes a graph of maintenance levels versus maintenance costs (see Fig. 7.1). These costs normally gradually decrease as equipment is bedded in, stay steady for the projected lifetime of equipment and then start to show a distinct climb again as equipment reaches the end of its useful life.

The hierarchy of controls offers a useful approach firstly to the risk avoidance approach to risk management. Choosing the option of new equipment or process offers the possibility of eliminating a hazard. The risk reduction approach allows substitution of the existing hazard in a process with a lesser hazard, or purchasing equipment in which engineering and design reduce the risk from a particular hazard which is present in the existing equipment or process.

The ultimate decision to purchase a new piece of equipment or process, rather than refurbish, will depend on weighing up not only these safety-related issues, including the estimated costs if out-of-date equipment fails, but also the financial implications. These would include negative aspects such as design and purchase costs, downtime – an issue in refit or replacement – and retraining.

Positive financial aspects of new equipment include reduced downtime (hopefully), reduced risk of failure or injury, higher production levels and better product quality.

Some costs would also be recouped from sale of the unwanted equipment for further use, scrap or parts.

Whilst most Robens-style OHS legislation makes some reference to the 'manufacturer's duty of care' in providing a safe product, there is also a duty of care for organizations to eliminate hazards where practicable. The application of risk management principles to the purchase of new products is one method to limit the introduction of new hazards into the organization.

The risk assessment can be performed by the use of complex tools such as fault tree analysis; by the use of the Kinney and Wiruth approach (see Appendix to Chapter 4); the use of a matrix such as that in AS/NZ 4360; or by simple checklists. The assessment must, however, take into consideration the full life of the component. To help with the successful implementation of risk assessment in purchasing, the following points must be considered:

- purchasing policy developed with reference to safety assessment, legislative requirements and national or international standards
- a risk assessment checklist to suit the technical requirements of the organization
- functional workplace trials to be performed by the users
- the purchasing/supply committee to provide ongoing product monitoring and quality assurance.

When purchasing with health and safety in mind, remember that 'active safety' involves the factors which prevent an accident. 'Passive safety' covers factors which limit injury if there is an accident.

Planning integration of the new risk management strategy with the organization's quality management (QM) programme

Identifying the links between the new strategy and the QM programme

Quality management (QM)

The concept of quality management is synonymous with an organization striving to be the best in everything it does. It is the management plan to achieve continuous improvement in the quality of the performance, both internal and external, of all processes and services, and of products.

The continuous improvement cycle, which is fundamental to the heart of QM, is conceptually the same as the risk management process. The two models can be compared as follows:

Table 14.1 QM versus risk management

QM	Risk management
Understand the process	Identification
Plan improvement changes	Analysis
	Assessment
Carry out changes	Treatment
Follow up results of changes	Monitoring
Accept changes if positive	

Within an organization which has adopted the QM philosophy, the safety processes will be continually reviewed to refine and optimize their performance.

Developing a policy and procedures linking the occupational risk strategy with the QM programme

If an organization has committed itself to introducing and applying QM, then this assumes the commitment has become part of the policy of the organization.

Procedures can then be developed so that the occupational risk strategy is smoothly integrated with the QM approach.

Table 14.1 shows how firstly it is necessary to gain a good understanding of the processes employed in the workplace so as to identify the risks – occupational and non-occupational.

The procedure to deal with risk should then allow for analysis of how the risk is created and an assessment or evaluation of its extent.

Assessing risk treatment options and applying them are the next steps and this fits within the framework of planning and carrying out changes envisaged by the QM approach.

The last part of the procedure linking QM with risk management involves a further common step. The follow up of the improvements or changes uses tools such as monitoring, inspecting, or auditing.

If these indicate improved results from the risk treatment, then the approach being used is continued. If not, the QM process, basing identification of risk on fully understanding the process, commences again.

A negative result from monitoring or follow up may suggest that some part of the system or process is not as well understood as was thought.

An example of this could be that new, more fully automated equipment has been introduced which it was thought would allow operators to work longer hours because of less required mental demand. However, monitoring indicates accidents and production errors have increased. A rethink may indicate that the system or process assumptions were wrong. Fatigue from longer working hours combined with boredom from reduced mental demand may have reduced operator alertness and ability to respond correctly to process problems or risks.

Elements of effective quality management (QM) and safety

These include:

- Management commitment
  Clear direction through policies and procedures is needed. The responsibility for safety starts with line management. Workplace standards must be high and monitored to ensure they are achieving their goals. Understanding that management commitment plays a key role is crucial.
- Continuous improvement
  All activities which influence the organization must be monitored and the performance measured. Improvements will be made based on that performance. The result of the improvements is then evaluated – hence the term continuous improvement. All safety performance indicators, both positive and negative, can be used to refine the safety system.
- Reducing barriers
  To optimize organizational potential all internal departments must liaise and improve communication channels. Teamwork is essential to maximize the resources available. Understanding the supply and demand sides of all internal and external customers is essential to streamline processes. Discussion and consultation over safety issues will include all parties affected and call upon specialist skills in the organization.
- Measurement and variation
  Process outcome measurements, production rates, loss rates, and positive and negative occupational performance indicators are all organizational parameters which can be measured. The reason measurement is important in the QM process is because it provides quantitative analysis of performance. Knowing the expected value or result of the measurement is important, as any variation from the expected result will be the starting point for further analysis of the cause of the variation. In relation to safety performance, benchmarking can be done across industry or internally against organizational safety performance indicators.
- Employee importance
  The QM process of using teams to solve problems and improve performance should utilize the involvement of all employees. As employees are the key to making a process work then their potential must be maximized by supporting their ideas and allowing them to make, or at least decisions, an approach strongly favoured by the International Labour Organization. Health and safety legislation generally is also orientated around consultation and communication, both of which fit in well with the QM problem-solving philosophy.

Further reading

Bailey, C.W. and Petersen, D. (1989). Using Perception Surveys to Assess Safety System Effectiveness. *Professional Safety*, Feb. 1989.
Balian, J. (1995). Are You Relying on Unreliable Software? *The Health and Safety Practitioner*, **13(11)**, 34–6.
Blewett, V. and Shaw, A. (1995/6). OHS Best Practice Column. An eleven part series in the *Journal of Occupational Health and Safety – Aust. NZ* from **11(1)**, 15–19 to **12(6)**, 731–7.
Blewett, V. and Shaw, A. (1996). Quality Occupational Health and Safety? *Journal of Occupational Health and Safety – Aust. NZ*, **12(4)**, 481–7.
British Standards Institute. (1999). *British Standard 18001 OHS Management Systems (Requirements), 18002 (Guidelines)*. London: British Standards Institute (developed with 13 other standards bodies). See also Canadian Standards Institute *OHSAS 18001:1999*.
Brumale, S. and McDowall, J. (1999). Integrated Management Systems. *The Quality Magazine (Australia)*, **8(2)**, 52–8.
Budworth, N. (1996). *Indicators of Performance in Safety Management. The Safety and Health Practitioner*, **14(11)**, 23–9.
Butyn, S. (2002). Harnessing Change. (Risk reduction at Bombardier). *Accident Prevention (Canada)*, November/December, p. 20.
Cameron, I. (1997). A Social Learning Approach to the Practice of Safety Management. *The Safety and Health Practitioner*, **15(3)**, 26–32.
Chemical Industry Association. (1979). *A Guide to Hazard and Operability Studies*. London: Chemical Industry Association.
Coonen, R. (1995). Benchmarking – A Continuous Improvement Process. *The Safety and Health Practitioner*, **13(10)**, 18–21.
Crittall, J. and de Plevitz, L. (1997). Best Practice in Managing Contractors: The OHS Obligations of Principals. *Journal of Occupational Health and Safety – Aust. NZ*, **13(4)**, 353–60.
Dawkins, P. and Littler, S. (2001). *Downsizing. Is it Good for Australia?* Melbourne: Melbourne Institute of Applied Economic and Social Research.
Department of Industry Science and Resources. (2000). *Guidelines for Preparation and Submission of Safety Cases*. Canberra: DISR.
Diaz, D. (2000). The Lone Danger (perils of working alone). *Safeguard (New Zealand)*, May/June, pp. 20–4.
Doctor, S. (1997). Safety – How Do You Rate? *Australian Safety News*, **68(8)**, 26–31.
Drucker, P. (1995). *Managing in a Time of Great Change*. Oxford: Butterworth-Heinemann.
Emmett, E. and Hickling, C. (1995). Integrating Management Systems and Risk Management Approaches. *Journal of Occupational Health and Safety – Aust. NZ*, **11(6)**, 617–24.
Eves, D. (1995). Health and Safety Beyond the Millennium. *The Safety and Health Practitioner*, **13(4)**, 13–17.
Fox, A. (1971). *A Sociology of Work in Industry*, London: Collier Macmillan.

Glendon, I. and Booth, R. (1995*)*. Measuring Management Performance in OHS. *Journal of Occupational Health and Safety – Aust. NZ*, **11(6)**, 559–65.

Glendon, I. (1995). Safety Auditing. *Journal of Occupational Health and Safety – Aust. NZ*, **11(6)**, 569–75.

Hopkins, A. (1995). *Making Safety Work – Getting Management Commitment to OHS*. Sydney: Allen and Unwin.

Hopkins, P. (1995). A New Approach to Risk Management. *Health and Safety at Work*, **17(10)**, 8–10.

Institution of Chemical Engineers. (1992). *Major Hazards Onshore and Offshore*. New York, London: Hemisphere Publishing. (Taylor and Francis).

Jay, A. (1997). Methods to Measure and Improve Health and Safety Performance, **15(1)**, 27–9.

Kase, W. and Wiese, K.J. (1990). *Safety Auditing for Loss Control*. New York: Van Nostrand Reinhold.

Kletz, T. (1996). Organisations Have No Memory. *The Safety and Health Practitioner*, **14(12)**, 16–18.

Lamian, S. (1995). The Role of Cost–Benefit Analysis in Formulating Strategies for Improving Health and Safety at Work. *The Safety and Health Practitioner*, **13(1)**, 18–20.

Mayhew, C. and Quinlan, M. (1997). The Management of OHS where Subcontractors are Employed. *Journal of Occupational Health and Safety – Aust. NZ*, **13(2)**, 161–9.

Murphy, S. (1996). Managing the Risk. *Health and Safety at Work*, **18(7)**, 18–20.

National Safety Council (US). (2000). *Accident Prevention Manual for Business and Industry*. 12th edn. Itasca, Ill: National Safety Council.

OECD. (1996). *Trade and Labour Standards. A Review of the Issues*. Paris: OECD.

O'Leary, D. (2001). Inplementing COMAH in a Multisite Organisation. *The Safety and Health Practitioner*, **19(1)**, 31–3.

Quinlan, M. and Bohle, P. (2000). *Managing Occupational Health and Safety in Australia. A Multidisciplinary Approach*. 2nd edn. Melbourne: Macmillan.

Quinlan, M. (2000). Downsizing, Outsourcing, Organisational Restructuring and Occupational Health and Safety: Evidence, Risk Factors and Remedies. *Proceedings Minesafe International Conference*, 4–8 September. Perth, Chamber of Minerals and Energy of Western Australia, 393–413.

Senge, P. (1990). *The Fifth Discipline*. New York: Doubleday.

Sengenberger, W. and Campbell, D. (1994). *International Labour Standards and Economic Interdependence*. Geneva: ILO.

Simpson, I. and Gardner, D. (2001). Using OHS PPIs to Monitor Corporate OHS Strategies. *Journal of Occupational Health and Safety – Aust. NZ*, **17(2)**, 125–34.

Stainaker, C.K. (2002). Making the Transition from Start-up to Normal Operations. *Professional Safety*, **47(11)**, 14–17.

Standards Australia. (2001). AS/NZS 4804. *OHS Management Systems – General Guidelines on Principles, Systems and Supporting Techniques*. Sydney: Standards Australia.

Standards Australia. (2000). AS 4801. *OHS Management Systems – Specification with Guidance for Use*. Sydney: Standards Australia.

Stephenson, J. (1991). *System Safety 2000*. New York: Van Nostrand Reinhold.
Taylor, G.A. (1995). Trade, Trauma and Travail. *Safety in Australia*, **18(1)**, 13–18.
Taylor, G.A. (2001). Proceed with Care. *National Safety (Australia)*, **72(8)**, 42–6 (re. procedures).
Thomen, J. (1991). *Leadership in Safety Management*. Canada: John Wiley.
Veevers, A. and Hussein, AZMO. (1999). Getting Information from Reliability Data. *The Quality Magazine (Australia)*, **8(2)**, 60–4.
Waring, A. (2000). Risk Management and Corporate Governance. *The Safety and Health Practitioner*, **18(10)**, 19–20.
Waterman, L. (1995). Safety Practitioners and the Quality Revolution. *The Safety and Health Practitioner*, **13(4)**, 61–2.
Whiting, J. (1995). Proof Positive. *Australian Safety News*, **66(9)**, 34–42.
Whiting, J. (1995). Suggested PPI's. *Australian Safety News*, **66(1)**, 50–1. (re. positive safety performance indicators).
Worksafe Australia. (1996). *Control of Major Hazard Facilities*. (NOHSC: 1014, 2016). Canberra: AGPS.

Activities

1. Choose a procedure from your organization that doesn't have a quality aspect and rewrite it incorporating quality principles.
2. Within your selected workplace, describe those elements of the OHS programme which are currently running.
3. Prepare a list of intended problems you perceive would occur if you introduced either random blood tests or a fitness programme in a selected workplace. How would you address these problems?
4. Select a suitable particular part of a workplace and identify the OHS standards which are applied. Are they adequate? Compare existing work practices with the standards. Comment.
5. Select a reasonably recent writer on management and explain how his or her views could influence the approach to health and safety management.

Information sources

Some worldwide web addresses on OHS

acc.org.nz – NZ Accident Organization
acrsp.ca – Association of Canadian Registered Safety Professionals
amsa.gov.au – Australian Maritime Safety Authority
ansi.org – American National Standards Institute
aposho.org – Asia Pacific OSH Organization
arpansa.gov.au – Australian Radiation Protection and Nuclear Science Agency
asosh.org – Asian Society of OSH
asse.org – American Society of Safety Engineers
awcbc.org – Association of Workers' Compensation Boards of Canada
bcsp.org – Board of Certified Safety Professionals (US)
canoe-kayak.org – Minnesota Canoe Association (recreation safety)
cas.org – Chemical Abstracts Service
ccohs.ca – Canadian Centre for OHS
cdc.gov/niosh – National Institute for OSH (US)
cenorm.be – European Committee for Standardization
china-tradenet.com – Chinese Laws and Regulations
cie.co.at – International Commission on Illumination
comcare.gov.au – Australian federal employees OHS administration
crsrehab.gov.au – Commonwealth Rehabilitation Service, Australia
csa.ca – Canadian Standards Association
csse.org – Canadian Society of Safety Engineering
dgms.net (India) – Directorate General of Mines Safety, India
dme.gov.za – Department of Minerals and Energy, South Africa
doh.gov.ph – Department of Health, Philippines
doir.wa.gov.au – Department of Industry and Resources (mines), Western Australia
dol.govt.nz – Department of Labour, New Zealand
dosh.gov.my – Department of OSH, Malaysia
dot.gov – Department of Transportation (USA)
dsd.go.th – Department of Skills Development, Thailand
eng.moph.go.th/safety – Ministry of Public Health, Thailand

entemp.ie – Department of Enterprise Trade and Employment, Ireland
env-sol.com/solutions/tscasara.html – Solutions Software Corporation (US Acts – TSCA and SARA)
europe.osha.eu.int (no www – includes 150 word translation software) – European Agency for Safety and Health at Work
flightsafety.org – Flight Safety Foundation
fs.fed.us – USOA Forest Service (USA)
government.go.ug – Government of Uganda
gov.on.ca./LAB – Ontario Ministry of Labour
gov.mb.ca – Manitoba Government
gov.sk.ca – Saskatchewan Government
gnb.ca, gov.nf.ca, gov.ns.ca, gov.nt.ca, gov.nv.ca, gov.pe.ca, gov.yk.ca – other Canadian province government sites, then search for the department handling labour affairs
hse.gov.uk – Health and Safety Executive (UK)
iapa.on.ca – Industrial Accident Prevention Association (Ontario, Canada)
icrp.org – International Commission on Radiological Protection
iec.ch – International Electrotechnical Commission
iesna.org – Illuminating Engineering Society of North America
ifap.asn.au – Industrial Foundation for Accident Prevention (Australia)
ilocarib.org.tt – Caribbean Office of the ILO with information on Caribbean labour departments
ilo.org – International Labour Organization (ILO)
ilo.org/public/english/protection/safework/cis/legosh/ – ILO Database on Countries and their OSH legislation
information.prevention.issa.int (no www) – International Social Security Association
iso.org – International Standards Organization
labour.gov.lk – Department of Labour, Sri Lanka
labour.go.ke – Department of Labour, Kenya
labour.gov.pk – Department of Labour, Pakistan
labour.gov.za – Department of Labour, South Africa
labour.gov.zm – Department of Labour, Zambia
labour.nic.in – Ministry of Labour, India
leat.or.tz – Lawyers' Environmental Action Team, Tanzania
mhlw.go.jp – Ministry of Health Labour and Welfare, Japan
mines.nic.in – Ministry of Mines, India
minerals.nsw.gov.au – Department of Mineral Resources, New South Wales
moftec.gov.cn – Ministry of Foreign Trade and Economic Cooperation, China
mom.gov.sg – Ministry of Manpower, Singapore
msha.gov – Mine Safety and Health Administration (USA)
myregs.com/dotrspa – Hazardous Materials Safety Regulations and Interpretations (USA)
nfpa.org – National Fire Protection Association (USA)
nigeria.gov.ng – Government of Nigeria
nist.gov – National Institute of Standards and Technology (US Government)
nohsc.gov.au – National OHS Commission, Australia

Information sources

nrm.qld.gov.au – Department of Natural Resources, Mines and Energy, Queensland
nsc.org – National Safety Council, USA
nt.gov.au/wha – WorkSafe, Northern Territory, Australia
occhealthnursing.net – Graduate Students and Faculty, University of Iowa, Occupational Health Nursing (useful site on occupational health nursing)
osha.gov – Occupational Safety and Health Administration (USA)
oshabulletin.com – OSHA Bulletin (USA)
oshc.dole.gov.ph – Department of Labour and Employment, Philippines
safetyline.wa.gov.au – Worksafe, Western Australia, Consumer and Employment Protection
safetylit.org – Injury Research and Prevention Literature Update
rospa.org.uk – Royal Society for the Prevention of Accidents (UK)
safety-council.org (Canada) – Canadian Safety Council
scc.ca – Standards Council of Canada
sheilapantry.com – Sheila Pantry Associates Ltd (former head of information at UK HSE) information site
sia.org.au – Safety Institute of Australia
travail.gouv.qc.ca – Department of Labour, Quebec (in French)
ttl.fi – Finnish Institute of Occupational Health (English option)
tuc.org.uk – Trades Union Congress (UK)
turva.me.tut.fi (no www) – Institute of Occupational Safety Engineering, Finland
ul.com – Underwriters Laboratories (USA)
unece.org – United Nations Economic Commission for Europe
who.int – World Health Organization
whs.gov.ab.ca – Workplace Health and Safety, Alberta
whs.qld.gov.au – Workplace Health and Safety, Queensland
workcover.act.gov.au – WorkCover, Australian Capital Territory
workcover.nsw.gov.au – WorkCover, New South Wales
workcover.vic.gov.au – Victorian WorkCover Authority, Victoria
workcover.sa.gov.au – Workcover Corporation of South Australia
worksafebc.com – Worksafe, British Columbia
worldlii.org – World Legal Information Institute
wsa.tas.gov.au – Workplace Standards, Tasmania
yahoo.com/health – general Yahoo health search site
and www4.law.cornell.edu/uscode – Cornell University site for US law

Index

AAS, 425–426
Absorption, 303, 304
Acceptability (of PPE), 158
Accidents, 5, 200
    causation, 192–194
    causes, 12, 195, 208
    classes, 25, 233
    costs, 21–24, 25, 26–29
    investigation, 201–216
    investigation team leader, 203
    models, 192
    prevention, 8, 191
    reporting, 20, 202, 207
Accidents investigation team, 203
Accountability, 66, 68
Accuracy, 82, 423
ACGIH *see* American Conference of Governmental Industrial Hygienists
Acid, 354
Acoustic control, 404–405
Acting on the message, 82, 400
Action plan, 410, 461
ACTRAC, 288, 296
Acts of parliament, 98
Acute toxin, 306
Additivity, 306
Administrative controls, 373
Adult learning, 518, 520
Adventure tourism, 130
Aerobic capacity, 473
Aerosol, 353, 414–415

Aerospace industry, 249
Agenda, 90
Aggressive behaviour, 87
Agreements, 60, 100, 175
Agricola, 299
Airborne chemical hazards, 149, 375
ALARA, 445
Alarm/s, 287, 289, 295
Alcohol, 336–338
Allergy, 313
Alternating current, 255
American Conference of Governmental Industrial Hygienists, 7, 138, 325
Analogue, 476, 477
Analysis, 14, 20, 229, 425, 484
Anger, 87, 236
Anthrax, 136
Anthropometry, 450, 454–455
Appreciation, 33, 73, 102, 155
Arlidge, 300
Arson, 290–291
AS/NZS 4360, 575
AS 4801, 545, 568
AS/NZS 4804, 544, 555, 556, 568
Asbestos, 6, 56, 249, 308, 309, 415
Asphyxiants, 309–310
Assertive behaviour, 88
Assessment, 143, 372, 388, 466, 522
    (hazards), 134, 138
    criteria, 139, 188, 296

Assessment (contd.)
   guidelines, 134
   methods, 520–521
Atomic weight, 350
Atoms, 346, 350, 426
Attention deficit, 250
Attitudes, 50, 86, 139, 299, 518
Audit, 72, 187, 189, 549
Auditing, 72, 185, 225, 544
Australian Standards, 284, 419, 433
Authority, 66, 142, 218
Automation, 249, 254
Awards, 100, 558, 563

Back, 254, 467
   injury, 321, 465, 470–471
   strain, 40, 459
Backrest, 457, 481, 482
Bali, 235, 253
Bar charts, 48
Barriers to:
   communication, 87–88
   learning, 519
Base, 354
Bathing patients, 484
Bathtub curve, 248, 574
Bed and breakfast, 291
Behaviour, 527
   model, 240
Behaviour-based safety, 76–78
Belief models, 50
Benches, 483
Benchmarked, 503
Benchmarking, 3, 60, 70, 577
Bends, 136, 316, 359
Bernstein, 132
Best practice, 71–73, 497
Bhopal, 52, 72, 384
Biological exposure indices, 325, 417
Biological hazards, 157, 316
Biological monitoring, 417
Biomechanical principles, 462, 464
Biomechanics, 462, 463
Bird Accident Ratio Triangle, 25, 26

Bird and Loftus, 12, 16
Births, marriages and deaths register, 301
Blame, 81, 95, 199, 202, 205, 206
Blood flow, 464, 481
Blood supply, 316, 464, 480
Bloom's Taxonomy, 514, 515
Brainstorming, 146, 517
Breivik, 104, 131–132
Bridger, 474, 476
Britain, 2, 114, 490
Budget planning, 62–64
Building classifications, 285
Building Code of Australia, 281
Building regulations, 280–282
Building security, 291
Bulk, dangerous goods, 377, 380
Bulk container, 377
Bulletproof vests, 238
Bullying, 236, 336
Bunding, 369, 382
Bureaucratic structures, 562
Bus driving, 236
Business plan, 507, 543

Canada, 4, 5, 105, 106, 109
Canadian Centre for Occupational
   Health and Safety, 122, 217, 581
Cancer, 310
Captive groups, 340
Carban in pulp gold, 388
Carbon monoxide, 387
Carcinogen, 334, 388, 422
Carpal tunnel syndrome, 322, 458
Case histories, 384
Case studies, 153, 510
Causes of cancer, 311, 312
Causes of fire, 275
CBT, 514, 521, 525
CCOHS, 107, 681
CEO, 71, 142, 566
Chains, 268
Chairs, 450, 454, 482
Challenger, 51, 384
Chance failure, 248
Change, 213, 558, 561, 564, 565, 567

Change management, 217, 250, 547, 558
Charcoal tube, 418
Checklist, 135, 178, 179, 222
Chemical bonds, 440
Chemical compounds, 349–350
Chemical contaminants, 149, 425, 429
Chemical control measure, 373
Chemical reactions, 272, 273, 355–356
Chemical receipt, 376
Chemical register, 45, 369
Chemicals, 346, 351
  disposal, 374
  documentation, 352
  emergency plan, 225–226
  employee duties, 374
Cherrypicker, 130
Child labour, 565
Children, 236, 287, 290, 350, 457
Chlorine, 309, 347, 349, 351, 357
Chronic poison, 305
CI *see* continuous improvement
CIE, 581, 100, 411
Citations, 115, 117
Civil and criminal jurisdictions, 97
Classification of hazardous substances, 360, 363
Clip art, 48
Codes of practice, 4, 99, 109, 171
Cognition, 19, 82, 201
Cold stress, 321
Cold weather gear, 159
Colorimetric tubes, 417
Colour and contrast, 413
Colour vision, 476
Comcare, 581
Comfort, 85, 158, 159
Commitment, 144, 152
  commission, 2, 100, 329
Commission, 325, 329
  commissioning, 553
Common law, 20, 95–96, 172
  liability, 7, 29, 491
Communicating risk, 159–161
Communication, 78–79
  skills, 78, 519
Compartmentation, 283
Compensable injury, 227, 490
Compensation, 227
Competencies, 522, 534
Competency-based training, 514, 525
Competition, 225, 228, 565
Compliance, 72, 172, 549
Comprehension, 84, 514
Compressed gases, 265
Computer, 48, 49, 562
Computer terminal, 450
Confined spaces, 181, 271, 309, 315
Congress, 96, 98
Construction, 98, 285, 286
Consultation, 4, 20, 105, 341
Continental shelf, 115
Contingency preparedness, 555
Continuous improvement, 544, 550, 569, 576, 577
Contract mining, 75
Contractor, 74, 191
Contractor safety, 74–75
Contributory factors, 192
Control, 373, 386, 402–403, 475
Control measures, 257, 373, 429
Control options, 156
Conveyors, 271
Cooper, 300
Corporate health indicators, 342
Corporate veil, 4, 108
Corrosion, 265, 268, 358, 359
Cost–benefit, 571, 573–574
Counselling, 52, 78, 237, 330, 550
Court, 96–97, 172
Covello, 161
Cranes, 266–267
Credibility transfer, 159
Crisis change, 564, 566
Critical behaviours, 76, 77
Crystalline structure, 358
Cultural, 49
Cultural backgrounds, 340
Cultural barriers, 49
Culture, 545

Cumulative trauma disorder, 321, 464
Customer contact, 238

Daily noise dose, 407
Danger tag, 521
Dangerous goods, 359–360
Decision making, 33, 139, 392
Default notice, 57, 62
Defences, 7, 9, 10, 99
Delegated legislation, 99
Delegates, 42, 67
Deming, 544
Departments of Labour, 2
Depression, 335, 496
Dermatitis, 136, 313, 314
Design:
  configurations, 404
  criteria, 475, 476
  of display or controls, 484
  scenarios, 457
Detector tube, 417, 418
Detectors, 287, 288, 357, 384, 418
Detoxification, 305
Digital, 476, 477, 481
Direct current, 255
Direct reading instruments, 418
Directed dilution, 430
Disability, 164, 340, 466
Discs, 62, 465, 468, 480
Dispatch, 41, 376
Displays, 445, 475, 476, 477, 518
Disposal, 374
Disputes, 55, 493, 494
DNA, 311, 312, 351, 440
DND, 407
Doctor, 5, 9, 147, 458
Document holders, 482
Documentation, 376–380, 502
  of inspections, 174–177, 178
Dose, 300, 305, 443, 445
Dose-response, 305
Drills, 280, 292, 453
Drucker, 563
Drugs, 336–338

Drug testing, 338
Duration, 21, 22
  rate, 21, 23
Dust, 307, 353
Duty of care, 4, 100–102, 105, 491, 575
Dynamic anthropometric data, 454, 455, 456

Early failure, 248
Early fire hazard indices, 283
Earthing, 53, 256, 257, 258
Economics, 131
Ecstasy, 337
Effective communication, 4, 78
Effective meetings, 89
Effective presentations, 83–87
Effects of injury, 496
Egger, 341
Egress, 287, 284
EIP *see* emergency information panel
Electric shock, 254, 256
Electrical control measure, 257
Electrical dangers, 256
Electrical hazards, 256
Electrical safety, 254–259
Electrical safety management, 256
Electrocution/s quality, 53
Electromagnetic fields, 442
Electromagnetic radiation, effects on eye, 54, 323, 324
Elements, 347–348, 349, 350
Elevating work platforms, 250
Elimination, 44, 144, 237, 403
Emergency, 390, 392
Emergency control functions, 288
Emergency equipment, dangerous goods, 366
Emergency information panel, 376, 385
Emergency lighting, 284
Emergency procedures, 289, 366
Emergency procedures guides, 236, 371
Emergency Services Manifest, 378, 369
Emergency training, 292, 293
Employee assistance programmes, 80, 336, 338

Index

Employee benefits, 342
Employee induction, 221
Employee representation, 88, 111, 115, 189
  responsibility, 68, 142, 293–295
Empowerment, 53, 559
Energy:
  cost, 472, 474
  damage model, 193–194
  exchange, 13, 252
  expenditure, 472, 474
Engineering controls, 40, 389, 403
Environment, 30, 43
Environmental factor, 32, 453
Environmental management, 70, 76, 544, 577
Epidemiology, 12, 301
Equal opportunity, 20, 79
Equipment, 30, 42, 264, 571, 572–573
Equivalent aerodynamic diameter, 307, 414
Ergonomics, 32–33, 43, 340, 404–405, 413–414, 438–439, 449–486
  design cycle, 454, 455
Error, 15, 18, 193, 201, 213, 214, 250, 253–254, 484
Error control, 214, 249, 251
Error management, 74, 216, 249
European Commission, 118
European Union (EU), 96, 118, 173, 302, 361, 381
Evacuation plans, 391–392
Evaluation procedures, 557–558
Evaporation, 318, 434
Event tree analysis, 192–193
Events, 6, 16, 29, 192, 202, 206, 520
EWP *see* elevating work platform
Executive summary, 48
Exemption limit, 362
Exit signs, 284
Exits, 282, 284, 287, 292
Explosion, 225, 256, 350, 357, 383
Explosion dampers, 247
Explosives, 272

Exposure standard, 149, 375, 416–417
External consultants, 160
External environment, 543, 562
Extinguishers, 187, 188, 272–280
Extinguishing gases, 278
Eye, 30, 159, 315, 323, 410, 412, 440, 441
Eyewash station, 375

Facility description, 51, 571
Factories inspectorate, 127, 129
Factories legislation, 105
Failure to danger, 247
Failure to safety, 247
Falls, 12, 53, 157, 250, 322
Family courts, 236
Farr, 300
Fatigue, 43, 73, 136, 200, 328, 337, 463–464, 577
Fault finding, 108, 195, 212,
Fault tree analysis, 192, 575
Feedback, 19, 200, 455, 520, 526, 528
Fibres, 268–269, 307, 308–309, 419
Field testing, 146–147
Financial resources, 142, 448, 563
Fire, 225, 272, 273, 274
Fire alarms, 351
Fire and explosion, 47, 223, 256–258
Fire and housekeeping, 275–276, 280, 281
Fire brigade, 295
Fire extinguishers, 187, 188, 276–278
Fire hydrants, 284, 287
Fire load, 285–286
Fire prevention, 247, 275, 292–297
Fire protection, 88, 222, 247
Fire resistance, 276, 282, 285
Fire safety, 173, 280, 286–288, 297
Fire safety managers, 280
Fire safety survey, 288
Fire tetrahedron, 273
Fire-fighting equipment, 284, 296, 391
Fire-resisting construction, 282
First aid, 29, 111, 187, 226, 390
First-line supervisor, 218

Five Star, 555
Flammable, 258, 276, 315, 351
Flammable liquids, 274, 276, 277, 292
Flashpoint, 274, 352
Flixborough, 51, 209
Floor warden, 296
Floors, 455
Flowchart, 49, 57, 423
Flow control, 247, 379
Flues, 282
FM 200, 278
FMEA, 51
Fog, 354
Footstool, 482
Foreign firms, 562
Formal safety assessment, 51
Freight container, 377, 380
Frequency Rate, 21–23
FRL, fire resistance level, 281, 283, 571
Fume, 35, 43, 171, 268, 307, 353

Gantt chart, 49
Gas, 136, 256, 264, 274, 277, 282, 353, 354
Gas liquid chromatograph, 418, 427–428
Gas solution, 354, 359
Gases and vapours, 415, 354
General dilution, 373, 430
General duty, 4, 115, 374
Gilbreth, 332
Glare, 410, 413, 414
GLC *see* gas liquid chromatograph
Globalization, 562
Glossary of dangerous goods terms, 209
Goals, 45, 47, 54, 58, 64, 224, 235, 339
Golden rule, 98
Gordon, 12
Graphical information, 48
Greenhow, 300
Ground fault, 247
Ground stability, 249
Guidance note, 100, 172, 362, 385

Haddon, 13, 14, 15
Hamilton, 300

Handicap, 466
Handling steel, 370
Handtools, 484
Hawthorne effect, 559
Hazards, 5, 15, 30–31, 39, 43, 130, 132–133, 136, 150, 216
 assessment, 139, 142
 control, 39, 40, 134, 139, 172, 249, 259–264
 evaluation, 133, 139, 144
 identification, 16, 71, 133–139
 management, 39, 72, 77, 133, 249, 251, 252
 monitoring, 146, 370, 420
Hazardous conditions, 10, 179
Hazardous substance, 107, 116, 346
Hazchem, 280, 364, 366, 386
HAZOP, 51
Health, 5
Health and safety committee, 42, 55, 60, 83, 88, 106, 113–114, 142–143
Health and Safety Executive, 191, 548, 582
Health and safety plan, 71, 217–226
Health and safety policy, 64, 108, 111, 217
Health and safety procedures, 220
Health and safety programme, 555, 556
Health and safety promotion, 67
Health and safety representative, 56, 57, 60
Health belief model, 50
Health care cost reduction, 314–342
Health promotion, 338, 340–342
Health risk appraisal, 345
Health risks, 363, 548
Health, statistics, 21
Health status, 254, 300, 340
Health surveillance, 45, 347, 370
Hearing conservation, 407, 409
Heat balance equation, 317, 436
Heat stress, 35, 223
Heat transfer, 434
Hegney–Lawson System Risk Model, 77, 212, 251

Index

Heinrich, 10–12, 25, 76, 233
Heroin, 337
Herzberg, 478
Hierarchy of controls, 77, 78, 373
High performance liquid
    chromatograph, 418
Hong Kong, 123
Hookah, 387
Horizontal elutriator, 414
Housekeeping, 42, 179, 275
Housekeeping practices, 275
HPLC, 418, 428
HSE, 138, 419, 582
Human element, 289
Human error, 15, 200, 484–485
Human factors engineering, 32, 449
Human inertia, 563
Human limitations, 199
Human machine model, 453
Human resources, 41, 146
Human Rights and Equal Opportunity
    Commission, 359
Humidity, 317, 434
Hunter, 300

IARC, 312
ICNAS, 134
ICRP, 100, 445
IEC, 100
IEC 61508, 247, 572
IFAP, 582
Ignition source, 258, 275, 282
Illumination, 32, 180, 413
ILO, 9, 301, 582
ILO conventions, 7, 302
Improvement notice, 4, 97, 107,
    112, 113, 115
Impure chemicals, 353
Incentive schemes, 77, 224
Incentives, 3, 302, 558, 561
Incidence rate, 21, 24, 505
Indirect accident costs, 26
Induction programmes, 75, 297
Induction training, 511, 515
Industrial chemicals notification, 134

Industrial Foundation for Accident
    Prevention, 582
Industrial hygiene, 422
Industrial relations, 42, 52, 56, 479
Industrial trucks, 48
Information capacity, 35
Information sources, 134, 371, 181–183
Information systems, 504
Informed choice, 52
Infra-red, 322, 441
Infra-red spectrophotometer, 428
Injury, 5, 11, 15
Injury classes, 25
Injury control, 249, 261
Injury experience, 21, 64, 253
Injury management, 497, 499–501
Injury measurement, 71
Injury prevention, 18, 157, 253, 259
Inorganic compounds, 349
In-running nip, 260
Insecticide, 310
Inspection, 174
Inspection team, 174
Inspector, 62
Instrumentation, 311, 410
Insulation, 194, 247, 256, 257, 281,
    283, 319
Insurance, 131–132, 491, 494
Insurance providers, 491
Insurer, 40, 64, 227, 503–504, 505
Intangible benefits, 342
Integrated management systems,
    544–545
International conventions, 301
International Labour Organisation,
    9, 301
International Social Security
    Association, 8, 582
Internet, 561, 562, 566
Interpretation, 98
Intervention, 495
Intervention strategies, 230
Interview techniques, 205
Interviewing, 79, 83, 205
Interviews, 79–81

Index

Intimidation, 3, 236
Investigation, 201, 202
Ions, 349
ISO, 70, 100
ISO 14000, 70, 217, 544, 554
ISO 9000, 70, 217, 544, 554
Isocyanates, 72, 314, 383
Isolation, 259, 373, 403
Isotopes, 349, 443
ISRS/DNV, 555

Jacks, 269–270, 272
Job analysis, 471
Job descriptions, 478, 513
Job design, 32–33, 340, 451–452, 460, 477–480
Job diagnostic survey, 478
Job enlargement, 479
Job enrichment, 479
Job performance sampling, 170
Job rotation, 479
Job safety analysis, 135
Job-centred, 477
Johnson, 17
Johnson, Lyndon, 14

Karasek, 328, 479
Knee-chair, 482
Knowledge, 17, 60–61, 527
KPI's, 76
Kyphosis, 481

Label, 360
Labelling, 45, 173, 353, 360–361, 381, 476
Labour hire firms, 74, 75
Labour relations, 52–54
Ladder, 12, 164, 250
L_{Aeq}, 406
Lapses, 485, 529
Lasers, 441
Lead poisoning, 136, 300, 333
Leadership, 546
Leadership styles, 546
Learning outcome, 296

Learning styles, 518–519
Legge, 300
Legionnaire's disease, 48, 317
Legislation on OHS issues, 55, 59
LEL *see* lower explosive limit
Lesson plans, 511, 512
Life cycle, 247, 250
Lifestyle, 164, 326, 332, 340
Lift installation, 284
Lift truck, 483
Lifting devices, 169, 176, 266
Light duties, 80, 226
Lighting, 410
Lighting measurements, 411
Line management, 143
Liquid, 353
Listening, 52–83
Literal rule, 98
Literature review, 502
Local exhaust ventilation, 383, 431
Local government, 52, 368, 375
Local toxin, 306
Locking, 289, 291, 292
Locking systems, 291–292
Lock-out, 76, 264
Long distance haulage, 336
Longitude, 246
Lordosis, 481
Loss control, 39, 45–46
Loss control management, 149
Lost-time injuries, 22, 73
Lower explosive limit, 274, 352
LTI *see* lost-time injury
LTIFR statistics, 48, 232
Lumbar, 467
Lumbar spine, 467
Lumbar support, 481
Lump sum, 493
Lung, 307
Lung diseases, 209, 307

Machine design, 261, 262
Machine guarding, 110, 176, 259–264
Machine safety checklists, 263, 264
Machinery, 42, 259–264, 571–575

Maintenance, 264, 271, 413
Maintenance programme, 54
Major hazards, 368, 369, 390, 392
Malaysia, 124
Management, 17, 19, 68
Management culture, 545–547
Management review, 225, 554, 568–571
Management strategy, 38–39, 211, 215
Management theories, 559
Manager, 61
Manual handling, 461
Manual handling procedures, 471–472
Mapping organizations, 547
Marking, dangerous goods, 360, 376, 377
Maslow, 478
Mass spectrometer, 428
Matches, 281, 292
Material handling equipment, 270–271
Material safety data sheet, 44, 112
Materials, 30, 42, 259, 357–358
Matrix-style, 331, 547
McDonald, 25
McMichael, 341
Measuring airborne contaminants, 374
Measuring instruments, 444–445, 474, 411
Measuring risk, 151–153
Mechanical, 33
Mechanical machines, 33
Mechanization, 254
Media, 50, 54, 141, 160
Medical practitioner, 2, 490
Medically treated injuries, 228
Meeting arrangements, 88–91
Meeting documentation, 90–92
Meetings, 47–49, 78–91
Memory, 34
Mental fitness, 335
Mercury, 136, 311, 352, 426
Message, 78, 82, 84, 85, 159, 160
Metals, 136, 310, 347, 356–357
Microprocessor, 477
Microwave oven, 440, 442
Microwave oven checking, 442

Middle management, 70, 562
Mine Safety and Health Administration, 95, 103, 582
Mines, 102, 423, 433
Mini-cyclone, 414
Mining, 99, 247, 258
Ministry of Labour, 124, 125
Minutes, 90
Mischief rule, 98
Mist, 354
Mistakes, 49, 450
Monitoring, 21, 58, 116, 133, 157, 374
Morbidity, 208, 339
Mortality, 208, 339
Moss, 514, 517, 518, 523
Motivation, 19, 131, 451
MSDS, 134, 352, 371, 381, 383
MSHA, 98, 115
Multi-skilling, 329, 479
Muscle groups, 481
Muscular activity, 472
Musculoskeletal damage, 322

National uniformity, 106–107
National Institute of Occupational Safety and Health, 114, 116, 468, 581
National Occupational Health and Safety Commission, 363, 511
National Plant Standard, 272
National standards, 100, 172
National uniformity, 106, 107
Natural fibre ropes, 269
Needs assessment, 236–237, 514, 548
Negative dominance, 159
Negligence, 7, 100
Network, 560
Neutral fault, 247
New Zealand, 57, 107
NFPA, 363
Nightclub fire, 289
NIOSH see National Institute of Occupational Safety and Health
No detectable damage, 208, 209, 251
No fault, 491

NOHSC *see* National Occupational Health and Safety Commission
Noise, 35, 159, 327, 400
Noise contour diagram, 407
Noise controls, 146, 402
Noise dose meter, 405, 407
Noise sampling, 420, 421
Non-assertive behaviour, 87
Normal distribution, 230
Norman, 484, 525, 527, 528
Note taking, 81
Notice of meeting, 90
Nursing home, 289, 292
Nutrition, 333

Occupational disease, 9, 218
Occupational health, 299–303
Occupational hygiene survey, 415–416
Occupational Overuse Syndrome, 200, 458–462
Occupational Safety and Health Administration, 103, 114, 583
Occupational safety and health bodies, 8, 9
Offshore, 132, 333
Off-the-job, 223, 550
OHS agencies, 117
OHS competencies, 518, 534
OHS issues, 52, 54–56
OHS management system, 217, 543, 550, 568
OHS plan, 49, 551, 556
OHS policy, 60, 546, 556
OHS programme, 548, 556
Oil, 277, 313
Oil rig disaster, 50
On-the-job, 27, 515, 527
OOS *see* occupational overuse syndrome
Operating and capital cost, 63
Operator certification, 264
Organic compounds, 249–250, 427
Organic solvent, 354, 355
Organization charts, 547
Organizational culture, 99, 148

Organizational structure, 6, 61, 547, 556
Organized resistance, 563
Origins of risk concepts, 130–132
OSH competencies, 511, 518, 534
OSHA, 103, 114, 115
Otis, 246
Outcomes, 251
Overseas suppliers, 562
Owen, 300
Oxidizer, 347
Oxygen levels, 393

Packaging, dangerous goods, 360, 364
Packing group, 351, 365
Pain and suffering, 29, 97
Paracelsus, 299, 305
Pareto principle, 173
Parliament, 95, 96, 99
Particle behaviour, 307
Passive smoking, 333
Passport, 76
Pause gymnastics, 460
PDCA cycle, 544
Peel, 300
Penetration, in building, 281, 283
Perception, 33, 159
Perception of risk, 5, 73
Perception surveys, 548
Percival, 300
Performance indicators, 73–74, 187
Permit to work, 509, 553
Personal health, 339, 343
Personal hearing protection, 409
Personal protective equipment, 157–159, 389
Person-centred, 477
Person–machine interface, 200
Person–machine system, 475
Personnel selection, 513
Petersen, 16–17
Petroleum, 107, 315
Physical fitness, 335
Physical hazards TLVs, 442, 416
Pie chart, 48, 506

Pipeline, 107, 379
Piper Alpha, 50, 51, 209, 384
Placarding, 360, 362, 365
Plan evaluation, 517
Plan implementation, 55, 220, 226
Plant, 42, 264
Polynuclear aromatic hydrocarbons, 300, 312
Population, 455
Portable fire extinguishers, 186, 278
Positive performance indicator, 103, 558
Posture, 460, 471, 472, 480
Potential barriers to change, 561–564
Pott, 300
Power, 270
Power trucks, 270
Powerpoint, 48, 84
PPI, 558
Precedence, 100, 114
Precision, 14, 320, 527
Pregnant, 312, 334
Premium, 29
Premium rates, 27, 504
Prescribed amount, 498
Prescriptive laws, 253
Presence-sensing, 247
Presentation, 83
Pressure vessels, 264–265
Prevention, 8–9. 10, 114, 191, 199–200
Primary, 361
Prison, 52, 118, 289
Procedures, 19
Process industries, 292, 297
Productivity, 141
Professional assistance, 147
Programme, 72
Programme elements, 72
Programme evaluation, 524
Programme implementation, 53
Programmable systems, 247, 248
Prohibition notice, 4, 59, 107
Protective mechanisms, 307–308
Provisional improvement notice, 56, 62
Psychological, 43, 137

Psychological barriers, 563
Psychological causes, 136
Psychological trauma, 235
Psychomotor, 475
Pure chemicals, 352
Pyromania, 290

QA, 544
QM *see* quality management
Quality assurance, 544
Quality management, 68, 575–756
Questioning, 81, 521

Radiant heat, 318, 434
Radiant heaters, 281
Radiation, 440, 441, 442, 444
Ramazzini, 300, 450
Ramp-up, 248
Rasmussen, 17, 18, 484, 529
Reach distance, 455, 483
Reactions, 350, 356
Reactivity, 356
Reason, 201, 485, 486, 529
Reasonably practicable, 105
Received message, 82
Recency, 160
Reciprocating trap, 260
Recognition, 300, 302, 303
Record keeping, 374
Recreation, 6, 463
Recruitment, 19, 20
Regulations, 7, 95, 99, 118, 255
  type, 109
Rehabilitation, 80, 226, 495, 499, 502–503
Rehabilitation coordinator, 498, 499, 500
Rehabilitation plan, 500
Rehabilitation programme, 498, 499–500
Relative risk, 549
Reliability, 247, 456
Remedy, 461
Repetitive Strain Injury, 253, 458
Reporting, 87
Reporting back, 87

Reporting of occupational diseases, 301
Reporting results, 437–438
Resistance, 255
Resolution of issues, 106, 112
Resources, 47, 142, 552
Respiratory disease, 339
Response, 305, 390
Responsibility, 552
Revealed fault, 247
Risk, 5
Risk communication, 49, 208
Risk control, 43, 44, 133, 155–157
Risk evaluation, 40, 153
Risk intervention strategies, 230
Risk management, 39, 40–41, 144, 155
Risk retention, 40
Risk transfer, 40
Robbery, 236
Robens, 103–110
Robens-style, 4, 7
Rockbolts, 247
Royal Society, 127, 583
RSI, 253, 458
Rule of three, 159
Rust-belt industry, 562
Rust-belt, 562

Sadler, 300
Safe system, 53, 168
Safety, 5
Safety and health committee, 47, 114
Safety and health representative, 114
Safety arrest mechanism, 246
Safety audit, 170, 185, 186, 187, 189
Safety behaviour, 518, 548
Safety case, 50–51
Safety coordinator, 219
Safety culture, 18, 303
Safety culture surveys, 558, 567
Safety dogs, 246
Safety equipment, 249, 375, 385
Safety equipment, chemicals, 364, 375
Safety initiatives, 54, 62, 63, 68, 94, 186, 208, 558, 561
Safety integrity level, 247

Safety management, 16, 38–92, 256
Safety management system, 217, 247
Safety MAP, 190
Safety parameters, 183
Safety precedence sequence, 35–36, 139, 156
Safety programmes, 53, 555
Safety survey, 170, 177–178, 288
Safety technology, 246–250, 253, 254
Sampling, 324–325, 416, 417–419, 420–425, 448
Sanders, 240, 244, 481, 487
Sanitary wear, 482
Sanitation, 2
SARS, 26, 253
Scaffolding, 56, 109, 123, 264
SCBA *see* self-contained breathing apparatus
Scenarios, 51, 453, 457
Scissorlift, 250
School teachers, 236
Screenings, 325, 326, 364, 374
Scrotal cancer, 300
Seat depth, 482
Seat belt, 271
Seated person, 455, 468
Seated posture, 480–481
Secondary barriers to change, 563
Secondary container, 379, 382, 386
Security, 63, 237, 291–292, 550
Selected duties, 500
Self-regulation, 4, 103–110, 172, 175
Self-contained breathing apparatus, 158, 389, 390, 433
Senge, 562
Senior manager, 143
Separation, fire, 281, 283, 286
Separation distance, 368, 369
Setting, 20, 168–173, 234, 263
Severity Rate, 21, 26, 225
Sexuality, 29, 154, 290, 303, 327, 332, 480, 493
Shearing trap, 260
Shielding, 156, 446

Shifts, 136, 138, 336
Shiftwork, 331–332
Shop floor, 41, 112, 198, 199
Shopping centres, fire in, 289, 292, 549, 568
Shotcrete, 247
Shrink-wrap pallets, 380
Site safety plan, 218
Sitting, 321, 465, 480, 481, 483
Skewed distribution, 230
Skills, 60, 74, 78, 295–296, 466, 513, 517
Skills enhancement, 524
Skills-rules-knowledge, 17–18, 484
Skin disorders, 209, 313, 319
Skin exposure, 313, 419
Skylarking, 329
Slings, 54, 266, 267, 269, 271, 272
Slips, 101
Smoke, 215, 282, 284, 288, 353
Smoke doors, 283
Smoking, 414, 419, 172, 275, 281, 304, 333
Software, 49, 247, 248, 250, 289
Solid, 353, 356
Solvents, 56, 313–315, 354
Sorting operations, 484
Sound, 400–402
  pressure, 400, 401
Sound pressure level, 400–401
Sound pressure level meter, 405–407
Sound waves, 400
South Africa, 105, 126, 172, 542, 581
Special-needs groups, fire, 289, 290
Spill collection, 382–383, 391
Spine, 321, 439, 462, 480
Spiritual health, 340
Spreader, 269
Sprinkler, 51, 281, 283, 284, 286, 288
Squab, 456, 481, 482
SRK, see skills-rules-knowledge
Stacking, 369, 465, 483
Staff training, 53, 63, 237, 369
Staff turnover, 342, 559, 574

Staffing levels, 55, 57, 59, 236, 328, 460
Stakeholders, 59
Standard deviation, 230, 231, 426, 454
Standard of care, 101
Static anthropometric data, 456
Static discharge, 370
Static loading, 459, 481
Statistical, 26
Statistics, 21–25, 227–234
Statute law, 52, 67, 95
Steel, 353, 357, 358, 388, 389
Storage facilities, 482–483
Storage of chemicals, 176, 180, 362, 375
Storey, 285, 286, 288, 291, 292
Stowage, dangerous goods, 363, 364, 377
Strategic changes, 558–561
Strategic planning, 27, 545
Stress, 328–329, 463, 466
Stress factors, back injury, 466
Stressors, 303, 307, 324, 326–336
Strobe effects, 413
Structural adequacy, 281, 282
Structural defects, 179–180
Structural integrity, 281, 282, 286
Structure of shifts, 59
Subcontractor, 74, 76, 399
Substance abuse, 80, 337
Substitution, 44
Successful outcomes, 457, 526
Supervisor, 16, 28, 61, 67, 142, 175, 218
Surgeons, 9, 50
Synergism, 305, 306
Synthesis, 514, 515
Synthetic ropes, 268
Systematic failure, 247, 248
System of work, 72, 105, 206, 211, 571
System reliability, 188
System risk model, 77, 212, 251
System yielding, 250
Systemic toxin, 306
Systems engineering, 250–254

Index

Tag-out, 75, 379, 555
Talks, 84
Tangible benefits, 342, 584
Target group, 461
Target organs, 303, 304, 345, 372
Task analysis, 477
Task design, 458, 459
Task redesign, 467
Task rotation, 474
Task selection, 474
Taxonomy, Bloom's, 514
Taylor, F.W., 322, 478
Team leader, 144, 472
Technical report, 47
Technical safety, 51
Technological change, 169, 250, 461
Tenders, 29, 76, 566
Tendonitis, 459
Tenosynovitis, 136, 322, 458
Tetrahydrocannabinol, 336
Textile mills, 300
Thackrah, 300
Thermal comfort indices, 435
Time management, 197, 329
Toilet pedestals, 450, 484
Tool design, 458, 459
Toolbox meeting, 61, 68, 224
Total loss control, 18–20
Tourism, 103, 104, 132
Toxic gases, 309–310
Toxic metals, 310, 426
Toxicology, 134, 299, 305
Toxins, 303, 310
Trade unions, 8–9
Trainee analysis, 514
Training course, 54, 62, 219, 509, 516
Training needs analysis, 46, 511, 513–520
Training objectives, 513, 514
Training programme, 292–297, 509–512, 513, 569
Training session, 516–518, 523–524
Transmitting the instructions, 294
Transport of dangerous goods, 367
Trauma, 9, 235, 253

Travel distance, 288
Tripartite, 57, 59, 61, 103
Trips, 101
Trust, 160, 161
Twisting, 250, 450
Typing pools, 461, 562
Tyre blow-out, 250, 252, 253

UEL *see* upper explosive limit
Ultraviolet, 323, 368, 440, 441
UN Recommendations, 351, 359, 360, 361, 363, 369, 377
Uncertainty, 98
Understanding risk, 151–155
Understanding the system, 197
Unit load, 380
United Kingdom, 450–451
United States, 114–118
Upper explosive limit, 274, 352

Validity, 229
Vapour, 353, 415
Vapour density, 352
Vapour pressure, 352, 354–355
VDT, 458
VDU, 43, 414, 458, 475
Vehicle requirements, dangerous goods, 364
Ventilation, 173, 174, 368, 430
Vertebrae, 462, 465, 480, 481, 484
Vibration, 315–316, 400, 404–405, 438–439
Violations, 95
Violence, 235–238
Virus, 253, 313, 351
Visionary CEO, 71, 142, 566
Vocational rehabilitation, 490, 501
Voice, 82
Voltage, 255
Voluntarily assumed risk, 50, 52
Voluntary assumption of risk, 7, 103

walk-through survey, 415, 416, 461
Walsh–Healy Act, 114
Warden, 289, 293, 294, 295, 296

Waste disposal, 43, 374–375, 390
Wearout failure, 248
Weather conditions, 233, 453
Welding, 197, 315, 387–388
Wellness, 337, 338–342
What if, 51
Wigglesworth, 14, 15, 16
Williams-Steiger Occupational Safety and Health Act, 114
Wire ropes, 268, 272
Witnesses, 112, 202, 204–206
Work environment, 30–36, 149–150, 172–173, 178, 188–189, 340
Work from home, 74
Work instructions, 525, 543, 544, 567
Work placement, 45, 47
Work practices, 43, 432–433
Work-related illness, 495, 498
Work-related injury, 227, 489, 494, 495, 498, 504
Work system, 32, 196–199, 202, 461
Worker's compensation data, 134, 235, 504, 505
Workers' compensation and rehabilitation, 29, 489, 490–508
Workplace agreement, 60
Workplace climate, 316–320, 343
Workplace health improvement programme, 339
Workplace health promotion, 338, 340, 341, 342
Workplace injuries, 21–29, 495, 504
Workplace Relations Ministers, 339, 341, 342
Workplace system, 39, 43, 44, 77
Workplace wellness, 337, 338–342
Worksafe Australia, 461, 466, 467
Worksite health promotion, 341
Work station, 200, 404, 407, 412, 413
World Trade Agreement, 302
World Trade Center, 235, 253, 291, 556

X-ray diffraction, 429

Yielding systems, 250
Young persons, 300
Your body, 411
Your voice, 86–87, 567

Zoonoses, 316